The Dynamics and Environmental Context
of Aeolian Sedimentary Systems

Geological Society Special Publications
Series Editor J. BROOKS

GEOLOGICAL SOCIETY SPECIAL PUBLICATION NO 72

The Dynamics and Environmental Context of Aeolian Sedimentary Systems

EDITED BY

KENNETH PYE

Postgraduate Research Institute for Sedimentology,
University of Reading

1993

Published by

The Geological Society

London

THE GEOLOGICAL SOCIETY

The Society was found in 1807 as the Geological Society of London and is the oldest geological society in the world. It received its Royal Charter in 1825 for the purpose of 'investigating the mineral structure of the Earth'. The Society is Britain's national society for geology with a Fellowship of 6965 (1991). It has countryside coverage and approximately 1000 members reside overseas. The Society is responsible for all aspects of the geological sciences including professional matters. The Society has its own publishing house which produces the Society's international journals, books and maps, and which acts as the European distributor for publications of the American Association of Petroleum Geologists.

Fellowship is open to those holding a recognized honours degree in geology or cognate subject and who have at least two years relevant postgraduate experience, or who have not less than six years relevant experience in geology or a cognate subject. A Fellow who has not less than five years relevant postgraduate experience in the practice of geology may apply for validation and, subject to approval, may be able to use the designatory letters C. Geol (Chartered Geologist).

Further information about the Society is available from the Membership Manager, The Geological Society, Burlington House, Piccadilly, London W1V 0JU, UK.

Published by The Geological Society from:
The Geological Society Publishing House
Unit 7
Brassmill Enterprise Centre
Brassmill Lane
Bath BA1 3JN
UK
(*Orders*: Tel. 0225 445046
        Fax 0225 442836)

First published 1993

British Library Cataloguing in Publication Data

A catalogue record for this book is available from the British Library
ISBN 0-903317-88-5

Distributors

USA
   AAPG Bookstore
   PO Box 979
   Tulsa
   Oklahoma 74101-0979
   USA
(*Orders*: Tel. (918)584–2555
         Fax (918)584–0469)

Australia
   Australian Mineral Foundation
   63 Conyngham St
   Glenside
   South Australia 5065
   Australia
(*Orders*: Tel. (08)379–0444
         Fax (08)379–4634)

India
   Affiliated East–West Press PVT Ltd
   G-1/16 Ansari Road
   New Delhi 110 002
   India
(*Orders*: Tel. (11)327–9113
         Fax (11)331–2830)

Japan
   Kanda Book Trading Co.
   Tanikawa Building
   3-2 Kanda Surugadai
   Chiyoda-Ku
   Tokyo 101
   Japan
(*Orders*: Tel. (03)3255–3497
         Fax (03)3255–3495)

Typeset by EJS Chemical Composition, Midsomer Norton, Bath, Avon

Contents

Temperate and cold climate continental dunes

Dust and loess

Preface

This volume arises from a two-day international symposium held at the Geological Society of London on 22–23 October 1991. The meeting was convened by K. Pye of Reading University and sponsored jointly by the British Sedimentological Research Group and the British Geomorphological Research Group. It brought together approximately one hundred researchers with interests in aeolian processes and environments, both ancient and modern. The aim of the meeting was to provide an opportunity to discuss recent advances in understanding of the environmental controls on aeolian sediment transport processes, dune morphodynamics, and dunefield evolution. The selection of twenty-four papers included in this volume address a wide range of issues, ranging from short-term experimental studies of individual grain movement and grain-bed collisions during aeolian transport to long-term climatic, eustatic and tectonic controls on the development of sand seas. Consideration is given to warm continental desert dunefields, cold climate and temperate continental aeolian environments, coastal dunes, aeolian dust transport and loess formation.

Several of the papers report recent advances in the development of methods for dating late Quaternary aeolian deposits. These methods make it possible, for the first time, to rigorously test hypotheses which relate major phases of aeolian activity to changes in climate, sea-level, and anthropogenic disturbance. Thick sequences of aeolian sand, loess and palaeosols potentially provide some of the most complete and detailed evidence of the nature of environmental changes which have affected continental areas during the Quaternary and earlier geological periods. Similarly, dust deposits in ocean sediments can record important information about changes in continental surface conditions and global atmospheric processes. The papers in this volume indicate that much has been learned in the past decade about the relationships between climate, sea-level, aeolian transport and deposition, although it is not yet possible to claim a full understanding.

The editor is grateful to many individuals who helped in the organization of the meeting and subsequent processing of the manuscripts for publication. In particular, the assistance of the following in refereeing papers is gratefully acknowledged: J. R. L. Allen, J. R. Burkinshaw, G. Butterfield, D. J. Carruthers, R. W. G. Carter, L. Clemmensen, E. Derbyshire, G. Kocurek, N. Lancaster, T. Littmann, I. Livingstone, D. Loope, B. Maher, K. Rasmussen, H. Rendell, R. Sarre, S. Stokes, D. S. G. Thomas, H. Tsoar, A. Warren, P. Wilson, B. B. Willetts, A. G. Wintle, P. Worsley, and V. P. Wright. Final processing of the manuscripts was ably handled by Joanna Cooke of the Geological Society Publishing House.

K. Pye

Introduction: the nature and significance of aeolian sedimentary systems

K. PYE

*Postgraduate Research Institute for Sedimentology, University of Reading,
Whiteknights, Reading RG6 2AB, UK*

The past 20 years have seen a major growth of interest in aeolian processes, landforms and sediments, as witnessed by a growing number of published papers in journals, conference proceedings and edited volumes (e.g. see the collections in Morales 1979; McKee 1979; Péwé 1981; Brookfield & Ahlbrandt 1983; Barndorff-Nielsen et al. 1985; Nickling 1986; Liu 1987; Eden & Furkert 1988; Kocurek 1988; Hesp & Fryberger 1988; Gimingham et al. 1989; Okuda et al. 1989; Leinen & Sarnthein 1989; Nordstrom et al. 1990; Barndorff-Nielsen & Willetts 1991a,b; Kocurek 1991; Carter et al. 1992; Pye & Lancaster 1993. A number of monographs dealing with sand dunes, dunefields, dust and loess have also been published (Greeley & Iversen 1985; Pye 1987; Lancaster 1989; Pye & Tsoar 1990; Coudé-Gaussen 1991). Several factors have contributed to this increased level of interest, including a desire to achieve a better understanding of recent aeolian environments as analogues for ancient aeolian hydrocarbon reservoirs (e.g. Glennie 1972; Fryberger et al. 1983) and other planets where aeolian processes play an important role (e.g. Breed & Grow 1979; Greeley & Iversen 1985), the requirement for the development of improved methods of desertification and sand control in arid and semi-arid regions (Khalaf 1989; Watson 1990), particularly in view of the predicted effects of global climate change (Houghton et al. 1990), the identification of coastal dune systems as an important recreational and conservation resource which is under threat from rising sea-level and human activities (ver der Meulen 1989; Bakker et al. 1990; van der Meulen et al. 1991), and the recognition that aeolian sedimentary sequences both on land and in the oceans potentially preserve a detailed record of changes in regional climate and global atmospheric transport during the Quaternary and earlier times (e.g. Rea et al. 1985; Rea & Leinen 1988; Kukla et al. 1988; Beget & Hawkins 1989; Ruddiman et al. 1989; Pye & Zhou 1989; Clemens & Prell 1990; Sirocko et al. 1991; Sirocko & Lange 1991).

The evidence provided by aeolian sediments and bedforms provides a useful means of testing the predictions of global circulation models for different times in previous earth history (e.g. Kutzbach & Guetter 1986; Parrish & Peterson 1988). However, successful interpretation and use of the aeolian sedimentary record in this way requires that the nature of aeolian processes, and the environmental controls on them, are adequately understood. In simple terms, the formation of aeolian sedimentary sequences has only three basic requirements: (1) a source of sediment; (2) sufficient wind energy to sort and transport the sediment; and (3) a suitable location where part or all of the sediment can accumulate. In practice, a wide range of environmental factors may influence the distribution, morphology and sedimentological character of aeolian deposits. These include tectonic factors, which affect weathering and erosion rates, influence the nature of the wind regime, and control the location, form and preservation potential of the resulting deposits. Climate also exercises a strong control on sediment production rates and the potential for aeolian transport, both directly and indirectly through its influence on vegetation cover and groundwater levels. In near-coastal areas, sea-level fluctuations may also have a strong influence, either by initiating phases of transgressive dune activity associated with shoreline changes, or by triggering widespread changes in groundwater level (Chan & Kocurek 1988). Finally, human activities play an ever-increasing role, being responsible under different circumstances both for destabilization of partially vegetated dunes and for stabilization and control of active systems.

Against this background, the papers included in this volume have been grouped into five main sections. The first section contains four papers dealing with aeolian grain mechanics and dune morphodynamics. Although the basic physics of aeolian sediment transport have been relatively well understood for more than 50 years, and many of the fundamental principles elucidated in Bagnold's (1941) *The Physics of Blown Sand and Desert Dunes* remain basically valid, recent

From Pye, K. (ed.), 1993, *The Dynamics and Environmental Context of Aeolian Sedimentary Systems.*
Geological Society Special Publication No. 72, pp. 1–4.

laboratory experimental and numerical modelling work has allowed the refinement of the basic model, as discussed in **McEwan & Willett's** paper in Section I. Difficulties remain, however, in taking full account of the effects of turbulence at different spatial and temporal scales. In this connection, gain function analysis promises to be a useful new approach, as described in the paper by **Hardisty**. An important challenge also remains to relate models of aeolian grain transport over flat or uniformly sloping beds to real field situations with complex terrain, where wind speed profiles may deviate markedly from the theoretical logarithmic profile. The paper by **Burkinshaw** *et al.* presents data which illustrate the variability of wind profiles over a reversing transverse dune in the Alexandra coastal dunefield of South Africa, while **Wiggs** illustrates the problems involved in evaluating shear velocity from wind profile measurements in the context of a dynamic dune in Oman.

Section II contains seven papers which deal primarily with the development and sedimentary history of desert dunefields, including the determination of sand chronostratigraphy. The first paper in this section by **Wintle** provides an overview of recent advances in luminescence dating of aeolian sands. The methodologies of thermoluminescence and optical luminescence dating are outlined, together with outstanding problems and promising areas for future development. The following three papers, by **Edwards**, **Rendell** *et al.* and **Stokes & Breed**, provide case studies of the application of luminescence dating to the evolutionary history of dunes in the Kelso area of California, the northern Negev Desert, and northeastern Arizona, respectively. Although some procedural problems remain, and there are currently too few dates available to draw firm conclusions about the number, timing and causation of dune episodes in these areas, the preliminary results are encouraging.

The paper by **Livingstone & Thomas** provides a discussion of the environmental factors which govern the morphological development and dynamics of linear dunes in southern Africa. It is pointed out that the presence of vegetation cover does not in itself imply that these features are entirely relict. Some sand movement is possible under conditions of partial vegetation cover, and episodic sand transport in response to climate fluctuations on time scales of 10–100 years should not be ignored.

The final two papers in this section provide illustrations of aeolian-dominated desert sedimentary sequences in the rock record. **Crabaugh & Kocurek** describe in detail the sedimentary architecture of the Jurassic Entrada Sandstone in the Utah area, USA, and discuss the evidence which suggests that large-scale aeolian deflation and accumulation events in this area were controlled principally by changes in groundwater levels which in turn were related to movements in sea-level. Careful evaluation of sedimentary structures is also used by **Chakraborty & Chaudhuri** as the basis for their interpretation that the Precambrian Mancheral Quartzite of northern India is principally composed of thin aeolian sand layers which accumulated under conditions of a high water table within a broad river floodplain environment.

Section III contains seven papers dealing with coastal dunes. The first paper in this section, by **Psuty**, considers the relationship between recent beach changes at Perdido Key, Florida and the development of foredune morphology. On a related theme, the relationship between coastal wind regime, sediment transport and foredune development at Tentsmuir, Fife, is considered by **Wal & McManus**, while **Carter & Wilson** provide an overview of the location, morphology, and historical development of coastal dune sand accumulations in Ireland.

The complex pattern of sand transport and mixing of sediment from different sources around the Baix Empordà, eastern Spain, is documented by **Cros & Serra**, while **Pye & Neal** present the results of a stratigraphic drilling and radiocarbon-dating investigation which have allowed the development of a model for the development of the late Holocene Sefton barrier dune complex in northwest England.

The final two papers in this section, by **Gardner & McLaren** and **McLaren**, consider the nature and significance of post-depositional cementation and diagenesis within carbonate-rich coastal dune sands from different parts of the world. They demonstrate that a simple model of progressive diagenesis in carbonate aeolianites cannot be supported, and that palaeoenvironmental interpretations based on degree of diagenetic alteration need to be made with caution.

Section IV contains two papers relating to aeolian deposits in temperate and cold continental environments. The first paper, by **Koster** *et al.*, examines the genesis and sedimentological character of late Holocene drift sands in northwest Europe which have formed by aeolian reworking of older Pleistocene sands. The second paper, by **Seppälä**, provides a rare account of climbing and falling dunes of Holocene age in Finnish Lapland.

The final section of the book includes three papers dealing with aeolian dust and loess.

Coudé-Gaussen & Rognon present a comparative analysis of the sedimentary character and palaeoclimatic significance of dust deposits in the Canary Islands and southern Morocco. They clearly illustrate that the dust deposits in the two areas have different source regions and were deposited at different times in response to opposite shifts in climatic belts during the late Pleistocene and Holocene.

Li & Zhou present new data relating to the age and origin of loess deposits in coastal areas of northeastern China, suggesting that much of the material was probably derived from the floor of the Bohai Sea during times of glacially lowered sea-level. Finally, **Rolph et al.** discuss recent results relating to the magnetic mineralogy of the loess section near Lanzhou, on the western edge of the Central Loess Plateau, and discuss its relationship to the depositional history of the loess sequence.

Overall the papers bear witness to the high degree of current activity in aeolian research, and will undoubtedly contribute to the formulation of new research programmes.

References

BAGNOLD, R. A. 1941. *The Physics of Blown Sand and Desert Dunes.* Methuen, London.

BAKKER, TH. W., JUNGERIUS, P. D. & KLIJN, J. A. (eds) 1990. *Dunes of the European Coasts.* Catena Supplement No. 18, Catena Verlag, Cremlingen.

BARNDORFF-NIELSEN, O. E., MØLLER, J. T., RASMUSSEN, K. R. & WILLETTS, B. B. (eds) 1985. *Proceedings of the International Workshop on the Physics of Blown Sand, Aarhus University, 26–31 May 1985.* Department of Theoretical Statistics, Institute of Mathematics, University of Aarhus, Memoirs No. 8, Volumes 1, 2 & 3.

—— & WILLETTS, B. B. (eds) 1991a. *Aeolian Grain Transport 1. Mechanics.* Acta Mechanica Supplementum 1, Springer Verlag, Vienna.

—— & —— (eds) 1991b. *Aeolian Grain Transport 2. The Erosional Environment.* Acta Mechanica Supplementum 2, Springer Verlag, Vienna.

BEGET, J. E. & HAWKINS, D. B. 1989. Influence of orbital parameters on Pleistocene loess deposition in central Alaska. *Nature*, **337**, 151–153.

BREED, C. S. & GROW, T. 1979. Morphology and distribution of dunes in sand seas observed by remote sensing. *In*: MCKEE, E. D. (ed.) *A Study of Global Sand Seas.* U.S. G.S. Professional Paper **1052**, 253–302.

BROOKFIELD, M. E. & AHLBRANDT, T. S. (eds) 1983. *Eolian Sediments and Processes.* Elsevier, Amsterdam.

CARTER, R. W. G., CURTIS, T. G. F. & SHEEHY-SKEFFINGTON, M. J. (eds) 1992. *Coastal Dunes. Geomorphology, Ecology and Management for Conservation.* Balkema, Rotterdam.

CHAN, M. A. & KOCUREK, G. 1988. Complexities in eolian and marine interactions: processes and eustatic controls on erg development. *Sedimentary Geology*, **56**, 283–300.

CLEMENS, S. C. & PRELL, W. L. 1990. Late Pleistocene variability of Arabian Sea summer monsoon winds and continental aridity: eolian records from the lithogenic component of deep sea sediments. *Paleoceanography*, **5**, 109–145.

COUDÉ-GAUSSEN, G. 1991. *Les Poussieres Sahariennes.* John Libbey Eurotext, Montrouge, France.

EDEN, D. N. & FURKERT, R. J. (eds) 1988. *Loess.* Balkema, Rotterdam.

FRYBERGER, S. G., AL-SARI, A. M. & CLISHAM, T. J. 1983. Eolian dune, interdune, sand sheet, and siliciclastic sabkha sediments of an offshore prograding sand sea, Dharan area, Saudi Arabia. *American Association of Petroleum Geologists Bulletin*, **67**, 280–312.

GIMINGHAM, C. H., RITCHIE, W., WILLETTS, B. B. & WILLIS, A. J. (eds) 1989. *Coastal Sand Dunes.* Proceedings of the Royal Society of Edinburgh Section B (Biological Sciences) Volume **96**.

GLENNIE, K. W. 1972. Permian Rotliegendes of northwest Europe interpreted in light of modern desert sedimentation studies. *American Association of Petroleum Geologists Bulletin*, **56**, 1048–71.

GREELEY, R. & IVERSEN, J. D. 1985. *Wind As A Geological Process.* Cambridge Planetary Science Series 4, Cambridge University Press, Cambridge.

HESP, P. A. & FRYBERGER, S. G. (eds) 1988. *Special Issue on Eolian Sediments.* Sedimentary Geology Volume **55**.

HOUGHTON, J. T., JENKINS, G. J. & EPHRAMUS, J. J. (eds) 1990. *Climate Change. The IPCC Scientific Assessment.* Cambridge University Press, Cambridge.

KHALAF, F. I. 1989. Desertification and aeolian processes in the Kuwait Desert. *Journal of Arid Environments*, **16**, 125–145.

KOCUREK, G. (ed.) 1988. *Special Issue: Late Paleozoic and Mesozoic Eolian Deposits of the Western Interior of the United States.* Sedimentary Geology Volume **86**.

—— 1991. Interpretation of ancient eolian sand dunes. *Annual Reviews of Earth and Planetary Science*, **19**, 43–75.

KUKLA, G., HELLER, F., LIU, X. M., XU, T. C., LIU, T. S. & AN, Z. S. 1988. Pleistocene climates in China dated by magnetic susceptibility. *Geology*, **16**, 811–814.

KUTZBACH, J. E. & GUETTER, P. J. 1986. The influence of changing orbital parameters and surface boundary conditions on climate simulations for the past 18,000 years. *Journal of Atmospheric Sciences*, **43**, 1726–1759.

LANCASTER, N. 1989. *The Namib Sand Sea.* Balkema, Rotterdam.

LEINEN, M. & SARNTHEIN, M. (eds) 1989. *Paleoclimatology and Palaeometeorology: Modern and Past Patterns of Global Atmospheric Transport.* NATO ASI Series C: Mathematical and Physical Sciences, Volume **282**, Kluwer, Dordrecht.

LIU TUNGSHENG (ed.) 1987. *Aspects of Loess Research.* China Ocean Press, Beijing.

McKEE, E. D. (ed.) 1979. *A Study of Global Sand Seas.* United States Geological Survey Professional Paper No. 1052.

MEULEN, VAN DER, F. (ed.) 1989. *Perspectives in Coastal Dune Management.* SPB Academic Publishing, The Hague.

——, WITTER, J. V. & RITCHIE, W. (eds) 1991. *Impact of Climatic Change on Coastal Dune Landscapes of Europe.* Landscape Ecology Volume 6 (1/2), SPB Academic Publishing, The Hague.

MORALES, C. (ed.) 1979. *Saharan Dust—Mobilization, Transport, Deposition.* Wiley, New York.

NICKLING, W. G. (ed.) 1986. *Aeolian Geomorpology.* Binghampton Symposia in Geomorphology International Series No. 17, Allen & Unwin, Boston.

NORDSTROM, K. F., PSUTY, N. P. & CARTER, R. W. G. (eds) 1990. *Coastal Dunes Form and Process.* Wiley, Chichester.

OKUDA, S., RAPP, A. & ZHANG, L. (eds) 1991. *Loess. Geomorphological Hazards and Processes.* Catena Supplement 20, Catena Verlag, Cremlingen.

PARRISH, J. T. & PETERSON, F. 1988. Wind directions predicted from global circulation models and wind directions determined from eolian sandstones of the western United States — a comparison. *Sedimentary Geology,* **56**, 261–282.

PÉWÉ, T. L. (ed.) 1981. *Desert Dust.* Geological Society of America Special Paper 186.

PYE, K. 1987. *Aeolian Dust and Dust Deposits.* Academic Press, London.

—— & LANCASTER, N. (eds) 1993. *Aeolian Sediments Ancient and Modern.* International Association of Sedimentologists Special Publication No. **16**, Blackwell Scientific Publications, Oxford.

—— & TSOAR, H. 1990. *Aeolian Sand and Sand Dunes.* Unwin Hyman, London.

—— & ZHOU, L. P. 1989. Late Pleistocene and Holocene aeolian dust deposition in North China and the northwest Pacific Ocean. *Palaeogeography, Palaeoclimatology, Palaeoecology,* **73**, 11–23.

REA, D. K. & LEINEN, M. 1988. Asian aridity and the zonal westerlies: late Pleistocene and Holocene record of eolian deposition in the northwest Pacific Ocean. *Palaeogeography, Palaeoclimatology, Palaeoecology,* **66**, 1–8.

——, —— & JANECEK, T. R. 1985. Geologic approach to long-term history of atmospheric circulation. *Science,* **227**, 721–725.

RUDDIMAN, W. F., SARNTHEIN, M., BACKMAN, J., BALDAUF, J. G., CURRY, W., DUPONT, L. M., JANECEK, T., POKRAS, E. M., RAYMO, M. E., STABELL, B., STEIN, R. & TIEDEMANN, R. 1989. Late Miocene to Pleistocene evolution of climate in Africa and the Low-latitude Atlantic: overview of Leg 108 results. *Proceedings of the Ocean Drilling Project, Scientific Results,* **108**, 463–484.

SIROCKO, F. & LANGE, H. 1991. Clay mineral accumulation rates in the Arabian Sea during the late Quaternary. *Marine Geology,* **97**, 105–119.

——, SARNTHEIN, M., LANGE, H. & ERLENKEUSER, H. 1991. Atmospheric summer circulation and coastal upwelling in the Arabian Sea during the Holocene and the last glaciation. *Quaternary Research,* **36**, 72–93.

WATSON, A. 1990. The control of blowing sand and mobile desert dunes. *In:* GOUDIE, A. S. (ed.) *Techniques for Desert Reclamation.* Wiley, Chichester, 35–85.

Aeolian mechanics and dune morphodynamics

Sand transport by wind: a review of the current conceptual model

IAN K. McEWAN & BRIAN B. WILLETTS

Department of Engineering, University of Aberdeen, Aberdeen, AB9 2UE, UK

Abstract: Rapid progress has been made in the last decade towards a more comprehensive model of wind blown sand transport extending and sharpening Bagnold's classic model in significant aspects. This paper reviews the current physical model and attempts to indicate future directions of research.

The currently accepted physical model of aeolian sand transport reduces the sand transport system to four distinct sub-processes: aerodynamic entrainment; the trajectory of the wind driven sand grains; the grain/bed collision; and the modification of the wind by the driven sand. The isolation and separate treatment of these sub-processes has been an important factor in the recent rapid development of aeolian sand transport mechanics. It is, however, their interaction that produces the rich behaviour of the system. Anderson & Haff (1991) and McEwan & Willetts (1991) have synthesized the four sub-processes and constructed full saltation models which follow the system from incipience to steady-state saltation. These computer simulations provide a stern test for the physical model as the results calculated can be compared to experimental observations from field and wind tunnel. The models compare well with available data; thus we can have some confidence that the physical model is realistic. Moreover, these computer simulations have become a powerful investigative tool and have highlighted areas where our understanding is deficient.

In a recent review paper, Anderson *et al.* (1991) noted that theoretical work in aeolian sand transport was on the point of outstripping its experimental database. This remark stems from the very rapid progress made during the last five years in the numerical modelling of the wind blown sand system. Four models have been described in the literature: Anderson & Haff (1988, 1991), Werner (1990), Sorensen (1991) and McEwan & Willetts (1991). These models descend from the analytical model of Owen (1964), but because of the relatively massive computational resources available they are able to relax a number of the assumptions forced on Owen in order to make his calculation tractable. However, a further, more significant, distinction exists between Owen (1964) and the present generation of models over the nature of the self-regulatory mechanism present in the system. Owen postulated that the equilibrium of the sand transport rate was maintained by the shear stress at the surface being limited to a value corresponding to the entrainment threshold of the sand. Thus, if the sand transport rate decreased below equilibrium then the surface shear stress increased and more gains were entrained by the wind. However, if the sand transport rate increased above its equilibrium value then the surface shear stress fell, fewer grains were entrained by the wind and the transport rate returned to equilibrium. The basis for this hypothesis is the assumption that aero-dynamic entrainment is the dominant mechanism for sustaining the grain cloud.

However, an important piece of evidence was omitted from this picture. Bagnold (1941) defined two thresholds, the fluid and the impact thresholds for sand transport. A wind of a certain strength is required to initiate sand transport over a loose dry surface: this wind strength was termed the *fluid threshold*. However, if the wind strength decreases after sand transport has begun, it can be sustained at wind strengths below the fluid threshold. Thus another threshold, termed the *impact threshold*, is observed below which sand transport ceases. Bagnold (1941) noted that the shear velocity describing this wind strength was some 70–80% of that which describes the fluid threshold. The suggestion must be that if sand transport can occur at wind strengths below the fluid threshold when aerodynamic entrainment is expected to be minimal then it should be expected that impact entrainment (the dislodgement of bed grains at collision) occurs over the whole range of shear velocities. In this way, doubt is cast on the feedback mechanism proposed by Owen (1964).

In the accepted physical model of wind blown sand transport, equilibrium occurs because the wind is retarded sufficiently so that it is saturated and its capacity to transport more sand is limited. An important ingredient in this equilibrium is the role of the grain/bed collision: the

From Pye, K. (ed.), 1993, *The Dynamics and Environmental Context of Aeolian Sedimentary Systems.*
Geological Society Special Publication No. 72, pp. 7–16.

wind deceleration means that grains hit the surface with reduced velocities and hence fewer grains are ejected and the numbers of grains participating in saltation is limited. This feedback mechanism is quite different to the one proposed by Owen (1964) because the numbers of grains being transported is determined through the medium of the grain/bed collision.

Numerical modelling of sand transport

A common feature which distinguishes the current generation of computer simulations from previous attempts is that they are firmly based on the physics of the system. In particular, each model includes a representation of the grain/bed collision derived from experimental data. The explicit inclusion of the grain/bed collision is a significant refinement on previous models and is indicative of the attempt made to faithfully represent the physics of the system. However, there are significant differences between the published numerical models. The numerical models of Anderson & Haff (1988, 1991) and McEwan & Willetts (1991) are based in the time domain and thus are able to provide insight into the temporal behaviour of the system; the models of Werner (1990) and Sorensen (1991) are steady-state models. Fig. 1 shows a schematic representation of the temporal models of McEwan & Willetts (1991) and Anderson & Haff (1988, 1991). The system is reduced into four separate sub-processes (aerodynamic entrainment, grain trajectory, grain/bed collision and wind modification). In keeping with the spirit of these models each of these sub-processes will be discussed in turn: both from a modelling perspective and from a wider viewpoint.

Aerodynamic entrainment

In the models of Anderson & Haff (1991) and McEwan & Willetts (1991) aerodynamic entrainment is modelled by an excess shear stress rule:

$$N_a = \alpha \, (\tau_a(0) - \tau_t), \qquad (1)$$

where N_a is the number of aerodynamically entrained grains leaving the bed per unit area, α is a constant (m s kg^{-1}), $\tau_a(0)$ is the fluid shear stress at the surface, and τ_t is the fluid threshold.

McEwan & Willetts (1991) have noted that the model of Anderson & Haff (1991) is likely to underestimate the shear stress at the bed. Thus, the conclusion that at steady state aerodynamic entrainment has a limited role, is based on insufficient evidence. The model of McEwan & Willetts (1991) predicts some entrainment at steady state; again however, their calculated value of steady-state bed shear stress is questionable. Thus, there is some discrepancy in the numerical modelling predictions about the role of aerodynamic entrainment during steady-state sand transport. The openness of this question is further emphasized by closer inspection of the validity of the aerodynamic entrainment rule used in the numerical models. Williams *et al.* (1990) and Willetts *et al.* (1991) examined aerodynamic entrainment near the leading edge of a sand deposit. Both studies indicate that the turbulent structure of the flow is significant in the entrainment of grains. Both studies concluded that it is the peak short-term values of shear stress near the bed which determine the level of aerodynamic entrainment. The aerodynamic entrainment rule used by Anderson & Haff (1991) and McEwan & Willetts (1991) (see Equation 1) is based on mean flow character-

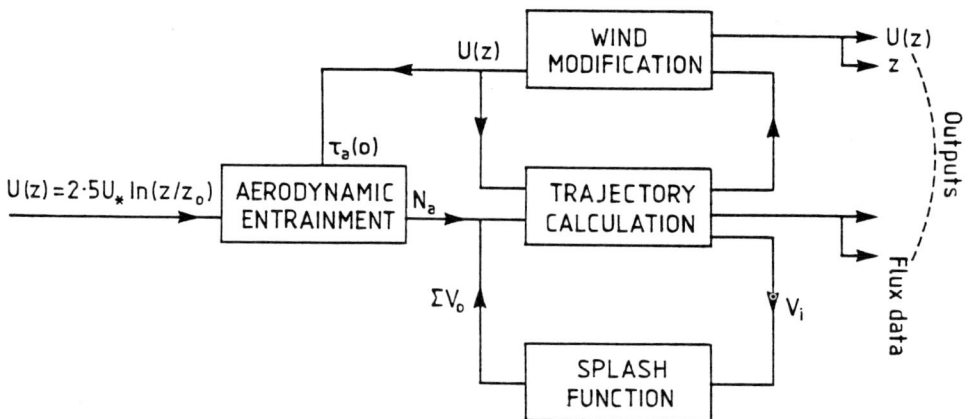

Fig. 1. A schematic diagram linking the four saltation sub-processes.

istics and so is likely to be an oversimplification of the process. Thus, more experimental work is required before the precise role of aerodynamic entrainment at steady state can be resolved.

The trajectory calculation

Saltation is the hopping or dancing of sand grains over a surface. Sharpening this definition, saltation is a flight or trajectory of a sand grain acted on by gravity and the wind. If the initial velocity vector of a grain leaving the surface and the forces acting on the grain throughout its flight are known, then the trajectory of that grain may be calculated. This calculation has been reliably performed in many previous studies of saltation (e.g. Owen 1964; White & Schulz 1977; Tsuchiya 1969a,b, 1970; Reizes 1978; Hunt & Nalpanis 1985; Jensen & Sorensen 1986; Ungar & Haff 1987; Anderson & Hallet 1986; Werner 1990; McEwan & Willetts 1991).

It is widely accepted that there are four forces which govern a saltation trajectory:

(1) force due to gravity;
(2) aerodynamic drag;
(3) Magnus effect;
(4) aerodynamic lift.

To calculate the force due to gravity, the grain mass needs to be known. This is calculated by defining an equivalent diameter sphere, as proposed by Bagnold, which has the same terminal velocity of fall in air as the grain itself. Bagnold determined that the equivalent diameter was given by the sieve diameter multiplied by a shape factor which, for the sand used in his experiments, was 0.75. Owen (1980) stated that the range of shape factors was 0.6 to 0.8. Therefore, knowing the equivalent diameter and the density, the volume and the mass of the grain may be calculated.

The aerodynamic drag has been calculated in many studies by assuming that the drag may be represented by the drag on the equivalent diameter sphere, for instance in Owen (1964). The drag on a grain is specified by a drag coefficient which is a function of Reynolds number. A number of empirical expressions are available to calculate the drag coefficient: Morsi & Alexander (1972), Schiller & Nauman (1933) and White (1974). The Reynolds number is dependent on the relative velocity between the grain and the air and so the drag coefficient depends on both the horizontal and the vertical velocities. As a result, the equations of motion in the horizontal and vertical direction cannot be decoupled (solved separately) and so, in practice, must be solved by computer. Owen (1964)

evaded this difficulty by taking the drag coefficient to depend linearly on the relative velocity, thus enabling the decoupling of the equations of motion. Sorensen (1991) also employed this assumption in his analytical model of saltation.

The Magnus effect results from the pressure difference above and below a spinning projectile. It was first noted in 1671 by G. T. Walker in trying to account for the slice of a tennis ball (Swanson 1961). Chepil (1945a) noted that grains appeared to spin at 200–1000 rev. s^{-1}. White & Schulz (1977) were the first to incorporate the Magnus effect into a trajectory calculation. They estimated the rotation of glass spheres by fitting the theoretical trajectories to photographically observed trajectories, and obtained initial spin rates of 100–300 rev. s^{-1}. They found that incorporating the spin in the trajectory calculation resulted in an approximate 10% increase in the height of the hops. However, there remains some doubt about this procedure as there is no check on the spin rates calculated. Anderson & Hallet (1986) noted that for a grain with topspin, the Magnus force will accelerate the grain upward as long as the grain's forward velocity is less than that of the wind. The same grain will receive a downwind acceleration on its ascent, but be decelerated on its descent. Other investigators have found the Magnus effect to be a second order force. Hunt & Nalpanis (1985), Jensen & Sorensen (1986) and Werner (1990) did not include the Magnus effect in their equations of motion. Owen (1980) speculated that the impacting grain will receive a spin on colliding with the surface but that this spin will be quickly damped by friction. Willetts (1987) suggested that the Magnus effect may have been more influential on the glass spheres used by White & Schulz (1977) than on quartz sand grains.

Aerodynamic lift acts on a particle in high shear flows (Saffman 1964) because of differences in fluid velocity above and below the particle. Thus, it is similar to the Magnus effect as they both result from a pressure difference over the grain. Aerodynamic lift will be most effective in influencing grain motion near the boundary where the velocity gradient is highest. Further from the surface this force diminishes considerably. Willetts & Murray (1981) and Naddeh (1984) have studied lift on a sphere near a solid boundary. They have both shown that the lift force decays rapidly as the gap ratio increases; the gap ratio is the ratio of distance from the wall to the sphere diameter.

This trajectory model is based on a number of assumptions. The grains are assumed to respond to the mean flow velocity only. This has been shown to be a valid assumption for grains larger than 100 μm (Owen 1980; Anderson 1987). Grains smaller than this will tend to move in suspension, responding to small scale turbulence and hence the trajectories will be stochastic rather than deterministic, as they are in pure saltation. In practice, there will be no definite demarcation between saltation and suspension. Jensen & Sorensen (1982, 1986) and Hunt & Nalpanis (1985) have used the term *modified* saltation to describe this intermediate transport mode. A further assumption is that the grains in the trajectory calculation are assumed to be spherical: the fluid drag is calculated on the basis that the grains are spheres. This may be a reasonable assumption for compact grains but Rice (1991) has shown that the trajectories of platy sand grains are different. Nevertheless, the trajectory calculation provides a useful model of the grain trajectories. It is worthwhile to note that a satisfactory model is obtained by including only weight and fluid drag. The forces due to the Magnus effect and aerodynamic lift are small in comparison and so may be neglected.

Grain/bed collision

Until recently, the grain/bed collision had received little attention in the literature. White & Schulz (1977) used the grain equations of motion to extrapolate direct mid-air observations of trajectories back to the surface. They found rebound angles of impacting grains were less than the 90° suggested by Bagnold (1941) and Chepil (1945b). Rumpel (1985) studied the collision process through a simple geometric model of colliding discs and also challenged the notion that grains left the surface travelling vertically. Willetts & Rice (1985, 1986a, 1986b, 1989) made high speed films of collisions. Previous attempts at high speed photography had found observation in the vicinity of the surface impossible because the high concentration of grains obscured the camera view. Willetts & Rice (1985) raised a small section of bed to make direct observation possible. They confirmed the

finding of White & Schulz (1977) and Rumpel (1985) that the rebound angle was not 90°. Other experimental studies of the grain/bed collision process include those by Mitha *et al.* (1986) who used stroboscopic photography of the impact of steel ball bearings and those by Werner (1987) who fired single sand grains into a sand surface and used stroboscopic photography to record the result. An alternative approach was taken by Werner & Haff (1988). They constructed a two-dimensional numerical simulation of the collision in which the grains were modelled as discs. A bed of regularly packed discs was formed and a single disc of the same size was fired into the surface, and the equations of motion of each disc in the system were integrated forward in time. A large number of such simulations was made for different beds and different impacts and the results were compiled into a statistical function which represents the grain/bed collision. Encouragingly, the results of this numerical simulation were in broad agreement with the results found by conventional experimental observation.

In typical collision, shown schematically in Fig. 2, a grain strikes the bed at a velocity V_1 and at an angle (α_1) of 10–16° and ricochets at a velocity, V_3, and a steeper angle (α_3) of between 20–40°, retaining some 60% of its impact speed. One or more surface grains may be ejected from the bed. These ejected grains tend to have a speed which is an order of magnitude less than that of the impact grain. Their angle of ascent may vary from 0–180° but the mean is roughly 50° from the forward direction.

Ungar & Haff (1987) put forward a simplified model of the system which included the grain/bed collision. They introduced the term *splash function* to describe the stochastic law (Anderson *et al.* 1991) required to predict the outcome of an impact. Thus collision was recognized as being an integral part of the wind blown sand system. The experimental work on collision has produced a stock of data which allows a limited parameterization of its properties. It is on this stock of experimental data that the current generation of numerical models are founded.

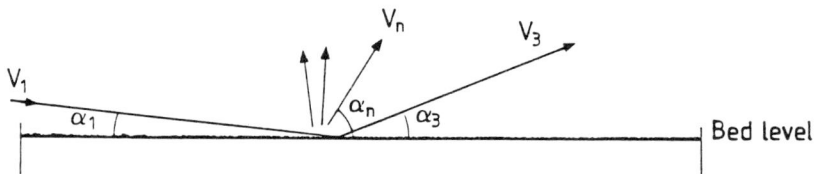

Fig. 2. A schematic view of a typical collision.

Wind modification of the grain cloud

The wind modification can be considered to have two components. Firstly, the wind is decelerated by the moving sand and, secondly, the fluid shear stress is modified by the grains so that it is no longer constant with height. The resulting differential stresses act as a force on the wind. This follows Owen (1964) who introduced the notion of grain-borne shear stress which is defined as the force exerted on a column of air of unit base area extending from that height and upwards.

Consider the equilibrium of the body of air of unit base area shown in Fig. 3. Assuming that

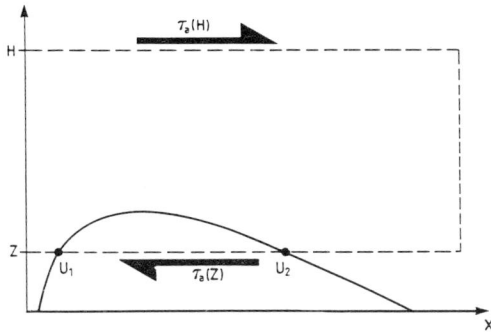

Fig. 3. Grain-borne shear stress. The dashed box represents a control volume of air, the pressure gradient is zero (dp/dx = 0) and the grain accelerates from horizontal velocity u_1 to u_2 in time t_{12} across height, z.

there is no horizontal pressure gradient (dp/dx = 0) then the forces on the air mass are the shear stress at height, H, which is above the grain layer, the shear stress at height, z, below the air mass and the drag exerted by the grain, of mass m_g, while being accelerated from u_1 to u_2 in time t_{12} while above height, z. For equilibrium of the air mass:

$$[\tau_a(H) - \tau_a(z)] + m_g(u_2 - u_1)/t_{12} = 0. \quad (2)$$

Thus, the mean shear stresses at heights H and z are no longer equal, as they would be in the absence of grains. Instead, this single grain trajectory has modified the shear stress at different heights by differing amounts which depend on the distribution of the grain's acceleration with height over its flight. The difference between $\tau_a(H)$ and $\tau_a(z)$ is equal to $\tau_g(z)$ which is the grain-borne shear stress at height z.

In nature the interaction between the grains and the wind is somewhat more complex. For a start, there are many trajectories and their influence must be summed; however, the qualitative argument presented above still holds true. Natural winds are also turbulent, but the above argument again holds if it is restricted to the mean flow conditions.

Figure 4 shows the stresses calculated by the computer simulation of McEwan & Willetts (1991). This confirms Owen's (1964) prediction of complementary stresses in the grain layer (NB the fluid and the grain-borne stresses sum to a

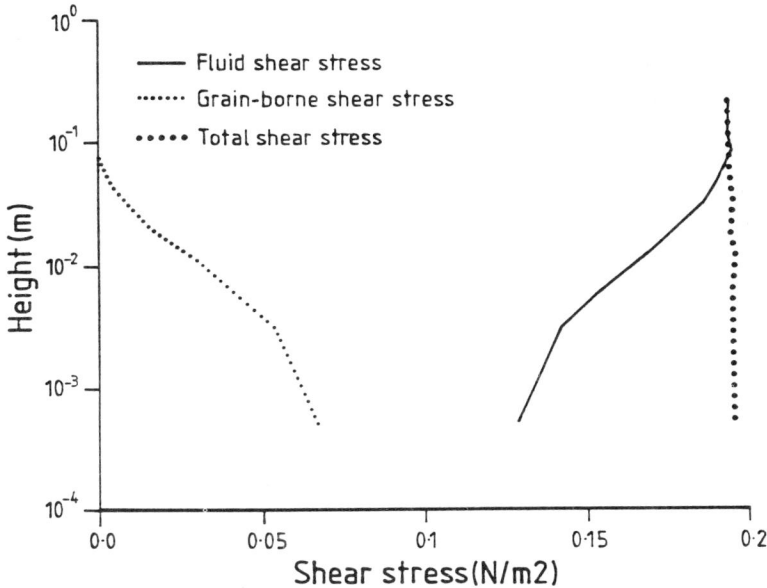

Fig. 4. The fluid, grain-borne and total stress profiles. The total stress is found from summing the fluid and grain-borne stresses. Note the complementary nature of these stresses.

constant total stress with height). Prandtl's mixing length can be used to relate the shear stress, height from the wall and the velocity gradient such that

$$\tau_a(z) = \varrho \, (\kappa z \, du/dz)^2, \qquad (3)$$

where $\tau_a(z)$ is the fluid shear stress at height, z; ϱ is air density ($= 1.23$ kg m^{-3} at 20°C); κ is von Karman's constant ($= 0.4$); du/dz is the velocity gradient.

Rearranging equation 3 produces

$$u(z) = 1/(\kappa\sqrt{\varrho}) \int_{z_0}^{z} \tau_a(z)/z \, dz; \qquad (4)$$

where z_0 is the aerodynamic roughness of the fixed sand bed.

Thus, the velocity profile in the grain layer is a function of the shear stress profile. If, as in the absence of grains, $\tau_a(z)$ is a constant, then Equation 4 reduces to the pure logarithmic wind velocity profile. However, if $\tau_a(z)$ is a function of height then the wind velocity profile departs from a straight line in semi-logarithmic space. Therefore, the distinctive curvature of measured wind velocity profiles is a result of the distribution of grain activity with height, or more rigorously the grain-borne shear stress profile. If the grain-borne shear stress is a maximum at the surface and zero above the saltation layer, as it must be by definition, then Equation 4 predicts that the grain modified wind profile must be convex upwards.

In a very thorough re-analysis of experimentally measured wind profiles which involved replotting many profiles on the same scale so that a visual comparison could be made, Gerety (1985) cast doubt on Bagnold's notion of a focal point at a certain height and velocity, finding instead stronger evidence for a focal zone.

Gerety concluded that 'the existence and dimensions of a kink in the velocity profile depend on how the log-law was drawn through the data, particularly on the forcing of a straight line through a presumed common focal point.'

Gerety suggested that some level in the saltation layer was the height below which grains significantly alter the flow and that above this height the grain concentrations are sufficiently low so that they do not contribute significantly to the momentum. From the experimental wind profiles, Gerety considered that the region below 2–3 cm, in the low gradient section of the profile, is where the grains significantly alter the flow and that above this height the wind profile could be described by the logarithmic law of the wall. This interpretation of experimentally

measured wind profiles is in agreement with the preceding theoretical discussion.

Owen (1964), in his first hypothesis, stated that the saltation layer behaves as an increased aerodynamic roughness to the flow outside it. He supposed that the thickness of the saltation layer was related to the mean vertical lift-off velocity of the grains, which in turn was proportional to a characteristic flow velocity u_*. Thus, the additional roughness length caused by the grain cloud is given by,

$$z_0' = C_0 \, u_*^2/2g, \qquad (5)$$

where C_0 is an empirical constant.

This has a much stronger physical basis than Bagnold's concept of a focus. Owen's relationship predicts that the roughness length during saltation will increase with u_*.

Figure 5 shows some logarithmic velocity rays given by the relationship:

$$u(z) = u_*/\kappa \, \ln \, (z/z_0), \qquad (6)$$

where $z_0' = C_0 \, u_*^2/2g$

Inevitably the velocity profiles intersect in a psuedo-focal zone. However, no clearly defined point emerges nor should one be expected. Rasmussen & Mikkelsen (1991) conducted a similar analysis and again found no firm evidence for a single focal point. Therefore, even if experimentally measured velocity rays pass through a focal zone, there is no sound physical basis to force them to pass through a single

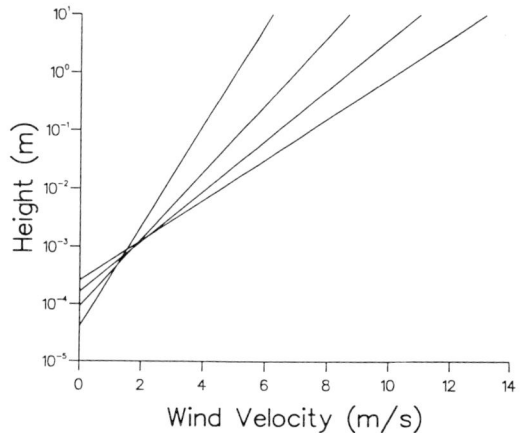

Fig. 5. Wind velocity profiles. The roughness length z_0 is calculated from Owen's roughness relationship; thus for a given u_* the intercept is found and so the profile may be calculated. Note the profiles do not cross at a definite focal point as suggested by Bagnold (1941).

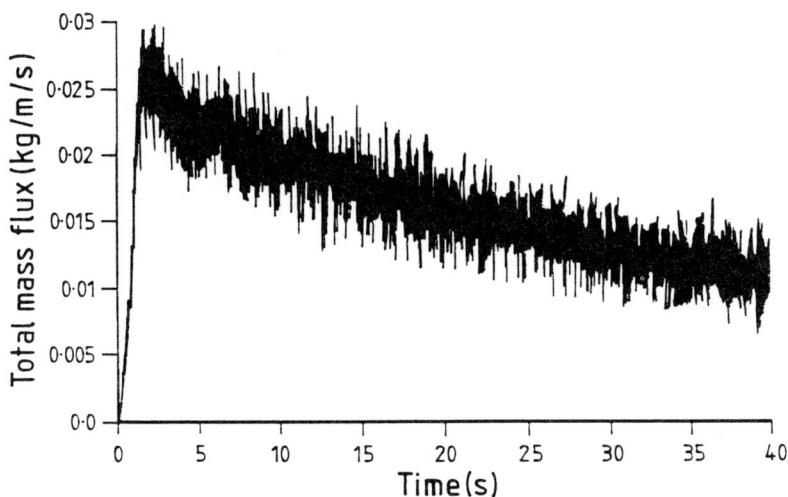

Fig. 6. The total mass flux against time, as calculated by the model of McEwan & Willetts (1991).

point. Instead the important physical parameter is the intercept or the effective roughness which, with experimentally measured profiles, can be compared with Owen's roughness relationship.

However, some problems remain. The value of the empirical constant in Equation 5 is well defined for wind tunnel work ($C_0 = 0.02$) but its value in the field is somewhat uncertain. Rasmussen *et al.* (1985) report values of 0.14 and 0.18 for field experiments at Foveran, Scotland and Hantsholm, Denmark. In theory, if a value for C_0 is known, the shear velocity may be found from one velocity measurement. This is an attractive proposition but until such a time as the behaviour of the constant C_0 is adequately defined for field conditions, such a practice is not advisable and subject to error. Moreover, Owen's roughness relationship may only apply in the case of a well-developed boundary layer and so its use is uncertain in non-equilibrium field conditions. Nevertheless, it represents a significant refinement over the use of Bagnold's focus and should be preferentially adopted.

The physics of the system

An important element in [6] understanding the physics of the wind blown sand system is to establish how the ingredients of the system, the transported grains, the wind and the surface, interact to produce a steady state. This interaction is well described by the numerical models of Anderson & Haff (1988, 1991) and McEwan & Willetts (1991) and thus some results from the

latter numerical model will be used to discuss the self-regulatory nature of the system. Figure 6 shows the transport rate variation over a time period of 40 s from the onset of saltation calculated by the model of McEwan & Willetts (1991). The transport rate grows very rapidly during the first second until it reaches a maximum. It then decays, initially rapidly, then more slowly, to a presumed equilibrium which, for this simulation, is less than half its peak value. This decay to a steady state appears to take place over a time period of the order of several tens of seconds.

Figures 7 & 8 show the response of the shear velocity and the effective roughness, respectively, over the same 40 s simulation period. The shear velocity was calculated from the gradient of the wind profile above 8 cm and the effective

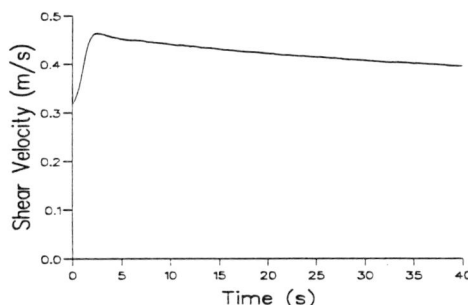

Fig. 7. The shear velocity against time, as calculated by the model of McEwan & Willetts (1991).

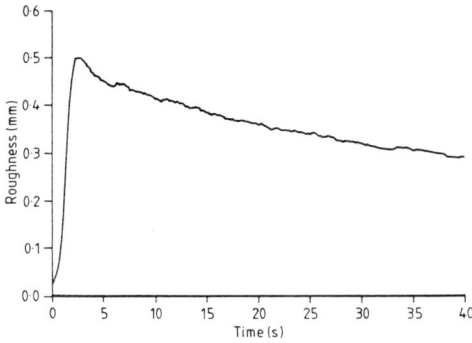

Fig. 8. The effective aerodynamic roughness length against time, as calculated by the model of McEwan & Willetts (1991).

roughness was calculated from projecting an intercept of this gradient down to the height where the wind velocity would apparently be zero. The shear velocity, initially 0.32 m s^{-1}, rises rapidly in the first second. This rise is in phase with the initial increase in transport rate. Then, the shear velocity and the roughness decay from their maximum values to a presumed equilibrium attained after several tens of seconds. Again, it is in phase with the development of the transport rate in time.

Thus, the equilibrium of the saltation system can be considered as a two-stage process. Firstly, the saltating cloud responds rapidly (less than one second) to a change in shear velocity. This estimate of the system's response time agrees well with experiments by Butterfield (1991) and numerical modelling by Anderson & Haff (1991). Secondly, the boundary layer must adjust to the increased roughness imposed on it by the saltation cloud. It does this by increasing its gradient (or shear stress). However, this new shear stress is not in equilibrium with the upper boundary layer (or outer flow condition) and so a second equilibrium must be sought. This is achieved by decreasing the shear stress over a period of the order of several tens of seconds. The first component of the steady state can be considered as the wind and the saltation cloud reaching an equilibrium, and the second component as the new surface roughness and the wind profile reaching equilibrium. It is only when this second condition is attained, over a much longer period relative to the first condition, that the saltation system can be considered to be in equilibrium.

The large difference between the maximum transport rate and the eventual equilibrium value occurs because of the highly non-linear

relationship between transport rate and the shear velocity.

A crucial implication of this result is that in natural or field conditions sand transport flow may never be in complete equilibrium with the flow conditions. Sand transport rate measurements under field conditions do not indicate an equilibrium transport rate but rather an average of a sequence of temporally adjusting flows. A further important implication is that because of the increase in surface roughness, the shear velocity during sand transport is not equal to the shear velocity before transport began. During saltation the shear velocity, measured above the grain layer, must increase because of the additional roughness imposed on the flow.

A further significant feature of the numerical models is that they spontaneously recover the non-linear relationship between the sand transport rate and shear velocity. Figure 9 was constructed from the results of four model simulations each of 25 s duration using the model of McEwan & Willetts (1991). The gradient of this line in the logarithmic space is 3.0, thus the behaviour of the model is in close agreement with experimental observations. A slight cautionary note should be added here because the exact value of the exponent determined by the simulation will depend somewhat on the nature of the splash function (Rice (1991) found that the exact splash function was dependent on the shape of the sand). Thus the very close agreement found between the models of McEwan & Willetts (1991) and Bagnold (1941)

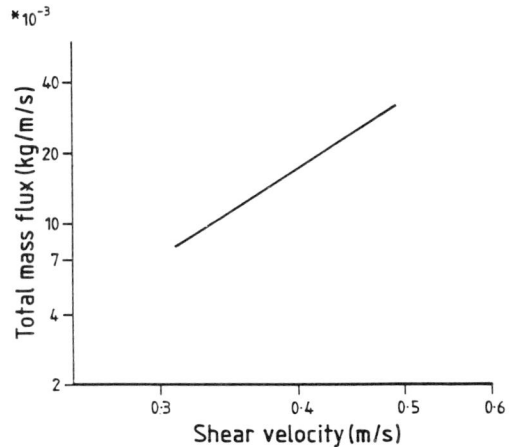

Fig. 9. The variation of total mass flux with shear velocity as calculated by the model of McEwan & Willetts (1991). The mass flux was calculated at $t = 25 \text{ s}$ for four shear velocities; $u_* = 0.31, 0.38, 0.44$ and 0.49 m s^{-1}.

should not be seen as a complete endorsement of Bagnold's cubic transport relationship. The non-linearity is not explicitly input into the model; rather the model, constructed from the fundamental sub-processes spontaneously reproduces this behaviour. The model is a system of forces which is free to find equilibrium. Therefore the conclusion must be that as that equilibrium corresponds closely with the one observed experimentally, then the physics on which the models are based is sound and represents a major advance in our understanding of aeolian processes.

Concluding remarks

Rapid progress has been made in the study of the fundamental processes involved in aeolian sand transport. The main impetus for this advance has been in computer simulation. Models have been used to study aeolian processes on almost all scales from the grain/bed collision to the full saltation system. Other significant developments have been made in experimental work. It has been recognized that factors such as sand trap efficiency and unrepresentative wind velocity measurements have contributed to large discrepancies between many data sets. Rasmussen & Mikkelsen (1992) compared the efficiency of several types of sand trap and found major discrepancies between measured and actual transport rates. Investigators should be aware of the potential inefficiency of sand traps in measurement and this factor should be considered in all measured data.

Moreover, Rasmussen & Mikkelsen (1992) also suggested that inaccurate wind velocity measurements are a potential cause of scatter in wind tunnel data. They reanalysed Kawamura's (1951) experimental data and found that his wind profiles were inconsistent with Owen's roughness relationship which suggests that the experiments were conducted under non-equilibrium conditions. Thus Kawamura's transport equation, derived semi-empirically from his data, is likely to overestimate the sand transport rate.

It should also be recognised that most sand transport rate formulae have been developed under uniform conditions in a wind tunnel. To extrapolate their use to field conditions may be unjustified. Features such as the spatial variations in sand transport (sand snakes) resulting from the larger scale variations in turbulence cannot be reproduced in wind tunnels. Thus, an important research objective should be an investigation of the influence of unsteady flow conditions on sand transport.

A further urgent short-term research objective should be a quantitative analysis of the interaction of different grain sizes in the aeolian sediment system. Specifically the role of surface sorting and ripple formation in determining the sand transport rate should be studied. The formation of ripples appears to cause a decay in the transport rate over the order of tens of minutes (Rasmussen & Mikkelsen 1991; Butterfield 1991) and thus may represent an important process in determining sand transport rate. Moreover this experimental study should be supported by extension of existing numerical models to accommodate multiple grain sizes. The main obstacle to this development of multiple grain size models is the lack of data about the size distribution of the grains ejected from the bed during collision. This issue needs to be addressed in order to gain insight into the interaction of several grain sizes in determining the transport rate.

References

ANDERSON, R. S. 1987. Eolian sediment transport as a stochastic process: The effects of a fluctuating wind on particle trajectories. *Journal of Geology*, **95**, 497–512.

—— & HAFF, P. K. 1988. Simulation of aeolian saltation. *Science*, **241**, 820–823.

—— & —— 1991. Wind modification and bed response during saltation of sand in air. *Acta Mechanica, Supplementum*, **1**, 21–52.

—— & HALLET, B. 1986. Sediment transport by wind: Toward a general model. *Geological Society of America Bulletin*, **97**, 523–535.

——, SORENSEN, M. & WILLETTS, B. B. 1991. A review of recent progress in our understanding of aeolian sediment transport. *Acta Mechanica Supplementum*, **1**, 1–20.

BAGNOLD, R. A. 1941. *The Physics of Blown Sand and Desert Dunes*. Methuen, London.

BUTTERFIELD, G. R. 1991. Grain transport rates in steady and unsteady turbulent airflows. *Acta Mechanica, Supplementum*, **1**, 97–122.

CHEPIL, W. S. 1945a. Dynamics of wind erosion: 1. Nature of movement of soil by wind. *Soil Science*, **60**, 305–320.

—— 1945b. Dynamics of wind erosion: 2. Initiation of soil movement. *Soil Science*, **60**, 397–411.

GERETY, K. M. 1985. Problems with determination of u_* from wind-velocity profiles measured in experiments with saltation. *In*: BARNDORFF-NIELSEN, O. E. & WILLETTS, B. B. (eds) *Proceedings International Workshop on the Physics of Blown Sand*. Memoir No. 8. Dept Theoretical Statistics, Aarhus University, Denmark, 271–300.

HUNT, J. C. R. & NALPANIS, P. 1985. Saltating and suspended particles over flat and sloping surfaces. *In*: BARNDORFF-NIELSEN, O. E. & WILLETTS, B. B. (eds) *Proceedings International*

Workshop on the Physics of Blown Sand. Memoir No. 8. Dept of Theoretical Statistics, Aarhus University, Denmark, 37–66.

JENSEN, J. L. & SORENSEN, M. 1982. On the mathematical modelling of aeolian saltation. *Euromech 156: Mechanics of sediment transport, Istanbul*, 65–72.

—— & —— 1986. Estimation of some aeolian saltation transport parameters: A re-analysis of Williams' data. *Sedimentology*, **33**, 547–558.

KAWAMURA, R. 1951. Study of sand movement. *Reports of Physical Sciences Research Institute of Tokyo University*, **5** (3–4), 95–112 (in Japanese). Available in English as *National Aeronautic and Space Administration (NASA) Technical Translation F-14.*

McEWAN, I. K. & WILLETTS, B. B. 1991. Numerical model of the saltation cloud. *Acta Mechanica, Supplementum*, **1**, 53–66.

MITHA, S., TRAN, M. Q., WERNER, B. T. & HAFF, P. K. 1986. The grain-bed impact process in aeolian saltation. *Acta Mechanica*, **63**, 267–278.

MORSI, S. A. & ALEXANDER, A. J. 1972. An investigation of particle trajectories in two-phase flow systems. *Journal of Fluid Mechanics*, **55**, 193–208.

NADDEH, K. 1984. *Lift forces on spheres at Reynolds numbers between 80 and 25000.* MSc Thesis, Aberdeen University.

OWEN, P. R. 1964. Saltation of uniform grains in air. *Journal of Fluid Mechanics*, **20**(2), 225–242.

—— 1980. *The Physics of Sand Movement.* Autumn Course on Physics of Flow in the Oceans, Atmosphere and Deserts. International Centre for Theoretical Physics, Trieste.

RASMUSSEN, K. R. & MIKKELSEN, H. E. 1991. Wind tunnel observations of aeolian transport rates. *Acta Mechanica, Supplementum*, **1**, 135–144.

—— & —— 1992. The transport rate profile and the efficiency of sand traps. *Sedimentology* (in press).

——, SORENSEN, M. & WILLETTS, B. B. (1985). Measurement of saltation and wind strength on beaches. *In*: BARNDORFF-NIELSEN, O. E. & WILLETTS, B. B. (eds) *Proceedings of International Workshop on the Physics of Blown Sand.* Memoir No. 8, Dept of Theoretical Statistics, Aarhus University, Denmark, 301–326.

REIZES, J. A. 1978. Numerical study of continuous saltation. *Journal of the Hydraulics Division, Proceedings American Society of Civil Engineers*, **104** (HY9), 1305–1321.

RICE, M. A. 1991. Grain shape effects on aeolian sediment transport. *Acta Mechanica, Supplementum*, **1**, 159–166.

RUMPEL, D. A. 1985. Successive aeolian saltation: studies of idealized collisions. *Sedimentology*, **32**, 267–280.

SAFFMAN, P. G. 1964. The lift on a small sphere in a slow shear flow. *Journal of Fluid Mechanics*, **22**, 385–400.

SCHILLER, L. & NAUMAN, A. 1933. *Z. Ver. Dent. Ing.*, **77**, 318.

SORENSEN, M. 1991. An analytic model of wind-blown sand transport. *Acta Mechanica, Supplementum*, **1**, 67–82.

SWANSON, W. M. 1961. The Magnus Effect: A summary of investigations to date. *Transactions American Society of Mechanical Engineers. Journal of Basic Engineering*, **83**, 461–470.

TSUCHIYA, Y. 1969a. Mechanics of the successive saltation of a sand particle on a granular bed in a turbulent stream. *Bulletin Disaster Prevention Research Institute, Kyoto University*, **19** (1), no. 152, 31–44.

—— 1969b. On the mechanics of saltation of a spherical sand particle on a turbulent stream. Proceedings 13th Congress of the International Association for Hydraulic Research. *Bulletin Disaster Prevention Research Institute, Kyoto University*, **19**(1), no. 152, 52–570.

—— 1970. Successive saltation of a sand grain by wind. *Coastal Engineering Conference American Society of Civil Engineers*, **III**, 1417–1427.

UNGAR, J. E. & HAFF, P. K. 1987. Steady state saltation in air. *Sedimentology*, **34**, 289–299.

WERNER, B. 1987. *A physical model of wind blown sand transport.* PhD Thesis, California Institute of Technology.

—— 1990. A steady state model of wind-blown sand transport. *Journal of Geology*, **98**, 1–17.

—— & HAFF, P. K. 1988. The impact process in aeolian saltation: Two dimensional simulations. *Sedimentology*, **35**, 189–196.

WHITE, B. R. & SCHULZ, J. C. 1977. Magnus effect in saltation. *Journal of Fluid Mechanics*, **81** (3), 497–512.

WHITE, F. 1974. *Viscous Flow.* McGraw Hill, New York.

WILLETTS, B. B. 1987. Particulate transport in the atmosphere. *In*: STEFFEN, W. L. & DENMEAD, O. T. *Flow and Transport in the Natural Environment: Advances and Applications.* Springer-Verlag, 300–311.

——, McEWAN, I. K. & RICE, M. A. 1991. Initiation of motion of quartz sand grains. *Acta Mechanica, Supplementum*, **1**, 123–134.

—— & MURRAY, C. G. 1981. Lift exerted on stationary spheres in turbulent flow. *Journal of Fluid Mechanics*, **105**, 487–505.

—— & RICE, M. A. 1985. Inter-saltation collisions. *In*: BARNDORFF-NIELSEN, O. E. & WILLETTS, B. B. (eds) *Proceedings International Workshop on the Physics of Blown Sand.* Memoir No. 8. Dept of Theoretical Statistics, Aarhus University, Denmark, 83–100.

—— & —— 1986a. Collisions in aeolian saltation. *Acta Mechanica*, **63**, 255–265.

—— & —— 1986b. Collisions in aeolian transport: the saltation/creep link. *In*: NICKLING, W. G. (ed.) *Aeolian Geomorphology.* Allen and Unwin, Boston, 1–17.

—— & —— 1989. Collisions of quartz grains with a sand bed: the influence of incident angle. *Earth Surface Processes and Landforms*, **14**, 719–730.

WILLIAMS, J. J., BUTTERFIELD, G. R. & CLARK, D. G. 1990. Rates of aerodynamic entrainment in a developing boundary layer. *Sedimentology*, **37**, 1037–1048.

Gain function analysis of sand transport in a turbulent air flow

JACK HARDISTY, HELEN L. ROUSE & SAMANTHA HART

School of Geography & Earth Resources, The University of Hull HU6 7RX, UK

Abstract: Recent work on the modelling of sand transport processes in the wind is concentrating on the effects of turbulence and, in particular, on flow accelerations and particle inertia. An approach is developed here in which the problem is cast in the frequency domain using the particle momentum equation which balances grain inertia against resultant steady and acceleration related fluid forces. A transport gain function is defined which relates the normalized transport spectrum to the normalized, turbulent flow spectrum. Experiments are reported in which the gain function was determined using a hot wire anemometer and a new pin impact probe to make high-frequency measurements of the velocity and sand transport profiles within the saltation layer in a small laboratory wind tunnel. The results were subjected to frequency analysis and the corresponding wind and transport spectra were obtained. Preliminary calculations suggest that the transport gain function is reasonably flat over the range up to 0.4 Hz and, therefore, models based upon gain function analysis may more properly account for sand transport within a turbulent boundary layer.

The problem of modelling the rate at which sand is transported across coastal and desert dune fields has been studied by many workers since the pioneering experiments of Bagnold (1941) and reviews are reported in recent texts such as Pye & Tsoar (1990). The transport of sediment in the aeolian environment is one aspect of the more general 'two-phase' flow problem (Bagnold 1966) and the following is based upon recent developments in a number of different environments. All concern the effect of some type of geophysical boundary layer on the movement or otherwise of non-cohesive particles in the range from fine sands to coarse gravels. Recent work is concentrating on the effects of turbulence and, in particular, on the flow accelerations and particle inertia terms which must be introduced when unsteady flow parameters are introduced into the model.

The grain momentum model

One approach which seeks explicitly to include the acceleration and intertial terms is the *grain momentum equation* which attempts to account for all of the forces acting on the grains and to balance the resultant against change in the grain momentum. That is, the formulation is based upon Newton's Second Law, force = mass × acceleration. The approach includes unsteady flow effects since it accounts for the instantaneous force values due to flow turbulence. This approach is reported by, for example, Hinze (1975 p. 463), Ogawa (1988 p. 170). Lasek (1991 p. 204), Hardisty (1990 p. 172, 1991 p. 727)

and in a different form by Madsen (1991 p. 18). Generally, forces due to pressure drag (F_p), viscous drag (F_v), added mass (F_a), the history integral (F_h), gravity (F_g), the pressure gradient (F_{pg}) and solid friction (F_s) are considered; thus:

$$m_s \frac{du_s}{dt} = F_g - F_p + F_v + F_a + F_h + F_{pg} + F_s \tag{1}$$

where m_s and u_s are the mass and velocity of the solid grain(s) respectively. In more detail, Hardisty (1991) gives:

$$\frac{4}{3} \pi r^3 \varrho_s \frac{du_s}{dt} = \frac{4}{3} \pi r^3 (\varrho_s - \varrho_f) g$$

$$- C_D \frac{1}{2} \pi r^2 \varrho_f |u_r| u_r - C_A \frac{4}{3} \pi r^3 \varrho_f \frac{du_r}{dt}$$

$$- C_H r^2 \sqrt{\pi \varrho_f} \mu \int_0^t \frac{du_r/dt'}{\sqrt{t - t'}} dt'$$

$$+ \frac{4}{3} \pi r^3 \varrho_f \frac{du_f}{dt} \tag{2}$$

where C_D, C_A and C_H are the drag coefficients, ϱ_s and ϱ_f are the solid and fluid densities, r is the particle radius, u_s, u_f and u_r are the solid, fluid and relative velocities respectively and the solid friction term is not represented. Hardisty (1991) demonstrated that, due to the inclusion of the acceleration related effects of the added mass and history integral forces, the model can adequately describe the behaviour of a single, spherical grain in a highly unsteady flow. However, he noted that, since flow must be described

From Pye, K. (ed.), 1993, *The Dynamics and Environmental Context of Aeolian Sedimentary Systems.*
Geological Society Special Publication No. 72, pp. 17–23.

at each particle and since the acceleration and inertial terms require information on the history of the flow and turbulence is, by definition, a random phenomenon, a general transport function is unlikely to emerge from this analysis.

Turbulent gain function modelling

The preceding model was based upon analyses in the time domain and demands information on the unique history of flow parameters in a random flow field. An alternative, perhaps, is to utilize spectral analysis techniques in addressing the two-phase problem. Spectral analysis has already proven useful in the study of geophysical boundary layers (e.g. Kaimal *et al.* (1972) for atmospheric systems and Bowden (1962), Heathershaw (1976), Soulsby (1977), Anwar (1981), Shiono & West (1987) and Williams *et al.* (1989) for marine currents). It is not unreasonable, therefore, to recast the problem in the frequency domain provided that there is a firm theoretical basis for so doing and that the technology is available to obtain sufficiently high-frequency measurements of the process. The technological aspects are discussed below and here gain function analysis is used to propose an alternative sediment transport model which may account for the turbulent effects. In particular, we utilize the spectral gain function (cf. Jenkins & Watts 1969), $G(f)$, which relates an input spectrum, in the present case the wind velocity spectrum, $u(f)$, to the output spectrum, in the present case the sand transport rate spectrum, $i(f)$:

$$i(f) = G^2(f) \, u(f) \qquad (3)$$

Utilizing normalized spectra (e.g. Soulsby 1977) the gain function is simply the root of the ratio of the corresponding transport amplitudes to the flow amplitudes at each frequency:

$$G(f_n) = \sqrt{ \frac{ni(f_n)}{\int_0^n i(f_n) \, dn} \, \frac{\int_0^n u(f_n) \, dn}{nu(f_n)}} \qquad (4)$$

where f_n is the appropriate frequency and n is the number of the frequency bin. The advantage of this approach is that, since spectral analysis explicitly defines the amplitude and period of the constituent sinusoids, the acceleration and inertial terms (i.e. the differentials) are explicit and, in principle, a theoretical spectral gain function can be obtained by combining the momentum equation (Eq. 2) with the gain

function (Eq. 4). Preliminary results of such numerical analyses are reported in Hardisty (in press). Here we present preliminary results from experiments designed to determine the empirical form of the gain function defined by Eq. 4.

Experimental technique

A series of experiments were conducted in the small wind tunnel which has been described by Hardisty & Whitehouse (1988). The tunnel has a working section 100 cm in length, 4 cm wide and 8 cm high. The tunnel sides and top were fabricated from perspex and two fans delivered air into the parallel sided working section. The wind tunnel was set up on a smoothed sand surface in the laboratory and wind speed was measured at three heights (1, 3 and 5 cm above the surface) using an E.T.A. 3000 hot wire anemometer manufactured by Airflow Developments Ltd (Fig. 1). Although thermoanemometers permit relatively high-frequency measurements of turbulent wind flow characteristics, there have been few attempts to make comparable and simultaneous measurements of the transport response. Fryberger *et al.* (1979) and Butterfield (1991) used load cell devices whilst Stockton & Gillette (1990) used momentum sensors. Here, a small impact pin-probe was constructed which consisted of a 10 mm needle mounted on the surface of a piezo-electric crystal and mounted with the needle horizontal and oriented across the tunnel at the same heights as the anemometer's sensor. Signals from the probe were amplified, low pass filtered to remove high frequency noise and rms-dc con-

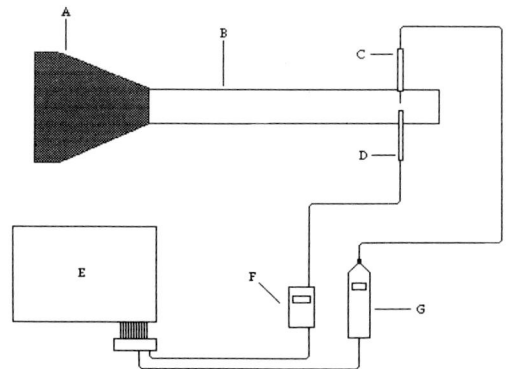

Fig. 1. Experimental arrangement showing, in plan view, the air pump unit (A), the wind tunnel (B), impact and thermoanemometer probes (C and D) and signal processing units (G and F) and the microcomputer and analogue to digital board (E).

verted by a standard Castle Instruments GA101 and GA701 sound level meter. The rectified probe output was calibrated in the wind tunnel by comparison with mean transport rates measured with a trap, similar to that shown in Bagnold's (1941) fig. 24 which extended from the sand surface to the roof of the tunnel. Analogue outputs from the anemometer and probe amplifier were led to a National Instruments 16 channel NB-M10-16H analogue to digital board. The board was mounted in a Macintosh II microcomputer and was controlled and interogated by a logging program written in the DataFlow language. The system permitted rapid and simultaneous measurements to be made of the velocity profile and impact probe voltages for a range of competent wind speeds in the tunnel. A series of experiments was carried out in each of which a record of wind speed and probe voltage was measured at 2 Hz over 128 s. The results were calibrated using the manufacturers data for the anemo-

meter and by comparing the intergrated probe signals over the depth of the tunnel with the total transport collected in a combined bedload and saltation trap during each experiment.

Results

The mean transport rate profiles for three of the sets of experiments are shown in Fig. 2, confirming that the mean concentration decreases with height above the bed in broad agreement with a power law. Two typical datasets were then identified for further analysis and are shown in Fig. 3. The data are plotted as corresponding time series over the duration of the experimental runs of the wind velocity and the saltation transport rate. It is apparent that although there is a broad agreement between trends in the wind velocity and transport rate time series, there appears to be little correlation between the instantaneous values at this scale.

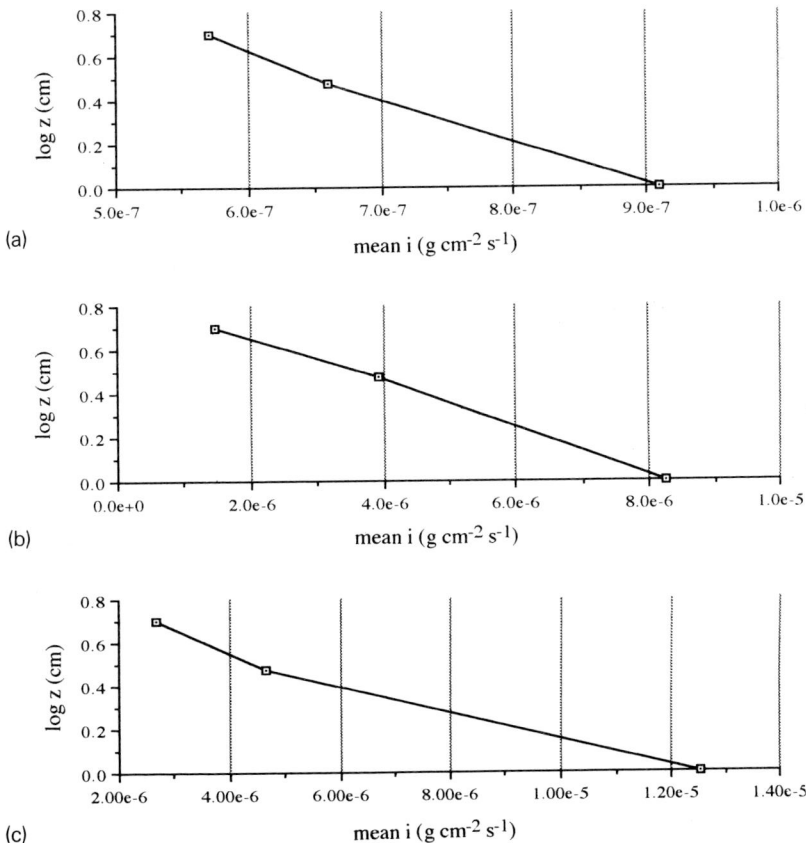

Fig. 2. Mean transport rate profiles for (**a**) record 7.5, (**b**) record 10.5 and (**c**) record 12.5, where the increasing record reference numbers indicate increasing mean wind speed.

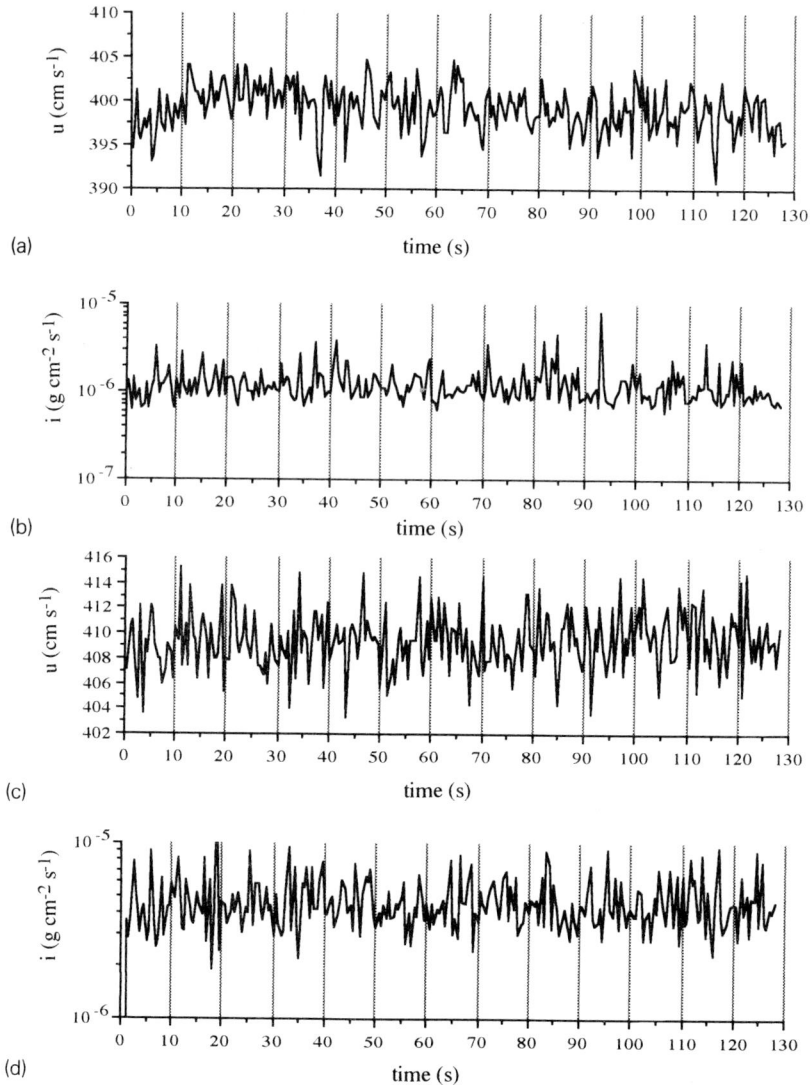

Fig. 3. Time series of wind speed (**a** and **c**) and transport rate (**b** and **d**) for record 9.5 zl (**a** and **b**) and 12.5 z2 (**c** and **d**).

Analysis

The calibrated and de-trended time series for each experiment were subjected to spectral analysis through Hanning window and a standard FFT routine. Three way ensemble and frequency domain averaging was applied to each record and the results are shown as power spectra in Figs 4 and 5. The wind-speed spectra appear to be reasonable showing the now well-documented decay with frequency. The new transport spectra appear not unreasonable, also showing a broadband shape with a relatively low frequency peak and the decay at the higher ranges. The wind-speed and transport spectra were, therefore, normalized by dividing each by the spectral value at an (arbitrary) frequency of 0.1 Hz and the gain functions (Eq. 3) were derived by plotting the square root of the resulting transport spectra divided by the corresponding wind speed spectra at each frequency as shown in Figs 4c and 5c.

(a)

frequency (Hz)

(b)

frequency (hz)

(c)

frequency (Hz)

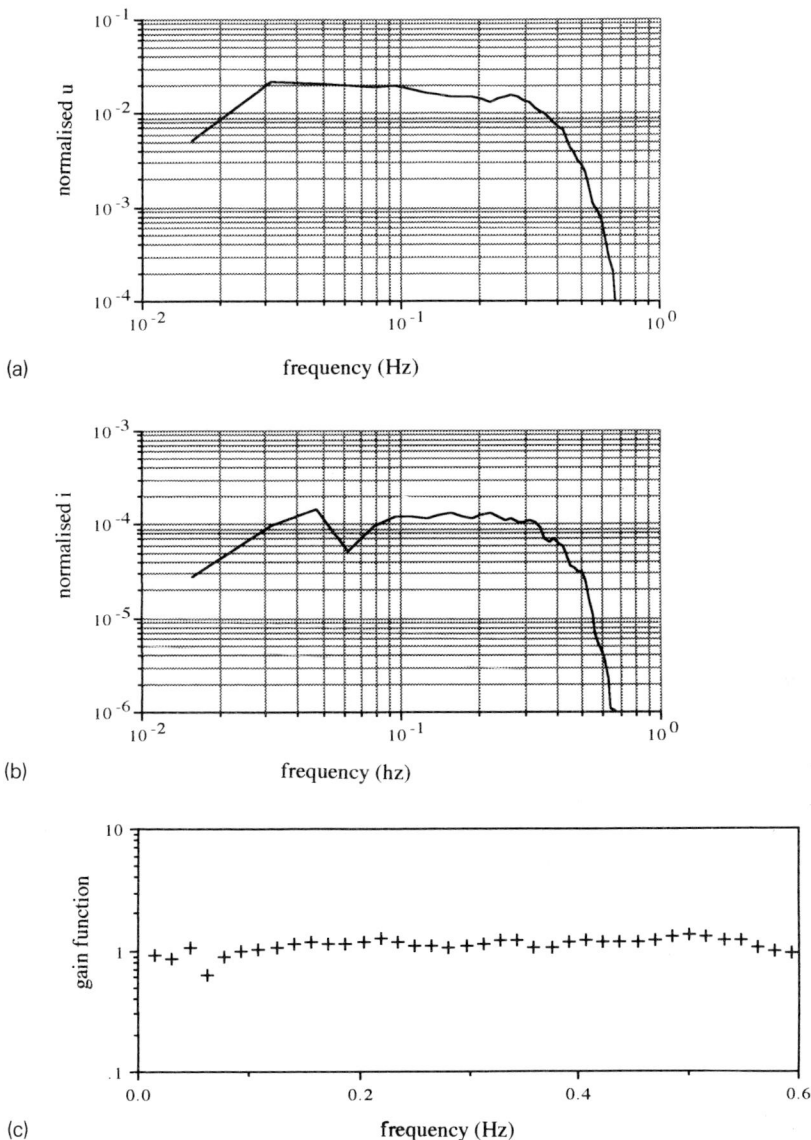

Fig. 4. Power spectra of (**a**) wind speed, (**b**) transport rate and (**c**) normalized gain function for record 9.5 z1.

Conclusion and discussion

The results appear to suggest that the sand transport gain function, as defined by Eq. 3 is remarkably flat with frequency and, using the normalizing procedures outlined above, varies between about 0.5 and 1.5. Clearly, such variation is very small when compared with the range of more than two orders of magnitude in both the windspeed and transport rate spectra and, therefore, we conclude that the use of gain function analysis may offer a useful technique for the quantification of sand transport rates in unsteady, turbulent winds. It appears that, for the fine-grained, sand-sized sediment examined in the present experiments, the normalized transport spectrum can be reasonably predicted using a unity gain function from the normalized, turbulent wind spectrum. The conclusion may provide some insight into a second geomorphological problem. Thornes (1990), writing about the apparent self similarities in geomorphological landform systems, com-

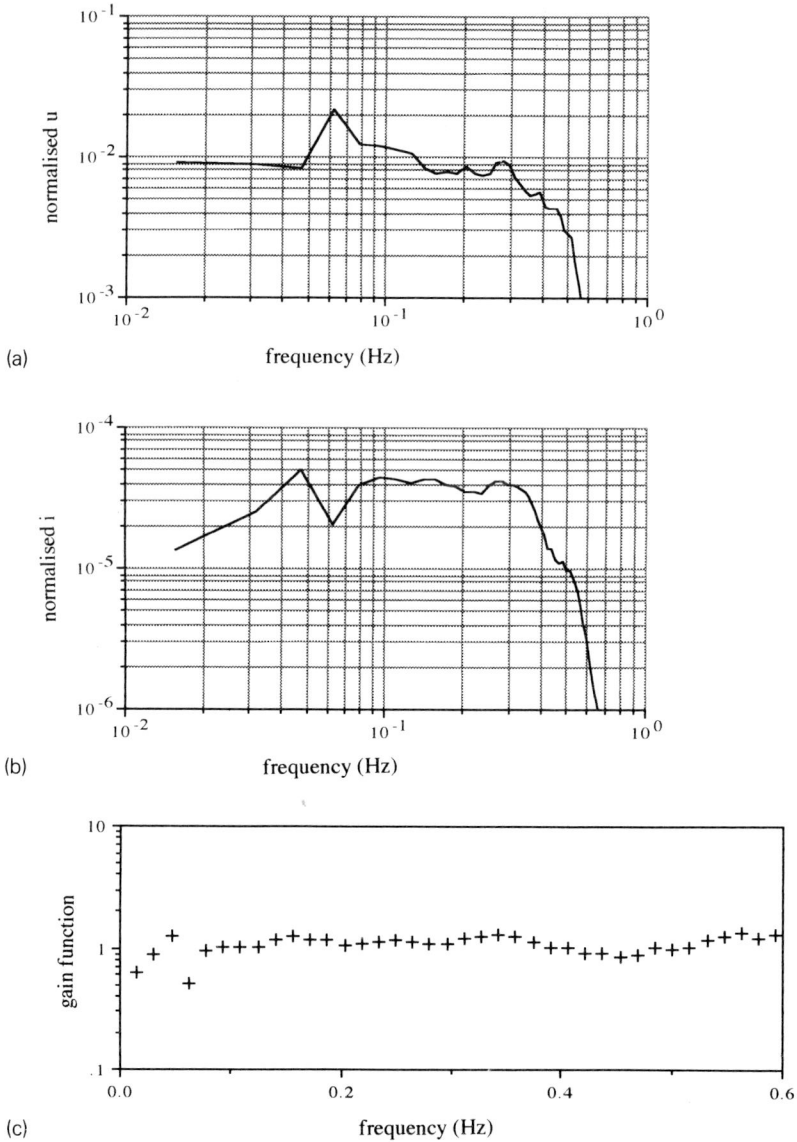

(a)

(b)

(c)

Fig. 5. Power spectra of (**a**) wind speed, (**b**) transport rate and (**c**) normalized gain function for record 12.5 z2.

mented on the spatial coupling between landscape surface forms and the erosive mechanisms revealed by the relationships between spectral energy density and wavenumber in the size spectrum of landscape features. He noted that any such coupling might be non-linear so that the resulting rate coefficient is a function of successive wavenumbers and would then be based upon the same basic proposition as the Kolmogorov–Obhukov cascade. That is to say

the model assumes the same energy cascade as is utilized at present for the structure of energy dissipation in turbulent fluid flow (cf. Huntley 1988). Thornes noted that this process would require a means of transforming a temporally varying erosional force (in the present case the wind speed) into a spatially self-similar landform. It is possible that the type of gain function analysis described above provides the elements of such a transformation, and the well known self similarity of aeolian structures (e.g. Anderson (1990), ripple to megaripples to various scales of dune dimension) may provide a suitable natural laboratory for appropriate wave number analyses.

References

ANDERSON, R. S. 1990. Aeolian ripples as examples of self organisation in geomorphological systems. *Earth Science Reviews*, **29**, 77–96.

ANWAR, H. O. 1981. A study of the turbulence structure in a tidal flow. *Estuarine, Coastal and Shelf Science*, **13**, 373–387.

BAGNOLD, R. A. 1941. *The Physics of Blown Sand and Desert Dunes*. Methuen, London.

—— 1966. *An approach to the sediment transport problem from general physics*. US Geological Survey Professional Paper, **422-I**.

BOWDEN, K. F. 1962. Measurement of turbulence near the seabed in a tidal current. *Journal of Geophysical Research*, **67**, 3181–3186.

BUTTERFIELD, G. R. 1991. Grain transport rates in steady and unsteady turbulent airflows. *Acta Mechanica, Supplement*, **1**, 97–122.

FRYBERGER, S. G., AHLBRANDT, T. S. & ANDREWS, S. 1979. Origin, sedimentary features and significance of low-angle eolian sand sheet deposits, Great Sand Dunes National Monument and vicinity, Colorado. *Journal of Sedimentary Petrology*, **49**, 733–746.

HARDISTY, J. 1990. *Beaches: Form and Process*. Unwin-Hyman, London.

—— 1991. Bedload transport under low frequency waves. *Proceedings of Coastal Sediments '91*. American Society of Civil Engineers, New York, 726–733.

—— in press. Frequency analysis of sand transport in a turbulent air flow. *In*: CLIFFORD, N. J., HARDISTY, J. & FRENCH, J. (eds) *Turbulence: perspectives on sediment transport*. Wiley, Chichester.

—— & WHITEHOUSE, R. J. S. 1988. Evidence for a new sand transport process from experiments on Saharan dunes. *Nature*, **332**, 532–534.

HEATHERSHAW, A. D. 1976. Measurements of turbulence in the Irish Sea benthic boundary layer. *In*:

McCAVE, I. N. (ed.) *The Benthic Boundary Layer*. Plenum Press, New York, 11–31.

HINZE, J. O. 1975. *Turbulence*. 2nd Ed. McGraw-Hill, New York.

HUNTLEY, D. A. 1988. A modified inertial dissipation method for estimating seabed stresses at low Reynolds numbers with application to wave current boundary layer measurements. *Journal of Physical Oceanography*, **18**, 339–362.

JENKINS, G. M. & WATTS, D. G. 1969. *Spectral Analysis and its Applications*. Holden-Day, San Francisco.

KAIMAL, J. C., WYNGAARD, J. C., IZUMI, Y. & COAT, O. R. 1972. Spectral characteristics of surface layer turbulence. *Quarterly Journal of the Royal Meteorological Society*, **98**, 563–589.

LASEK, A. 1991. On some problems of particles' behaviour in shear flow. *In*: SOULSBY, R. & BETTESS, R. (eds) *Euromech 262 — Sand Transport in Rivers, Estuaries and the Sea*. Balkema, Rotterdam, 203–207.

MADSEN, O. S. 1991. Mechanics of cohesionless sediment transport in coastal waters. *Proceedings of Coastal Sediments '91*. American Society of Civil Engineers, New York, 15–27.

OGAWA, Y. 1988. Mechanism of sediment transport. *In*: HORIKAWA, K. (ed.) *Nearshore Dynamics and Coastal Processes*. University of Tokyo Press, Tokyo, 167–193.

PYE, K. & TSOAR, H. 1990. *Aeolian Sand and Sand Dunes*. Unwin Hyman, London.

SHIONO, K. & WEST, J. R. 1987. Turbulent perturbations of velocity in the Conwy Estuary. *Estuarine and Coastal Shelf Science*, **25**, 533–553.

SOULSBY, R. L. 1977. Similarity scaling of turbulence spectra in marine and atmospheric boundary layers. *Journal of Physical Oceanography*, **7**, 934–937.

STOCKTON, P. & GILLETTE, D. A. 1990. Field measurements of the sheltering effect of vegetation on erodible land surfaces. *Land Degradation and Rehabilitation*, **2**, 77–85.

THORNES, J. B. 1990. Big rills have little rills. *Nature*, **345**, 764–765.

WILLIAMS, J. J., THORNE, P. D. & HEATHERSHAW, A. D. 1989. Measurements of turbulence in the benthic boundary layer over a gravel bed. *Sedimentology*, **36**, 959–971.

Wind-speed profiles over a reversing transverse dune

JENNIFER R. BURKINSHAW, WERNER K. ILLENBERGER
& IZAK C. RUST

*Institute for Coastal Research, Geology Department, University of Port Elizabeth,
PO Box 1600, Port Elizabeth 6000, South Africa*

Abstract: Wind-speed profiles were measured within 1.5 m of the surface on the windward slope of a 7 m high transverse dune that reverses seasonally with respect to summer easterly and winter northwesterly winds. The profiles were measured along section lines normal to the dune trend at different stages of dune reversal. An independent weather station recorded relatively undisturbed flow at 6 m elevation above the surface.

Flow deceleration at the base of the windward slope and flow acceleration against the middle to upper slope generate non-logarithmic wind-speed profiles. The shape of the dune and the strength of the wind play a major role in determining the behaviour of the wind-speed profile. Airflow compression is enhanced by steeper slopes, resulting in high surface shear stress that causes rapid erosion of the windward slope.

The wind-speed profile can be divided into three zones: a surface shear stress layer, an amplification layer and a recovery layer. It seems that the shear stress that governs sand transport is developed within 5 cm of the surface, in the surface shear stress layer. Future work should concentrate on measurements within this zone.

Detailed measurements of wind-speed profiles over aeolian bedforms are rare. Prior to 1988 most of the work was hampered by lack of equipment. In many studies only one anemometer was used per station, and assumptions, based on tenuous data, were made about the behaviour of wind-speed profiles. Several detailed studies have now improved our knowledge of flow behaviour over aeolian bedforms (e.g. Hesp *et al.* 1989; Mikkelsen 1989; Butterfield 1991; Wiggs this volume), but many more measurements need to be made before the dynamic nature of the aeolian environment can be modelled reliably.

The Karman–Prandtl law for turbulent flow has been extended for measurement over mobile surfaces by Bagnold (1941) who found that for a naturally graded rippled sand (grain size 0.25 mm), threshold speed is 4 m s^{-1} at z_0 of 1 cm (Bagnold 1941 p. 69). In addition, he established that the sand-transporting capacity of the wind is proportional to the cube of the shear velocity. In this paper, we assume that this relationship holds.

It is well established in fluid dynamic theory that the projection of an obstacle, such as a sand dune, into the atmospheric boundary layer results in divergence of streamlines at the dune base and convergence towards the dune crest, causing an acceleration of flow on the upstream (windward) side of the form. Flow separation occurs at the crest and the separated layer re-attaches itself to the surface some distance downstream of the separation. Airflow behaviour controls the erosion, transport and deposition of sand on a dune surface, and ultimately, the formation of sand dunes.

Wind-tunnel investigation of airflow over a range of forward-facing escarpments of different slopes has yielded the height of recovery of airflow from the influence of the escarpment to be three times the escarpment height under strong wind conditions (Bowen & Lindley 1977). Field measurements by Bowen & Lindley (1974, *in* Hunt 1980) recorded a jet flow at the top of a vertical escarpment.

Because airflow streamlines first diverge and then converge on the windward slope, wind speed will not increase linearly with respect to the logarithm of the height above the surface as predicted by Karman–Prandtl theory. Mulligan (1988) confirmed the non-logarithmic nature of wind-speed profiles measured within 1.6 m of the surface on the windward slope of a 5 m high dune, and found that the behaviour of the wind-speed profile varied according to slope position. However, the nature of the response of the wind-speed profile to various dune shapes of varying size has not been investigated, although some wind-speed profiles over hills have been published (e.g. Bradley 1980, 1983).

Equilibrium transverse dune forms typically have an asymmetric shape, with a stoss slope of 10–15° and a lee slope of 33° (Cooke & Warren

From Pye, K. (ed.), 1993, *The Dynamics and Environmental Context of Aeolian Sedimentary Systems.*
Geological Society Special Publication No. 72, pp. 25–36.

1973). Winds approaching a dune at 90° experience most compression. The effective slope gradient is reduced for winds approaching a dune obliquely. We will only deal with wind approaching the dune at 90°.

We have measured wind-speed profiles within 150 cm of the surface on the windward flanks of a 7 m high reversing transverse dune during two seasonally opposing wind regimes. Burkinshaw & Rust (in press) showed that wind speed measured at 6 cm above the dune surface correlates with sand flux over the surface, and is very responsive to the surface gradient and to changes in surface gradient. Here, we discuss the entire wind-speed profile which adjusts progressively as the dune shape is modified towards attaining an equilibrium shape with respect to the prevailing winds.

Our measurements are a first step towards developing an airflow model that would allow the prediction of sand movement over a transverse sand dune and adjacent interdune area. We relate our data to some aspects of the model by Hunt *et al.* (1988).

Study site and methods

The study dune was chosen from a 10 km long strip of small reversing transverse dunes located behind the beach east of the Sundays River

mouth, along the seaward edge of the western end of the Alexandria coastal dunefield, South Africa (Figs 1 & 2). The dunes average 150–300 m in length, have a crest to crest spacing of about 80 m and a maximum crest height of about 7 m.

The dunefield is subjected to a trimodal wind regime (Illenberger 1988; Fig. 1). The dominant onshore southwesterlies blow all year round; the easterlies blow onshore during summer, and the northwesterlies blow offshore during winter. The latter two winds blow transverse to the study dune axis (Fig. 3), and cause the morphology of the dune to reverse seasonally.

Wind-speed measurements were made during the early and middle stages of the summer easterly wind season, and during the middle to late stage of the winter northwesterly wind season. Different equipment was at out disposal for each season's measurements. The procedure in both cases was to record the airflow on the windward slope from the base to the crest of the dune, positioned on a section line, using vertical arrays of anemometers.

Typically, wind speed was measured over one-minute intervals and averaged over a minimum period of 20 minutes, except when surface conditions changed rapidly, in which case a reduced period of 1 to 2 minutes was used.

Conventionally, anemometers are rigged at a

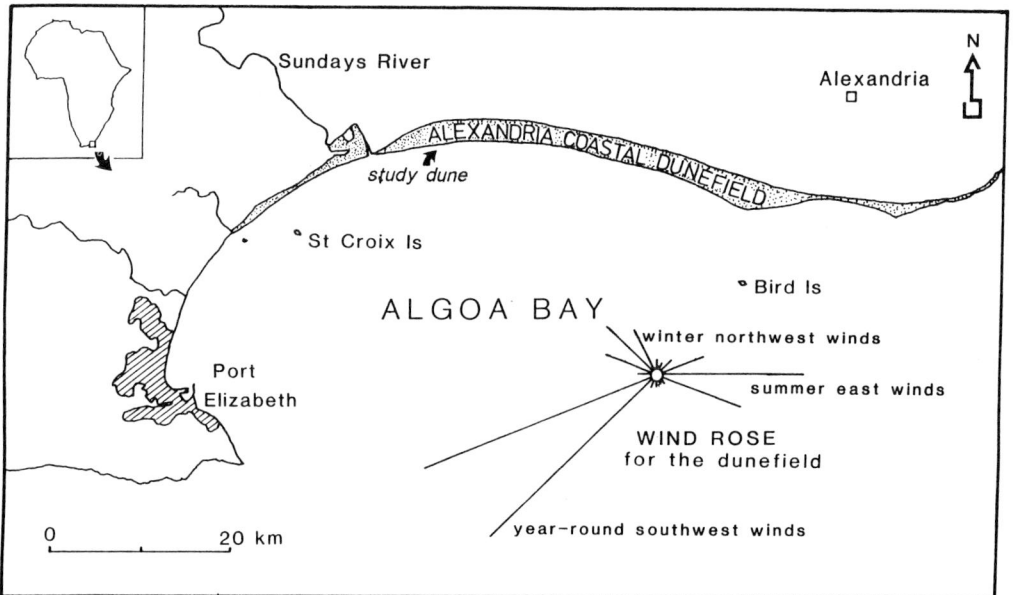

Fig. 1. Locality map of the Alexandria coastal dunefield including a stylized windrose for the dunefield (modified from Illenberger 1988).

Fig. 2. High-angle oblique aerial photograph of the Alexandria coastal dunefield, looking east. The strip of small reversing transverse dunes along the coast is delineated. The study dune is in the central portion of this strip.

logarithmic spacing above the surface in accordance with the logarithmic wind-speed profile. In practice, as the wind speed profile over a dune surface is non-logarithmic, and as the number of anemometers was limited, arbitrary heights were chosen on an approximately logarithmic vertical spacing.

We consider our measurements to be sufficiently close to the surface for thermal buoyancy effects to be neglected. Rasmussen *et al.* (1985) speculate that, for measurements made within 1 m of the dune surface, stability effects are insignificant compared to measurement error. In addition, Greeley & Iversen (1985) assume that the turbulence of winds that exceed the threshold velocity for sand movement mixes air sufficiently to result in neutral conditions.

In addition to measuring airflow over the dune, a 6 m control station monitored, as far as possible, the relatively undisturbed airflow at 3 and 6 m, as opposed to the locally accelerated airflow close to the dune surface. The reference station was rigged up at the beach end of the dune, where the crestal height was about half the maximum dune height, during measurement of the summer easterlies, and on the beach during the winter northwesterlies. During measurement sessions in the winter wind season, four additional anemometers were rigged up to a height of 1.5 m at the beach site next to the

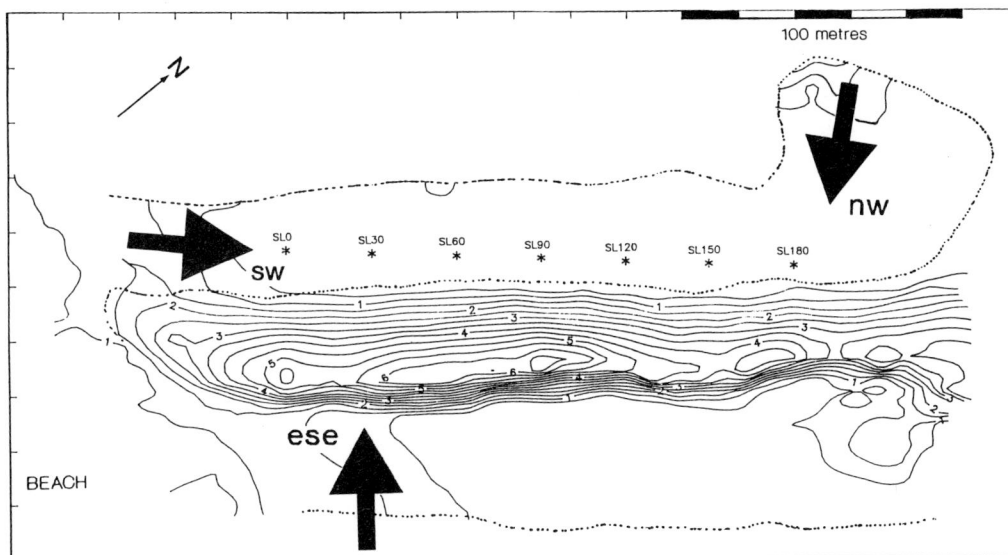

Fig. 3. Contour map of the study dune showing the orientation of the dune crest with respect to the prevailing winds. SL0, SL30, SL60, etc. are section lines positioned at 0 m, 30 m, 60 m, etc. from the southwestern end of the dune. Contour interval is 0.5 m.

reference station, completing measurement of the profile from 16 cm to 6 m above the surface.

Anemometers were calibrated using two different methods: (i) instruments were cross-calibrated in the field and (ii) selected instruments were calibrated in a wind tunnel at the University of Cape Town. During field cross-calibration, it was found that the results for a single anemometer could vary by up to 4%. The inconsistencies were probably due to variations in the rigging despite considerable care taken in setting up the configuration. Such inconsistencies could easily have been duplicated in the field. Consequently, the error in wind-speed measurement is taken to be 4%.

The dune was surveyed with a theodolite at regular intervals along seven section lines transverse to the dune axis (Fig. 3). The amount of erosion and deposition and local gradient measurements were recorded at erosion pins staked at 1–2 m intervals along each section line.

Results and interpretation

The data are presented in terms of the status of the dune with respect to the prevailing wind. A dune profile is considered to be in equilibrium when the stoss slope angle is 10–15°, and the shape of the dune does not change under the prevailing winds (Fig. 4). The extreme case of non-equilibrium occurs when a slipface becomes the upwind surface of the dune, in which case the stoss slope angle would be 33°. A slope which is halfway in the process of being converted from a slipface to a stoss slope, is in semi-equilibrium with respect to the wind.

Equilibrium profile

By the time the winter wind season measurements were made, the dune had almost reached equilibrium. Several sets of wind-speed profiles were measured on the west flank of the dune during light winds (typically 7–9 m s^{-1} at 6 m).

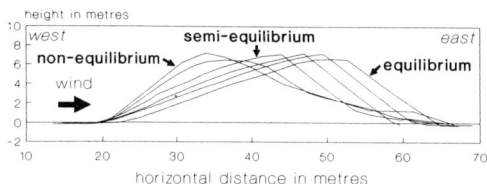

Thirty-five anemometers manufactured in-house, using Aandera style cups (rotating cup diameter of 15 cm, vertical axis 10 cm), were available for measurement of the winter wind-speed profile. Usually 4 vertical arrays of either 4 or 5 anemometers each were deployed simultaneously, spaced equidistantly on a section line from the base of the dune to the dune crest. Anemometers were deployed at heights of 10 cm, 30 cm, 60 cm and 150 cm above the surface.

Data from 26 June 1989 measured over a 20 minute period along section line SL60 are reported here. The reference station recorded an average wind speed of 9.7 m s^{-1} at 6 m height on the beach. The wind approached the dune at an angle of 82° with respect to the dune crest line. The windward slope of the study dune was close to attaining an equilibrium shape with respect to the prevailing northwest winds (Fig. 5). Typical gradient measurements were 13–14° on the lower to mid-slope; the dune gradient decreased towards the crest.

The graphs of the wind-speed profile are non-linear when plotted on a log scale (Fig. 6). Wind speed increases with height above the dune surface, and also progressively upslope at all levels in the arrays, except close to the surface at the brink, where airflow decelerates within 10 cm of the surface in the rounded crestal region, but recovers its speed within 20 cm of the surface. Maximum shearing occurs within 70 cm of the surface. Above this height the wind-speed profile shows little increase in velocity with increase in height.

Wind speed at the base of the dune is considerably slower than the reference profile, and the largest rate of acceleration occurs from the base of the dune to the next measurement position, 13 m upslope. The basal profile differs from those measured upslope in that shear velocity

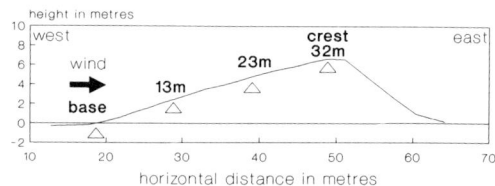

Fig. 5. Profile of section line SL60 on 7 July 1989, 10 days after the wind-speed measurements presented in Fig. 6. Triangles indicate the approximate locations of anemometer arrays on the windward slope, given in metres upslope from the dune base. The dune shape is almost in equilibrium with the northwest wind; stoss slope angle is 11–14°.

Fig. 4. Comparison of dune profiles for section line SL90, showing the different stages of reversal from March–September 1989, during the 1989 winter wind season.

height in m above surface

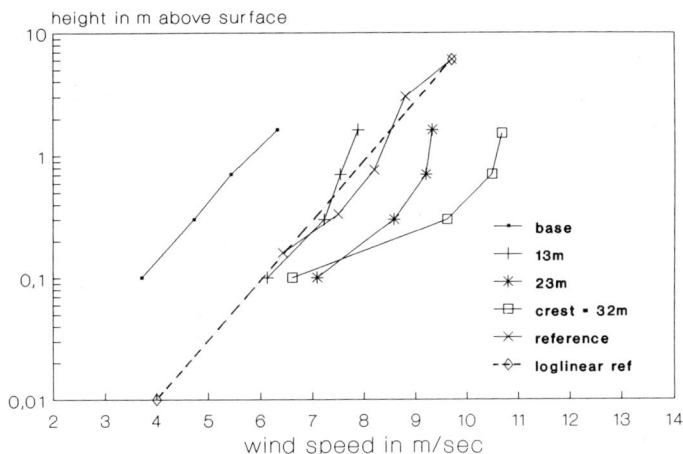

Fig. 6. Equilibrium wind-speed profiles measured along SL60 on 26 June 1989. The reference profile was measured on the beach. Legend indicates the station locations in metres upslope from the dune base; dune profile is shown in Fig. 5.

tends to increase towards the top of the wind-speed profile.

The reference wind-speed profile recorded on the beach is non-linear when plotted on a log scale, as compared with the linear profile produced by projecting the 6 m wind speed down to Bagnold's (1941 p. 69) threshold values of 4 m s^{-1} at z_0 of 1 cm (Fig. 6). Factors which may have distorted the reference wind-speed profile are the beach gradient, changes in surface roughness upwind of the station, and the presence of the upwind dune.

Wind strength on the beach is comparable to the strength recorded on the lower to middle slope of the dune. At 16 cm above the surface the reference wind speed is slower than the wind speed measured 13 m upslope from the dune base; but above 16 cm, the reference wind speed is faster. Wind speed measured on the upper slope and at the crest is faster than that of reference profile at all levels.

Non-equilibrium profile

Measurements were made at the start of the summer wind season when profiles of very light winds blowing against the east-facing slipface were recorded using 5 or 6 Rimco miniature

height in m above surface

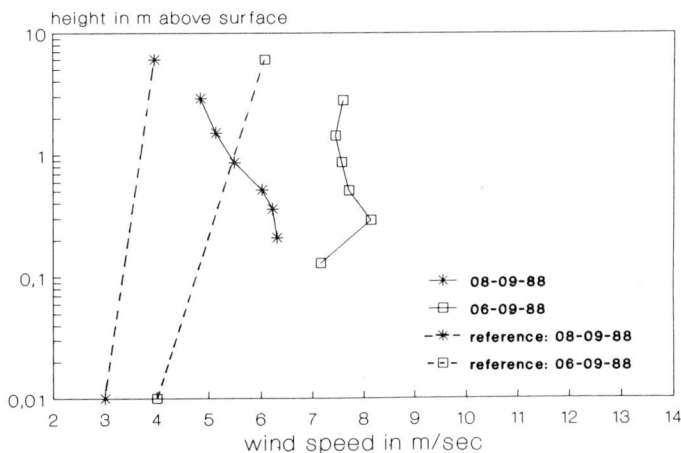

Fig. 7. Examples of two non-equilibrium wind-speed profiles with near-surface jet developed, measured during September 1988. Values for reference wind speed at 6 m are interpolated (see text). The near-surface wind-speed gradients of the non-equilibrium profiles are approximately three times that of the reference profile.

anemometers (rotating cup diameter 10 cm) in an array 2.8 m high, stationed at the top of the slipface. On 6 and 8 September 1988, the fastest wind speed (6–8 m s^{-1}) was recorded at 20–50 cm height above the surface with wind speed decelerating higher up in the wind-speed profile to give an inverted wind-speed profile or jet close to the surface (Fig. 7). This results in extremely high shear velocity over the surface, calculated to be roughly three times that of the reference profile (Fig. 7). As reference profile measurements were not measured at the dune site on this occasion, values of the reference profile are interpolated from a nearby weather station on the coast 3 km away, and are included in the graph to illustrate the extent of the flow acceleration compared to the low shear velocity of the prevailing light winds.

The results reflect extreme streamline compression against the top of the slipface as the wind approaches the steep 33° slope. The decrease in wind speed higher in the wind-speed profile is probably a reaction to extreme compression close to the surface. The jet in the airflow results in a sand plume being transported several metres downwind of the brink. Airflow separates in the downwind region.

Semi-equilibrium profile

This set of measurements was made midway through the summer wind season on the east-facing dune flank along section line SL60; the dataset of 17 November 1988 is reported here. The east flank of the dune was in the process of being reversed from a regular avalanche slipface to a stoss slope (Fig. 8); typical gradients of the lower dune slope were 20–24°. The slope started rounding off towards a flattened crest half-way upslope. The wind approached the dune crest line at almost exactly 90°.

At this time, eight Rimco microanemometers

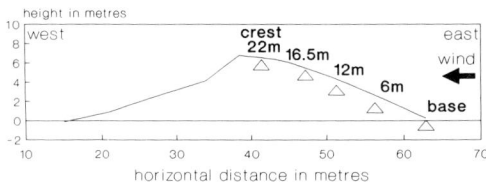

Fig. 8. Profile of section line SL60, 18 November 1988. Triangles indicate the approximate locations of anemometer arrays on the windward slope, given in metres upslope from the dune base. The dune shape is in semi-equilibrium with the east wind; stoss slope angle = 20–24°.

were available for wind-speed measurements. Two vertical arrays of four anemometers each rigged at average heights of 6 cm, 20 cm, 60 cm and 140 cm above the surface were deployed simultaneously. One array was retained at the crest as a control while the other array was moved to different locations downslope along a section line.

Hourly wind-speed averages of 14–16 m s^{-1} were recorded at 6 m by the reference station at the lower (4 m dune height) beach end of the dune. Wind speed dropped to 10 m s^{-1} at the end of the period of measurement so wind-speed data are normalized with respect to the data measured at the top of the crestal array to enable simultaneous comparison of the wind-speed profiles.

The semi-equilibrium wind-speed profiles of 17 November 1988 (Fig. 9) do not show the same systematic increase of wind speed upslope as those recorded under near-equilibrium conditions in winter. At 6 cm above the surface, airflow accelerates rapidly from the base of the dune to the middle slope (12 m from the base), where the highest wind speed and most erosion were recorded. The high rate of acceleration is attributed to enhanced streamline compression against the steeper lower slope of the dune. Further upslope, airflow decelerates at 6 cm above the surface at a slope distance of 16.5 m from the dune base. Further deceleration occurs at the crest, where 25 cm of sand was deposited during a 270 minute period. This reduction in wind speed describes incipient airflow separation close to the surface in response to a rapidly reducing gradient on the upper slope and in the crestal region. Above 6 cm, airflow accelerates resulting in large wind-speed gradients between 6 cm and 20 cm above the surface. Wind speed at the crest attains its maximum speed-up with respect to the reference profile at 60 cm above the surface. There is little increase in speed between 60 cm and 140 cm above the surface at the crest.

The fastest wind speed on this occasion was measured at the top of the array stationed 16.5 m upslope from the dune base, where airflow accelerates uncharacteristically between 60 cm and 140 cm, shown as a kink in the wind-speed profile (Fig. 9). We speculate that this may be a secondary airflow response to the extreme acceleration experienced between 5 and 20 cm above the surface under strong wind conditions. However, we have recorded this feature only once; typically, the fastest speed is recorded at the top of the crestal array.

At the base of the dune shear velocity increases towards the top of the wind-speed pro-

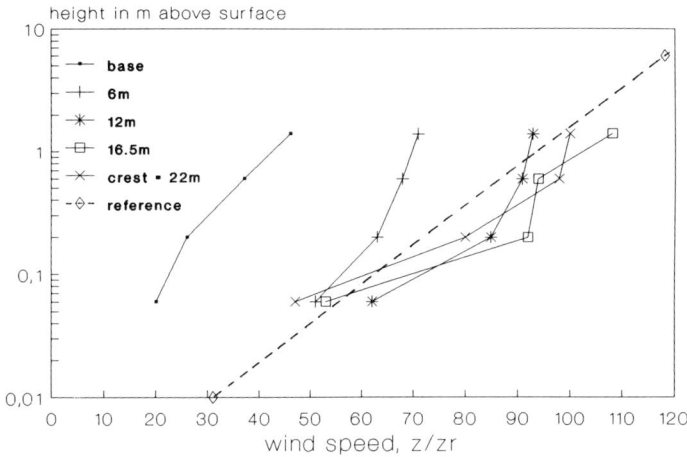

Fig. 9. Semi-equilibrium wind-speed profiles measured on 17 November 1988. Data are normalized with respect to wind speed measured at the top of the crestal array ($zr = 10$–14 m s^{-1}). Legend indicates the station locations in metres upslope from the dune base; dune profile is shown in Fig. 8. The reference profile was measured at the seaward end of the dune, at a site where the crest height was 4 m, about half the normal dune height (7 m).

file, as wind speed recovers from reduced flow close to the surface upstream of the dune.

Wind speed measured at 6 m above the surface at the reference station is extrapolated to threshold values assuming a logarithmic profile (Fig. 9). Comparison with the wind-speed profiles measured over the dune shows that the wind speed measured at the base of the dune and 6 m upslope from the dune base is slower than the approximation of the speed of the reference profile at all levels. At 12 m slope distance from the base, wind speed is accelerated with respect to the reference profile from 6 cm to approximately 80 cm above the surface. There is a zone of low wind-speed gradient between 60 and 140 cm, and this could be compensation for flow acceleration and crowding of the streamlines within 60 cm of the surface. Wind speed measured further upslope, at 16.5 m slope distance from the dune base, and at the crest, is faster than that of the reference profile. We acknowledge that the reference profile is not a true representation of undisturbed upwind conditions, and probably represents accelerated flow. Nevertheless, it appears that higher than 1.5 m above the dune surface, the wind-speed gradient would have to increase for wind speed to recover to undisturbed conditions.

Discussion

Wind-speed profile model

The seasonal reversal of a reversing transverse dune demonstrates the feedback relationship between flow and form as morphology tends towards an equilibrium shape with respect to the prevailing winds of each season (Fig. 4). Enhanced streamline compression against the steep slope at the start of the seasonal reversal results in the upper slope being modified as the wind pushes the dune over. Only once an equilibrium shape is attained does the dune start to move downwind, maintaining its shape (Fig. 4).

Airflow on the windward slope of the dune deviates from the logarithmic wind-speed profile for all three dune states under discussion. Our current data consist of wind-speed profiles measured for light winds recorded under equilibrium and non-equilibrium conditions, and for strong winds under semi-equilibrium conditions. It is necessary to differentiate between strong and gentle winds as the wind-speed profile probably behaves differently under strong wind conditions, and sand is eroded at the base of the dune only in strong wind. In our conceptual model (Fig. 10) we discuss the wind-speed profile at the crest, mid-slope and base of the windward slope for the different stages of dune reversal.

For the non-equilibrium state when the wind approaches a 33° slope at an angle of 90°, extreme compression of flow occurs close to the surface at the top of the slipface (Fig. 10a). The effective region of compression of airflow appears to depend on the strength of the wind. Under light winds, compression is confined to the brink edge. Under stronger winds, a larger area downslope of the brink is eroded. (This may be a flow response to the already reduced

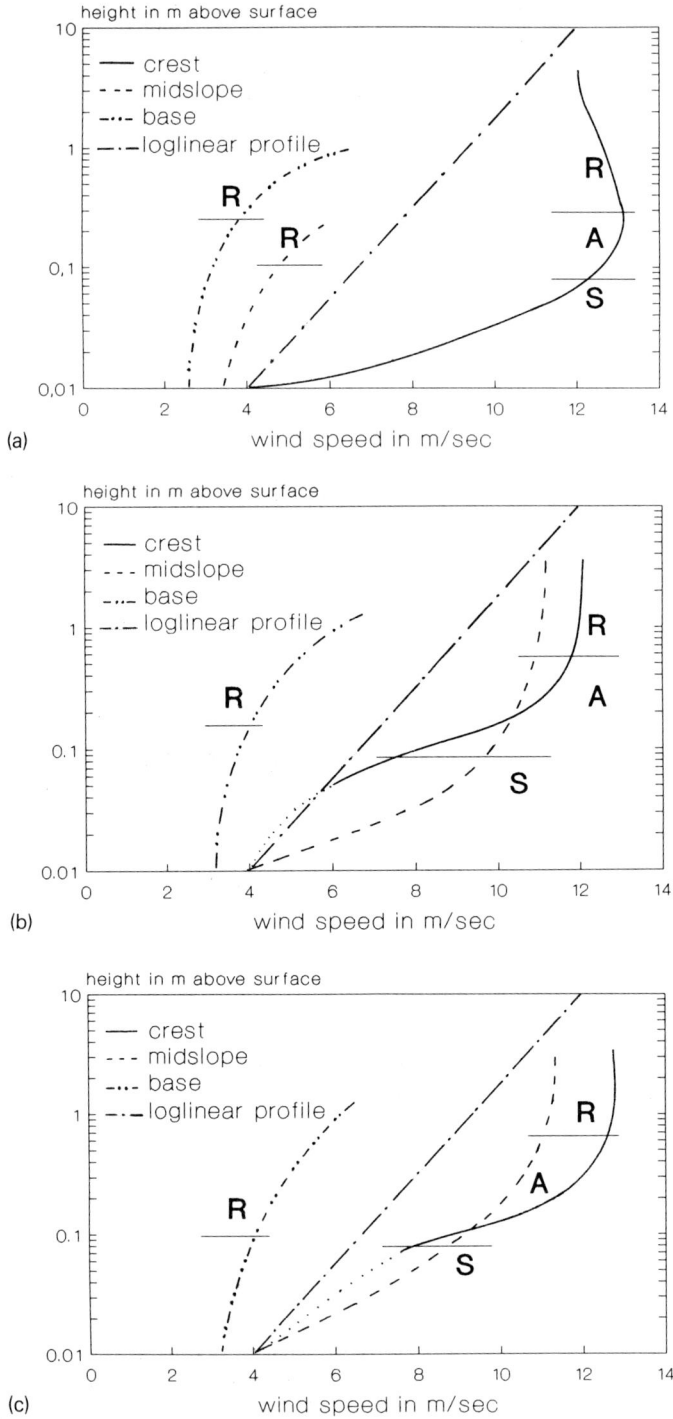

Fig. 10. Conceptual wind-speed model for different stages of dune reversal: (**a**) non-equilibrium, (**b**) semi-equilibrium and (**c**) equilibrium. S, A and R denote the surface shear stress, amplification and recovery layers respectively.

gradient at the brink edge.) Amplification is so extreme that wind speed decreases higher up in the wind-speed profile until approximately 2 to 3 m above the surface (one third of the dune height, the limit of our measurements). This extreme amplification, with shear velocity three times the undisturbed shear velocity (Fig. 7), can cause a 30-fold increase in sand transport rate (sand transport is proportional to the cube of the shear velocity), compared to the transport rate of the undisturbed airflow. No sand will be eroded from the lower slope region or the dune base, even under strong wind conditions, because the upwind topographic step that the dune presents to the wind causes flow divergence in this region.

As the process of reversal progresses, the midslope region becomes more accessible to approaching winds, and the wind-speed profile close to the surface in the midslope area changes from flow divergence to accelerated flow (Fig. 10b). The attack is focussed further downslope as illustrated by the data of 17 November 1988 (Fig. 9) and Fig. 4. Surface shear stress is enhanced in the mid-slope area, where maximum erosion occurs. Under strong wind conditions shearing would increase, and erosion of the non-equilibrium shape would be enhanced, extending lower down the slope than in gentle winds. The slope gradient alters under continued erosion. Incipient flow separation can occur on the upper slope and in the crestal area when the gradient is considerably reduced in contrast to the steeper lower slope. However, this is very much a surface effect and wind speed accelerates rapidly within 20 cm of the surface. Also, due to the feedback relationship between the dune form and airflow, a region of incipient flow separation will be a short-lived feature when it is accompanied by deposition, as occurred on 17 November 1988.

Under equilibrium conditions (Fig. 10c) the surface gradient is such that shear stress is predicted to increase systematically upslope maintaining dune curvature (Bagnold 1941). The consistent acceleration of wind speed measured at 10 cm above the surface on the windward slope from the dune base towards the crest (as on the 26 of June 1989) probably means that shear velocity close to the surface increases upslope; correspondingly, an increasing volume of sand is eroded from the slope and transported up the slope, remaining in transport until the crestal region. Incipient airflow separation, with some sand deposition, can occur at the crest as also reported by Lancaster (1987) and Mulligan (1988), but complete airflow separation occurs

at the brink, and all the sand transported upslope is dropped here.

Our measurements during equilibrium conditions only relate to light winds. Wind speed at the base of the dune was below the threshold for sand transport. We predict that for strong wind the same systematic increase of wind speed would occur upslope, and the wind-speed profile would exhibit much stronger shearing close to the surface. Under strong winds, the effects of flow disturbance may be felt much higher in the wind-speed profile.

Wind-tunnel measurements and theoretical modelling confirm the reduction of airflow at the base of a dune (Tsoar 1985). Wind-speed profile behaviour at the dune base for semi-equilibrium and equilibrium dune states suggests that airflow starts to recover from the topographic effect of the dune within 20–30 cm above the surface, depending on the gradient of the lower slope. However, despite flow reduction close to the surface, sand is eroded from the dune base under strong wind conditions. The mechanism for this erosion remains to be accounted for (see Wiggs, this volume). For the non-equilibrium dune state airflow recovery would start closer to the surface in the midslope region than at the dune base.

The wind-speed profile on the upper windward slope can be divided into 3 zones (Fig. 10). The thicknesses of these zones probably vary with strength of wind and depend on the degree of equilibrium/disequilibrium of the dune. Surface friction, surface irregularities, surface gradient, and change of surface gradient affect the wind-speed profile within 5–6 cm of the surface, the 'surface shear stress' layer. Increased surface gradient results in increased surface shear stress within this layer. Fluctuation of velocity and shear stress in this layer can be related to changes in sediment flux on the surface (Burkinshaw & Rust in press). Subtle changes in gradient can lead to incipient flow separation in this layer resulting in drastically reduced surface shear velocity.

Higher up the wind-speed profile, airflow recovers from surface effects and slope effects dominate in the 'amplification' layer. Maximum speed-up relative to the reference profiles occurs between 20 cm and 70 cm above the surface for a 7 m high dune. The height of this layer depends on the dune shape, and is closer to the surface for steeper windward slopes. Large wind-speed gradients are experienced in this layer but flow behaviour in this layer does not relate to sand transport on the dune surface.

Above this zone, in the 'recovery' layer, wind

speed either decreases over a certain height interval in response to acceleration in the amplification layer, and reverts to the 'undisturbed' profile; or in the case of an equilibrium dune form, wind speed increases slowly to recover to the 'undisturbed' profile.

It is difficult to predict the height of recovery of the airflow from the influence of the dune, but it may be of the order of three times the dune height. We think that under lighter wind conditions, this height could be lower. Airflow may recover quicker on the lower slope where acceleration is minimal, perhaps within twice the height of the dune. Recovery height would probably reach its maximum above the crest where maximum flow acceleration occurs.

It must be noted that despite the fact that it is difficult to get a reliable, undisturbed profile as field locations are seldom horizontally homogeneous with respect to roughness parameters (Rasmussen *et al.* 1985), wind-speed profiles measured over a dune need to be related to a profile where airflow is relatively undisturbed. Mulligan (1988) related his non-logarithmic wind-speed profiles to logarithmic profiles projected from the top of each measured profile to threshold values of Bagnold. Butterfield (1991) used the same technique. Such projected logarithmic profiles are unlikely to represent the shear velocity of the undisturbed flow, unless the measured wind speed profile has already recovered from the disturbance to the flow.

Application of our airflow data to existing numerical models and amplification

Several attempts have been made to model and predict turbulent flow behaviour over a topographic obstacle. The linear analytical model of Jackson & Hunt (1975) has proved very popular despite its limitations. Hunt *et al.* (1988) modified and extended the 1975 theory to give greater resolution of flow behaviour in the surface layer for an increased range of upwind profiles.

The Hunt *et al.* (1988) model is sensitive to h/L for low symmetric hills, where the length scale, L, is the half length of the hill at its half-height. L is sometimes regarded as the length of the stoss slope (Lancaster 1985; Jensen 1983), which is a more appropriate assumption for an asymmetric dune. The heights of the inner region, l, the inner surface layer, l_s, and maximum compression, h_{max}, of the Hunt *et al.* (1988) model are all sensitive to L. Assuming L = stoss slope length, we have computed the various layer parameters using the model formulae applied by

Table 1. *Computation of parameters relating to the Hunt et al. (1988) model for equilibrium and semi-equilibrium dune states, using* L = *stoss slope length*

HLR parameters in m	Semi-equilibrium 17-11-88	Equilibrium 26-06-89
h	6.8	6.5
L	24	31
l	1.5	2
l_s	0.12	0.14
h_{max}	0.50	0.67
h/L	0.29	0.21

Rasmussen (1989; Table 1). Values of l, l_s and h_{max} are reduced for the semi-equilibrium dune reflecting increased compression of flow against the steeper slope.

Our measurements have been made within the shear stress layer of the inner region of the Hunt *et al.* (1988) model, except for our lowermost anemometer which was within the inner surface layer. Calculated values of l_s appear to correspond approximately to the top of our surface shear stress layer. If this is the case, then measurements need to be made within the inner surface layer to determine shear stress close to the surface, as this is where we think sand transport is controlled. The values of h_{max} correspond to the top of the 'amplification' layer of our model.

The fractional speedup ratio (Jackson & Hunt 1975) calculated at 70 cm height above the crest for the equilibrium dune shape (Table 2) is approximately 1.7 h/L, which is in fairly good agreement with Jackson & Hunt's prediction. But for the semi-equilibrium dune shape, the maximum value of the ratio at 20 cm above the surface on the upper slope is 1.1 h/L on the upper slope, which can be attributed to the steeper dune surface. Fractional speedup does not reflect the sand transporting capacity of the wind, and amplification should be considered a secondary influence in this respect, as suggested by Watson (1987).

Despite the inherent limitations of the Hunt *et al.* (1988) model, it seems that parameters related to the eroding stoss slope can be calculated successfully for an equilibrium dune shape. Our breakdown of airflow behaviour (Fig. 10) is similar to that of the Hunt *et al.* (1988) model but we see the behaviour of the airflow being affected by h/L rather than L, and that for sand dunes the effect of slope gradient on shear stress close to the surface is more relevant than the amplification of the flow.

Table 2. *Calculation of the fractional speedup ratio for equilibrium and semi-equilibrium dune states*

Semi-equilibrium dune state (17-11-88)

Height above the surface (cm)	Slope distance (m)				
	22 (crest)	16.5	12	6	0
140	− 0.02	0.14	− 0.08	− 0.25	− 0.48
60	0.13	0.12	0.03	− 0.19	− 0.54
20	0.13	0.33	0.18	− 0.11	− 0.55
6	− 0.14	− 0.01	0.14	− 0.10	− 0.65

Equilibrium dune state (26-06-89)

Height above the surface (cm)	Slope distance (m)			
	32 (crest)	23	13	2
160	0.26	0.09	− 0.07	− 0.25
70	0.36	0.19	− 0.02	− 0.29
30	0.35	0.21	0.01	− 0.33
10	0.08	0.20	0.03	− 0.37

Wind-speed measurement close to the surface

At present, our determination of shear velocity is dependent on measurement of wind speed close to the dune surface, which is fraught with difficulties. Cup anemometers are physically too large to obtain an accurate measure of wind speed close to the surface, and more specialized equipment should be used. Butterfield (1991), using steel-clad hotwire anemometers, identified approximately logarithmic segments in the upper saltation layer within 20 cm of the surface, but the effect of saltating sand grains on wind speed measurements has yet to be established. Rasmussen *et al.* (1985) suggest that reliable measurements can be made up to within 10 cm of the surface using hot-wire anemometers. The results of Burkinshaw & Rust (in press) indicate that sand transport correlates well with fluctuation in wind speed in the lowermost 5–6 cm; ultimately shear velocity may have to be determined indirectly.

Threshold values still need to be reliably established. Rasmussen (1989) estimates the roughness value of saltating sand surfaces to be of the order of 0.1–1.5 mm, in contrast with Bagnold's roughness value of 1 cm.

Conclusions

Our measurements give useful insights into wind behaviour over a dune surface that is changing to an equilibrium shape. The extent of the variation in wind-speed profiles on the windward slope depends mainly on the slope of the surface, which varies with different approach angles of the wind, and the degree of equilibrium of the dune with the wind.

We divide the wind-speed profile of the windward slope into three layers: the surface shear stress layer where shear stress relates to sand transport; the amplification layer where flow acceleration is a maximum with respect to the undisturbed wind; and the recovery layer where airflow recovers to the undisturbed flow. The height of these various layers is probably a function of the dune gradient, dune height and also the strength of the wind.

An accurate measure of shear velocity close to the surface is required for calculation of sand transport over a dune. We believe that measurements should be made within 5 cm of the surface with specialized equipment. The effect of saltation on wind speed measurements would have to be taken into account. Owing to the difficulties of setting up instrumentation close to the surface (within 10 cm), techniques such as monitoring erosion pins can be used to indirectly determine shear stress close to the surface, assuming that Bagnold's transport equation is valid.

An adequate predictive model of airflow and surface shear stress variations over aeolian bedforms, in which changing dune shape and flow separation are taken into account, has yet to be developed. Models need to be supported by field measurements; in particular, measurements close to the surface need to be verified. More measurements need to be made across the entire dune profile under varying wind regimes, particularly strong winds.

We are grateful to the referees for their useful comments. The Department of Environment Affairs, Republic of South Africa, funded this study.

References

BAGNOLD, R. A. 1941. *The Physics of Blown Sand and Desert Dunes.* Methuen, London.

BOWEN, A. J. & LINDLEY, D. 1977. A wind-tunnel investigation of the wind speed turbulence characteristics close to the ground over various escarpment shapes. *Boundary-Layer Meteorology*, **12**, 259–271.

BRADLEY, E. F. 1980. An experimental study of the profiles of wind speed, shearing stress and turbulence at the crest of a large hill. *Quarterly Journal of the Royal Meteorological Society*, **106**, 101–123.

—— 1983. The influence of thermal stability and angle of incidence on the acceleration of wind upslope. *Journal of Wind Engineering and Industrial Aerodynamics*, **15**, 231–242.

BURKINSHAW, J. R. & RUST, I. C. (in press). Aeolian dynamics on the windward slope of a reversing transverse dune, Alexandria coastal dunefield, South Africa. *In*: PYE, K. & LANCASTER, N. (eds) *Aeolian Sediment Ancient and Modern.* International Association of Sedimentologists Special Publication, **16**, Blackwell Scientific Oxford.

BUTTERFIELD, G. R. 1991. Grain transport rates in steady and unsteady turbulent airflows. *In*: BARNDORFF-NIELSEN, O. E. & WILLETTS, B. B. (eds) *Aeolian Grain Transport. Acta Mechanica (Supplementum)* **1**, 97–122.

COOKE, R. U. & WARREN, A. 1973. *Geomorphology in Deserts.* University of California Press, Berkeley and Los Angeles.

GREELEY, R. & IVERSEN, J. D. 1985. *Wind as a Geological Process.* Cambridge University Press, Cambridge.

HESP, P. A., ILLENBERGER, W. K., RUST, I. C., MCLACHLAN, A. & HYDE, R. 1989. Some aspects of transgressive dune geomorphology and dynamics, south coast, South Africa. *Zeitschrift für Geomorphologie Supplement Band*, **73**, 111–123.

HUNT, J. C. R. 1980. Wind over hills. *In*: WIJNGAARD, J. C. (ed.) *Workshop on the Planetary Boundary Layer.* American Meteorological Society, Boston, 107–149.

——, LEIBOVICH, S. & RICHARDS, K. J. 1988. Turbulent shear flows over low hills. *Quarterly Journal of the Royal Meteorological Society*, **114**, 1435–1470.

ILLENBERGER, W. K. 1988. The dunes of the Alexandria coastal dunefield, Algoa Bay, South Africa. *South African Journal of Geology*, **91**, 381–390.

JACKSON, P. S. & HUNT, J. C. R. 1975. Turbulent wind flow over a low hill. *Quarterly Journal of the Royal Meteorological Society*, **101**, 929–955.

JENSEN, N. O. 1983. Escarpment induced flow perturbations, a comparison of measurements and theory. *Journal of Wind Engineering and Industrial Aerodynamics*, **15**, 243–251.

LANCASTER, N. 1985. Variations in wind velocity and sand transport on the windward flanks of desert sand dunes. *Sedimentology*, **34**, 511–520.

—— 1987. Variations in wind velocity and sand transport on the windward flanks of desert dunes. Reply. *Sedimentology*, **34**, 511–520.

MIKKELSEN, H. E. 1989. *Wind Flow and Sediment Transport Over a Low Coastal Dune.* Geoskrifter Nr 32, Aarhus Universitet.

MULLIGAN, K. R. 1988. Velocity profiles measured on the windward slope of a transverse dune. *Earth Surface Processes and Landforms*, **13**, 573–582.

RASMUSSEN, K. R. 1989. Some aspects of flow over coastal dunes. *Proceedings of the Royal Society of Edinburgh*, **96B**, 129–147.

——, SORENSEN, M. & WILLETS, B. B. 1985. Measurement of saltation and wind strength on beaches. *In*: BARNDORFF-NIELSEN, O. E., MOLLER, J. T., RASMUSSEN, K. R. & WILLETTS, B. B. (eds) *Proceedings of International Workshop on the Physics of Blown Sand.* Institute of Mathematics, Aarhus, Volume **2**, 301–326.

TSOAR, H. 1985. Profiles analysis of sand dunes and their steady state signification. *Geografiska Annaler*, **67A**, 47–59.

WATSON, A. 1987. Variations in wind velocity and sand transport on the windward flanks of desert sand dunes. Discussion. *Sedimentology*, **34**, 511–520.

WIGGS, G. F. S. 1993. An integrated study of desert dune dynamics. *This volume* 37–46.

Desert dune dynamics and the evaluation of shear velocity: an integrated approach

GILES F. S. WIGGS

Department of Geography, University College London,
26 Bedford Way, London WC1H 0AP, UK
(Present address: Department of Geography, Coventry University,
Coventry CV1 5FB, UK)

Abstract: Few empirical measurements have been undertaken of shear velocity, sand transport or dune morphology in the upwind basal regions of dunes. This study compares field measurements in this region from a barchan dune in Oman with calculations from a complex flow model using linearised equations (FLOWSTAR) and measurements in a wind tunnel.

The calculations of shear velocity from wind velocity observations in the field and those predicted by the FLOWSTAR model reflect the observations of previous studies in exhibiting the widely reported upwind reduction in shear velocity. Despite these observations, measurements of sand transport did not reduce at the base of the dune. The implications of these findings for the predictive power of typical saltation flux equations and the dynamics of the dune itself are discussed.

Wind tunnel modelling using a pulse wire probe suggests that the conventional methods of shear velocity derivation are inadequate. The pulse wire measurements exhibited no reduction in shear at the toe. An argument is presented to explain the maintenance of shear velocity in this position. It is suggested that the reduction in shear stress due to the fall in wind velocity is counteracted by an increase in shear stress due to streamline curvature. The adequacy of field methods for determining shear velocity in this respect is considered.

Many studies have been undertaken of the interaction of dune form and wind flow. These investigations have utilized field observations (Lancaster 1987; Mulligan 1988; Livingstone 1986), wind tunnels (Lai & Wu 1978; Tsoar 1985; Tsoar *et al.* 1985) and mathematical models (Howard *et al.* 1978; Howard & Walmsley 1985; Wippermann & Gross 1986) to examine the distribution of wind flow, sand transport and erosion/deposition over transverse and linear dunes. Most have concentrated on the mid-windward slopes and crests/brinks of dunes, pin-pointing the zones of maximum shear velocity, sand transport and erosion (Lancaster 1985; Watson 1987). Little emphasis has been placed on the variations in shear velocity near the toe and just upwind of dunes, despite the widely reported existence of an interesting anomaly in this zone.

Most studies have demonstrated that as wind approaches a dune its velocity and calculated shear velocity are reduced at the toe and just upwind, before increasing up the windward slope. If it is assumed that sand transport is proportional to the third power of shear velocity (u_*^3), as it is in all the commonly used sand transport equations (e.g. Lettau & Lettau 1978), and it is also assumed that the wind is fully saturated with sand as it approaches the dune, then a reduction in shear velocity at the toe of the dune should result in sand deposition. Under these circumstances the toe of a dune would advance upwind while the brink moved downwind. Such a situation is not observed in the field.

This paper describes variations in shear velocity around the toe of a barchan dune and shows that standard methods of evaluating shear velocity are inadequate in explaining the morphology of the dune. The study compares field measurements with surface shear stress predictions from a new mathematical model (FLOWSTAR) and measurements over a scale model of the dune in a wind tunnel using a pulse wire probe. The differences between the measurements are discussed with regard to the methods of evaluating shear velocity and the implications for the morpho-dynamics at the toe of dunes.

Methods

Field Study

The study dune was an 80 m long and 9.6 m high

From Pye, K. (ed.), 1993, *The Dynamics and Environmental Context of Aeolian Sedimentary Systems.*
Geological Society Special Publication No. 72, pp. 37–46.

Fig. 1. The study dune. Wind direction from the right.

unvegetated barchan (Fig. 1) on a gravel surface in the eastern part of the Sultanate of Oman (Fig. 2). The area was chosen because it is subject to a consistently strong southwesterly monsoonal wind regime (the Kharif) in the summer months. The dune itself was selected because, despite the existence of a small (2 m) high dune approximately 80 m windward, other conditions upwind were relatively undisturbed. The dune was small enough to be in a quasi-equilibrium with the wind regime, but large enough to offer an inner layer of depth 1.5–2.0 m (as defined by Jackson & Hunt 1975) which is sufficiently deep for the insertion of an array of cup anemometers. Measurement within the inner layer is necessary because it is here that the changes in shear stress and turbulent structure caused by the intrusion of the dune into the airflow are greatest, and where feedback mechanisms between the airflow and the dune surface can be assumed to be important (Taylor *et al.* 1987).

The objective was to measure wind velocity along the dune centre-line in order to calculate shear velocity (u_*). The potential sand transport capacity (Q_{pot}) calculated from the shear velocity was then compared to measurements of actual sand transport (Q_{act}) and erosion/deposition at the toe. Vertical and horizontal wind velocity profiles were taken along and upwind of the centre-line. This was achieved with arrays of cup anemometers erected 10–15 m apart and arranged at heights of 0.25 m,

0.35 m, 0.6 m and 1.0 m above the surface (Fig. 3). The anemometers were connected to Grant Squirrel 1600 series data loggers on which were recorded average windspeeds over periods of one minute. Synchronous with the wind velocity measurements, sand transport was measured using a simple Bagnold-type sand trap (similar to the Aarhus design and shown in Fig. 3) and erosion/deposition was monitored with erosion pins. A reference array of anemometers was erected 50 m upwind of the dune centre-line allowing wind and sand flux measurements to be normalized.

The data presented in this paper were collected in a two-minute period along the centre-line of the dune in August 1990. The short duration of the measurement period is governed by the inaccuracy of the sand traps over longer periods in windy conditions. Many other recording runs were made and show the same pattern as those reported here.

The calculation of shear velocity from wind velocity profiles is problematic (Gerety 1985; Mulligan 1988). A typical method used over horizontal surfaces is the calculation of a regression line through velocity profile data. However, it has been shown that due to the differential speed-up of airflow over low hills such as sand dunes the vertical velocity structure becomes non-logarithmic (Mulligan 1988; Finnigan *et al.* 1990). Velocity profiles measured at the toe of the study dune also revealed a non-logarithmic character, prohibiting the use of regression

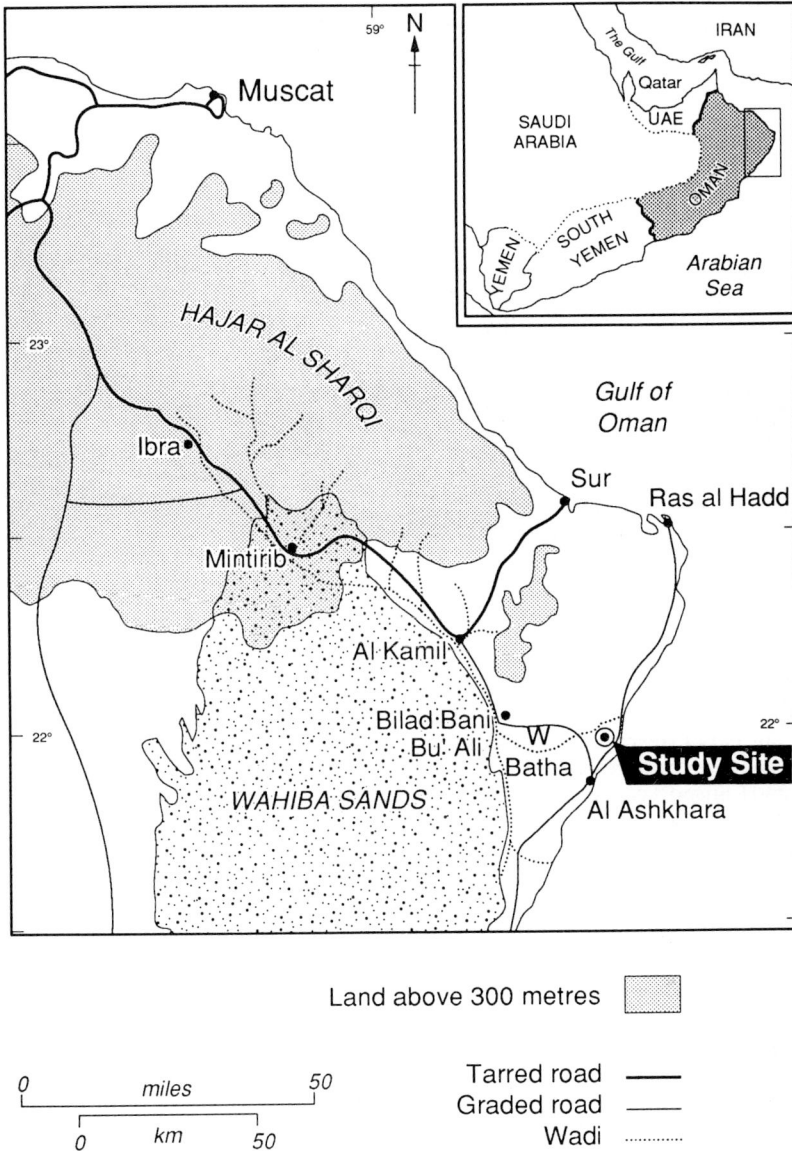

Fig. 2. Map of the eastern part of Oman showing the relationship between the study site and the Wahiba Sand Sea.

analysis to calculate shear velocity. To overcome this problem it has been suggested (Mulligan 1988) that shear velocity should be calculated from a single velocity measurement close to the surface using the expression derived by Bagnold (1941):

$$u_* = \frac{\kappa}{\ln(z/z_0)}(U_z - U_t) \qquad (1)$$

where u_* = shear velocity, κ = von Karman's constant (0.4), z = measurement height, z_0 = aerodynamic roughness height, U_z = velocity at height z, U_t = threshold of grain entrainment.

The values of z_0 and U_t define the focus of the velocity profile under sand-driving conditions and can be calculated from the mean grain diameter of the sand using the mathematical relationships developed by Bagnold (1941 p. 105). This method of calculating u_* assumes that z_0 and U_t are spatially constant quantities,

Fig. 3. Vertical array of anemometers at the crest of the dune and the position of the sand traps.

a supposition criticized by Gerety (1985). Despite this criticism, the method continues to be used to determine shear velocity and it is employed in this paper so that comparisons can be made with contemporary research.

Potential sand transport capacity can be determined using the equation of Lettau & Lettau (1978):

$$Q_{pot} = C \left(\frac{d}{D} \right)^{1/2} \frac{\varrho}{g} (u_* - u_{*t}) u_*^2 \qquad (2)$$

where Q_{pot} = potential sand transport, C = constant (4.2), d = mean grain diameter of the sand, D = standard grain diameter, ϱ = air density, g = gravitational acceleration, u_* = shear velocity, u_{*t} = threshold shear velocity.

Equation (2) was used in this study first because it has emerged as one of the more commonly used expressions for determining potential sand transport in the literature (Howard *et al.* 1978; Weng *et al.* 1991), and secondly because regression analysis of measured sand transport on calculated sand transport in the present study showed that it had a higher predictive power than many other expressions ($R^2 = 0.8$). Using Equations (1) and (2), the shear velocity and potential sand transport capacity for the field velocity measurements were calculated and compared to the measured sand transport rate (Fig. 4). It can be seen from Fig. 4 that the predicted potential sand transport rate (Q_{pot}) reduced to a minimum of 12 g m^{-1}s^{-1} at the toe (from an upwind value of 23 g m^{-1}

s^{-1}), coincident with a fall in calculated shear velocity from 0.43 m s^{-1} to 0.36 m s^{-1}. In contrast, the measured transport rate (Q_{act}) remained fairly constant. Figure 5 shows the perturbations in the measured and calculated quantities, highlighting these differences. The reduction in u_* at the toe of about 15% resulted in a reduction in potential sand transport of nearly 50%. Any model based on u_* relation-

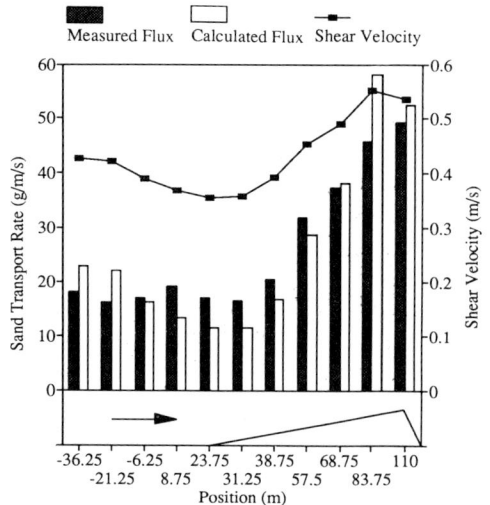

Fig. 4. The calculated shear velocity and potential sand flux compared to the measured flux on the centre line of the dune.

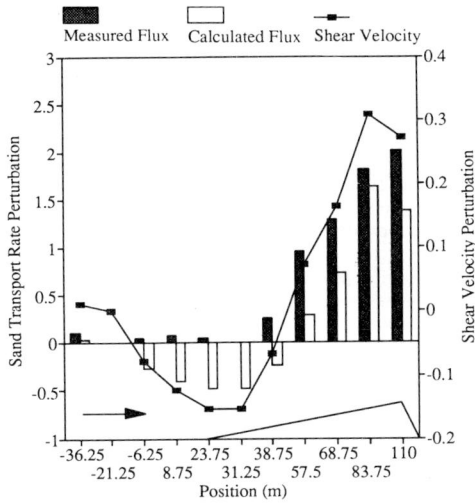

Fig. 5. Perturbations in shear velocity and potential sand flux compared to that of the measured flux.

Mathematical modelling

Available mathematical models do not help to explain the anomaly. The FLOWSTAR model, developed by Cambridge Environmental Research Consultants Ltd (CERC), is a PC-based computer model which can predict the velocity and shear velocity field over shapes of low slope. It is based on the model of Hunt *et al.* (1988) and gives a more accurate description of airflow close to the surface than previous models (Weng *et al.* 1991). The method of calculation is to use an analytical solution in terms of the Fourier transforms of the topography and the velocity and shear stress fields which are inverted to calculate the actual flow variables at a point. The Hunt *et al.* (1988) analysis for the shear stress perturbations, solving the approxi-

ships of this type would predict a piling up of sand at the toe, but Fig. 5 shows that actual sand transport did not reduce from its upwind rate and visual observation suggested no deposition of sand in this area.

It is important to be aware of the errors that are involved in this type of calculation and measurement. The sample period was only two minutes, sand traps of this type are notoriously unreliable and the wind may not have been sand-saturated. Nonetheless, many profiles of this type were carried out in different wind conditions and places on the dune and the trend is for the relationships described here to be repeated.

mate equation was:

$$\tau = \frac{2H/L}{U^2(l)} \delta[1 + \delta(2\ln\kappa + 4\gamma + 1 + \iota\pi)] \qquad (3)$$

where τ = shear stress perturbation, H = hill height, L = hill half length, U = upstream velocity, σ = normalized pressure perturbation, γ = Euler's constant (0.57721).

The FLOWSTAR model uses a higher-order solution of Equation (3) and since it does not assume a logarithmic velocity profile over the surface, its prediction of u_* over sand dunes is likely to be more exact than the field measurements (Weng *et al.* 1991). Weng *et al.* (1991) give more detail of the program and its analysis.

Figure 6 shows shear stress calculated by FLOWSTAR along the centre-line of the prototype dune. Like the field measurements, FLOWSTAR predicts a 10–15% drop in shear velocity at the toe. Using Equation (2) to determine sand transport capacity, the calculated reduction in u_* results in a decline in potential sand flux of about 35%. Hence the problem of a decreasing u_* and potential sand flux, but an apparent maintenance of actual sand transport remains.

There are two possible explanations of the anomaly. First, if the drop in shear velocity at the toe of the dune is real, then some other process must be operating which maintains sand transport. Various possible processes are presented by Cooke *et al.* (1992). The most promising of these possibilities is a change in roughness between the dune and interdune, causing an increase in u_* which would offset the apparent decrease due to streamline divergence. The argument is that owing to increased sand transport on the dune compared to the interdune, the apparent roughness (z_0) increases,

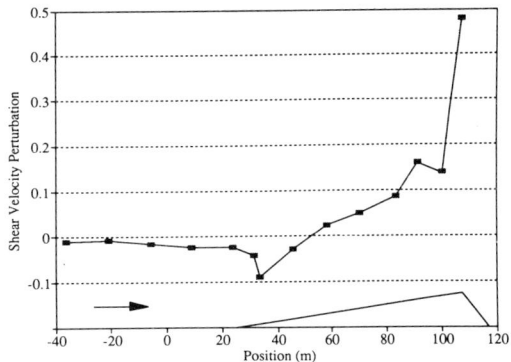

Fig. 6. Surface shear stress along the centre-line of the dune as calculated by FLOWSTAR.

hence increasing u_*. However, if the wind across the gravel interdune is fully saturated with sand, then transgression onto a sand-laden surface (the dune) would more likely cause a decrease in z_0, due to the weaker saltation rebound, and this would lead to a reduction in u_*. Furthermore, Figs 4 and 5 reveal that the apparent drop in u_* occurs upwind of the toe of the dune, before any potential change associated with the roughness height.

A second explanation is that the methods of shear stress derivation in the field and mathematical model are inadequate. There may be no reduction in shear velocity at the toe of the dune.

Wind tunnel modelling

In order to test this latter hypothesis, further measurements were undertaken on a 1 : 200 scale model of the field dune in a wind tunnel at Surrey University. The wind tunnel was set-up as shown in Fig. 7. The logarithmic part of the Atmospheric Boundary Layer was reconstructed with a mixing mesh and barrier arrangement (Cook 1977). This resulted in similar roughness, shear velocity and turbulence scale to the field prototype. A comparison of field and wind tunnel similarity parameters is shown in Table 1.

Vertical and horizontal gradients of velocity and Reynolds stress were measured along the centre-line of the model using single and cross hot-wire probes. Further shear stress measurements were undertaken using a pulse wire probe (Fig. 8), which is designed specifically to make measurements in highly turbulent regions where

Table 1. *Comparison of field and wind tunnel similarity parameters*

Parameter	Field value	Laboratory value
Reynolds number ($\times 10^6$)	4.4	0.0032
Half length (m)	40.6	0.2
Dune height (m)	9.6	0.048
Inner layer depth (m)	2.0	0.01
H/z_0	1890	1580
Shear velocity, u_* (m s^{-1})	c. 0.42	0.37

traditional hot-wire probes cannot be used because of their lack of sensitivity to angular changes in the velocity vector normal to the wire axis. The principal behind pulse wire anemometry (PWA) is the measurement of the 'time-of-flight' of a heat tracer generated as a pulse in one wire and detected by a second sensor wire. The velocity is deduced from the reciprocal of the flight time. The method has been adapted for measurements in the near-wall region where it is the only available technique that can measure the fluctuating component of wall shear stress. Its design allows it to be inserted through holes drilled in the tunnel floor, thus enabling measurements at the dune surface. Detailed descriptions of the development, use and calibration of the probe are given by Bradbury & Castro (1971), Castro *et al.* (1987), Handford & Bradshaw (1989) and Castro & Dianat (1990). A recent review is given by Castro (1991).

The measurements of Reynolds stress (-$\bar{u}\bar{w}$) using the traditional cross-wire probe technique

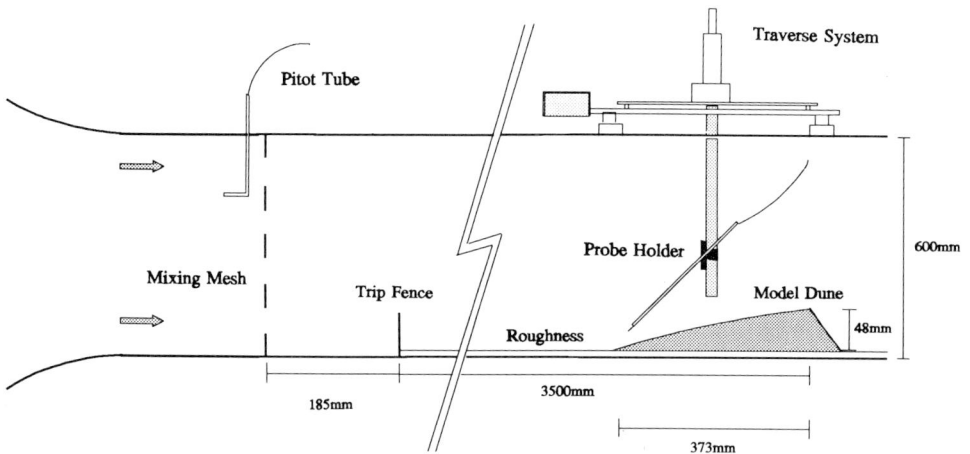

Fig. 7. The wind tunnel, showing positioning of mixing mesh, trip fence and dune model.

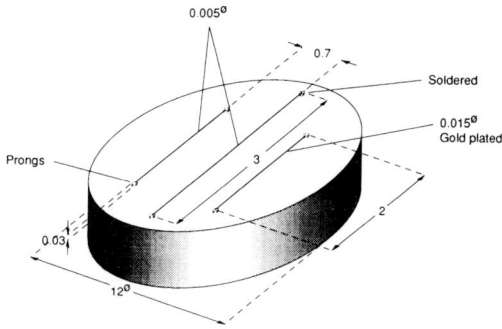

Fig. 8. The top of the pulse wire skin friction probe.

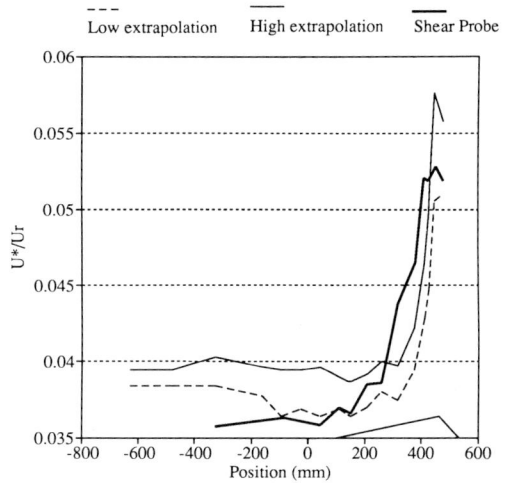

Fig. 10. Comparison of the derived surface shear stress from the cross-wire measurements (high and low extrapolations) and the surface shear stress measured with the pulse-wire probe.

require correction for the effects of turbulence and streamline angle (with respect to the probe axis calibration direction) before surface shear stress can be derived. Figure 9 indicates that the correction for streamline angle (which affects the angle of approach of the flow onto the probe) is considerable. This is despite the angular difference leading to the largest error being only 10°. This is important because the correction factors introduce a source of error into the final analysis of shear stress. The surface shear stresses extrapolated from the corrected Reynolds stress profiles above the surface are shown in Fig. 10. The minimum and maximum limits are a result of the method of extrapolation. Neither of the limits suggests a signifi-

cant reduction in surface shear stress at the toe of the dune, but the actual value may lie anywhere between these limits, with a reduction of about 7% a possibility. However, this maximum possible reduction in shear is half that calculated from the field and predicted by FLOWSTAR.

Comparing the measurements from the cross-wire method to the pulse wire probe measurements (Fig. 10) it can be seen that the trend of the surface shear stress is similar. However, the pulse wire probe measurements suggest no reduction in surface shear stress at the toe of the dune. They are likely to be more accurate than the cross-wire measurements because they require no correction and are actually taken at the surface of the model, hence avoiding errors due to extrapolation.

Discussion

Converting the measurements of surface shear stress measured in the wind tunnel and predicted by FLOWSTAR into sand flux perturbations using equation (2), allows them to be compared directly to the field measurements of sand transport (Fig. 11).

Figure 11 indicates that the perturbation in sand flux calculated from the pulse wire probe measurements in the wind tunnel most closely describes the measured sand flux from the field (Q_{act}). All the other techniques of determining shear stress result in a reduction in sand transport of between 35% and 65% at the toe (at 24 m

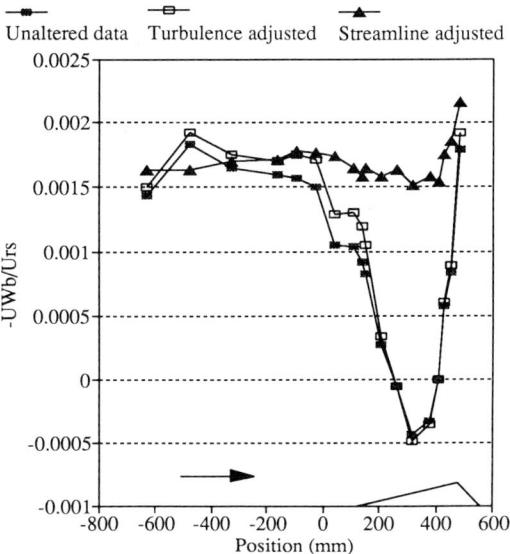

Fig. 9. The effect of correcting for turbulence intensity and streamline angle on measured values of Reynolds stress (-ūw̄) along the centre-line.

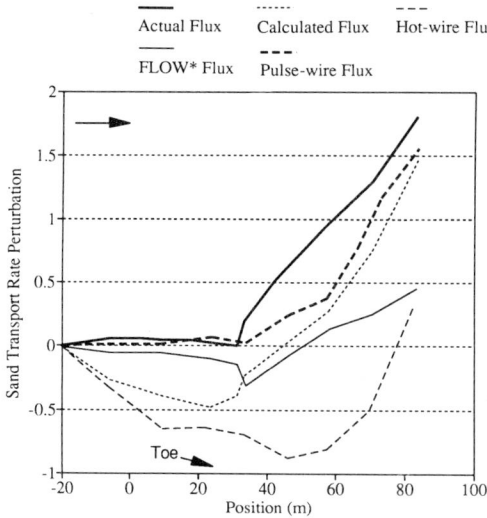

Fig. 11. Comparison of the measured sand transport rate perturbation at the toe of the dune (actual flux) and those calculated from various methods of shear stress derivation.

in Fig. 11). The flux perturbation calculated from the FLOWSTAR model more closely describes the actual flux (Q_{act}) than the potential flux calculated from the field wind velocity measurements. A similar trend in flux perturbation to the field calculated flux is shown by the hot-wire flux on Fig. 11. This line represents the flux calculated from the wind tunnel velocity measurements using the same method of evaluating u_* as the field analysis (Equation (1)). Figure 11 suggests that the methods of determination of shear stress which imply a reduction in sand flux at the toe are incorrect.

The field method of evaluating shear stress (Equation (2)) relies heavily on the assumption of z_0 constancy and wind speed measurement as a determinant. Such a dependence will always result in a reduction in u_* at the toe because of the divergence of streamlines in this region. Hence, u_* calculated from the wind tunnel velocity measurements using Equation (1), exhibits the same trend (Fig. 11). Shear stress, however, is not controlled by wind velocity alone.

One factor which only the pulse wire probe takes into account in its determination of shear stress is streamline curvature. The sensitivity of airflow over hills to streamline curvature has been emphasized by Zeman & Jensen (1987), Gong & Ibbetson (1989) and Finnigan et al. (1990). At the toe of a dune or low hill there is concave curvature, with turbulent structures of

high velocity being conveyed into regions of lower velocity. This causes instability in the flow and leads to an increase in shear stress. It is postulated that the increase in shear stress owing to concave streamline curvature at the toe of the dune counteracts the effects of the reduction in shear stress caused by the drop in wind velocity. The additional stresses resulting from such curvature are not apparent in velocity profiles, and this means that they are not accounted for in the field method of calculating u_*. Nor are such effects accounted for in the FLOWSTAR calculations. The additional stresses are, however, apparent in Reynolds stress profiles and so would be accounted for by the cross-wire and pulse wire probe measurements.

The fact that the cross-wire measurements demonstrate a smaller reduction in shear stress at the toe than the field and FLOWSTAR results (7% as opposed to 15%) but more of a reduction than the pulse wire measurements (Fig. 10), may be explained by the inverse relationship between streamline curvature and height (Finnigan et al. 1990). The cross-wire Reynolds stress measurements could not be taken at the dune surface because of the bulk of the probe. Hence, they were subject to less curvature, with the result that they show a small decline in surface shear stress.

The degree of airflow destabilisation due to streamline curvature is clearly a function of the morphology of the windward slope of the dune. It is conceivable, therefore, that the form of the windward slope near the toe is partly governed by the curvature required to counteract the effects of shear stress reduction in that region.

Conclusion

The results of this study suggest the following.

(1) Commonly used methods of calculating shear stress and potential sand transport do not explain the morphological dynamics in the basal regions of dunes. This is because the method relies too heavily on variations in wind velocity and hence predicts a reduction in shear stress at the toe, implying sand deposition. The difference between the change in calculated sand transport and actual sand transport may be as much as 50%.

(2) Measurements using a pulse wire probe in a wind tunnel demonstrate no reduction in shear stress at the toe. The preservation of u_* in this region reflects actual sand transport and dune morphology.

(3) The differences in the evaluation of surface shear stress by each method may be due to the effects of streamline curvature which destabilizes the airflow structure and tends to increase shear stress at the surface. Such an effect is apparent in Reynolds stress profiles and hence only measured by the wind tunnel techniques in this study.

(4) The interaction between the form of the windward slope, the degree of streamline curvature and the additional stresses imposed may be an important process which has so far been ignored in dune research.

This project was funded by a Natural Environment Research Council (UK) Research Studentship held at the Department of Geography, University College London. Permission to work in Oman was granted by the Diwan of Royal Court Affairs and invaluable assistance was provided by George Gamlen at Sultan Qaboos University. The FLOWSTAR model was generously provided by Cambridge Environmental Research Consultants Ltd. I am grateful to Adrian Chappell, Mark Ashby, Richard Ledger and Neil Scoble for their unstinting field assistance and I also thank Guy Baker in the Drawing Office at University College London. Helpful comments on a draft of this paper from Andrew Warren, Ian Livingstone and Ian Castro were greatly appreciated.

References

BAGNOLD, R. A. 1941. *The Physics of Blown Sand and Desert Dunes*. Methuen, London.

BRADBURY, L. J. S. & CASTRO, I. P. 1971. A pulse wire technique for turbulence measurements. *Journal of Fluid Mechanics*, **49**, 657–691.

CASTRO, I. P. 1991. Pulsed wire anemometry. *Second World Conference on Experimental Methods in Heat Transfer, Fluid Mechanics and Thermodynamics, Dubrovnik*.

—— & DIANAT, M. 1990. Pulsed wire anemometry near walls. *Experiments in Fluids*, **8**, 343–352.

——, DIANAT, M. & BRADBURY, L. J. S. 1987. The pulsed wire skin-friction measurement technique. *Turbulent Shear Flows*, **5**, 278–290.

COOK, N. J. 1977. Wind tunnel simulation of the adiabatic atmospheric boundary layer by roughness, barrier and mixing device methods. *Building Research Establishment, Paper 6*, 157–176.

COOKE, R. U., WARREN, A. & GOUDIE, A. 1992. *Desert Geomorphology*. University College London Press (in press).

FINNIGAN, J. J., RAUPACH, M. R., BRADLEY, E. F. & ALDIS, G. K. 1990. A wind tunnel study of turbulent flow over a two-dimensional ridge. *Boundary-Layer Meteorology*, **50**, 277–317.

GERETY, K. M. 1985. Problems with determination of u_* from wind-velocity profiles measured in experiments with saltation. *In*: BARNDORFF-NIELSEN, O. E., MØLLER, J. T., RASMUSSEN, K. R. & WILLETTS, B. B. (eds) *Proceedings of International Workshop on the Physics of Blown Sand*. Department of Theoretical Statistics, Aarhus University, Denmark, Memoirs **8**, 2, 271–300.

GONG, W. & IBBETSON, A. 1989. A wind tunnel study of turbulent flow over model hills. *Boundary-Layer Meteorology*, **49**, 113–148.

HANDFORD, M. & BRADSHAW, P. 1989. The pulsed wire anemometer. *Experiments in Fluids*, **7**, 125–132.

HOWARD, A. D. & WALMSLEY, J. L. 1985. Simulation model of isolated dune sculpture by wind. *In*: BARNDORFF-NIELSEN, O. E., MØLLER, J. T., RASMUSSEN, K. R. & WILLETTS, B. B. (eds) *Proceedings of International Workshop on the Physics of Blown Sand*. Department of Theoretical Statistics, Aarhus University, Denmark, Memoirs **8**, 2, 377–392.

——, MORTON, J. B. & GAD-EL-HAK, M. 1978. Sand transport model of barchan dune equilibrium. *Sedimentology*, **25**, 307–338.

HUNT, J. C. R., LEIBOVICH, J. B. & RICHARDS, K. J. 1988. Turbulent shear flow over hills. *Quarterly Journal of the Royal Meteorological Society*, **114**, 1435–1470.

JACKSON, P. S. & HUNT, J. C. R. 1975. Turbulent wind flow over a low hill. *Quarterly Journal of the Royal Meteorological Society*, **101**, 929–955.

LAI, J. & WU, J. 1978. *Wind Erosion and Deposition along a Coastal Sand Dune*. Sea Grant Program, University of Delaware, Report DEL-SG-10-78.

LANCASTER, N. 1985. Variations in wind velocity and sand transport on the windward flanks of desert sand dunes. *Sedimentology*, **32**, 581–593.

—— 1987. Reply: Variations in wind velocity and sand transport on the windward flanks of desert sand dunes. *Sedimentology*, **34**, 511–520.

LETTAU, K. & LETTAU, H. H. 1978. Experimental and micrometeorological field studies on dune migration. *In*: LETTAU, H. H. & LETTAU, K. (eds) *Exploring the World's Driest Climate*. University of Wisconsin-Madison, Institute for Environmental Studies, Report **101**, 110–147.

LIVINGSTONE, I. 1986. Geomorphological significance of windflow patterns over a Namib linear dune. *In*: NICKLING, W. G. (ed.) *Aeolian Geomorphology*. Allen & Unwin, Boston, 97–112.

MULLIGAN, K. R. 1988. Velocity profiles measured on the windward slope of a transverse dune. *Earth Surface Processes and Landforms*, **13**, 573–582.

RASMUSSEN, K. R., SØRENSEN, M. & WILLETTS, B. B. 1985. Measurement of saltation and wind strength on beaches. *In*: BARNDORFF-NIELSEN, O. E., MØLLER, J. T., RASMUSSEN, K. R. & WILLETTS, B. B. (eds) *Proceedings of International Workshop on the Physics of Blown Sand*. Department of Theoretical Statistics, Aarhus University, Denmark, Memoirs **8**, 2, 301–325.

TAYLOR, P. A., MASON, P. J. & BRADLEY, E. F. 1987. Boundary-layer flow over low hills: a review. *Boundary-Layer Meteorology*, **39**, 107–132.

TSOAR, H. 1985. Profile analysis of sand dunes and their steady state signification. *Geografiska Annaler*, **67A**, 47–59.

Tsoar, H., Rasmussen, K. R., Sørensen, M. & Willetts, B. B. 1985. Laboratory studies of flow over dunes. *In*: Barndorff-Nielsen, O. E., Møller, J. T., Rasmussen, K. R. & Willetts, B. B. (eds) *Proceedings of International Workshop on the Physics of Blown Sand*. Department of Theoretical Statistics, Aarhus University, Denmark, Memoirs **8**, 2, 327–349.

Watson, A. 1987. Discussion: Variations in wind velocity and sand transport on the windward flanks of desert sand dunes. *Sedimentology*, **34**, 511–520.

Weng, W. S., Hunt, J. C. R., Carruthers, D. J., Warren, A., Wiggs, G. F. S., Livingstone, I. & Castro, I. 1991. Air flow and sand transport over sand dunes. *Acta Mechanica* [Suppl] **2**, 1–22.

Wippermann, F. K. & Gross, G. 1986. The wind-induced shaping and migration of an isolated dune: a numerical experiment. *Boundary-Layer Meteorology*, **36**, 319–334.

Zeman, O. & Jensen, N. O. 1987. Modification of turbulence characteristics in flow over hills. *Quarterly Journal of the Royal Meteorological Society*, **113**, 55–80.

Desert dunefields

Luminescence dating of aeolian sands: an overview

ANN G. WINTLE

Institute of Earth Studies, University College of Wales,
Aberystwyth, Dyfed SY23 3DB, UK

Abstract: Aeolian sands are ideal materials for the application of luminescence techniques. Their primary mineral constituents, quartz and feldspars, exhibit a variety of luminescence properties and can be separated for mineral-specific measurements. Individual grains are usually well exposed to light prior to deposition thus enabling their signal to be zeroed at the time of interest.

Luminescence is produced either by heating the grains to produce thermoluminescence (TL) or by stimulating at room temperature with wavelengths specific for each mineral component. Only light-sensitive signals are observed in measurements of optically stimulated luminescence (OSL). For feldspars, infra-red stimulated luminescence (IRSL) can be obtained using IR-emitting diodes.

Luminescence dating techniques can be applied to inland sand seas, small dunefields, coastal dune systems or periglacial sand sheets. Natural variation in radioactive content in different geographical areas can be used to advantage — sands with a high radioactive content, as found in areas dominated by granitic rocks, are best suited for applications to Holocene dunefields (such as Kelso, California), whereas sands with a low radioactive content, as found in areas containing little feldspar-bearing rock and with much recycled sediment (e.g. Australia), are best suited to applications spanning the last 100 ka.

There is a great need to provide a time frame for aeolian deposits. This may be on the scale of hundreds of years to ascertain recent aeolian activity, in particular the movement of desert margins which might relate to anthropogenic activity: on the scale of thousands of years to provide the background against which to assess possible anthropogenic activity; and on the scale of tens to hundreds of thousands of years to study changes relating to external climatic forcing. Although radiocarbon dating can be used for the last 40 ka, there are many sedimentary deposits which do not contain suitable organic remains. It is, therefore, advantageous to develop a method which dates the sediment itself. Luminescence techniques provide such a method as they are applied to the quartz or feldspar grains that make up the bulk of the sediment.

Luminescence dating

The fundamental basis of luminescence dating is the increase in the number of trapped electrons within the crystals that results from their exposure to ionizing radiation. The ionizing radiation is provided in sediments by the decay of radioactive nuclides in the uranium and thorium decay chains and the decay of ^{40}K, with minor contributions from ^{87}Rb and cosmic rays.

The trapped electrons can then be released either by the application of heat which results in the release of light (thermoluminescence, TL) or by the application of stimulating light which results in the release of light (photostimulated luminescence, PL; also known as optically stimulated luminescence, OSL). In both processes the luminescence is produced by the released electrons recombining within the crystal lattice at defects known as luminescence centres.

For this phenomenon to be useful as a dating method, it is necessary for there to be a zeroing mechanism which empties out the previously stored electrons at the time of interest. For TL it has long been realized that heating would act as a zeroing mechanism and this led to the development of TL dating for pottery and other heated materials. Not until 1979 was it thought that sunlight exposure would be an effective zeroing mechanism (Wintle & Huntley 1979).

The basic equation is the same for all the methods

$$\text{Age (a)} = \frac{\text{Equivalent dose (Gy)}}{\text{Dose rate (Gy a}^{-1})}$$

Luminescence measurements are used to ascertain the radiation dose to which the grains have been exposed since the zeroing event. The natural luminescence signal is compared with that induced by known doses of laboratory radiation from an artificial beta or gamma

From Pye, K. (ed.), 1993, *The Dynamics and Environmental Context of Aeolian Sedimentary Systems.*
Geological Society Special Publication No. 72, pp. 49–58.

source. The dose required to produce a signal which matches the natural signal is known as the equivalent dose (ED); the term equivalent is used because the natural radiation does not derive from one type of radiation but from a mixture of alpha, beta and gamma radiation from the radioactive decay processes. The SI unit of absorbed dose is the gray (Gy).

To obtain the age, the ED needs to be divided by the dose rate (in grays per annum) and this is obtained by measuring the radioactive content of the sediments. This is achieved either by direct measurements of the gamma activity in the field, and the beta and alpha activity in the laboratory, or by indirect measurement involving element analysis of K, U and Th by techniques such as atomic absorption and neutron activation analysis. The advantages and disadvantages of various methods of radioactivity analysis can be found elsewhere (Aitken 1985).

Mineral luminescence

Many minerals give luminescence signals. These include quartz and feldspars, as well as zircon and calcite. As with most analytical techniques it is always advantageous to separate the mineral species and isolate the luminescence signal from a particular mineral. Concentration of particular minerals can be achieved for grains above 100 μm using heavy liquids, with a specific gravity of 2.62 and 2.58 to separate quartz, sodium and potassium feldspars, respectively. Grains are sieved to select a grain size range before settling in sodium polytungstate solutions (e.g. 100–150 μm or 100–300 μm). Further purification may be achieved by magnetic separation and selective etching (e.g. using HF to remove feldspars from the quartz fraction).

It is also possible to enhance the signal from particular luminescence centres within a given mineral. Different impurity atoms in the crystal lattice are the cause of different luminescence centres and light from these centres can be observed by selecting the appropriate wavelengths with a coloured glass filter placed in front of the detector. Huntley et al. (1988b) have demonstrated that quartz TL has two main emission bands, one in the blue and the other in the red; plagioclase feldspars emit in the green and red (relating to Mn and Fe atoms substituting for Ca and Al in the crystal lattice); and K feldspars emit in the violet and near-UV (Huntley et al. 1988a). Additional spectral studies of TL emission from various minerals have been reported by Prescott et al. (1990). Emission spectra for optically stimulated luminescence from feldspars have been found to be similar to the TL spectra (Huntley et al. 1989). On the other hand, luminescence from quartz stimulated by either an argon laser (Huntley et al. 1989) or a krypton laser (647 nm) (Huntley et al. 1991) gives a single emission band around 365 nm, a wavelength region not seen in high temperature quartz TL spectra. The most likely emission region for stimulated luminescence in quartz is 360–380 nm. This region is characteristic of the 110°C TL peak for quartz and a link has been reported between this peak and the OSL signal on the basis of experiments designed to induce sensitivity changes (Stoneham & Stokes 1991). Spectral measurements are used to choose the appropriate coloured glass filters for dating applications.

Thermoluminescence dating

For most sediments the signals from quartz and feldspars are the most useful. Both mineral types produce TL when the crystals are heated. When mineral grains are heated, as in the manufacture of pottery, all the electron traps are emptied resulting in a zero luminescence signal for a recent pot. However, in the case of grains making up a sediment, sunlight empties a large fraction of the potential TL signal, but does not reduce it to zero. This results in a residual TL signal at deposition which is difficult to mimic in laboratory bleaching experiments; the exact light conditions (wavelengths, intensity and duration) at deposition are not easily duplicated, though for aeolian sediments long exposures to unfiltered sunlight may be assumed to have occurred.

Because TL was the first method to be developed, many more dating studies appear in the literature (for earlier reviews see Singhvi & Mejdahl 1985; Berger 1988). Recent applications will be discussed within a framework which first considers TL dating of quartz in arid environments and presents the main limitation of the method for older samples (the filling of the electron traps) and then considers sands in temperate areas which are the result of earlier periglacial or marine activity. These sediments also contain potassium feldspars, which permit the dating of much older deposits because of the higher saturation level found in these minerals. The importance of choosing an appropriate optical filter is demonstrated in several studies on sands with independent chronological information. For sands deposited before the last interglacial, the upper age limit for feldspars is determined by the long-term thermal stability.

TL dating case studies

The first dating of aeolian sand deposits was reported by Singhvi *et al.* (1982) when they applied the method to dune sands from India. Instead of using sand-sized grains, they used the fine silt fraction (4–11 μm), an approach which could have led to an incorrect age if these grains had filtered down through the dune from an overlying deposit (Pye 1982). In a deliberate attempt to put a lower age limit on a coastal dune from northern California, Berger *et al.* (1991) selected sands capped by later aeolian deposits. They used the TL from the fine silt fraction within the lower dune unit to date these dunes. Comparison with radiocarbon dates at two sections indicated that the grains had been well exposed to light prior to deposition.

More recently, Chawla *et al.* (1992) have used 90–150 μm quartz grains to date three sand profiles in western Rajasthan. They demonstrated by obtaining several dates of about 14 ka that sand was mobile in the Thar desert at that time. Older dates of about 40 ka were obtained at the base of one of the sections. They were consistent with radiocarbon dates from the same section, but have large errors because the natural TL appears to be close to saturation. The regenerated TL growth curve soon becomes non-linear (Fig. 1), at first apparently saturating at a total dose level of 100 Gy and then increasing rapidly with further laboratory irradiation. However, no such rapid increase is found when the unbleached sample is irradiated in the laboratory (Chawla & Singhvi 1989). It thus appears that 40 ka is the maximum age that can be obtained for these sands, where the dose rate is of the order of 3 ± 0.5 Gy ka^{-1}).

In spite of this non-linear response, TL dating of quartz grains has been applied extensively in Australia, where the dose rates for aeolian deposits are often only a third of those above. The lower dose rate is a result of the intense weathering of potential sedimentary material prior to deposition, a process that has left a sand which is rich in quartz. This quartz is well exposed to light immediately prior to deposition and is likely to have experienced multiple cycles of erosion and deposition before reaching its current position.

TL dating methods have been applied to coarse-grain quartz from samples collected from siliceous coastal dune fields in northern Australia (Lees *et al.* 1990). The bulk samples had quartz contents above 99% and annual dose rates of around 1 Gy ka^{-1} or less. Three phases of dune emplacement were recognized on the basis of the TL measurements: the late Pleistocene; early Holocene; and late Holocene. Some control was provided by associated radiocarbon dates.

A limited number of TL dates have also been obtained for coastal dunes from the eastern coast of Australia (Tejan-Kella *et al.* 1990), including one over 700 ka which indicated an unusually high TL saturation level, as well as dose rates as low as 0.33 Gy ka^{-1}. Hutton *et al.* (1984) also reported dates for coastal dune sands.

Further south, TL has been applied to the aeolian sands on the coast of New South Wales (Bryant *et al.* 1990). Once again, the non-linear response resulted in large errors for TL age estimates over 130 ka, with dose rates ranging from 0.6–1.0 Gy ka^{-1}. Along the southern and western coasts of Australia, the coastal dunes are highly calcareous and also provide a low activity environment, with the possibility of separating the different dune-building phases associated with high sea-level stands back to about 700 ka (Huntley *et al.* 1985b).

In New Zealand, TL dating has been applied to 90–125 μm quartz grains from two dune sands in southwestern North Island (Shepherd & Price 1990). Comparison with radiocarbon dates on associated organic material indicated good agreement for the older sample (24.2 ± 3.7 ka), but a slight overestimate for the younger sample (3.0 ± 0.5 ka). Unlike some previously quoted dates, the surface residual level was not used as an indication of the original TL level in the dune sand at deposition. This demonstrates the diffi-

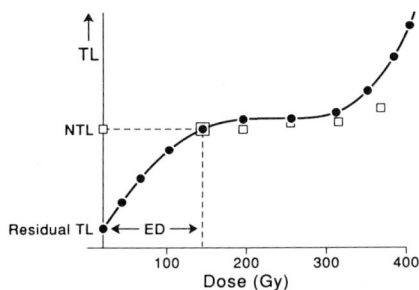

Fig. 1. Limitation due to saturation. ● Growth curve for quartz TL regenerated by laboratory beta radiation after bleaching with sunlight, showing apparent saturation followed by rapid growth above 300 Gy. □ Natural TL (NTL) and response to addition of laboratory dose (after Chawla *et al.* 1992). ED: equivalent dose.

culty in applying TL techniques to very young samples.

The use of a surface sample to determine the effective residual signal at deposition was developed by Readhead (1988) in his study of sands from the edge of the former Lake Mungo in western New South Wales. Good agreement was achieved with the age ranges expected on the basis of independent evidence. This study complemented that of Prescott (1983), who found that TL signals from surface samples of sand from Roonka, central Australia, could also be bleached in the laboratory after the surface coating had been etched away.

TL dating has also been reported for gypseous dunes which occur on the former shorelines of playa lakes in central Australia (Chen et al. 1990). At Lake Amadeus the quartz made up less than 20% of the bulk sand sample, but the low activity of the gypsum also resulted in low dose rates ranging from 0.5–1.5 Gy ka^{-1}. The TL ages obtained at two sites indicate that the dunes formed between 60 and 40 ka. Once again, non-linearity of response indicated that the method would be unsuitable for dating older dune material of this type.

Dose rates of around 1 Gy ka^{-1} were found for sand aprons at the foot of the Arnhem Land plateau in northern Australia. This enabled TL dating of quartz from sands containing artefacts. Several TL dates suggested that modern humans arrived in northern Australia about 50 000 years ago (Roberts et al. 1990). A surface sample gave an age of around 1 ka confirming the well-bleached nature of the sand grains as they lie on the surface of the sand apron. The TL ages at the two sites are in correct stratigraphic order and are in agreement with radiocarbon dates from the same section.

In North America there are many dunefields on the High Plains west of the Rocky Mountains. Some stabilized dunefields contain soils, which, on the basis of radiocarbon dating, have been shown to have developed during the Holocene. At a site in northern Colorado two samples were taken from buried A horizons, for which radiocarbon dates were also obtained (Forman & Maat 1990). The TL dates for the fine grains (4–11 μm) were in agreement with the radiocarbon dates and indicated that soil formation ceased about 8000 years ago when sand covered the land surface.

Besides the coastal and lacustrine dune systems, inland sand seas and small dunefields, the other aeolian sand deposits which have received attention are the periglacial sand sheets found at the edge of previously glaciated areas. These sands usually contain abundant feldspar,

and potassium feldspars are specifically separated for TL measurements alongside quartz. In Europe several studies have been carried out on aeolian sands in the eastern and southern parts of the Netherlands. Cover sands from the Late Glacial, about 10–14 ka, are easily distinguished from the overlying late Holocene drift sands by TL measurements (Dijkmans et al. 1992). However, compared with radiocarbon dates on peat layers within the sands, the TL ages for the potassium feldspars were too low by 20–40% (Dijkmans & Wintle 1991).

Similar underestimation of TL ages was found when using potassium feldspars from Late Glacial coversands in Denmark (Grün et al. 1989). However, in a parallel study in which a blue (rather than an ultraviolet) filter was used in the observation of the feldspar TL, ages were obtained which were in agreement with both the geological age estimates and the TL dates on quartz (Kolstrup et al. 1990). The quartz dose rates were about 1.2 Gy ka^{-1} and also showed saturation of the TL signal at a level similar to that reported for quartz from India and Australia.

It is currently hypothesized (Wintle & Duller 1991) that the age underestimation encountered when using an ultraviolet filter for large-grained potassium feldspars is related to the optical absorption properties of the feldspar grains. Other reasons have also been suggested (Grün et al. 1989; Rendell 1992).

TL studies using a blue filter indicated considerable aeolian activity in Denmark between 27 and 20 ka (Kolstrup & Houmark-Nielsen 1991; Kolstrup & Mejdahl 1986). Other parts of Scandinavia with TL-dated aeolian sands are Finland (Jungner 1987) and Sweden (Lundqvist & Mejdahl 1987; Mejdahl 1991).

The oldest dates for periglacial (predominantly fluvially deposited) sands are those for the Chelford Sands (Rendell et al. 1991). Dating has been carried out on the quartz fraction and on the feldspar-rich fraction (s.g. < 2.62), which constitutes only 1–2% of the mineral content. Good agreement was obtained between the age estimates for the feldspar fraction and that for the quartz. Ages increased systematically downsection and resulted in an estimate of 90–100 ka for the Chelford Interstadial organic horizon. Once again the dose rate to quartz grains was about 0.8 Gy ka^{-1}, and hence the quartz TL signal for the sample beneath the organic horizon was close to saturation and thus the limit of the technique.

This limitation has also given further impetus to the study of potassium feldspars, since they have been shown to have a higher saturation level. They were used to study two littoral dune sands from the island of Jersey in the English Channel (Balescu *et al.* 1991), which had been assigned on stratigraphic evidence to Oxygen Isotope Sub-stage 5d. The natural TL of the potassium feldspar separates was well below saturation, in spite of the dose rate being about three times higher than at Chelford. Both samples were unfortunately measured with an ultraviolet filter and the calculated ages were 40% lower than the expected ages. More recently, a blue filter has been used when re-measuring some of the same sands from the English Channel and this has given older ages which agree with the geological estimates (Balescu & Lamothe 1992; Balescu *et al.* 1992).

The upper age limit for potassium feldspars does not seem to be saturation. Instead long-term loss of signal as a result of thermal emptying at ambient temperature has been suggested (Mejdahl 1988). The effect of such loss is shown schematically in Fig. 2. The natural TL reaches a saturation level in the field environment which is lower than that observed in the laboratory. This would result in an underestimation of the ED. Mejdahl (1988) has suggested that corrections should be applied to ages over 100 ka, the exact amount depending upon the effective mean annual temperature experienced by the grains. This would depend upon the geographical area and the types of potassium feldspar being observed. The extent of thermal loss of signal can only be checked using samples with independent dating control.

Optically stimulated luminescence

In 1985, Huntley *et al.* (1985*a*) used an argon ion laser emitting green light at 514 nm to stimulate a signal from quartz grains and also from fine-grained sediment containing quartz and feldspar. The optically stimulated luminescence

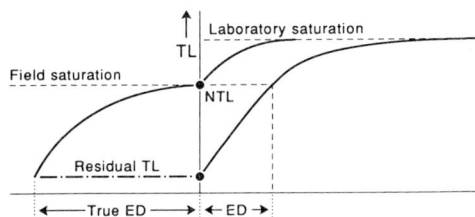

Fig. 2. Schematic representation of the growth of TL signal from feldspars over several hundred ka for which electron loss has occurred at ambient temperature. This behaviour results in the natural TL (NTL) level being lower than the saturation level obtained by laboratory irradiation.

(OSL) is observed while the laser is switched on and the signal decays as the electrons are emptied from the trap with laser exposure time. Because the stimulation is in the green, green luminescence cannot be observed but it is possible by careful selection of optical filters placed in front of the detector to observe emission from potassium feldspars in the violet (400 nm) and quartz in the ultraviolet (365 nm). Suitable filters attenuate the stimulating light by about a factor of 10^{16}.

The advantage of using this approach is that the electron trap that is sampled by the laser light is very sensitive to light and is rapidly bleached by sunlight. A 500 ka quartz sand had its OSL reduced to only 1% of its initial value after only 10 seconds exposure to direct sunlight (Godfrey-Smith *et al.* 1988), whereas no change was observed in the TL during the same exposure. In a parallel experiment they showed that bleaching also occurs under overcast conditions at a slower rate, proportional to the intensity of the diffuse solar radiation.

The disadvantage of the method is the cost of setting up a suitable laser system. Recently, green-emitting photodiodes (Galloway, pers. comm.) and lamps with narrow pass filters (Botter-Jensen & Duller 1992) have been tried as alternative stimulation sources.

The laboratory procedures used to determine the ED are similar to those for TL measurements and involve constructing curves of OSL signal as a function of added laboratory dose using multiple aliquots. Instead of heating the grains to observe the luminescence, the OSL is observed while the laser is switched on (Fig. 3b). A complication in the procedure is that the grains must be heated before measurement to erase any unstable luminescence signal (Aitken 1992; Rhodes 1988; Stokes 1992).

OSL dating case studies

The time range of OSL application to quartz was demonstrated in the first paper by Huntley *et al.* (1985*a*) when they found that the magnitude of the OSL signals from a series of beach dunes in South Australia could be used to distinguish dunes of different ages; the oldest dune was about 700 ka old on the basis of palaeomagnetic measurements. However, the saturation of the OSL signal is the limiting factor and it appears to occur at similar levels for the OSL and the TL.

In a programme designed to determine the depositional environments which are best suited to OSL measurements on quartz, Stokes (1992) has determined EDs on a number of modern or

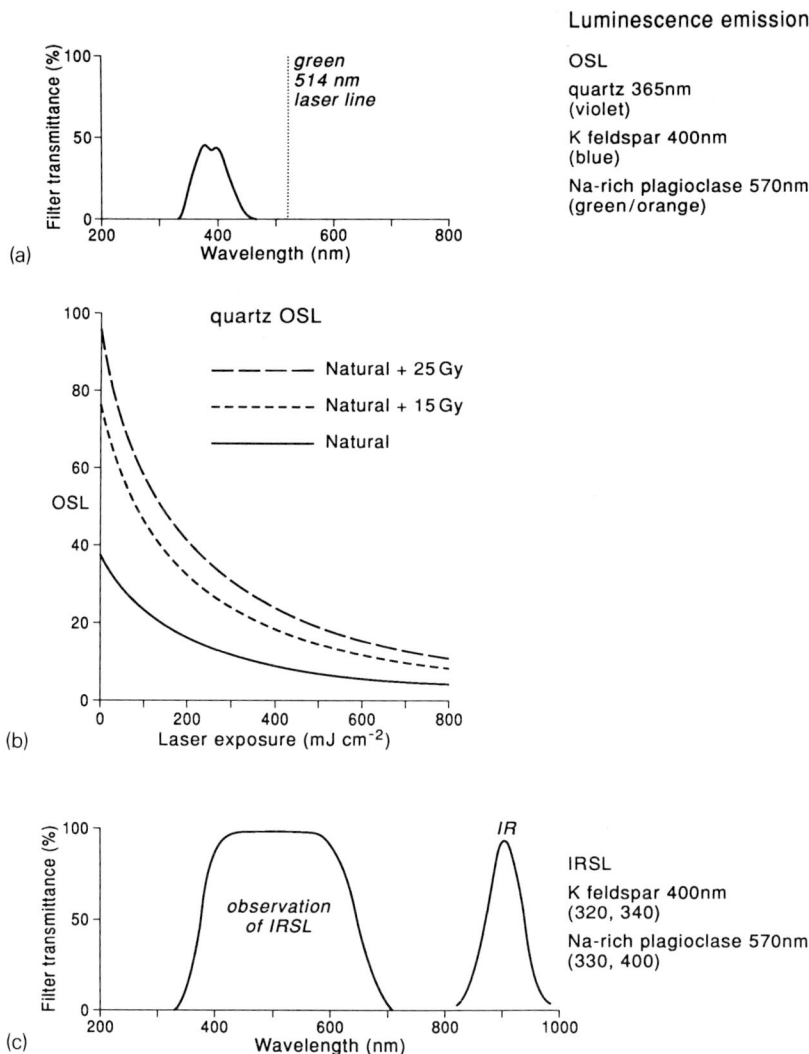

Fig. 3. (a) Schematic representation of argon laser stimulation line relative to transmission region for detecting filters used in OSL measurements. Emission peaks for main mineral species under optical stimulation. (b) Typical quartz OSL decay curves under continuous laser stimulation for natural and irradiated samples (after Smith *et al.* 1990). (c) Schematic representation of infra-red stimulation spectrum relative to transmission region for detecting filters used in IRSL measurements. Emission peaks for feldspars under IR stimulation (weaker peaks in brackets).

young samples. The aeolian sands, primarily from dunefields in the USA, gave lower EDs than coastal or fluvial dunes, indicating that the former were almost totally reset as a result of the grains experiencing multiple cycles of erosion and deposition. This would result in a long cumulative exposure to sunlight for each grain prior to its eventual incorporation in the sand deposit.

A very young age (46 ± 30 years) was ob-

tained for a dune sand from North Mali (Smith *et al.* 1990). In the same study an age of 95 ± 14 ka was obtained for sand from the base of the Dune du Pyla, France, and this was close to the age limit because the natural OSL was close to the saturation level. Stokes (1991) obtained agreement with radiocarbon dates for OSL dates on periglacial sand bracketing the Usselo Layer at a site in the Netherlands. His two results did not show the underestimation reported for the

TL of feldspars (Dijkmans & Wintle 1991). Rhodes (1988) has reported a series of quartz OSL dates on dune sands from archaeological sites at Hengistbury Head, Dorset, UK, and Chaperon Rouge, Morocco, where there was good agreement with other age estimates.

Relatively few applications of OSL dating of feldspars have been reported. A preliminary study of potassium feldspars separated from Holocene aeolian sediments yielded dates in agreement with the limited geological information; they were obtained using green light from a xenon lamp and monochromator (Hütt et al. 1988). Godfrey-Smith et al. (1988) studied the OSL stimulated by an argon laser from feldspars extracted from three non-aeolian sand deposits from Canada. The signals, from potassium and plagioclase feldspars, were a couple of orders of magnitude higher than those from quartz from the same sand units. Under laser stimulation the feldspar signal decayed more rapidly than that for quartz, but under bright sunlight bleaching conditions the feldspar OSL was reduced slightly more slowly than the quartz OSL, but still much more rapidly than the accompanying TL signal. OSL resulting from green stimulation has been detected for a wide variety of feldspars (Spooner 1992). In this study he showed that sanidine, oligoclase and labradorite, types of feldspar which show a short-term instability of the TL signal known as anomalous fading (Wintle 1973), also show anomalous fading of their OSL signals. Hence checks for anomalous fading need to be carried out when dating feldspars using stimulated luminescence techniques.

Infra-red stimulated luminescence

In their pioneering study, Hütt et al. (1988) used a xenon lamp and a monochromator to obtain the stimulation spectrum for a few feldspar samples. They found that not only green light, but also infra-red, around 800–900 nm, gave rise to an OSL signal detectable in the blue (400 nm). At the same time, Godfrey-Smith et al. (1988) were obtaining OSL from feldspars with infra-red lines (799 and 753 nm) from a krypton ion laser. These results were totally unexpected, but very welcome as it made it possible to construct a cheap and simple stimulation source based on infra-red diodes (Spooner et al. 1990). Such a diode array can be built into an existing TL reader (Botter-Jensen et al. 1991) and allow TL measurements to be made on the same sample discs after infra-red stimulated luminescence (IRSL) measurement.

Another advantage of using infra-red for stimulation is that it is possible to select a glass filter which can reject the stimulating wavelengths and pass a wide range of visible wavelengths (Fig. 3c). Huntley et al. (1991) have shown that emission spectra for feldspars are similar to their TL spectra and thus it is possible to use extra filters to select certain types of feldspar (Spooner 1992). Detailed studies of the charge transfer mechanisms which operate in feldspars when they are exposed to light of different wavelengths have been made by Bailiff & Poolton (1991). No IRSL has been reported from pure quartz (Spooner et al. 1990) and lack of response to IR stimulation is used routinely to check the purity of quartz samples prior to OSL measurements (Stokes 1992).

As with OSL, there is no built-in method of checking the long-term stability of the signal as there is in TL. For the latter, the higher the temperature at which the TL signal is emitted during the laboratory measurement, the greater the stability of the signal at ambient temperature. Hence, only the high temperature TL is used to obtain the ED. For IRSL preheat procedures, either short times at elevated temperatures or several weeks at ambient temperature have been developed empirically in an attempt to isolate a thermally stable signal.

A further advantage of the IRSL (and OSL) techniques is that it is possible to use only a short stimulation time (as little as 0.1 s) to obtain a measurable signal and yet cause less than a 1% drop in the overall signal. This means that discs can be measured many times either immediately after the first measurement or several weeks or months later. Hence, checks can be made for anamalous fading using the same set of discs. It also means that a dose response curve can be constructed for a single sample aliquot, leading to much higher precision in measuring the ED. This 'Single Aliquot' technique can also be extended to single grains of potassium feldspar provided that they have sufficient sensitivity; in this case if the potassium content can be determined for each grain, the age of the sediment will be able to be determined with greater accuracy. For this approach to be applied, measurement of a correction factor for the preheat procedure was developed (Duller 1991).

IRSL dating case studies

As yet very few examples of the application of IRSL to aeolian material have been published. Duller (1992) analysed a series of dune sands from raised marine terraces in southern North Island, New Zealand, using both IRSL and TL methods. Although no ages were presented, Duller showed that for these well-bleached materials both methods gave similar EDs. Samples ranged in age from modern to about 350 ka, and the similarity in EDs showed that saturation and thermal stability problems were affecting both the TL and IRSL signals in the same way. Problems with the measurement of the uranium and thorium contents of these grains precluded accurate calculation of the ages (Duller, pers. comm.), but ignoring this dose rate component, the ages were in agreement with the geological evidence for the last 150 ka.

Edwards (1993) presents a suite of results from the Mojave Desert, California, on Holocene dune sands. He mainly used the 'Single Aliquot' approach, though two samples were replicated using standard methods of ED determination (using both TL and IRSL) to check the method. For a modern surface sample the 'Single Aliquot' approach gave a very young age of 40 ± 17 years. The dose rate in this environment was very high (4.5–5 Gy ka^{-1}) due to the large quantity of potassium feldspars present in the sediment. However, the oldest sample dated gave an age of 4114 ± 334 years, and so no problems with saturation or thermal stability were observed.

There is a clear need to test the IRSL method further to establish its potential and limitations. The use of the 'Single Aliquot' approach may radically improve the speed and precision of dating for young samples.

Concluding remarks

Having presented an optimistic view of the new luminescence techniques based on optical stimulation, how should we view TL? For aeolian sediments which have been well exposed to light, TL still has a very important role to play. The sediment samples must still be prepared in the same way and hence the labour intensive part of the dating procedure is not avoided. Since the OSL measurement does not significantly affect the TL signal, it makes sense to measure the TL after the OSL signals have been made. This provides an internal check on both measurements.

OSL methods are advantageous for young samples and for those whose exposure to light at deposition may have been short. However, the limits for older samples are the same as those with the TL method. Quartz is limited by the linearity of the signal response and for feldspars the signal may not be stable over very long periods of time.

In summary, it seems that though there are many advantages to be gained from being able to sample the most light-sensitive electron traps by optical stimulation, the new techniques will be used alongside TL for many years to come.

TL and IRSL studies at Aberystwyth have been supported by NERC grants GR3/7242, GR9/333 and GR3/8190. The author wishes to thank G.A.T. Duller and S.R. Edwards for helpful comments. This is publication number 232 of the Institute of Earth Studies, UCW, Aberystwyth.

References

AITKEN, M. J. 1985. *Thermoluminescence Dating*. Academic Press, London.

——, 1992. Optical dating. *Quaternary Science Reviews*, **11**, 127–131.

BAILIFF, I. K. & POOLTON, N. R. J. 1991. Studies of charge transfer mechanisms in feldspars. *Nuclear Tracks and Radiation Measurements*, **18**, 111–118.

BALESCU, S. & LAMOTHE, M. 1992. The blue emission of K-feldspar coarse grains and its potential for overcoming TL age underestimation. *Quaternary Science Reviews*, **11**, 45–51.

——, PACKMAN, S. C. & WINTLE, A. G. 1991. Chronological separation of interglacial raised beaches from northwestern Europe using thermoluminescence. *Quaternary Research*, **35**, 91–102.

——, ——, & GRUN, R. 1992. Thermoluminescence dating of the Middle Pleistocene raised beach of Sangatte (Northern France). *Quaternary Research*, **37**, 390–396.

BERGER, G.W. 1988. Dating Quaternary events by luminescence. *In*: EASTERBROOK, D. J. (ed.) *Dating Quaternary Sediments*. Geological Society of America Special Paper, **227**, 13–50.

——, BURKE, R. M., CARVER, G. A. & EASTERBROOK, D. J. 1991. Test of the thermoluminescence dating with coastal sediments from northern California. *Chemical Geology*, **87**, 21–37.

BOTTER-JENSEN, L., DITLEFSEN, C. & MEJDAHL, V. 1991. Combined OSL (infrared) and TL studies of feldspars. *Nuclear Tracks and Radiation Measurements*, **18**, 257–263.

—— & DULLER, G. A. T. 1992. A new system for measuring OSL from quartz samples. *Nuclear Tracks and Radiation Measurements* (in press).

BRYANT, E. A., YOUNG, R. W., PRICE, D. M. & SHORT, S. A. 1990. Thermoluminescence and uranium–thorium chronologies of Pleistocene coastal landforms of the Illawarra region, New South Wales. *Australian Geographer*, **21**, 101–112.

CHAWLA, S., DHIR, R. P. & SINGHVI, A. K. 1992. Thermoluminescence chronology of sand profiles in the Thar Desert and their implications. *Quaternary Science Reviews*, **11**, 25–32.

—— & SINGHVI, A. K. 1989. Limits of quartz based thermoluminescence chronology: possible relevance of sensitivity changes. *Synopses from a workshop on Long and Short Range Limits in Luminescence Dating.* Research Laboratory for Archaeology and the History of Art. Oxford University. Occasional Publication No. 9, 85–90.

CHEN, X. Y., PRESCOTT, J. R. & HUTTON, J. T. 1990. Thermoluminescence dating on gypseous dunes of Lake Amadeus, central Australia. *Australian Journal of Earth Sciences,* 37, 93–101.

DIJKMANS, J. W. A., VAN MOURIK, J. M. & WINTLE, A. G. 1992. Thermoluminescence dating of aeolian sands from polycyclic soil profiles in the southern Netherlands. *Quaternary Science Reviews,* 11, 85–92.

—— & WINTLE, A. G. 1991. Methodological problems in thermoluminescence dating of Weichselian coversand and late Holocene drift sand from the Lutterzand area, E. Netherlands. *Geologie en Mijnbouw,* 70, 21–33.

DULLER, G. A. T. 1991. Equivalent dose determination using single aliquots. *Nuclear Tracks and Radiation Measurements,* 18, 371–378.

—— 1992. Comparison of equivalent doses determined by thermoluminescence and infrared stimulated luminescence for dune sands in New Zealand. *Quaternary Science Reviews,* 11, 39–43.

EDWARDS, S. R. 1993. Luminescence dating of sand from the Kelso Dunes, California. *In:* PYE, K. (ed.) *The Dynamics and Environmental Context of Aeolian Sedimentary Systems.* Geological Society, London, Special Publication, 72, 59–68.

FORMAN, S. L. & MAAT, P. 1990. Stratigraphic evidence for late Quaternary dune activity near Hudson on the Piedmont of northern Colorado. *Geology,* 18, 745–748.

GODFREY-SMITH, D. I., HUNTLEY, D. J. & CHEN, W. H. 1988. Optical dating studies of quartz and feldspar sediment extracts. *Quaternary Science Reviews,* 7, 373–380.

GRUN, R., PACKMAN, S. C. & PYE, K. 1989. Problems involved in TL dating of Danish coversands using K-feldspars. *Synopses from a workshop on Long and Short Range Limits in Luminescence Dating.* Research Laboratory for Archaeology and the History of Art. Oxford University. Occasional Publication No. 9, 13–18.

HUNTLEY, D. J., GODFREY-SMITH, D. I. & HASKELL, E. H. 1991. Light-induced emission spectra from some quartz and feldspars. *Nuclear Tracks and Radiation Measurements,* 18, 127–131.

——, —— & THEWALT, M. L. W. 1985a. Optical dating of sediments. *Nature,* 313, 105–107.

——, ——, —— & BERGER, G. W. 1988a. Thermoluminescence spectra of some mineral samples relevant to thermoluminescence dating. *Journal of Luminescence,* 39, 123–136.

——, ——, ——, PRESCOTT, J. R. & HUTTON, J. T. 1988b. Some quartz thermoluminescence spectra relevant to thermoluminescence dating. *Nuclear Tracks and Radiation Measurements,* 14, 27–33.

——, HUTTON, J. T. & PRESCOTT, J. R. 1985b. South Australian sand dunes: a TL sediment test sequence: preliminary results. *Nuclear Tracks and Radiation Measurements,* 10, 757–758.

——, McMULLAN, W. G., GODFREY-SMITH, D. I. & THEWALT, M. L. W. 1989. Time-dependent recombination spectra arising from optical ejection of trapped charges in feldspars. *Journal of Luminescence,* 44, 41–46.

HUTT, G., JAEK, I. & TCHONKA, J. 1988. Optical dating: K-feldspars optical response stimulation spectra. *Quaternary Science Reviews,* 7, 381–385.

HUTTON, J. T., PRESCOTT, J. R. & TWIDALE, C. R. 1984. Thermoluminescence dating of coastal dune sand related to a higher stand of Lake Woods, Northern Territory. *Australian Journal of Soil Research,* 22, 15–21.

JUNGNER, H. 1987. Thermoluminescence dating of sediments from Oulainen and Vimpeli, Ostrobothnia, Finland. *Boreas,* 16, 231–235.

KOLSTRUP, E., GRUN, R., MEJDAHL, V., PACKMAN, S. C. & WINTLE, A. G. 1990. Stratigraphy and thermoluminescence dating of Late Glacial cover sand in Denmark. *Journal of Quaternary Science,* 5, 207–224.

—— & HOUMARK-NIELSEN, M. 1991. Weichselian palaeoenvironments at Kobbelgard, Mon, Denmark. *Boreas,* 20, 169–182.

—— & MEJDAHL, V. 1986. Three frost wedge casts from Jutland (Denmark) and TL dating of their infill. *Boreas,* 15, 311–321.

LEES, B. G., LU, Y. C. & HEAD, J. 1990. Reconnaissance thermoluminescence dating of northern Australian coastal dune systems. *Quaternary Research,* 34, 169–185.

LUNDQVIST, J. & MEJDAHL, V. 1987. Thermoluminescence dating of eolian sediments in central Sweden. *Geologiska Föreningensi Stockholm Förhandlingar,* 109, 147–158.

MEJDAHL, V. 1988. Long-term stability of the TL signal in alkali feldspars. *Quaternary Science Reviews,* 7, 357–360.

—— 1991. Thermoluminescence dating of Late-Glacial sand sediments. *Nuclear Tracks and Radiation Measurements,* 18, 71–75.

PRESCOTT, J. R. 1983. Thermoluminescence dating of sand dunes at Roonka, South Australia. *Council of Europe Journal PACT,* 9, 505–512.

——, AKBER, R. A. & GARTIA, R. K. 1990. Three-dimensional thermoluminescence spectroscopy of minerals. *In:* COYNE, L. M., McKEEVER, S. W. S. & BLAKE, D. F. (eds) *Spectroscopic Characterization of Minerals and their Surfaces.* American Chemical Society Symposium Series 415, Washington, 180–189.

PYE, K. 1982. Thermoluminescence dating of sand dunes. *Nature,* 299, 376.

READHEAD, M. L. 1988. Thermoluminescence dating study of quartz in aeolian sediments from southeastern Australia. *Quaternary Science Reviews,* 7, 257–264.

RENDELL, H. M. 1992. A comparison of TL age estimates from different mineral fractions of sands. *Quaternary Science Reviews,* 11, 79–83.

——, WORSLEY, P., GREEN, F. & PARKS, D. 1991. Thermoluminescence dating of the Chelford Interstadial. *Earth and Planetary Science Letters,* 103, 182–189.

RHODES, E. J. 1988. Methodological considerations in the optical dating of quartz. *Quaternary Science Reviews,* **7**, 395–400.

ROBERTS, R. G., JONES, R. & SMITH, M. A. 1990. Thermoluminescence dating of a 50,000-year-old human occupation site in northern Australia. *Nature,* **345**, 153–156.

SHEPHERD, M. J. & PRICE, D. M. 1990. Thermoluminescence dating of late Quaternary dune sand, Manawatu/Horowhenua area, New Zealand: a comparison with ^{14}C age determinations. *New Zealand Journal of Geology and Geophysics,* **33**, 535–539.

SINGHVI, A. K. & MEJDAHL, V. 1985. Thermoluminescence dating of sediments. *Nuclear Tracks and Radiation Measurements,* **10**, 137–161.

——, SHARMA, Y. P. & AGRAWAL, D. P. 1982. Thermoluminescence dating of sand dunes in Rajasthan. *Nature,* **295**, 313–315.

SMITH, B. W., RHODES, E. J., STOKES, S., SPOONER, N. A. & AITKEN, M. J. 1990. Optical dating of sediments: initial quartz results from Oxford. *Archaeometry,* **32**, 19–31.

SPOONER, N. A. 1992. Optical dating: preliminary results on the anomalous fading of luminescence from feldspars. *Quaternary Science Reviews,* **11**, 139–145.

——, AITKEN, M. J., SMITH, B. W., FRANKS, M. & McELROY, C. 1990. Archaeological dating by infrared-stimulated luminescence using a diode array. *Radiation Protection Dosimetry,* **34**, 83–86.

STOKES, S. 1991. Quartz-based optical dating of Weichselian coversands from the eastern Netherlands. *Geologie en Mijnbouw,* **70**, 327–337.

—— 1992. Optical dating of young (modern) sediments using quartz: results from a selection of depositional environments. *Quaternary Science Reviews,* **11**, 153–159.

STONEHAM, D. & STOKES, S. 1991. An investigation of the relationship between the 100°C TL peak and optically stimulated luminescence in sedimentary quartz. *Nuclear Tracks and Radiation Measurements,* **18**, 119–123.

TEJAN-KELLA, M. S., CHITTLEBOROUGH, D. J., FITZPATRICK, R. W., THOMPSON, C. H., PRESCOTT, J. R. & HUTTON, J. T. 1990. Thermoluminescence dating of coastal sand dunes at Cooloola and North Stradbroke Island, Australia. *Australian Journal of Soil Research,* **28**, 465–481.

WINTLE, A. G. 1973. Anomalous fading of luminescence in mineral samples. *Nature,* **245**, 143–144.

—— & DULLER, G. A. T. 1991. The effect of optical absorption on luminescence dating. *Ancient TL,* **9**, 37–39.

—— & HUNTLEY, D. J. 1979. Thermoluminescence dating of a deep-sea sediment core. *Nature,* **279**, 710–712.

Luminescence dating of sand from the Kelso Dunes, California

STEPHEN R. EDWARDS

Institute of Earth Studies, University College of Wales, Aberystwyth, Dyfed SY23 3DB, UK

Abstract: A relatively new luminescence dating technique, infra-red stimulated luminescence (IRSL), has been used to date dune sand from the Kelso Dunes, eastern Mojave Desert, California. This is the first time the technique has been applied to desert aeolian sands. The IRSL technique measures the time since sediment grains were last exposed to light, by sampling a highly light-sensitive signal which reduces rapidly to a negligible level when so exposed. The technique can be used to date very young samples. Of the ten samples dated, the surface sample gave an age of 40 ± 17 years, while subsurface samples gave high-precision middle and late Holocene dates. These results, obtained from the upper 8 m of the sand mass, indicate significant aeolian activity during the later part of the Holocene. Clustering of IRSL ages from the data so far available suggests that the aeolian activity may have been episodic.

The principal aim of this study is to evaluate the relatively new luminescence dating technique of infra-red stimulated luminescence (IRSL) when applied to dune sand from a desert environment. The technique, in common with other luminescence techniques, measures the time since samples were last exposed to light. It is, therefore, likely to be suitable for determining the timing of periods of aeolian activity and dune formation. As with almost all desert dune areas, no direct, independent age control is available for the Kelso Dunes in the Mojave Desert; however, the dunefield was considered a good test area because there were areas with different dune orientation, indicating different periods of aeolian activity.

Geological background

The Kelso Dunes are located at latitude 34°34'N, longitude 115°43'W, in the eastern Mojave Desert, San Bernardino County, California, and comprise, in terms of sand volume and height, the largest dunefield in the Mojave Desert (Figs 1 & 2). The Mojave River, draining north from the San Bernardino Mountains, is the only through-flowing river of any consequence in the Mojave; today floodwaters from heavy precipitation in the river's source area only fill the river's terminal basin, Silver Lake playa, in exceptionally wet years. On emerging from Afton Canyon (Fig. 1), the Mojave River forms a plain of active alluviation from which a narrow aeolian sand sheet — the Devil's Playground — extends for about 55 km to the ESE. At its eastern end, the sand sheet merges into the Kelso Dunes. The source of sand for the dunes is thought to be the Mojave River alluvial plain (Sharp 1966); however, unconsolidated sands on the floors of the Soda and Silver Lake playas, exposed during former low lake level stands, may also have formed an important source (Fig. 1; Lancaster 1990).

The Kelso Dunes lie in the Kelso Valley, surrounded by mountains on their southern, eastern and northern sides. The dune mass is elliptically-shaped, covers roughly 175 km^2 and contains *c.* 4 km^3 of sand. The sand is well rounded, well sorted and rather coarse relative to typical dune sand, with 90% of grains between 0.25 and 0.50 mm in diameter.

Active dunes are mostly confined to the SE portion of the dunefield; within the active area, three nearly parallel, complex linear ridges, bearing N 65°E, constitute the largest dunes (Fig. 3). The southernmost of the linear ridges is the largest, being around 7 km long and rising nearly 170 m above the alluvial fan to the south (Fig. 2). Surrounding and partly superimposed upon the linear ridges is an irregular complex of medium-sized crescentic dunes.

Areas of sand stabilized by vegetation occur primarily on the lower slopes of the dune mass. In these areas, relatively stable and subdued crescentic dunes form strongly linear patterns of differing orientations (Fig. 3). A 25–35% vegetation cover, including large creosote bushes, is typical of the relatively stable areas.

Prevailing winds in the Kelso Dunes are from the WNW, but the work performed by these pervasive sand-transporting winds is locally counterbalanced by infrequent but strong orographically-controlled winds from other directions (Sharp 1966). Despite a considerable part of the dunefield being currently active, the Kelso Dunes are at the present time undergoing stabilization.

From Pye, K. (ed.), 1993, *The Dynamics and Environmental Context of Aeolian Sedimentary Systems.*
Geological Society Special Publication No. 72, pp. 59–68.

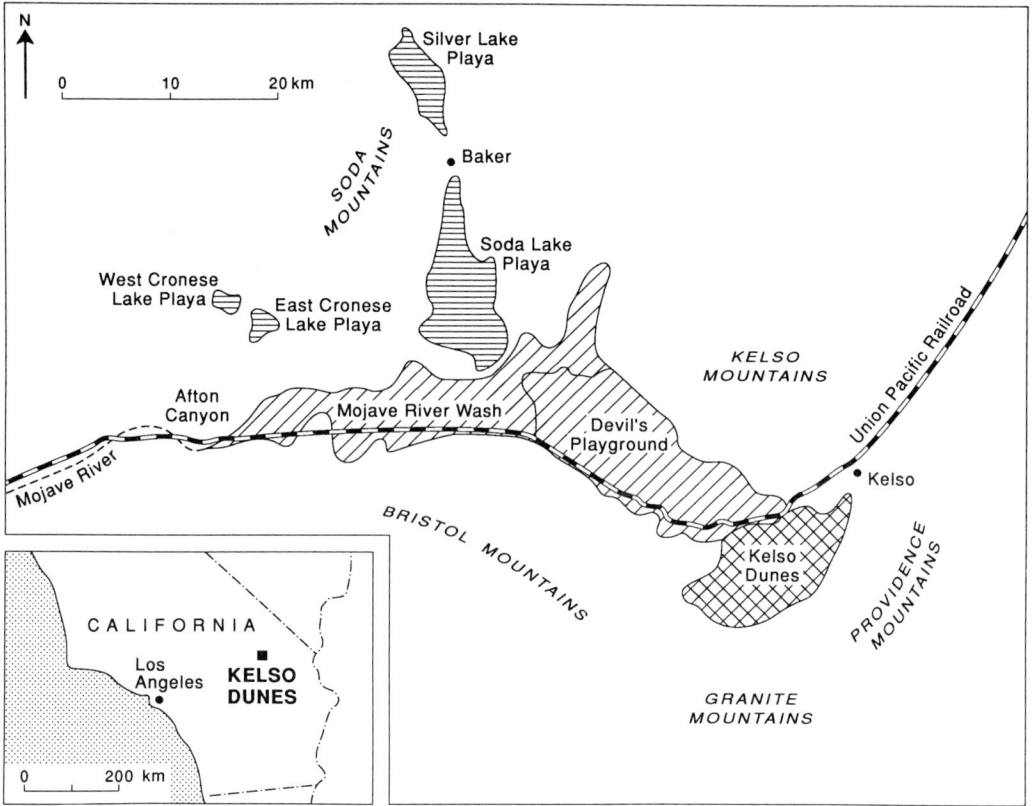

Fig. 1. Location map for the Kelso Dunes, eastern Mojave Desert, California.

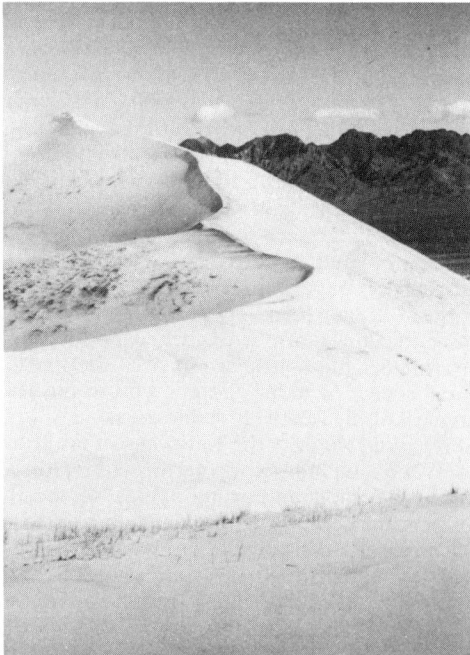

By studying Cottonwood Wash, a major wash which cuts through the Kelso Dunes, Smith (1967) concluded that the active dunes represent a relatively recent renewal of movement through rejuvenation of a once entirely stabilized surface; the Wash was seen to post-date a main dune-building phase in the Kelso Dunes and pre-date the more recent time of reactivation.

Sampling strategy

The objective of the sampling strategy was to obtain samples of representative dunes from areas with different dune orientation within the Kelso Dunes so as to maximize the possibility of collecting samples last exposed to light during

Fig. 2. The southernmost complex linear ridge in the Kelso Dunes, viewed from the NW. The Granite Mountains and an alluvial fan surface covered with regularly spaced creosote bushes can be seen in the background.

different dune-forming episodes. A Landsat image of the Kelso Dunes (Fig. 3) was used to guide the selection of sampling sites.

All but one set of samples were selected from areas of stabilized sand, the samples being obtained by augering to a depth of up to 8 m; one set of samples was taken from one of the currently active linear ridges. The use of an auger ensured that the likelihood of many samples being affected by recent reworking of dune crests would be minimized. Care was taken not to expose samples to any light on collection.

A brief description of the sites from which analysed samples were taken is given in Table 1.

Luminescence dating

Both the thermoluminescence (TL) and the IRSL techniques measure the time since sediments were last exposed to light, as such exposure removes the geologically acquired luminescence signal. Light, however, does not remove the entire TL signal, so that, in TL sediment dating, an assessment of the TL remaining after deposition, known as the 'residual', is required. Grains can generally be assumed to have had a long exposure to sunlight prior to burial in an environment such as that of the study area, thereby reducing the uncertainty in the residual value determined in the laboratory.

In IRSL dating, residuals need not be assessed, as the signal sampled is highly sensitive to light, rapidly reducing to a negligible level on exposure.

IRSL dates are here obtained on potassium feldspars; no IRSL signal has been observed from quartz (Spooner *et al.* 1990). For comparison, two samples were also dated using the

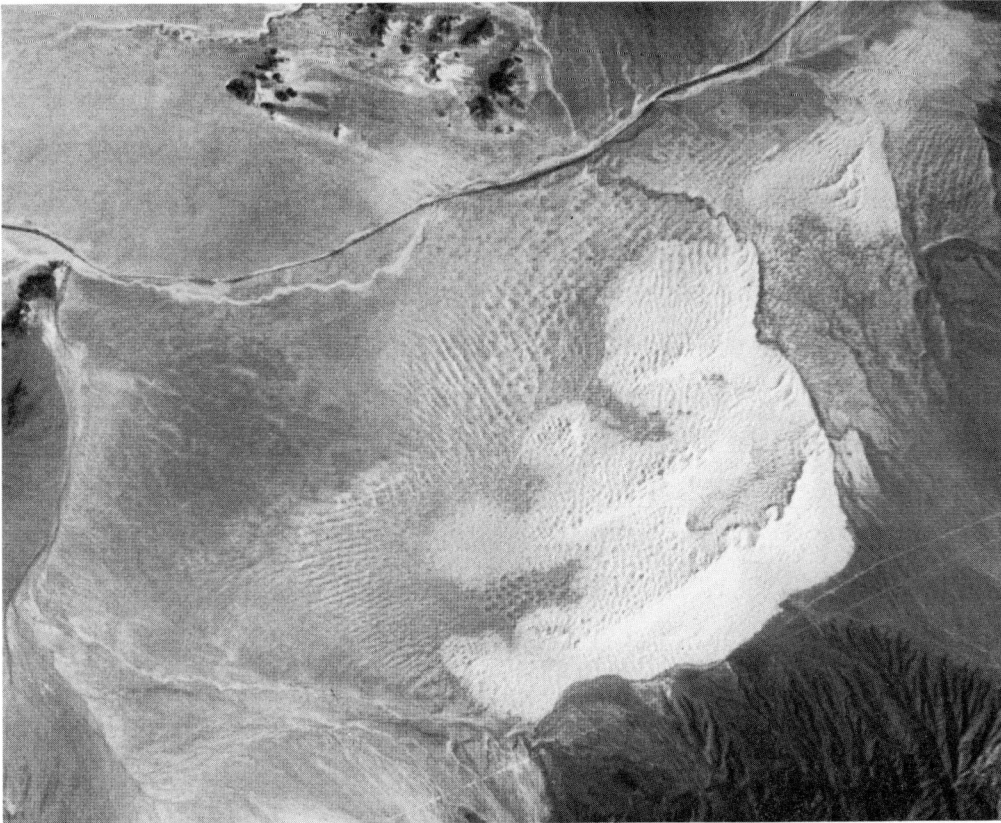

Fig. 3. Landsat image of the Kelso Dunes. Light-coloured areas indicate the presence of mobile sand and dark coloured areas of stabilized sand. Within the main active area, three parallel, complex linear ridges are clearly seen; elsewhere, stabilized crescentic dunes form strongly linear patterns of differing orientations. The northernmost Granite Mountains and a number of washes are visible; the Union Pacific railroad cuts across the image. Width of field of view is approximately 20 km.

Table 1. *Location of analysed samples within the Kelso Dunes**

Sample	Location
Kt	Sample collected from top 1 mm of surface within Unit II, an area of stabilized sand
K7	Sample located at 0.5 m depth within Unit IX, an area of stabilized sand in the NW part of the dunefield exhibiting a surface rhombic waffle pattern
K10	Sample located at 5 m depth within Unit IX, an area of stabilized sand in the NW part of the dunefield exhibiting a surface rhombic waffle pattern
K12	Sample located at 1.1 m depth within Unit XIII, an area of stabilized sand exhibiting crescentic dunes aligned NW–SE
K20	Sample located at 8 m depth from within a crestal hollow in the southernmost active linear ridge; the sample is located within Unit XI, an area of active sand
K21	Sample located at 3 m depth within Unit VII, an area of stabilized sand forming a hollow north of the southernmost linear ridge and mostly surrounded by active sand
K24	Sample located at 8 m depth within Unit VII, an area of stabilized sand forming a hollow north of the southernmost linear ridge and mostly surrounded by active sand
K29	Sample located at 8 m depth within Unit II, an area of stabilized sand in the NE part of the dunefield exhibiting crescentic dunes aligned NNE–SSW
K35	Sample located at 5 m depth within Unit XIV, an area of stabilized sand exhibiting a surface rhombic waffle pattern
K36	Sample located at 8 m depth within Unit XIV, an area of stabilized sand exhibiting a surface rhombic waffle pattern

* Table to be used in conjunction with Figs 3 & 5.

TL technique, the TL dates being likewise obtained on potassium feldspars; in TL dating, potassium feldspars hold a number of advantages over quartz as a dosimeter (see Wintle, this volume).

The age of a sample can be calculated from the age equation:

$$\text{Age} = \frac{\text{Equivalent dose}}{\text{Annual dose}} \qquad (1)$$

where: (1) the equivalent dose is the laboratory-administered dose necessary to replicate the natural TL; and (2) the annual dose is the radiation dose delivered annually since burial from the material surrounding the sample (the external dose) and from the sample itself (the internal dose).

The equivalent dose (ED) was determined using the Additive Dose method: a luminescence growth curve, establishing the sensitivity of a sample's luminescence response to dose, is obtained by adding known laboratory doses in addition to the sample's natural dose; the ED is then found by extrapolating the growth curve to its intercept with either: (i) the residual ($\geq$ 1 day's sunlight exposure or equivalent) TL level for TL dating; or (ii) the dose axis for IRSL dating.

IRSL signals were obtained by 'short shines' — short (0.1 s–1.0 s) exposures of an aliquot to IR stimulation. As the latent TL signal is not significantly reduced by a short shine, for the two samples on which TL dating was performed, multiple disc short shine IRSL dating was undertaken first on the prepared aliquots.

A new IRSL dating technique, the Single Aliquot technique, was used to determine sample ages for all the analysed samples. The technique, which is fast to apply and requires less sample than multiple disc techniques as it allows (subject to the qualification below) ED determinations to be obtained from single aliquots of a sample, has been described by Duller (1991). The technique uses the Additive Dose method, growth curves being constructed for single aliquots by the addition to such aliquots of dose increments. A pre-heat correction is required to raise the luminescence signal corresponding to each cumulative dose level to its true value; this correction is determined on separate undosed aliquots of each sample.

Sample preparation and luminescence measurements

Sample preparation was undertaken in subdued orange light. Samples were sieved to obtain the 180 μm to 211 μm grain size fraction; this fraction was then treated with 10% HCl to remove carbonates, and 30% H_2O_2 to remove organics. The alkali feldspar-rich fraction of each sample was obtained by separating out sub-samples of no more than 5 g using sodium poly-tungstate solution as the heavy liquid, set at a specific gravity of 2.62. The separated light fraction was itself separated using a sodium poly-tungstate solution set at a specific gravity of 2.58 so as to obtain the high potassium alkali feldspar-rich fraction. This light fraction was

Table 2.* Sample dose-rates

| Sample | External dose-rates derived from emission counting | | | | Internal dose-rates derived from laboratory element analysis | | Total dose-rate ($\mu Gy\,a^{-1}$) |
	α Dose-rate† ($\mu Gy\,a^{-1}$)	β Dose-rate‡ ($\mu Gy\,a^{-1}$)	γ Dose-rate§ ($\mu Gy\,a^{-1}$)	Cosmic dose-rate¶ ($\mu Gy\,a^{-1}$)	α Dose-rate‖ ($\mu Gy\,a^{-1}$)	β Dose-rate‖** ($\mu Gy\,a^{-1}$)	
Kt	27 ± 2	2680 ± 351	573 ± 40	290 ± 29	348 ± 24	817 ± 8	4735 ± 354
K7	43 ± 4	2592 ± 89	1148 ± 60	215 ± 22	194 ± 18	595 ± 6	4787 ± 111
K10	37 ± 3	2619 ± 71	1158 ± 61	116 ± 12	299 ± 19	600 ± 6	4829 ± 96
K12	25 ± 2	2637 ± 62	1209 ± 64	183 ± 18	215 ± 25	640 ± 6	4910 ± 94
K20	23 ± 2	2863 ± 45	1190 ± 63	85 ± 9	285 ± 18	650 ± 6	5096 ± 80
K21	29 ± 2	2504 ± 53	1164 ± 61	146 ± 15	282 ± 51	706 ± 7	4832 ± 97
K24	35 ± 2	2655 ± 53	1164 ± 61	85 ± 9	315 ± 21	676 ± 7	4930 ± 84
K29	41 ± 2	2734 ± 80	1087 ± 57	85 ± 9	232 ± 30	666 ± 6	4845 ± 103
K35	60 ± 6	2550 ± 635	1234 ± 65	116 ± 12	361 ± 22	451 ± 4	4772 ± 639
K36a	54 ± 3	2573 ± 86	1234 ± 65	85 ± 9	308 ± 34	632 ± 6	4885 ± 114
K36k	87 ± 4	2658 ± 89	1234 ± 65	85 ± 9	279 ± 17	414 ± 4	4757 ± 112

* Dose-rate calculations are for the relevant-sized grains from which 10 μm had been removed by etching; a water content of 2% and an α efficiency a of 0.2 were used. Errors associated with the α efficiency, grain size, layer removed by etching and water content have been set to zero.
† α Counting performed on a Daybreak 582 Thick Source Alpha Counter.
‡ β Counting performed on a 'SURRC' Thick Source Beta Counter.
§ γ Counting performed using a gamma-scintillometer in the field.
¶ Cosmic ray dose-rates adjusted for depth using the formula of Prescott & Hutton (1988).
‖ U and Th concentrations determined using an Inductively Coupled Plasma Mass Spectrometer.
** K concentrations determined using an atomic absorption technique.

then etched for 40 minutes in 10% HF to remove the outer alpha-irradiated layer of the grains, and then washed in 10% HCl to remove any fluorides. Finally, the etched sample was sieved through a 100 μm diameter mesh to remove fines created by the break-up of feldspars during etching. Monolayers of grains from each etched sample were mounted on clean 10 mm diameter aluminium discs.

Both the TL and the IRSL signals were measured on an automatic Risø Reader System (Bøtter-Jensen *et al.* 1991); the photomultiplier tube used was a type EMI 9635B, the luminescence from the grains being observed through Corning 7-59 and Schott BG-39 filters. A supply of argon provided an inert atmosphere within the Risø's glow oven/sample changer unit during measurement.

Around 60 discs for each of two samples (K10 and K24) were prepared for multiple disc IRSL and TL ED determinations. Laboratory irradiations were performed using a Daybreak ^{90}Sr/^{90}Y beta source; laboratory bleaching was performed by a Hönle SOL2 solar simulator, with a spectrum closely matching that of sunlight. Each aliquot was given a 0.1 s short shine so as to obtain a multiple disc IRSL ED to compare with the TL ED. TL glow curves (plots of TL versus temperature) were then obtained for each disc, up to a temperature of 450°C, using a ramp rate of 3°C s^{-1} for sample K10 and 5°C s^{-1} for sample K24. Second-glow normalization was used to normalize both the IRSL and the TL signals, TL EDs being calculated over the temperature range *c.* 250–420°C.

Eleven samples were analysed using the Single Aliquot technique: for each sample, ten discs were prepared, six for ED determinations, four for ascertaining the pre-heat correction.

Laboratory irradiations were performed using the Risø's ^{90}Sr/^{90}Y beta source. The IRSL was measured each time at 50°C for 1 s, an elevated temperature being used both to create a stable temperature for measurement and, together with the 'long' short shine chosen, to increase the IRSL signal (Duller & Wintle 1991) as the Kelso samples were expected to have low light levels. Six determinations of the ED were obtained for each sample.

Dosimetry

The external dose-rates were determined (i) by counting alpha, beta and gamma emissions and (ii) by determining the bulk sample abundances of U, Th and K. In this way a comparison was made between dose-rates obtained by direct (emission counting) and indirect (element abundance) techniques; the direct techniques were considered preferable (for details, see Edwards 1991). The internal dose-rates were also obtained by the determination of the abundances of U, Th and K in the separated mineral fractions. Results are given in Table 2.

The internal dose-rates form, on average, around 20% of the total dose-rates, the K contents being high (*c.* 10%) for all mineral separates; the total dose-rates are themselves very high (Table 2). Table 3 shows that the decay of ^{40}K accounts for about 90% of the external beta dose-rate; the dominance of K in contributing to the external beta dose-rate may be explained by the richly feldspathic nature of the bulk material: by weight, potassium feldspars account, on average, for one-fifth of the bulk samples. Conversely, only a very small proportion of each bulk sample is composed of silts and very fine sands.

Table 3. *Contributions from U, Th and K to the external beta dose-rate*

Sample	From U and Th* ($\mu Gy\ a^{-1}$)	β Dose-rate (%)	From K† ($\mu Gy\ a^{-1}$)	β Dose-rate (%)
Kt	206	8	2255	92
K7	270	11	2285	89
K10	260	11	2190	89
K12	182	7	2431	93
K20	151	6	2431	94
K21	200	8	2285	92
K24	235	9	2285	91
K29	271	11	2263	89
K35	314	11	2532	89
K36a	359	16	1948	84
K36k	373	16	2007	84

* U and Th concentrations determined from α-counts by using the 'pairs' technique.

† K concentrations determined using an atomic absorption technique.

Table 4. Sample dose-rates, EDs and ages

Sample	Total* dose-rate ($\mu Gy\,a^{-1}$)	Equivalent dose			Age			
		Single aliquot† IRSL (Gy)	Multiple disc† IRSL (Gy)	TL† (Gy)	Single aliquot IRSL (years)	Single aliquot–surface IRSL (years)	Multiple disc IRSL (years)	TL (years)
Kt	4735 ± 422	0.19 ± 0.08			40 ± 17	—		
K7	4787 ± 198	0.68 ± 0.20			142 ± 42	102 ± 45		
K10	4829 ± 223	3.52 ± 0.28	5.58 ± 0.53	4.53 ± 0.43	729 ± 67	689 ± 69	1156 ± 122	938 ± 99
K12	4910 ± 198	2.41 ± 0.18			491 ± 42	451 ± 45		
K20	5096 ± 215	0.53 ± 0.07			104 ± 14	64 ± 22		
K21	4832 ± 229	4.08 ± 0.26			844 ± 67	804 ± 69		
K24	4930 ± 232	19.52 ± 1.66	19.45 ± 1.64	20.21 ± 0.97	3959 ± 385	3919 ± 385	3945 ± 380	4099 ± 275
K29	4845 ± 213	19.93 ± 1.36			4114 ± 334	4074 ± 334		
K35	4772 ± 685	7.34 ± 0.23			1538 ± 226	1498 ± 227		
K36a	4885 ± 241	7.18 ± 0.62			1470 ± 146	1430 ± 147		
K36K	4757 ± 289	7.62 ± 0.30			1602 ± 116	1562 ± 117		

* Errors include those set to zero in Table 2 (see Table 2 * footnote).
† Single aliquot ED errors are at one standard deviation, and are associated with the central values of six ED determinations; all other errors are at one standard deviation and are associated only with the scatter on the naturals.

Results

The IRSL signals from Kelso samples were seen to be rapidly zeroed on exposure to light (Edwards 1991). The dose-rates, EDs and resultant ages for the analysed samples are given in Table 4, each sample's Single Aliquot ED being the arithmetic mean of six ED determinations. A representative Single Aliquot growth curve for one of the analysed samples is given in Fig. 4; each ED determination for this sample is also listed in the figure.

Table 4 and Fig. 4 illustrate the high precision that is obtained for Kelso sample EDs by the Single Aliquot technique. The table also shows that the multiple disc IRSL, TL and Single Aliquot ages for sample K24 are in excellent agreement; however, there is at best only reasonable agreement between the three techniques for sample K10. The difference between the multiple disc IRSL and TL ages for K10 may have been reduced by the application of a more prolonged laboratory bleach time; however, it is not clear why the Single Aliquot technique results in an underestimation of the K10 age relative to the other techniques, given that no similar underestimation occurs for sample K24. The issue is dealt with further by Edwards (1991), where it is shown not to be a significant problem and, therefore, does not invalidate the discussion which follows.

Figure 5 gives the Single Aliquot ages of

Single Aliquot EDs
on Sample K36a

Gy

6.86 ± 0.44
7.30 ± 0.08
6.59 ± 0.09
8.35 ± 0.22
7.08 ± 0.11
6.90 ± 0.16

Mean ED:
7.18 ± 0.62 Gy

Sample K36a

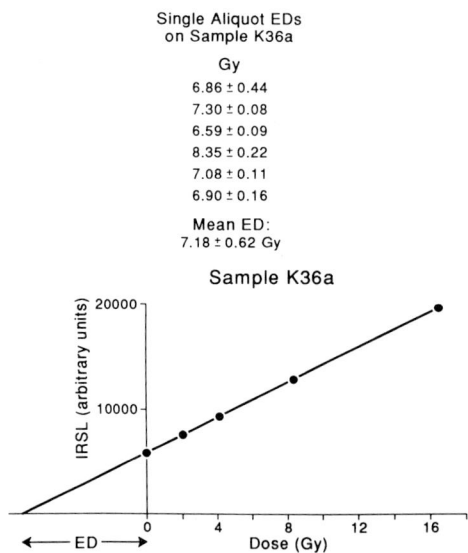

Using the jackknifing technique, the ED equals 7.08 ± 0.11 Gy

Fig. 4. A representative Single Aliquot growth curve and Single Aliquot ED determinations on a sample.

samples in relation to their location and depth; accepting these ages as valid (see above) it can be seen from the figure that sample ages increase with depth, there being no 'stratigraphical inversions' present, that the 'age' of the surface sample is very low, that the samples are comparatively young (all dates indicate middle and late Holocene ages) and that within this time range, despite there being a considerable spread of ages, some clustering of ages is nevertheless discernible.

Discussion

The very low 'age' of 40 ± 17 years for surface sediment shows that, in the field, a negligible residual level of luminescence is obtained for surface material, thus implying that the dating signal is reduced to negligible levels before burial — one of the key attractions of IRSL dating.

The luminescence dates so far obtained from subsurface samples are not inconsistent with the notion of episodic aeolian activity within the Kelso Dunes; the clustering of IRSL ages suggests possibly three main periods of late Holocene aeolian activity and deposition at c. 4000, c. 1500 and 450–800 years BP. If this is the case, the crescentic dunes of Unit II may have formed around 4000 years BP and the crescentic dunes of Unit XIV around 1500 years BP; extensive reworking of sands in Units VII, IX and XIII may have occurred in the period 450–800 years BP.

The currently active complex linear ridges are being shaped by transverse winds at the present day (Sharp 1966); the age of the K20 sample (64 ± 22 years), taken from the southernmost of these ridges, indicates that 8 m net thickness of sand has accumulated at this locality, by migration of superimposed dunes, in a very short period of time.

As luminescence dates provide an estimate of the time since sediments were last buried, and as dune areas may be subject to periods of re-mobilization, it may prove impossible using luminescence dating techniques to estimate the age of the Kelso Dunes as a whole. The absence of dates older than 4000 years BP from the upper 8 m of the Kelso sand mass may be due to dune areas having been extensively reworked prior to this date. Work carried out elsewhere in the eastern Mojave Desert is consistent with this interpretation: in the Cronese Basin, 55 km WNW of the Kelso Dunes, desert varnish developed on wind-grooved bedrock surfaces after 5000 years BP indicates increased aeolian activity prior to this date (Dorn et al. 1989)

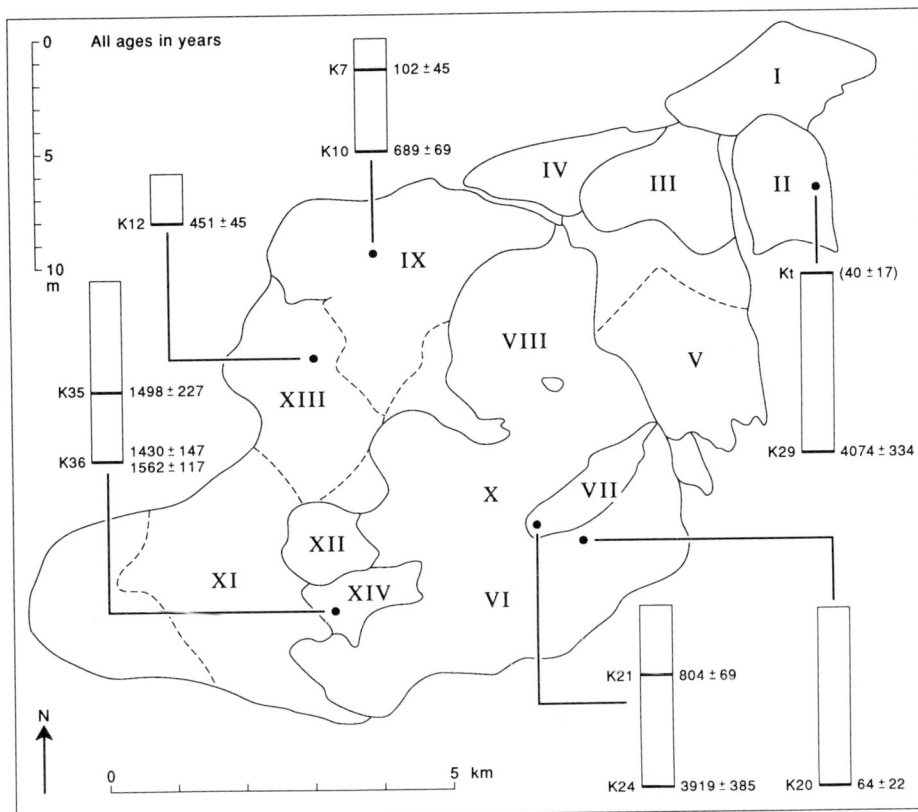

Fig. 5. Location, depths and Single Aliquot ages of samples.

while increased aridity is indicated for the period 5060–6800 years BP by palaeobotanical data obtained from packrat middens in the McCullough Range 100 km NE of Kelso (Spaulding 1991). Evidence is available for dune formation and aridity elsewhere in the southwest United States during the middle Holocene, e.g. the Great Plains (Gaylord 1990) and the southern High Plains (Holliday 1989).

Conclusions

The good agreement obtained between the multiple disc TL and IRSL techniques for two analysed samples demonstrates the validity and applicability of the latter technique in dating dune sand from a desert environment; agreement between the multiple disc IRSL and the Single Aliquot techniques was also good for one of the two samples. Eleven subsamples were dated using the Single Aliquot technique; the scatter between ED determinations on individual aliquots of each sample was low,

the precision offered by the technique correspondingly high. A very low 'age' of 40 ± 17 years was obtained for surface sediment. Luminescence dates obtained from the upper 8 m of the sand mass indicate significant aeolian activity during the later part of the Holocene, with clustering of IRSL ages from the data so far available suggesting that this aeolian activity may have been episodic. The IRSL dating technique appears to work well for desert dune sands and holds much promise in unravelling geological problems in the future.

The author wishes to thank Dr N. Lancaster of the Desert Research Institute, Reno, Nevada for extensive assistance in the field and for stimulating discussions; Dr H. M. Rendell of Sussex University for beta counting measurements; Dr A. G. Wintle for supervision of the project and G. A. T. Duller for generously providing software and practical help in the laboratory. The automated TL/IRSL reader was obtained with NERC grant GR3/8190. Fieldwork expenses were provided by NATO collaborative research grant 900151. This is publication number 236 of the Institute of Earth Studies, UCW, Aberystwyth.

References

BØTTER-JENSEN, L., DITLEFSEN, C. & MEJDAHL, V. 1991. Combined OSL (Infrared) and TL Studies of Feldspars. *Nuclear Tracks and Radiation Measurements*, **18**, 257–263.

DORN, R.I., JULL, A.J.T., DONAHUE, D. J., LINICK, T. W. & TOOLIN, L. J. 1989. Accelerator Mass Spectrometry Radiacarbon Dating of Rock Varnish. *Bulletin of the Geological Society of America*, **101**, 1363–1372.

DULLER, G. 1991. Equivalent Dose Determination using Single Aliquots. *Nuclear Tracks and Radiation Measurements*, **18**, 371–378.

—— & WINTLE, A. G. 1991. On Infrared Stimulated Luminescence at Elevated Temperatures. *Nuclear Tracks and Radiation Measurements*, **18**, 379–384.

EDWARDS, S. R. 1991. *Luminescence Dating of Sand from the Kelso Dunes, California*. Unpublished MSc Dissertation, University of Wales.

GAYLORD, D. R. 1990. Holocene Palaeoclimatic Fluctuations revealed from Dune and Interdune Strata in Wyoming. *Journal of Arid Environments*, **18**, 123–138.

HOLLIDAY, V. T. 1989. Middle Holocene Drought on the Southern High Plains. *Quaternary Research*, **31**, 74–82.

LANCASTER, N. 1990. Dune Morphology and Chronology, Kelso Dunes, Mojave Desert, California. *Geological Society of America Abstracts with Programs 22, 1990 Annual Meeting*, **A86**.

PRESCOTT, J. R. & HUTTON, J. T. 1988. Cosmic Ray and Gamma Ray Dosimetry for TL and ESR. *Nuclear Tracks and Radiation Measurements*, **14**, 223–227.

SHARP, R. P. 1966. Kelso Dunes, Mojave Desert, California. *Bulletin of the Geological Society of America*, **77**, 1045–1074.

SMITH, H. T. U. 1967. *Past Versus Present Wind Action in the Mojave Desert Region, California*. Air Force Cambridge Research Laboratories (USAF Bedford, Massachusetts) Publication AFCRL-67-0683.

SPAULDING, W. G. 1991. A Middle Holocene Vegetation Record from the Mojave Desert of North America and its Paleoclimatic Significance. *Quaternary Research*, **35**, 427–437.

SPOONER, N. A., AITKEN, M. J., SMITH, B. W., FRANKS, M. & McELROY, C. 1990. Archaeological Dating by Infrared Stimulated Luminescence using a Diode Array. *Radiation Protection Dosimetry*, **34**, 83–86.

WINTLE, A. G. 1993. Luminescence dating of aeolian sands — an overview. *In*. PYE, K. (ed.) *The Dynamics and Environmental Context of Aeolian Sedimentary Systems*. Geological Society, London, special publication, **72**, 49–58.

Thermoluminescence dating of periods of sand movement and linear dune formation in the northern Negev, Israel

H. M. RENDELL[1], A. YAIR[2] & H. TSOAR[3]

[1] *Geography Laboratory, University of Sussex, Brighton, UK*

[2] *Institute of Earth Science, Hebrew University of Jerusalem, Givat Ram Campus, Jerusalem, Israel*

[3] *Department of Geography, Ben-Gurion University of the Negev, Be'er Sheva, Israel*

Abstract: The dating of periods of sand movement and dune formation in the Negev has, until the present study, been based on the presence of stone artefacts on, within or beneath sand dunes, or upon ^{14}C dating of associated deposits. We present the results of a preliminary thermoluminescence (TL) dating study of sand and interdune-playa deposits from the Sede Hallamish area of the northern Negev. The results appear to indicate that: (a) sediments in the interdune area were deposited over a period from at least 43 ka to 9 ka and include several depositional cycles; and (b) the lower parts of the linear dunes were formed 6–10 ka and the dune flanks have been stabilized since then. These results are placed in the context of existing studies of the palaeoclimate of the Negev during the Late Pleistocene.

Until now the dating of sand movements in the Negev has been by association of the sand bodies with deposits or archaeological materials datable by [14]C (e.g. Magaritz 1986; Magaritz & Goodfriend 1987). The archaeological richness of the northern Negev and the problems with dating and palaeoenvironmental analysis are clearly demonstrated in the study by Goring-Morris & Goldberg (1990). They show that the uncertainties in the palaeoclimatic reconstruction reflect problems with some of the ^{14}C dates, and also the difficulty of establishing depositional sequences involving sand, and incorporating datable carbon or carbonates. The existing palaeoclimatic data involving the incursion of sand into the northern Negev and the establishment of linear dunes are summarized in Table 1.

Spreads of palaeolithic artefacts associated with sand dunes provide an approximate chronology. The validation of ages based on prehistoric implements by other dating methods is especially problematic in the younger dunes of the Nizzana area, in which the lack of diagnostic soil horizons does not even allow relative dating. Furthermore, organic materials or shells, which could be used for dating, are extremely uncommon in this sandy environment. Even where shells are present, the dating results may be problematic or erroneous (Goodfriend & Stipp 1983). Under such conditions, luminescence dating is probably one of the most promising methods for the establishment of the chronology of the sediments in the study area.

Thermoluminescence (TL) dating of sediments is based on the ability to date the last exposure to light of sediment grains, prior to burial by other grains. This dating technique, along with other allied luminescence methods, affords unique insight into the chronology of sedimentary processes, since the dating 'signal' is carried by the sand grains themselves.

TL dating involves an assessment of (a) the radiation dose absorbed by the sediment grains since their last exposure to light (the Equivalent Dose), and (b) the annual radiation dose. The TL age equation is simply:

$$\text{TL Age} = \frac{\text{Equivalent Dose (Gy}^*)}{\text{Annual Radiation Dose (Gy a}^{-1})}$$

The key assumption is that the TL 'clock' was effectively zeroed during the grains' last exposure to light. In reality all sediment samples exhibit a measurable unbleachable 'residual' TL signal and the establishment of this residual level can be problematic. Also, in the case of geologically 'young' samples, the residual tends to account for a significant proportion of the natural TL signal.

The study area

The present study involves the investigation of the chronological relationship between a linear dune and the adjacent interdune area in the Sede

---

* Gy: Grays, units of radiation dose.

From Pye, K. (ed.), 1993, *The Dynamics and Environmental Context of Aeolian Sedimentary Systems.* Geological Society Special Publication No. 72, pp. 69–74.

Table 1. *Summary of existing palaeoenvironmental data for the northern Negev*

Palaeoenvironmental information	Dating evidence	Reference
Linear dunes established Blown sand incursion *c.* 20 ka BP	Palaeolithic artefacts	Goldberg 1986
Sand incursion in Negev *c.* 20 ka BP	Palaeolithic artefacts and sites	Issar, Tsoar & Levin 1989
Linear dune incursion *c.* 20 ka BP Dunes stabilized 10–11 ka BP	Palaeolithic artefacts and sites ^{14}C dates*	Goring-Morris & Goldberg 1990
Aridity 18 ka BP Wetter period 14–10.5 ka BP (emphasizes steep environmental gradient at desert margin)	Oxygen isotope ^{14}C dates	Magaritz 1986
Aridity 10–11 ka BP Dune migration blocks emphemeral torrent (Nahal Sekher) Wetter conditions 8–9 ka BP	[14]C date for dune-dammed lake	Magaritz & Goodfriend 1987

* Based on dates for similar archaeological material outside the northern Negev.

Fig. 1. Location of the Sede Hallamish/Nizzana study area in the northern Negev (lat. 30°56′N; long. 34°22′E).

Hallamish/Nizzana area of the northern Negev (Fig. 1). The present mean annual rainfall in this area is in the range of 80–90 mm. The linear dune ridges are orientated approximately east–west and have partially vegetated flanks, with active sand movement confined to the crests (Fig. 2). The dune alignment of N 265°E exhibits a 20° deviation from the resultant vector of the wind direction data, but the alignment is parallel to the modal storm wind direction (Tsoar & Møller 1986). Yair (1990) recognizes two subunits within the dune ridges, namely the **active dune** comprising about 25% of dune area, with high slope angles and sparse vegetation, and the inactive **basal dune**, with gentle slopes and well vegetated, particularly on the northern sides.

The surfaces of the basal portions of these dunes are covered with a biological crust that

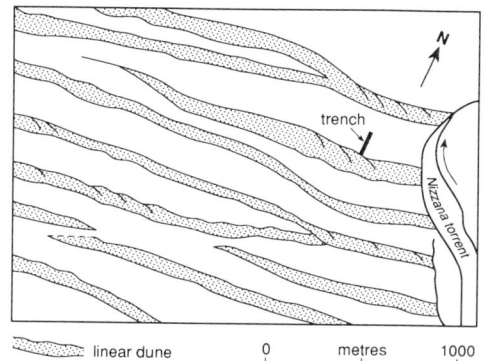

Fig. 2. Spatial distribution of linear dunes and interdune areas in the study area, and the location of the trench through the interdune sequence.

dramatically changes rainfall–runoff relationships (Yair 1990). The linear dunes are separated by interdune areas, comprising dissected and non-dissected playa surfaces, that have a complex irregular topography (Yair 1990). The interdune sediments comprise sands alternating with thin bands of horizontally bedded silt or silt/clay. Grain size analyses of the dune and interdune sediments, show that the dune sand is 98.3% sand size, with a median diameter in the range 250–300 μm, while an interdune silt/clay sample had a median diameter of 40 μm (Yair 1990).

Sampling and TL analysis

The location of the sampling points for TL dating is shown in the cross-section of the dune–interdune area in Fig. 3. The stratigraphic relationships were established in the field by trenching through the interdune area and by augering the adjacent dune. It was noted that the small playa deposit does not extend beneath the dune. At depth, both the dune and interdune sediments appear to be underlain by a loessic silt (shown as a dark horizon at the base of the boreholes in Fig. 3). Samples for TL analysis were taken from the dune crest, the basal dune flank and from several different levels, including two of the silt–clay horizons in the interdune sediments. A decision was made to try to separate out the feldspar fraction of the dune sand for the TL study, on the grounds that

feldspars not only tend to bleach more readily than quartz, but also that they are usually far more sensitive, as radiation dosimeters, than quartz grains.

The sand samples were prepared for TL analysis by sieving to separate the 250–125 μm size fraction, washing in 0.1 M HCl to remove carbonates, and heavy liquid separation (sg 2.62) to isolate the minor (2%) feldspar component in these samples. Samples were then etched for 40 minutes in 10% HF solution. Monolayers of grains were deposited on to clean 10 mm diameter rhodium-plated copper discs using silicone spray for adhesion.

The silt samples were prepared by pretreatment in 0.1 M HCl to remove carbonates and then separation of the 2–10 μm size fraction by sedimentation. Aliquots of this size fraction, in acetone suspension, were pipetted on to clean aluminium discs and allowed to evaporate to dryness at room temperature.

TL emission spectra were monitored through Schott UG11 (ultra-violet) and Chance-Pilkington HA3 (heat absorbing) filters using an EMI 9635Q photomultiplier tube. Heating rates were $150°$ min^{-1}, laboratory beta and alpha irradiations were given by ^{90}Sr and ^{241}Am sources, respectively. The choice of optical filters for feldspar dating is currently the subject of some debate. In particular, some workers (Wintle & Duller 1992; Balescu & Lamothe 1992) have argued that the measurement of UV emissions of feldspars results in the under-

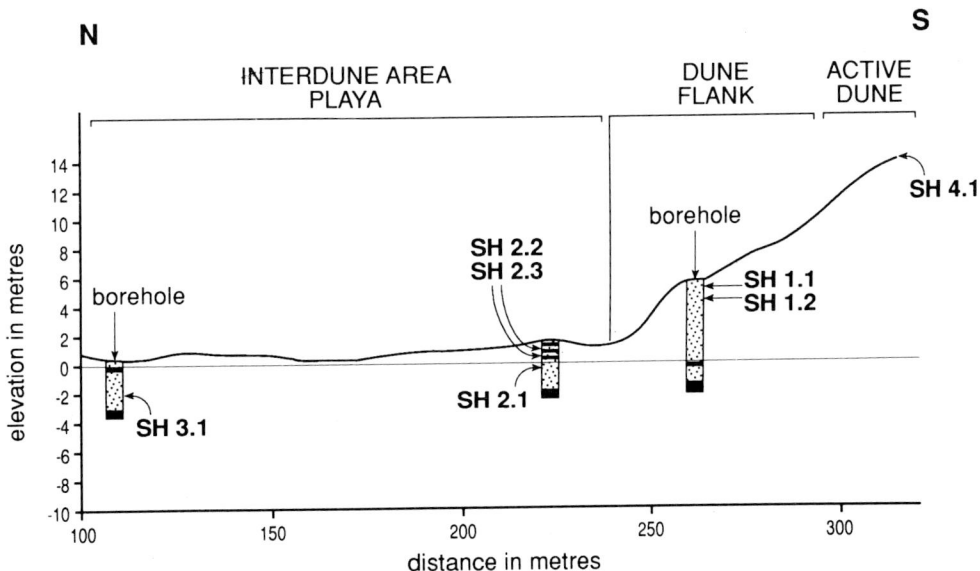

Fig. 3. Location of sampling points on the dune and in the interdune area.

estimation of TL ages. Rendell (1992) argues that problems with feldspar dating cannot be explained simply in terms of choice of filters, but rather that the basic assumptions underlying dosimetry calculations require re-examination. It is also clear from the data quoted by Balescu & Lamothe (1992, Fig. 5) that these effects are strongly sample-dependent.

Equivalent Doses were assessed using both the Additive Dose and Regeneration methods (Aitken 1985). In all cases, linear fits were applied to both the additive dose and the regeneration growth curves, using the method of Berger & Huntley (1986). Although the additive dose relationships showed some scatter, with correlation coefficients as low as 0.89–0.90, the regeneration fits were uniformly good, with correlation coefficients of 0.98–0.99. In the case of sample SH3.1, the amount of feldspar material separated was so small that it was possible to prepare only 14 discs for analysis. In all other cases the growth curves were based on data from at least 20 discs. The Equivalent Doses were assessed over the glow curve range 320°C to 380°C.

The annual radiation doses were calculated on the basis of data from a range of analytical techniques: indirect determination of uranium and thorium contents of bulk samples using sealed alpha counting of crushed samples; beta counting using uncrushed bulk material; direct determination of uranium and thorium contents of the bulk sample and the feldspar fraction by ICP-MS (inductively coupled plasma mass spectrometry) analysis of acid-digested samples; and measurement of the potassium content of both bulk sediment and the feldspar separates using AAS (atomic absorption spectrophotometry). The gamma doses received by the fine grain samples SH2.2 and SH2.3 were recalculated to take account of the gamma contribution from the sand layer above and below each of the silt–clay layers. The dose rates were calculated on the basis of internal and external doses, and allowance was made for dose attenuation within the large grains. The mean annual moisture content for the dune sand samples was estimated to be 2% dry weight, while that for the fine-grained interdune sediments was estimated to be 5% dry weight.

Results and discussion

The dosimetry data and TL results are summarized in Table 2. The TL results appear to indicate that:

(a) sediments in the interdune area, overlying the loessic silt, were deposited over a period from at least 43 ka to 9 ka and include several depositional cycles. There is a distinct erosional break between SH3.1 and the sequence of sediments including SH2.1;

(b) the linear dunes were in place by 6–10 ka and the dune flanks have been stabilized since then;

(c) the TL dates obtained for the SH2 series of interdune sediments point to relatively fast changes in environmental conditions and depositional processes during the period 15 ka–11 ka. Four distinct sedimentary units (two of them sandy and two silty-clay) with a total thickness of 1.2 m were deposited during this period. Only 0.4 m of sediment have been added since then, i.e. during most of the Holocene;

(d) given the topographic relationship between the dune and the interdune, and the fact that the fine-grained interdune sediments do not continue beneath the dune, we conclude that the dunes and the fine-grained interdune sediments are contemporary. A more detailed discussion of these age relationships must await further detailed sampling and analysis including sampling down from the dune crest and further sampling of the dune flanks.

The results also present some problems.

(a) The choice of the feldspar fraction for these preliminary measurements was determined by the fact that feldspars give larger TL signals per unit of radiation dose than does quartz, and thus it should be possible to obtain measurable TL signals even for very young samples. The feldspar content of the sand samples was very low (*c.* 2%) and the separation of a suitable quantity of grains for measurement was a major problem. As a first attempt, and owing to the paucity of feldspars, the feldspar fraction lighter than sg 2.62 was selected for analysis. Thus, the feldspar fraction is likely to contain both potassium and sodium feldspars and this poses dosimetry problems (see Rendell (1992) for discussion). A future study will have to be based on the selection of potassium feldspars (lighter than sg 2.58) in order to avoid this problem.

(b) The results for SH1.1 and 1.2 and SH2.2 and 2.3 show an age inversion with depth which makes no sense in terms of the stratigraphical relationships between these samples. In the case of the fine-grained samples SH2.2 and SH2.3, the TL age estimates have overlapping error bars, whereas this is not the case for the pair of dune samples SH1.1 and 1.2.

Table 2. *Dosimetry and TL age estimates*

Sample	U (ppm)	Th (ppm)	K (%)	External dose rate (mGy a⁻¹)	Internal dose rate (mGy a⁻¹)	Total dose rate (mGy a⁻¹)	Equivalent dose (Gy)		TL age (ka)	
							Additive dose	Regeneration	Additive dose	Regeneration
SH4.1	0.45 ± 0.07	0.90 ± 0.23	0.47 ± 0.05	0.947 ± 0.085	0.235 ± 0.035	1.182 ± 0.130	0.20 ± 0.02	0.10 ± 0.01	0.16 ± 0.02	0.08 ± 0.01
SH1.1	0.61 ± 0.07	1.14 ± 0.21	0.32 ± 0.05	0.898 ± 0.081	0.021 ± 0.003	0.919 ± 0.101	9.00 ± 0.90	9.17 ± 0.32	9.79 ± 1.00	9.98 ± 1.00
SH1.2	0.61 ± 0.07	1.21 ± 0.21	0.49 ± 0.05	0.935 ± 0.084	0.116 ± 0.017	1.051 ± 0.115	6.05 ± 0.41	7.20 ± 0.46	5.75 ± 0.70	6.85 ± 0.70
SH2.1	0.53 ± 0.11	1.94 ± 0.34	0.51 ± 0.05	1.111 ± 0.100	0.023 ± 0.003	1.134 ± 0.107	18.00 ± 2.05	17.14 ± 1.32	15.87 ± 1.79	15.11 ± 1.51
SH2.2	2.41 ± 0.22	5.51 ± 0.71	0.75 ± 0.05	nd	nd	3.271 ± 0.359	*	37.50 ± 5.00	*	11.46 ± 1.48
SH2.3	2.46 ± 0.23	3.99 ± 0.74	0.75 ± 0.05	nd	nd	2.871 ± 0.315	24.30 ± 4.10	27.68 ± 0.22	8.46 ± 1.42	9.64 ± 1.00
SH3.1	1.05 ± 0.13	1.64 ± 0.40	0.33 ± 0.05	1.026 ± 0.092	0.049 ± 0.007	1.075 ± 0.125	45.00 ± 4.50	43.77 ± 3.80	41.86 ± 4.20	43.06 ± 4.30

Notes: 'a' values for fine grained samples: SH2.2 0.204 ± 0.014; SH2.3 0.196 ± 0.007.
nd only total dose rate was determined for the fine grain samples SH2.2 and 2.3.
* owing to high scatter on additive dose, only Regeneration ED determined for SH2.2.

In the case of both pairs of samples we need to consider the potential for errors in both the dose rate estimates used and in the Equivalent Dose determinations. Although part of the problem concerning annual dose rate could probably have been resolved by making dosimetry measurements in situ, the main difficulty in estimating dose rate lies in the estimation of the water content history of the sediments since burial. In the case of the dune samples, however, differences in water content alone cannot explain the inverted ages. For example, even if we assume that sample SH1.2 was completely saturated with water throughout its history, this would decrease the dose rate by only 13% thus increasing the age estimate from 6.5 ka to 7.9 ka. It is hoped that the 'inversion' problems will be resolved by further work on both the luminescence measurements and the dosimetry, but at the moment the results still provide a conservative guide as to the age of the stabilized dune flanks and of the interdune sediments.

This preliminary TL chronology established for the Nizzana research site appears to be in reasonably good agreement with existing data as summarized in Table 1, and with Goring-Morris & Goldberg (1990; p.130) in particular, who conclude that, in the Nizzana area 'by 10,000 BP the present dune topography was in place, on the basis of prehistoric occupations near the crests'.

Conclusions

The results of this preliminary TL study of a linear dune/interdune area in the northern Negev have established the applicability of this particular dating technique for providing a chronological framework for sand movement and dune stabilization. These results indicate that sediments in the interdune area were deposited over a period from at least 43 ka until 9 ka and that there has been little further deposition during the rest of the Holocene, and that the dune flanks have been effectively stabilized for the last 6–10 ka. The results obtained are in reasonably good agreement with existing estimates based on archaeological evidence. Although the results of this small pilot study are encouraging, further work on new samples from this and other areas, preferably using optically stimulated luminescence in addition to conventional TL measurements, is required in order to confirm these results.

The study was conducted at the Nizzana Experimental Station of the Arid Ecosystems Research Centre of the Hebrew University of Jerusalem. We also wish to thank Ann Wintle for undertaking additional dosimetry measurements.

References

AITKEN, M. J. 1985. *Thermoluminescence Dating.* Academic Press, London.

BALESCU, S. & LAMOTHE, M. 1992. The blue TL emission of K-feldspar coarse grains and its potential for overcoming TL age underestimation. *Quaternary Science Reviews,* **11,** 45–51.

BERGER, G. W. & HUNTLEY, D. J. 1986. Linear regression of TL data. *Ancient TL,* **4,** 31–35.

GOLDBERG, P. 1986. Late Quaternary environmental history of the southern Levant. *Geoarchaeology,* **1,** 225–244.

GOODFRIEND, G. A. & STIPP, J. J. 1983. Limestone and the problem of radiocarbon dating of land-snail shell carbonate. *Geology,* **11,** 575–577.

GORING-MORRIS, A. N. & GOLDBERG, P. 1990. Late Quaternary dune incursions in the southern Levant: archaeology, chronology and palaeoenvironments. *Quaternary International,* **5,** 115–137.

ISSAR, A., TSOAR, H. & LEVIN, D. Climatic changes in Israel during historical times and their impact on hydrological, pedological and socio-economic systems. *In:* LEINEN, M & SARNTHEIN, M. (eds) *Paleoclimatology and Paleometeorology: Modern and Past Patterns of Global Atmospheric Transport,* Kluwer, Dordrecht, 525–541.

MAGARITZ, M. 1986. Environmental changes recorded in the Upper Pleistocene along the desert boundary, Southern Israel. *Palaeogeography, Palaeoclimatology, Palaeoecology,* **53,** 213–229.

——— & GOODFRIEND, G. A. 1987) Movement of the desert boundary in the Levant from Latest Pleistocene to Early Holocene. *In:* BERGER, W. H., LABEYRIE, L.D. (eds) *Abrupt Climatic Change.* Reidel, Dordrecht, 174–184.

RENDELL, H. M. 1992. A comparison of TL age estimates from different mineral fractions of sands. *Quaternary Science Reviews,* **11,** 79–84.

THOMAS, D. S. G., & TSOAR, H. 1990. The geomorphological role of vegetation in desert dune systems. *In:* THORNES, J. B. (ed.) *Vegetation and Erosion.* Wiley, Chichester, 471–489.

TSOAR, H. & MØLLER, J. T. 1986. The role of vegetation in the formation of linear sand dunes. *In:* NICKLING , W. G. (ed.) *Aeolian Geomorphology.* Allen & Unwin, Boston, 75–95.

WINTLE, A. G. and DULLER, G. A. 1992. The effect of optical absorption on luminescence dating. *Ancient TL* **9,** 37–39.

YAIR, A. 1990. Runoff generation in a sand area—the Nizzana Sands, Western Negev, Israel. *Earth Surface Processes and Landforms,* **15,** 597–609.

A chronostratigraphic re-evaluation of the Tusayan Dunes, Moenkopi Plateau and southern Ward Terrace, Northeastern Arizona

STEPHEN STOKES[1] & CAROL S. BREED[2]

[1] *Research Laboratory for Archaeology & the History of Art, and School of Geography, Oxford University, 6 Keble Road, Oxford, OX1 3QJ, UK*

[2] *US Geological Survey, Flagstaff, Arizona, Geologic Division, USA*

Abstract: Optical dating of the Tusayan Dunes of northeastern Arizona indicates that at least three distinct late Holocene phases of aeolian reactivation took place in the area, partly upon pre-existing, Pleistocene linear dunes. The reactivation phases occurred approximately 400, 2000–3000, and 4700 years ago. We hypothesize that the large discrepancy in age between the basal and upper parts of the linear dunes relates to the reactivation of only the linear ridge crests, leaving an undisturbed plinth. If so, linear dunes hold a potential for preserving long-term records of successive aeolian depositions. Evidence of regional alluvial-fill episodes and palaeoclimatic data for the area indicate that sediment supply, in addition to aridity and effective sand-transporting winds, is a key factor in determining the timing of aeolian reactivation.

Aeolian landforms and fluvial deposits are the most common surficial deposits over large areas of the southwestern United States (e.g. Reeves 1976; Ahlbrandt *et al.* 1983; Breed *et al.* 1984; Billingsley 1987*a*; Osterkamp *et al.* 1987; Holliday 1990; Wells *et al.* 1990; Madole in press). The deposits reflect a widespread regional response to an extended late Quaternary period of weathering, mass wasting, and sedimentary redistribution of detritus during a phase of relative dryness (e.g. Antevs 1962; Bryant & Holloway 1985). Interactions among climate, weathering, mass wasting, and fluvial and aeolian erosion and deposition are recorded in the Quaternary landscape of the Moenkopi Plateau, Ward Terrace, and Little Colorado River valley approximately 100 km northeast of Flagstaff and southwest of Black Mesa, Arizona (Fig. 1).

The entrenchment of the Little Colorado River and the retreat of cliffs along the eastern side of its valley as long ago as 2.4 Ma (Billingsley 1987*a*) resulted in the formation of a series of broad, sparsely vegetated, stepped terraces and stripped structural benches on which the southwest-trending Tusayan Washes (Hack 1942) flow in a direction opposite to the prevailing southwesterly wind (Figs 1 & 2). The highest-level erosion surface within the study area is the Moenkopi Plateau, at about 1660 m elevation; the lowest is the Little Colorado River floodplain at about 1300 m elevation. Regional precipitation is low (200 mm or less per annum) and

is distributed with bimodal maxima during the summer and winter, separated by autumn and spring intervals of reduced rainfall (Breed *et al.* 1984, fig. 16.1; Dean 1988). Average annual precipitation over a 10-year period of records at the US Geological Survey geometeorological station near Gold Spring, at the south edge of the Moenkopi Plateau (Fig. 2) was 147 mm.

Prevailing southwesterly winds, which have been observed to exceed 75 km hr^{-1} (Breed & Breed 1979), have entrained loose surficial material and deposited sand in dunes on terraces and pleateaux east of the Little Colorado River (Fig. 2). Typically, sand dunes are barchanoid and parabolic on the lower levels, including the floodplain, tributary washes, and Ward Terrace, and linear on the Moenkopi Plateau. The barchanoid and parabolic dunes are particularly abundant around a salient of the Red Rock Cliffs known as Paiute Trail Point (Fig. 2), where they climb to the highest level presently reached by active migrating dunes. No dunes now climb the 150 m-high Adeii Eechi Cliffs that rise from the Red Rock Cliff to the Moenkopi Plateau (Fig. 2).

Although the Little Colorado River is an obvious source of fluvial sediments derived from outside the local area, Billingsley (1987*a*) has shown that the dunes on Ward Terrace and the Moenkopi Plateau are built mostly of sand eroded from local outcrops within the drainage basins of the ephemeral Tusayan Washes. Within this area, therefore, the approximately

From Pye, K. (ed.), 1993, *The Dynamics and Environmental Context of Aeolian Sedimentary Systems.*
Geological Society Special Publication No. 72, pp. 75–90.

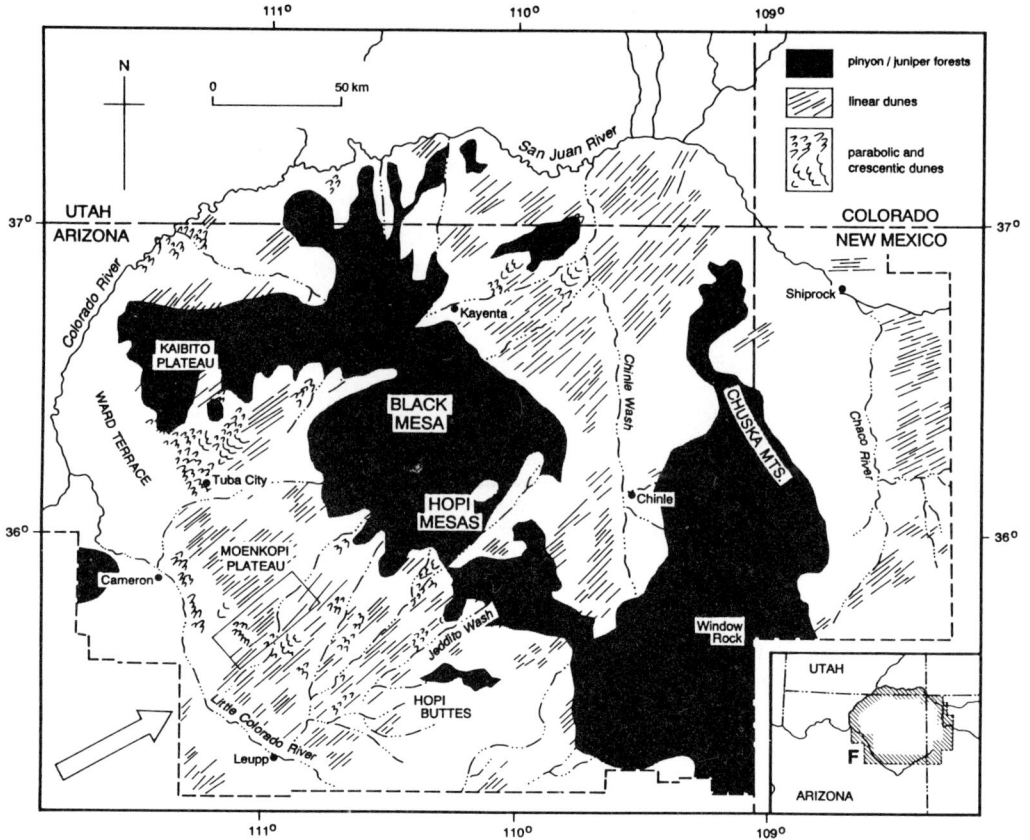

Fig. 1. Locality map depicting major physiographic units, and types and aerial extent of commonly occuring dunes in northeastern Arizona and vicinity (redrawn from Hack (1941)). Samples for optical dating taken from within small rectangular area between Moenkopi Plateau and Little Colorado River. Arrow at bottom left shows prevailing resultant wind direction for the area (after Breed & Breed (1979)). Regional context shown on small map, lower right (F: Flagstaff).

diametric opposition of fluvial and aeolian sediment transport pathways has resulted in the development of an almost closed depositional system (Breed *et al.* 1984). As such, the area is an ideal locality in which to observe the temporal response to fluctuations in late Quaternry climate of an aeolian depositional system in a semi-arid to arid environment.

The chronology of the aeolian deposits of Ward Terrace and the adjacent Moenkopi Plateau (herein collectively termed the Tusayan Dunes) is the subject of this study. In particular, we have revised the chronology of the stratigraphically upper parts of the dunes in the vicinity of Paiute Trail Point on Ward Terrace and the Moenkopi Plateau by this first application of the optical dating method to quartz sand grains in deserts of the American South-

west (Stokes *et al.* 1991). The optical dating method, a developmental absolute dating method closely related to thermoluminescence (TL) dating, allows the direct age evaluation of aeolian and other deposits over a time scale of at least 100 ka (Rhodes 1990; Smith *et al.* 1991; Stokes 1991, 1992*a*). The regional geological and geoarchaeological significance of the dunes (e.g. Hack 1941, 1942; Breed & Breed 1979; Breed *et al.* 1984), their relatively high quartz content (Billingsley 1987*a*), and good exposure and accessibility make them highly suited for quartz optical dating.

Before the presentation of new dates for some of the Tusayan Dunes and a discussion of the local and regional palaeoenvironmental implications, we give a brief review of previous investigations of the dunes, followed by an out-

Fig. 2. Oblique aerial photograph oriented eastwards across southwest part of area shown in Fig. 1. Active barchanoid dunes are visible within poorly vegetated ephemeral drainage channels that carry seasonal run-off to the Little Colorado River. Dunes migrate northeast in response to prevailing wind direction (Fig. 1), both within margins of poorly defined channels and braided plains and more widely across Ward Terrace. More subdued expression of inactive linear dunes can be observed on Moenkopi Plateau. (Source: US Air Force U-2 photograph, 1968).

line of the quartz optical dating method and the equipment and experimental procedures employed.

The Tusayan Dunes

The Tusayan Dunes are part of a much more extensive sand sea spanning areas of north-eastern Arizona, southeastern Utah, and north-western New Mexico (Fig. 1), as first systematically described and mapped by Hack (1941). Hack recognized three major dune types: transverse (barchanoid), parabolic, and longitudinal (linear), occurring commonly in geographical proximity (Fig. 1). Subsequent studies have described the morphology and chronology of these dunes (e.g. Breed & Breed 1979; Wells 1983; Breed et al. 1984; Billingsley 1987a, b; Wells et al. 1990) and of other dune fields to the

east and north in the Great Plains (e.g. Ahlbrandt et al. 1983; Muhs 1985; Gaylord 1990).

Most of the presently active dunes in north-eastern Arizona (southwest part of the area shown in Fig. 1) are barchanoid or parabolic features on the Little Colorado River flood-plain, in tributary washes, and on Ward Terrace (Fig. 2). These dunes may attain heights of 10 m and have slipfaces as long as 80 m. Sand is supplied to them via the braided Tusayan Washes that flow towards the Little Colorado River during the rainy season. Vegetated, palaeosol-capped, climbing barchanoid and parabolic types are preserved as relict dune features in topographically sheltered localities on Ward Terrace. Similarly relict features occur as linear dunes atop the Moenkopi Plateau, more than 150 m higher. Volumetrically minor sand sheets veneer the surfaces at both elevations (Billingsley 1987b).

Deposition of linear dunes on the Moenkopi Plateau occurred when sand grains were provided a route for saltation to that elevation via sand ramps, which were built by climbing dunes to the top of the plateau at some former time or times (Breed & Breed 1979; Billingsley 1987a). From this sand source linear sand ridges developed, which today extend downwind from gully heads and cliff points along the Adeii Erchi Cliff. Since the sand ramps were eroded away from the cliff head, the upwind ends of the linear dunes themselves have been severely deflated, exposing interiors partly cemented by caliche (calcrete). A thin veneer of white sand, eroded from bedrock at the cliff edge, mantles the dune crests. The linear dunes average more than 3.5 km in length and about 43 m in width. They range in height from 2–10 m (Breed & Breed 1979). Dune spacing varies, but is typically of the order of hundreds of metres. In some localities the dune ridges have been recently reactivated probably in response to overgrazing and trampling by domestic cattle.

The Tusayan Dunes consist primarily of quartz sand with varied quantities of silt and clay (up to 15% in basal parts of sampled linear dunes). The sand is yellowish red in most places (Table 1). The dunes may have trough and low-angle cross-bedding in their upper parts, grading downwards to a typically structureless sand, or they may be structureless throughout. Cross-beds are commonly preserved within the barchanoid dunes, whereas the linear dunes are commonly structureless except for the uppermost 0.3 m. Primary sedimentary structures have been obliterated wherever pedogenesis has taken place.

The age of the Tusayan Dunes has been the subject of several previous investigations. Hack (1941) suggested that all dune types in this region were formed contemporaneously, that they continue to form under present conditions, and that a continuum of aeolian deposition has taken place for an indeterminate but extended period, probably since the late Pleistocene. He based his age estimate on observed relations of falling dunes to alluvial deposits in Tsegi Canyon near Black Mesa and on the presence of stabilized longitudinal (linear) dunes beneath forests of pinyon and juniper at several localities that now receive more than 375 mm annual precipitation (Hack 1942). Hack also noted the superposition of barchanoid dunes upon formerly stable dwelling surfaces of Pueblo II-

and III-age archaeological sites in the Antelope Mesa area near Jeddito Wash (Fig. 1). He concluded that there have been regional phases of aeolian inactivity for at least some dune forms, and he suggested that the primary restraining factor for aeolian deposition was the resistance of the vegetation cover to erosion.

Stokes (1964) undertook a regional geomorphological reconnaissance of the linear dunes and concluded, on the basis of stream incision rates, that some of the dunes were formed as much as 1 Ma ago. Subsequently, Breed & Breed (1979) sectioned two of the linear dunes on the Moenkopi Plateau. They sampled a deposit of pyroclastic basaltic glass mixed with sand near the base of one of these dunes, and also analysed a similar deposit found interbedded with interdune sands on top of Paiute Trail Point. The chemistry, petrology and glass morphology of the volcanic ash deposits at both localities suggest their derivation from an eruption in the nearby San Francisco Volcanic Field. Probable correlative lavas within the San Francisco Field have been dated using the K–Ar method to approximately 0.46 ± 0.35 Ma ago (Breed & Breed 1979). They concluded, from this admittedly very broad date, that the linear dunes were already forming at least 100 000 years ago. Wells (1983) suggested (without supporting absolute chronological data) that the linear dunes were last significantly reactivated at the time of the last glacial maximum, approximately 18 ka ago. Billingsley (1987a) reconstructed the cliff retreat that formed the Moenkopi Plateau and surmized that the dunes on the erosion surface on the Moenkopi Plateau might have been deposited initially as early as late Tertiary (?Pliocene) time. He (Billingsley 1987b) mapped the dunes in the area from the Little Colorado River to the top of the Moenkopi Plateau, recognizing active and stabilized dunes of linear, barchanoid, parabolic, and complex types, topographically-controlled (climbing and falling) semi-active dunes, and active and stabilized sand sheets. Billingsley assigned ages to these units that range from Pleistocene to the present, based primarily on their stratigraphic and topographic relations and on the extent of their degree of consolidation, erosion and vegetation cover (Billingsley 1987b). Of particular interest is his recognition of apparently related units on the Moenkopi Plateau and on Ward Terrace, and his designation of stabilized linear dunes of Pliocene(?) and Pleistocene age as a base for younger, reactivated linear dunes on the Moenkopi Plateau.

Table 1. Locality and physical characteristics of the dune sands sampled for optical dating

Sample	Location	Elevation (m above sea level)	Sample depth (m)	Colour (Munsell Color Co. 1954)	Description
Inactive or semi-active dune samples					
$OX_{od}836/1$	35°55'N, 111°01'30"W	1768	1.7	5YR 5/6 yellowish red	Linear dune in road cut, approx. 5.5 km from cliff, head, limited soil development down to 30 cm, recent sand top 50 cm, no sedimentary structures.
$OX_{od}837/1$	35°46'40"N, 111°04'W	1658	2.0	5YR 5/6 yellowish red	Linear dune less approx. 1 km from cliff head, sample taken near locality formerly dug by Breed & Breed (1979), no sedimentary structures, limited soil development (30 cm).
$OX_{od}838/1$	35°46'35"N, 111°04'30"W	1653	1.4	5YR 6/6 yellowish yellow	Linear dune, semi-consolidated, no sedimentary structure overlain by very poorly developed soil (5–10 cm).
$OX_{od}839/1$	35°46'22"N, 111°04'40"W	1646	0.6	5YR 5/6 yellowish red	Moderately consolidated, structureless sand, well-developed (B_k) calcic soil horizon approx. 40 cm thick.
$OX_{od}840/1$	35°46'38"N, 111°10'30"W	1414	2.5	5YR 5/6 yellowish red	Dissected climbing barchanoid dune, medium- to large-scale low-angle cross-bedding (planar and trough), limited soil development (20–30 cm).
$OX_{od}842/1$	35°45'16"N, 111°9'W	1433	1.7	5YR 5/6 yellowish red	Dissected, trough cross-bedded (medium- to large-scale) climbing barchanoid dune, very poor soil development (5–10 cm).
Active dune samples					
$OX_{od}838B/1$	35°46'40"N, 111°04'W	1658	—	5YR 6/4 light reddish brown	Sand on unvegetated linear ridge crest on plateau.
$OX_{od}841/1$	35°44'40"N, 111°11'12"W	1372	—	5YR 6/4 light reddish brown	Saltation and traction load on barchan dune crest on Ward Terrace, no soil development, distinct active dune form.

Sampling of the Tusayan Dunes for optical dating

Field reconnaissance in 1989 resulted in the collection of eight samples for optical dating (Table 1, Fig. 3). Six were collected from within inactive dunes (vegetated, palaeosol-capped and weakly to moderately consolidated). Two were collected from surfaces of active aeolian deposition, one from a linear ridge on the Moenkopi Plateau and one from the crest of a barchan dune on Ward Terrace (Table 1). Our goals in sampling within the inactive dunes were to attempt to establish any trends in the timing of the formation of the linear dunes and any relations between the dunes on the plateau and those at lower elevation on Ward Terrace. The purpose of collecting the two samples from 'modern' contexts was to test for the completeness of resetting of the dating signal, and related effects, during aeolian transportation and deposition.

Field observation indicates the presence of at least three separate sub-units within the linear dunes, and at least two within the climbing dunes on Ward Terrace (Fig. 3). The units are recognized based on degree of soil development, subtle colour contrasts, preservation or absence of primary sedimentary structures, broad vegetation contrasts, and stratigraphic superposition (Table 1, Fig. 3).

At least one representative sample from each of the informally defined units was collected for optical dating using opaque PVC cylinders, driven into vertical exposures cut into the dunes. The samples were kept in darkness or subdued red light throughout storage, processing, and analysis, in order to preserve their record of time elapsed since they were last exposed to daylight.

Optical dating

Methodology

Optical dating is a luminescence dating method based on radiation exposure, closely related to thermoluminescence (TL) dating. Like the latter, it can determine the time that has elapsed since detrital mineral grains from sediments

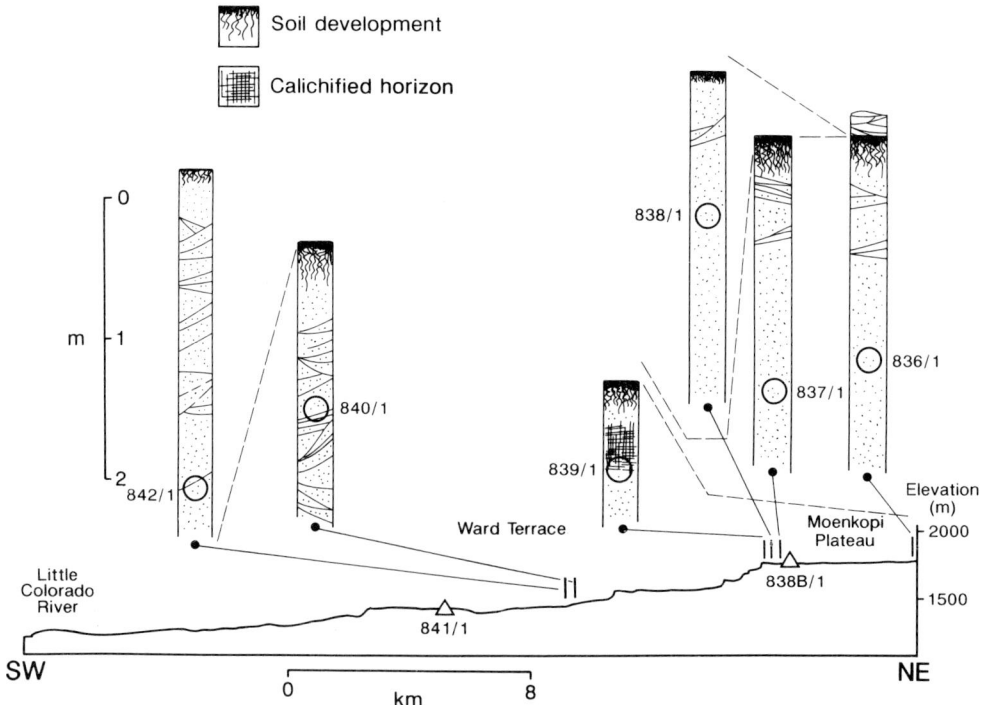

Fig. 3. Schematic SW–NE section showing sample locations, partial stratigraphic columns and inferred relative stratigraphic positions of the eight optically dated samples. See Table 1 for details. Inferred relative stratigraphic positions of units informally defined on Moenkopi Plateau and Ward Terrace are delimited by dotted lines between columns.

were last exposed to daylight. The quartz optical dating method was invented by Huntley *et al.* (1985) and has subsequently undergone extensive methodological development (e.g. Smith *et al.* 1986, 1991; Godfrey-Smith *et al.* 1988) and testing on a range of independently dated aeolian and other depositional environments (e.g. Rhodes 1990; Smith *et al.* 1990; Godfrey-Smith 1991; Stokes 1991, 1992*a*). The technique is now considered to be sufficiently well understood and soundly based to be applied to aeolian contexts that lack independent age constraint. Discussion within this paper is restricted to optical dating using quartz, and our methodological and other discussion do not apply to optical dating using coarse-grained feldspathic minerals, zircon, or fine-grained polymineralic mixtures.

With the exception of the zircon auto-regenerative technique, the essential basis of luminescence dating may be expressed by the equation:

$$Age = \frac{Palaeodose}{Dose\text{-}rate} \qquad (1)$$

The calculated age is the time since the grains were buried beyond reach of daylight (a minimum of a few millimetres), and became subjected to the low-level radiation from the surrounding sedimentary matrix and cosmic rays. The dose of radiation responsible for producing the luminescence signal is termed the palaeodose (measured in grays, Gy). The luminescence signal is produced due to the release of electrons from metastable trap locations within a crystal lattice; the initial emplacement of the electrons into traps being in response to radiation exposure. The dose-rate is an estimated measure of environmental radioactivity, resulting from: (a) the decay of radioisotopes (^{238}U, ^{232}Th, ^{40}K) within a radius of 30 cm surrounding the sample; and (b) a cosmic ray contribution.

The key advantage of the optical dating method over conventional TL dating of sediment is that it significantly reduces the complicating factors of residual signal levels at deposition (Stokes 1991, fig. 2). In optical dating, signal is observed only from optically sensitive electron traps; the electron traps most readily reset at deposition. The luminescence signal produced is termed optically stimulated luminescence (OSL). Other advantages of the optical dating method over TL include the speed of signal measurement, the avoidance of the TL-related phenomenon of black body radiation and thermal quenching, and the ability to

measure OSL at room temperature and in air (Huntley *et al.* 1985).

For a more detailed discussion of the methodology and principles of optical dating and its relation to TL, readers are referred to Aitken (1985, 1991), Godfrey-Smith *et al.* (1988), Rhodes (1990), Smith *et al.* (1986, 1990, 1991) and references therein.

Sample preparation

Routine separation procedures were used for the isolation of the coarse-grained (90–150 μm) quartz fraction (Aitken 1985; Smith *et al.* 1986; Stokes 1992*a*). A portion of each sample was wet sieved (retaining the 90–150 μm fraction), treated with hydrochloric acid and hydrogen peroxide, and density separated using a concentrated sodium polytungstate solution (density = 2.70 g cm^{-3}) to remove heavy minerals. Treatment with hydrofluoric acid resulted in concentration of the quartz grains. The prepared samples were dry sieved (retaining the 90–150 μm fraction) and mounted on 10 mm diameter stainless steel discs using a silicone spray (Silkospray*).

The optical excitation source

The excitation beam used for palaeodose evaluation was obtained from an argon-ion laser (Coherent* 2W) tuned to an emission wavelength of 514.5 nm. It was operated so that the power reaching the sample was approximately 47 mW cm^{-2}, and the samples were maintained at a temperature of 17°C. An EMI* photomultiplier tube with an 11 mm thickness of Corning* 7-51 filter and a 2 mm thick Schott* BG-39 filter (for red fluorescence rejection) were used to observe the luminescence of the samples.

Palaeodose evaluation

Palaeodose evaluation was undertaken using the so-called additive dose method that was initially developed for TL dating (Fleming 1970). Aliquots of sample were administered various doses of beta radiation (using a ^{90}Sr–^{90}Y source) and exposed to the laser for a period of 30 s. Final palaeodose estimates are based on the OSL integrated over durations of between 20 and 25 s of laser illumination (approximately 1.0–1.2 J cm^{-2}), although palaeodoses were calculated for successive intervals throughout

---

* Use of trade names does not imply endorsement by the US Geological Survey.

the duration of laser illumination. This so-called 'palaeodose plateau test' is generated to confirm that adequate sample bleaching (resetting) took place during deposition, to test for the possibility of accidental bleaching during sampling or processing, and to test for other factors relevant to the dating method (Aitken 1992).

A portion of the trapped electrons within quartz occupy light-sensitive sites which are not stable over geological time scales but are filled by laboratory irradiation. This portion must be removed, and for quartz this is achieved by pre-heating all sample aliquots at 160°C for 16 hours (Stokes 1992*a*).

Dose-rate evaluation

The dose-rate was evaluated by directly counting sample alpha activity using thick source alpha counting (Th- and U-derived dose) and measuring the potassium content using flame photometry. The thick source alpha counting was undertaken on approximately 3 g of unsealed, ground sample (50 minutes grinding in a ring mill), using ZnS scintillation screens whose active areas were 42 mm in diameter. Counts were also made on the same sample aliquots to test for radon emanation after they had been sealed and left for two weeks. Splits of the

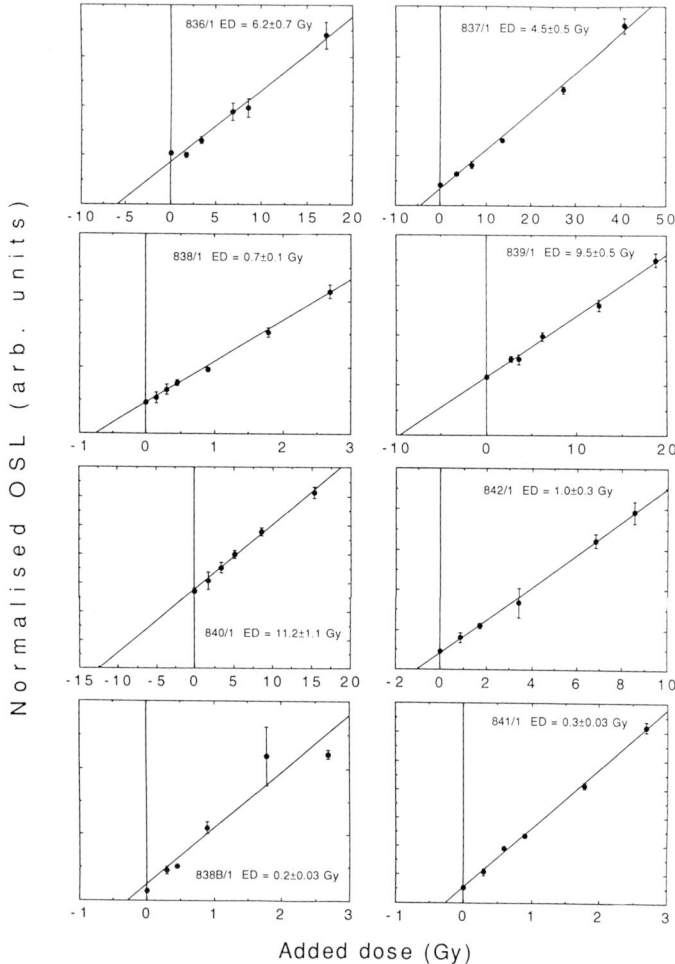

Fig. 4. Additive dose OSL growth curves for total integrated OSL over laser exposures of up to 1.2 J cm^{-2} for Tusayan Dune samples. A modified 'Equal Pre-Dose' normalization procedure was employed (Stokes, in prep). In this method samples are normalized by a test dose following the dating measurements and further radiation dosing and pre-heating. Prior to sample normalization, a background (scattered laser light and electronic noise) and herein termed 'long term OSL decay component' was subtracted from the total OSL signal. The OSL integrated over the period from 1.2–1.4 J cm^{-2} was used for this purpose. Six separate natural sample aliquots and four aliquots each for all additive dose points were used to generate the data points for the growth curves. Averaged scores and one-sigma errors are plotted.

crushed alpha counting samples (approximately 100 mg) were used for the flame photometry. The cosmic ray contribution was incorporated following the recommendations of Prescott & Hutton (1988). Correction factors to the dose-rate data have been incorporated to account for the attenuation of beta particles within the quartz grains and for the reduced overall beta dose caused by HF etching and moisture attenuation (Aitken 1985).

Results

Optical dates for the Tusayan Dunes

Growth of OSL in response to added radiation was linear, and as a result a linear least squares

fitting algorithm was used to extrapolate the data and calculate palaeodoses (e.g. Fig. 4). As the calculated palaeodoses did not vary dramatically with extended periods of laser exposure (Fig. 5), final palaeodose estimates are based on the total integrated OSL signals for the first 1.0 to 1.2 J cm^{-2} of laser exposure. Table 2 is a summary of the calculated palaeodose and dose-rate data, and the resulting age estimates.

The dates generated are significantly more recent than were anticipated from previous studies (e.g. Wells 1983) and suggest successive phases of middle to late Holocene aeolian deposition in the area. The most recent phase was approximately 400 years ago and resulted in the vertical accretion of the linear dunes of more than 1 m. This latest prehistoric mobilization

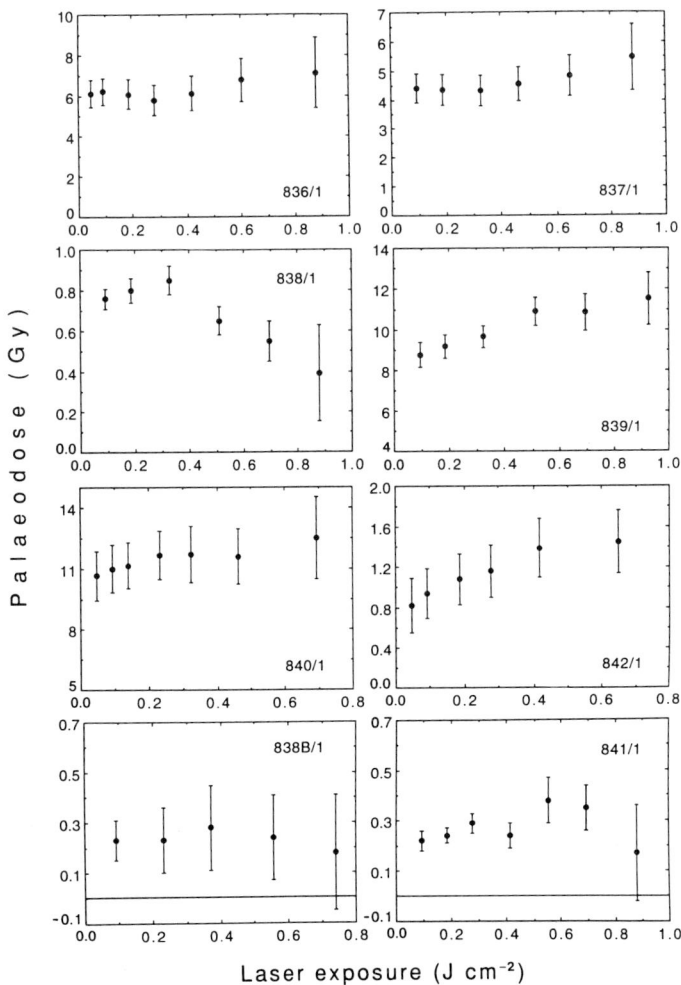

Fig. 5. Palaeodose plateaux for the Tusayan Dune samples. The palaeodoses were calculated for progressive durations of laser exposure. See text for details.

Table 2. *Summary of palaeodose and dose-rate data**

Sample	W†	F	C(t)‡ (ks⁻¹)	C(Th) (ks⁻¹)	K_2O (%)	Dose-rate (mGy a⁻¹)	Palaeodose (Gy)	Age (a)
Inactive or partly-active dune samples								
OX$_{od}$836/1	0.35	0.14 ± 0.03	3.70 ± 0.06	1.37 ± 0.18	1.7 ± 0.1	2.14 ± 0.16	6.16 ± 0.69	2900 ± 400
OX$_{od}$837/1	0.43	0.06 ± 0.01	4.26 ± 0.07	1.69 ± 0.20	1.5 ± 0.1	2.11 ± 0.15	4.45 ± 0.51	2100 ± 300
OX$_{od}$838/1	0.34	0.12 ± 0.02	2.59 ± 0.51	0.47 ± 0.11	1.6 ± 0.1	1.85 ± 0.14	0.74 ± 0.05	400 ± 40
OX$_{od}$839/1	0.47	0.07 ± 0.01	4.05 ± 0.06	1.35 ± 0.18	1.5 ± 0.1	2.03 ± 0.15	9.53 ± 0.54	4700 ± 400
OX$_{od}$840/1	0.38	0.07 ± 0.01	6.12 ± 0.09	1.92 ± 0.27	2.4 ± 0.1	3.11 ± 0.23	11.17 ± 1.10	3600 ± 400
OX$_{od}$842/1	0.37	0.03 ± 0.01	3.49 ± 0.06	1.65 ± 0.29	1.9 ± 0.1	2.35 ± 0.18	1.04 ± 0.25	440 ± 110
Active dune samples								
OX$_{od}$838B/1§	0.35	0.05 ± 0.01	3.81 ± 0.08	1.31 ± 0.22	2.6 ± 0.1	2.95 ± 0.22	0.24 ± 0.12	70 ± 40
OX$_{od}$841/1							0.26 ± 0.03	90 ± 30

* Calculated errors incorporate both random and systematics and are calculated after Aitken & Alldred (1972).
† W: saturation water content; F: fraction of W corresponding to the estimated average sample moisture content over the burial period.
‡ C(t), C(Th) are total and thorium alpha count-rates (unsealed), respectively. Sealed/unsealed counts ranged from 1.0 to 1.2. Count rates were converted to dose-rates after Aitken (1990). The given alpha count-rates correspond to equivalent concentrations of between 1.2–2.4 and 1.0–3.8 ppm of U and Th, respectively.
§ Dose-rate data were not directly calculated for this sample. Dose-rates for 841/1 were used.

event is preserved on both Ward Terrace and the Moenkopi Plateau, and it corresponds closely in time to the final abandonment of the region by the Pueblo people, in approximately AD 1400 (Euler *et al.* 1979; Euler 1988).

The age of at least the upper part of the linear dunes underlying the 400 year old deposits is well constrained to approximately 2000–3000 years ago, suggesting late Holocene linear dune vertical accretion rates of the order of metres per thousand years. A major regional phase of aeolian activity post-dating approximately 3000 years ago has been documented widely for the Great Plains (Ahlbrandt *et al.* 1983). The calculated age of 4700 years ago for the lower unit within the linear dunes possibly indicates a temporally distinct late middle Holocene phase of activity. Such an earlier timing of deposition for the lower unit is supported by its greater degree of pedo-genesis, in particular the development of a calcic B-horizon (Table 1). This unit may be correlative with the well-known carbonate-rich palaeosol that caps the Jeddito Formation of Pleistocene age on Black Mesa (Hack 1942; Karlstrom 1988, fig. 3.9).

The date of 3600 years ago for sample 840/1 falls between the first and second phases of aeolian activity recorded within the linear dunes and it could in fact represent an additional distinct episode of activity. We infer that the sample most probably correlates with, and forms the extension of, the period of activity 2000–3000 years ago. The apparently limited extent on the Moenkopi Plateau of the older phase (4700 years ago) may be only an artifact of our limited sampling, but it suggests that widespread preservation of climbing dunes of the older phase on the terrace below is not likely.

Except for the possibility of a latest Altithermal date (4700 ± 600 years ago) for the upwind, relict part of a linear dune on the Moenkopi Plateau, we have no evidence of aeolian deposition in the sampled area during the middle Holocene Altithermal. That time has been suggested by several authors to represent the period of greatest regional aridity during the Holocene (e.g. Holliday 1989, 1990; Spaulding 1991). Likewise, no dunes have been sampled that yield a last glacial maximum age. Nearby, however, at Black Mesa, Quaternary stratigraphic relations described by Hack (1942) and later by Karlstrom & Karlstrom (1986) and Karlstrom (1988) indicate aeolian deposition associated with the Altithermal. It is possible that a major hiatus, possibly spanning a hundred thousand years or more, separates the emplacement of basal (volcanic ash-bearing) sands from the middle to late Holocene multiple reactivation deposits in the linear dunes.

Several episodes of reworking of the upper parts of the linear dunes may have occurred during the hiatus noted above, of which a record of only the latest depositional sequence, superposed on much older basal sands, is preserved. How could the basal parts of the dunes be so old, if the upper parts have been so recently deposited? An answer may lie in a suggested mechanism of growth and migration of linear dunes of the sand ridge variety (Breed *et al.* 1984) as opposed to the growth and migration of the much better understood crescentic (transverse or barchanoid) dunes. Sand in barchanoid dunes is 'turned over' each time a dune migrates and reconstitutes itself in its entirety in a downwind direction, transverse to the effective wind. Thus the sand of barchanoid dunes is repeatedly exposed to sunlight during migratory episodes. Dating of these sands by quartz optical dating will reveal the age of the most recently deposited sand within the field (but not necessarily the age of the dunefield pattern, which has episodically been 'refilled' by migrating sand).

Linear dunes, on the other hand, are believed to grow by lengthwise extension in a direction parallel to their long axes and approximately parallel to the effective resultant wind direction. Linear dunes in semi-arid regions tend to have vegetated lower slopes, and they leave their flanks and basal parts behind as stabilized 'plinths' when the loose sand on their crests migrates downwind. Only the upper parts of the linear dunes are episodically reworked and replenished with sand that is transported down the length of the dunes ridges. Thus different parts of individual linear dunes may very well show great differences in age, just as indicated on Billingsley's (1987*b*) map and as suggested by the large age discrepancy between the lower and upper parts of the linear dune dated by K–Ar and quartz optical dating, respectively. Clearly, more sampling and dating of these dunes by all available methods will be necessary to resolve this question.

The two samples from active dunes yielded small, non-zero ages (Table 2). Concurrent studies at other localities have confirmed similar small, non-zero ages for a variety of aeolian depositional settings (Stokes 1992*a*). In principle, the ages quoted for the other samples may be in excess by the order of approximately 100 years, which is insignificant and falls within error limits except in respect of the two samples dating to approximately 400 years. However, on the basis of a hypothesis related to the phenomena of recuperation (Aitken & Smith 1988), no subtraction of residual has been made. It is noted that the subtraction of a 'residual-at-deposition' of the equivalent of approximately 100 years from these samples would result in ages that are inconsistent with the alternative evidence of increased aridity and possible late prehistoric dune mobilization (e.g. Euler *et al*. 1979; Euler 1988).

An alternative explanation for such an apparent palaeodose being present at, or immediately following, transportation and deposition was suggested by Aitken & Smith (1988). They attribute the recuperation of charge to rapidly bleachable trap sites from thermally unstable trap sites populated during the bleaching process. In the case of aeolian deposits, the extended (> hours) duration and frequently repeated episodes of transportation and surface exposure during deposition have been suggested to negate such recuperation processes at the order of magnitude initially suggested (Stokes, 1992*b*).

Discussion: regional comparisons of Holocene aeolian activity

A combination of radiocarbon dating of interdune organic deposits, soil chronostratigraphic studies, archaeology, and recently generated optical dates has resulted in a clearer picture of regional latest Quaternary aeolian deposition (Fig. 6). Generally synchronous, although not regionally simultaneous, aeolian deposition throughout the Great Plains and Southern High Plains indicates that the Holocene has been a very dynamic period of punctuated, rapid aeolian sedimentation. The most widespread and chronostratigraphically well-constrained phase of aeolian activity occured between approximately 3500 and 1500 years ago, corresponding closely in time to an interval of glacial retreat in the Rocky Mountains described by Benedict (1973). Benedict suggested that this period was a time of significant soil formation on the Colorado Front Range, possibly similar in intensity to the soil-forming phase of the earlier Altithermal. The only chronological evidence for aeolian deposition in the Great Plains during the middle Holocene, prior to approximately 5000 years ago is indirect. It was inferred from the intercalation of, and possible superposition of Cody Complex palaeoindian artifacts within and upon aeolian sands that pre-date the late Holocene depositional phase (Ahibrandt *et al*. 1983). Further investigations are required to

directly establish the age and extent of this initial phase of latest Pleistocene or Holocene activity.

In comparing glacial and other palaeoclimatic records across the region (Fig. 6), disagreement between the results of various studies is apparent. There is, however, a good correlation between phases of aeolian deposition as recorded in northeastern Arizona and regional episodes of alluvial filling. Such a relation may imply that reactivation was heavily dependent on sediment availability, as well as on such factors as aridity and the effectiveness of sand transporting winds.

If we consider palaeoclimatic evidence for regional aridity from a range of investigations based on pollen, pack-rat middens, and dendrochronology (Fig. 6), once more the regional pattern of late Holocene climatic change and aridity shows discrepancies. Many recent studies (e.g. Van Devender *et al*. 1984) suggest that late Holocene climates have been considerably more severe than the middle Holocene Altithermal. Such inferences are supported by the apparent absence of extensive middle Holocene aeolian deposits in the Great Plains and the recognition of widespread late Holocene activity there and in northeastern Arizona. The only recent study that supports a middle Holocene phase of greatest aridity is that of Spaulding (1991), for the Mohave Desert. Pack-rat midden evidence is, however, absent for most of the last 5000 years.

An abundant literature, based on archaeological, geological, and palaeoclimatic relations, provides a detailed palaeoenvironmental record for an area (Black Mesa) close to the Tusayan Dunes. The Quaternary history of this area was first outlined by Hack (1942) and later by several other investigators (summarized by Karlstrom 1986). According to Karlstrom (1988), 'the Black Mesa chronostratigraphic data support the classic interpretations of a major postglacial drought (the "Altithermal" of Antevs [1948]) culminating approximately 6000–5000 years ago, and followed by wetter and cooler climate (the "Little Ice Age") punctuated by secondary droughts.' Euler *et al*. (1979) interpreted the Black Mesa record as representative of the southern Colorado Plateau as a whole; they described evidence for four distinct phases of increased aridity in approximately the last 2500 years (Fig. 6). Two of their drought periods are coincident with aeolian reactivation as indicated by the optical dates on the Tusayan Dunes, and two are not. The two coincident aeolian depositional phases are also coincident with periods of alluviation and thus presumably with a more abundant sediment supply. The relation

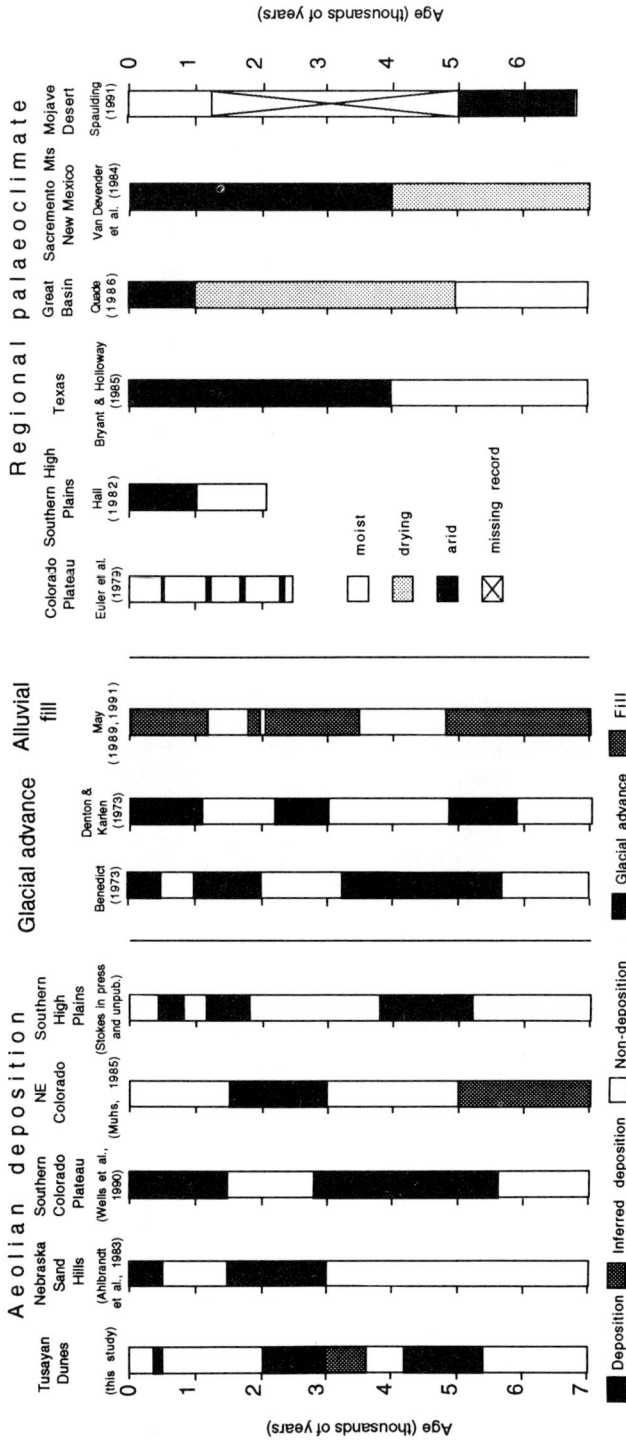

Fig. 6. Selected regional palaeoclimatic and palaeoenvironmental data.

supports the idea that sediment availability, in addition to aridity, may have been a driving force for the widespread mid- to late Holocene aeolian deposition recorded within the Tusayan Dunes. The apparent coincidence of alluviation with drought, however, rather than with moister conditions envisaged by Hack (1942) and by Karlstrom & Karlstrom (1986), raises some doubts regarding the correlation by some investigators of aeolian *non*-depositional intervals with episodes of drought in the South-west.

Conclusions

Optical dating of quartz sand from within the fixed (inactive or party active) Tusayan Dunes of the Moenkopi Plateau and Ward Terrace indicates that the dunes have been remobilized during at least three distinct middle to late Holocene epidodes at approximately 4700, 3500–2000 and 400 years ago. Given the presently accepted age for the base of the dunes as over 100 000 years, it is highly likely that many more reactivation/remobilization phases are recorded than have been sampled in this study.

The dunes are highly sensitive to minor shifts in the local environment that reflect changes in aridity and/or sediment supply. We hypothesize that reactivation events within the linear dunes primarily affect the ridge crests and upper flanks of the dunes, leaving lower (vegetated) plinth parts stable. The hypothesis provides a mechanical explanation for the difference in age between the lower and upper parts of the linear dunes. If it proves correct, the linear dunes (and possibly the trailing arms of some parabolic dunes) hold the promise of preserving much, if not all, of the history of activation of this dune field, in contrast to the barchanoid dunes, which can preserve only the record of their most recent reconstruction.

Regional comparisons of dated mid- to late Holocene aeolian reactivation events suggest that the High Plains erg system was highly dynamic during the period. The apparent non-synchronism of some local phases of reactivation requires further investigation but may imply localized changes in microclimate and/or sediment supply as being responsible for the initiation of reactivation. Such localized effects may not correspond to regional trends of deposition or aridity.

The authors wish to thank Jack McCauley, George Billingsley, Marlene Tuesink and Gerald Schaber for their assistance in locating various dune units for sampling on Ward Terrace and the Moenkopi Plateau. We thank Jim Swinehart for providing stimulating discussions related to the Late Holocene timing of alluviation. GB and JM provided valuable comments on an earlier version of this manuscript. We would also like to thank Diane Elder Anderson for her assistance in presenting some initial results of this work at the Annual Meeting of the Geological Society of America in October 1991, while Stokes & Breed attended the Geological Society Aeolian Conference in London. Logistical support for the fieldwork in Arizona was provided by the US Geological Survey, RLAHA, Pembroke College, and the Dudley Stamp Memorial Fund. This manuscript was greatly improved by the comments of M. J. Aitken (RLAHA), J. McCauley & G. Billingsley (USGS).

References

AHLBRANDT, T. S., SWINEHART, J. B. & MARONEY, D. G. 1983. The dynamic Holocene dune fields of the Great Plains and Rocky Mountain basins, U.S.A. *In*: BROOKFIELD, M. E. & AHLBRANDT, T. S. (eds) *Eolian Sediments and Processes.* Elsevier, Amsterdam, 379–406.

AITKEN, M. J. 1985. *Thermoluminescence Dating.* Academic Press, London.

—— 1990. Pairs precision required in alpha counting. *Ancient TL*. **8**(2), 12–14.

—— 1992. Optical dating. *Quaternary Science Reviews*, **11**, 127–131.

—— & ALLDRED, J. C. 1972. The assessment of error limits in thermoluminescent dating. *Archaeometry*, **14**, 257–267.

—— & SMITH, B. W. 1988. Optical dating: Recuperation after bleaching. *Quaternary Science Reviews*, **7**, 387–393.

ANTEVS, E. 1962. Late Quaternary climates in Arizona. *American Antiquity*, **28**,. 193–198.

BENEDICT, J. B. 1973. Chronology of cirque glaciation, Colorado Front Range. *Quaternary Research*. **3**, 584–599.

BILLINGSLEY, G. H. 1987a. Geology and Geomorphology of the Southwestern Moenokopi Plateau and southern Ward Terrace, Arizona. *U.S. Geological Survey Bulletin*, **1672**.

——. 1987b. Geologic map of the southwestern Moenkopi Plateau and (and a comparison with linear dunes on the Moenokopi Plateau). *In*: EL-BAZ, F. & WARNER, D. M. (eds) *Apollo–Soyuz Test Project Summary Science Reports, v. 2. Earth Observations and Photography*. US Aeronautics and Space Administration SP-412, 319–358.

BREED, C. S. & BREED, W. J. 1979. Dunes and other windforms of central Australia (and a comparison with linear dunes on the Moenkopi Plateau). *In*: EL BAZ, F. & WARNER, D. M. (eds) *Apollo–Soyuz Test Project Summary Science Reports, v. 2. Earth Observations and Photography*. US Aeronautics and Space Administration SP-412, 319–358.

——, McCAULEY, J. F., BREED, W. J., McCAULEY, C. K. & COTERA, A. S. Jr. 1984. Eolian (wind-formed) landscapes. *In*: SMILEY, T. L. *et al.* (eds)

Landscapes of Arizona — The Geological Story. Lanham, Md, University Press of America, New York, 359–413.

BRYANT, V. M. Jr. & HOLLOWAY, R. G. 1985. A late Quaternary palaeoenvironmental record of Texas: An overview of the pollen evidence. *In*: BRYANT, V. M. Jr. & HOLLOWAY, R. G. (eds) *Pollen Records of Late Quaternary North American Sediments*. American Association of Sedimentologists & Paleontologists Foundation, 39–70.

DEAN, J. S. 1988. Dendrochronology and palaeoenvironmental reconstruction on the Colorado Plateaus. *In*: GUMERMAN, G. J. (ed.) *The Anasazi in a Changing Environment*. School of American Research Advanced Seminar Series, Cambridge University Press, Cambridge, 119–167.

DENTON, G. H. & KARLEN, W. 1973. Holocene climatic variations — their pattern and possible cause. *Quaternary Research*, 3, 155–205.

EULER, R. C. 1988. Demography and cultural dynamics on the Colorado Plateau. *In*. GUMERMAN, G. J. (ed.) *The Anasazi in a Changing Environment*. School of American Research Advanced Seminar Series, Cambridge University Press, Cambridge, 192–229.

——, GUMERMAN, G. J., KARLSTROM, T. N. V., DEAN, J. S. & HEVLY, R. H. 1979. The Colorado Plateau: Cultural dynamics and palaeoenvironment. *Science*, 205, 1089–1011.

FLEMING, S. J. 1970. Thermoluminescent dating: Refinement of the quartz inclusion method. *Archaeometry*, 12, 133–145.

GAYLORD, D. R. 1990. Holocene palaeoclimatic fluctuations revealed from dune and interdune strata in Wyoming. *Journal of Arid Environments*, 18, 123–138.

GODFREY-SMITH, D. I. 1991. *Optical Dating of Sediments*. PhD thesis, Simon Fraser University, Barnaby, BC, Canada.

——, HUNTLEY, D. J. & CHEN. W.-H. 1988. Optical dating studies of quartz and feldspar sediment extracts. *Quaternary Science Reviews*, 7, 373–380.

HACK, J. T. 1941. Dunes of the Navajo Country. *Geographical Review*, 31, 240–263.

—— 1942. The changing physical environment of the Hopi Indians of Arizona: Report of the Awatovi Expedition: *Peabody Museum Papers. Harvard University*, 35(1).

HALL, S. A. 1982. Late Holocene paleoecology of the Southern High Plains. *Quaternary Research*, 17, 391–407.

HOLLIDAY, V. T. 1989. Middle Holocene drought on the Southern High Plains. *Quaternary Research*, 31, 74–82.

—— 1990. Soil and landscape evolution of eolian plains: the Southern High Plains of Texas and New Mexico. *Geomorphology*, 3, 489–515.

HUNTLEY, D. J., GODFREY-SMITH, D. I. & THELWART, M. L. W. 1985. Optical dating of sediments. *Nature*, 313, 105–107.

KARLSTROM, E. T. & KARLSTROM, T. N. V. 1986. Late Quaternary alluvial stratigraphy and soils of the Black Mesa — Little Colorado River areas, northern Arizona. *In*: NATIONS, J. D., CONWAY, C. M. & SWANN, G. A. (eds) *Geology of Central and Northern Arizona*. Geological Society of America, Rocky Mountain Section Guidebook, 71–91.

KARLSTROM, T. N. V. 1988. Alluvial chronology and hydrologic change of Black Mesa and nearby regions. *In*: GUMERMAN, G. J. (ed.) *The Anasazi in a Changing Environment*. School of American Research Advanced Seminar Series, Cambridge University Press, Cambridge, 45–91.

MADOLE, R. F. in press. Colorado Piedmont. *In*: WAYNE, W. J. (ed.) Chapter 14, Northern Great Plains. *In*: MORRISON, R. B. (ed.) *Quaternary Nonglacial Geology — Conterminous United States*. The Geology of North America, Geological Society of America, Boulder, volume K-2.

MAY, D. W. 1989. Holocene alluvial fills in the Loup Valley, Nebraska. *Quaternary Research* 32, 117–120.

—— 1991. Stratigraphic evidence of late-Quaternary alluviation, erosion and high-magnitude flooding in valleys on the Great Plains. *Abstracts with Programs. Geological Society of America Annual Meeting*. San Diego, California.

MUHS, D. R. 1985. Age and paleoclimatic significance of Holocene sand dunes in northeastern Colorado. *Annals of the Association of American Geographers*, 75, 566–582.

MUNSELL COLOR CO. 1954. Munsell Soil Color Charts. Baltimore, Maryland.

OSTERKAMP, W. R., FENTON,. M. M., GUSTAVSON, T. C., HADLEY, R. F., HOLLIDAY. V. T., MORRISON, R. B. & TOY 1987. Great Plains. *In:* GRAF. W. J. (ed.) *Geomorphic Systems of North America*. Geological Society of America, Centential Volume 2, Boulder, 163–202.

PRESCOTT, J. R. & HUTTON, J. T. 1988. Cosmic ray and gamma ray dosimetry for TL and ESR. *Nuclear Tracks and Radiation Measurements*, 14, 223–230.

QUADE, J. 1986. Late Quaternary environmental changes in the Upper Las Vegas Valley, Nevada. *Quaternary Research*, 26, 340–357.

REEVES, C. C. Jr. 1976. Quaternary stratigraphy e.g. geological history of the Southern High Plains, Texas and New Mexico. *In*: MAHANEY, W. C. (ed.) *Quaternary Stratigraphy of North America*. Dowden, Hutchinson and Ross, Stroudsberg, Pennsylvania, 213–234.

RHODES, E. J. 1990. *Optical Dating of Sediment*. D.Phil thesis, Oxford University, England.

SMITH. B. W., AITKEN, M. J., RHODES, E. J., ROBINSON, P. D. & GELDARD, D. M. 1986. Optical Dating: Methodological aspects. *Radiation Protection Dosimetry* 17, 229–233.

——, RHODES, E. J., STOKES, S. & SPOONER, N. A. 1991. Optical Dating of quartz. *Radiation Protection Dosimetry*, 34, 75–78.

——, ——, ——, —— & AITKEN, M. J. 1990. Optical dating of sediments: Initial quartz results from Oxford. *Archaeometry*, 32, 19–31.

SPAULDING, W. G. 1991. A middle Holocene vege-
tation record from the Mojave desert of North
America and its palaeoclimatic significance.
Quaternary Research, **35,** 427–437.

STOKES, S. 1991. Quartz-based optical dating of
Weichselian cover sands from the eastern
Netherlands. *Geologie en Minjbouw,* **70,** 327–
337.

—— 1992a. Optical dating of young (modern)
sediments using quartz: Results from a selection
of depositional environments. *Quaternary
Science Reviews,* **11,** 153–159.

—— 1992b. Optical dating of independently-dated
Late Quaternary eolian deposits from the
Southern High Plains. *Current Research in the
Pleistocene.*

—— in prep. OSL normalisation procedures for
quartz optical dating.

——, BREED, C. S. & ELDER, D. A. 1991. Holocene
maintainence of Pleistocene sand dunes in northeastern Arizona: Climatic implications of quartz
optical dating. *Abstracts with Programs.
Geological Society of America Annual Meeting,*
San Diego, California, A354.

STOKES, W. L. 1964. Incised, wind-aligned stream
patterns of the Colorado Plateau. *American
Journal of Science,* **262,** 808–816.

VAN DEVENDER, T. R., BETANCOURT, J. L. &
WIMBERLY, M. 1984. Biogeographic implications
of a packrat midden sequence from the
Sacramento Mountains, South-Central New
Mexico. *Quaternary Research,* **22,** 344–360.

WELLS, G. L. 1983. Late-glacial circulation over
central North America revealed by aeolian
features. *In*: STREET-PERROTT, F. A., BERAN. M.
& RATCLIFFE, R. (eds) *Variation in the Global
Water Budget.* Reidel, Dordrecht, 317–330.

WELLS, S. G., MCFADDEN, L. D. & SHULTZ, J. D. 1990.
Eolian landscape evolution and soil formation
in the Chaco dune field, southern Colorado
Plateau, New Mexico. *Geomorphology,* **3,**
517–546.

Modes of linear dune activity and their palaeoenvironmental significance: an evaluation with reference to southern African examples

IAN LIVINGSTONE[1] & DAVID S. G. THOMAS[2]

[1]*Department of Geography, Coventry University, Coventry CV1 5FB, UK*
[2]*Department of Geography, The University, Sheffield S10 2TN, UK*

Abstract: Supposedly relict continental desert sand dunes have frequently been used as palaeoenvironmental indicators. Often the argument has been, first, that vegetated dunes are fixed relicts, and second, that fixed dunes indicate past aridity. Recent work suggests that neither of these maxims is necessarily true. Using the example of the linear dunes of southern Africa, this paper examines the extent to which we can continue to support a simple assumption that vegetated dunes are inactive relicts. In particular, it addresses the problem of what constitutes activity on linear dunes, and suggests some hypotheses for investigating the 'grey' area between active, unvegetated dunes and densely wooded, inactive dunes. The concept of episodic activity is introduced in the context of partially vegetated linear dune forms.

Much has been made in the past of the use of continental desert sand dunes as palaeoenvironmental indicators (e.g. Sarnthein 1978). Simple rainfall measures, most frequently the mean annual precipitation, have been employed as a surrogate for all the environmental conditions controlling dune development, and vegetation cover has often been viewed as an indication of dune inactivity. Thus, the argument has been that vegetated dunes in areas of rainfall greater than, for example, 100–300 mm per annum (Goudie 1983, p. 72), are currently fixed by the vegetation cover, and that the dunes therefore testify to an episode or episodes of greater aridity in the past.

Most frequently these studies of relict dunes have considered dunes with a linear form (longitudinal dunes, seif dunes, sand ridges, etc.) for two reasons. First, linear dunes are the most common desert dune type (Fryberger & Goudie 1981). This is due, at least in part, to the fact that their mode of development predisposes them to preservation (see below). There are extensive areas of vegetated or partially vegetated linear dunes, notably in Australia, India and Africa, north and south of the equator. Second, there has been a parallel assumption that linear dunes are aligned with a single, uni-modal wind, and can therefore tell us not only about past rainfall but also about past wind direction. This is well illustrated in southern Africa where the unvegetated and supposedly active dunes of the coastal Namib Desert have been contrasted with the vegetated and allegedly inactive dunes of the interior Kalahari Desert (e.g. Goudie 1970).

Care must clearly be exercised in expressing opinions about past climatic regimes based on a landform the contemporary dynamics of which have yet to be fully elucidated. The key issues in the use of linear dunes as palaeoenvironmental indicators, as for any other landforms, are (1) how can we tell when a dune is relict?; and (2) if it is a relict feature, what can be inferred from the dune about the climate at the time of its formation?

In the light of advances in understanding both of dune dynamics and of past environmental changes, this paper aims to concentrate on the first of these issues, and in particular aims to examine the extent to which we can continue to support a simple assumption that vegetated linear dunes are necessarily geomorphologically inactive features.

The southern African context

Southern Africa provides an ideal area in which to address these problems because it incorporates a range of linear dunes stretching from the hyper-arid coastal Namib sand sea through the Kalahari Desert to more humid areas in the north and east of the Kalahari (Fig. 1).

The Namib

Along the Atlantic coast of southern Africa in the Namib Desert there are transverse dunes

From Pye, K. (ed.), 1993, *The Dynamics and Environmental Context of Aeolian Sedimentary Systems.*
Geological Society Special Publication No. 72, pp. 91–101.

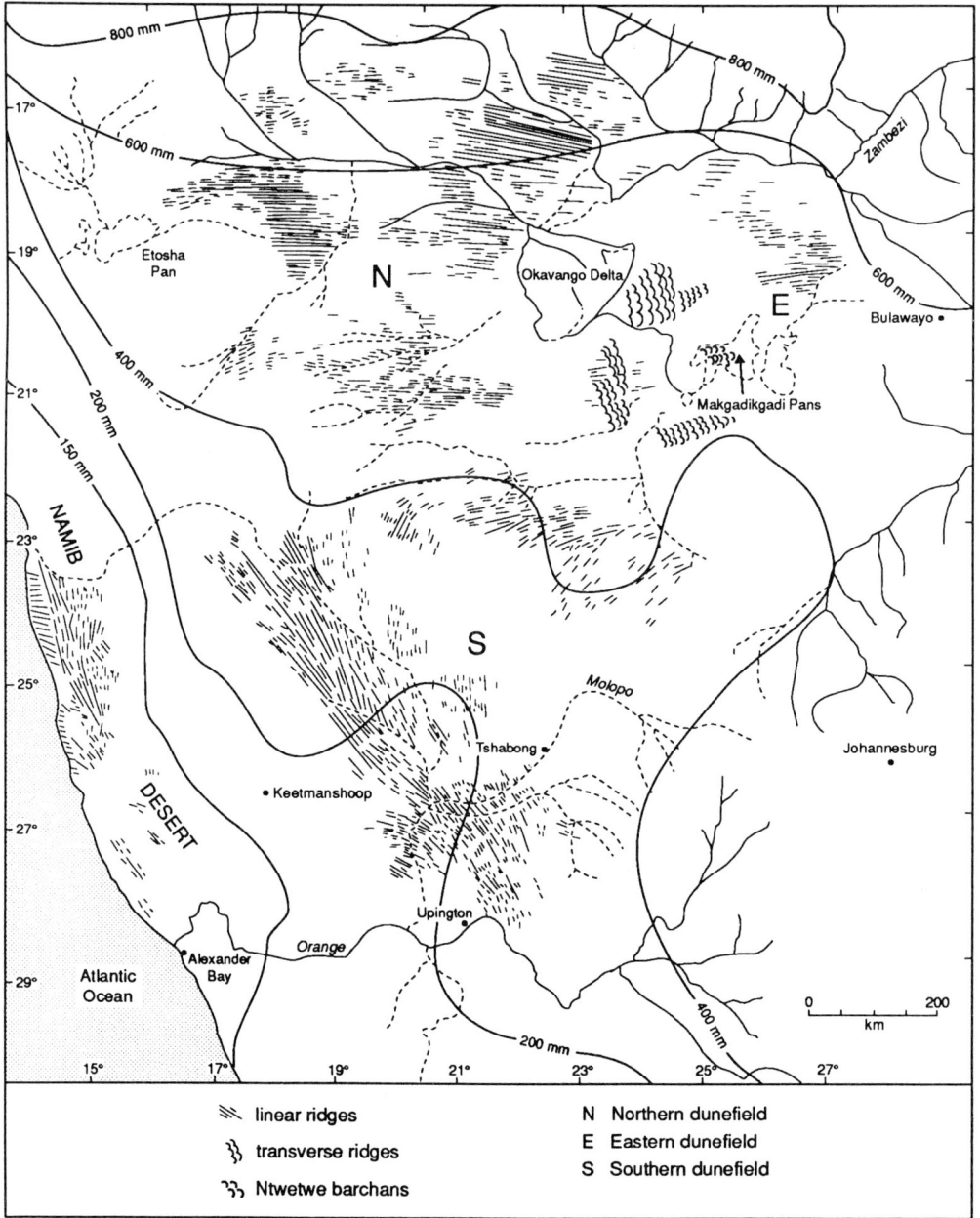

Fig. 1. Map showing isohyets and major linear dune locations and alignments in southern Africa.

formed by a broadly unimodal, southwesterly wind regime, while at the eastern fringe of the sand sea there are large star dunes created in a low energy, complex wind regime. However, the central sand sea is dominated by substantial, complex and compound linear dunes, formed in bimodal regimes. Wind regimes in the central area are described as 'intermediate to low energy bi-modal' by Breed et al. (1979, p. 346). Despite the sparsity of vegetation on these dunes (Fig. 2), Besler (1980) has argued that the Namib linear dunes are relicts of stronger

Fig. 2. View from the crest of a largely unvegetated complex linear dune in the hyper-arid Namib Desert. Evidence from monitoring programmes indicates that the surfaces of these dunes are active.

Pleistocene winds, following Glennie's (1983, 1985) argument for dunes of a similar size on the Arabian peninsula, but Lancaster (1989a), Livingstone (1989) and Ward (1984) all provide direct, empirical evidence of present-day activity. The linear dunes of the Namib sand sea are undoubtedly active at present, although it is probable that they are very long lived, and may have suffered periods of greater activity in the past when winds were stronger.

The Kalahari

Perhaps surprisingly, there has been a greater consensus that the Kalahari dunes are essentially fixed and inactive (Flint & Bond 1968; Grove 1969; Lancaster 1981; Thomas 1984). Although there are a few parabolic dunes, barchans and pan-depression fringing lunette dunes, the Kalahari dunefields virtually exclusively comprise linear forms (Fryberger & Goudie 1981), but under that heading they display a marked variety of morphologies, intra-dunefield relationships and vegetation covers, ranging from low, degraded, straight, wooded ridges in the northwest (Flint & Bond 1968) to straight and sinuous bifurcating dunes with variable vegetation cover in the southwest (Thomas 1988) (Figs 3 & 4).

All the Kalahari linear dunes, which can be divided into three systems on the basis of orientation and form (Lancaster 1981; Thomas 1984) (Fig. 1), have been regarded as relicts inherited from a period or periods of drier climate (less effective net precipitation), frequently associated with the last glacial maximum. As the 150 mm mean annual isohyet has been seen as

the threshold between active and inactive dunes in southern Africa today (e.g. Lancaster 1980), the inference made from such studies is that rainfall must have been below this figure at the time of dune formation.

Kalahari environmental change

There are many problems with such a general assumption of markedly drier late Pleistocene Kalahari climates (Shaw & Cooke 1986; Shaw et al. 1988). The most important developments suggest that late Quaternary climatic changes in the area should not necessarily be seen in terms of substantial shifts between wetter and drier periods, but as more subtle fluctuations around a semi-arid mean. Thus, pan depressions and at least some of the region's dry valleys can be explained in terms of development under groundwater control (Farr et al. 1982; Arad 1984; Shaw & De Vries 1988; Nash et al. 1993), rather than as clear cut indicators of climatic change. In the Kuruman valley, major suites of Holocene alluvium are not now viewed as indicators of wetter climates but as having analogues in the modern deposits that result from short-lived, episodic floods (Thomas & Shaw 1991). At Equus Cave on the southern Kalahari fringe, Klein et al. (1991) note that Late Pleistocene climatic changes may not have deviated substantially from the historical norm.

These developments have important implications for the palaeoenvironmental context into which the Kalahari dunes are placed. First, the variety and nature of vegetation communities on Kalahari dunes, the medium through which inactivity is viewed, needs to be

Fig. 3. A wooded linear dune bordering Sikumi Vlei in northwestern Zimbabwe. On a variety of criteria, including the particle size characteristics and the presence of dense *Baikaea plurijuga* woodland, there can be little doubt that such features are relicts.

Fig. 4. Linear dunes in the southwestern Kalahari. Often credited with relict status and a simple palaeoclimatic significance, such dunes display evidence of more complicated environmental relationships including spatial and temporal variations of surface sand movement.

examined in more detail, and its contribution to dune surface conditions and linear dune characteristics more fully assessed. Second, dune activity in the area needs to be considered in terms of rather more than simple, major, late Quaternary environmental shifts, because as yet the evidence for these, especially from the southwestern Kalahari, remains unproven. An important and often overlooked point is that there is no direct or indirect evidence that points to the Kalahari dunes having been either completely devegetated or resembling the generally plantless dunes of the Namib. Published, directly derived dates for periods of dune activity are also absent, while the relevance to dune development and activity of dates derived from soils and other landforms in the southwest Kalahari (Heine 1982; Lancaster 1987) is very marginal. Episodes of dune development often tend to be placed in 'windows' in an otherwise humid palaeoenvironmental sequence (Thomas & Shaw 1991), and ideas about how the dunes respond to climate change are only loosely inferred. A more appropriate point to start addressing dune activity in the region may therefore be through examining the overall nature of dune dynamics and activity, including the role of vegetation and the characteristics of present-day environmental conditions in the region.

Recent advances in dune studies

Over the past decade considerable advances have been made in the understanding of both dune dynamics and the controlling environmental parameters. There has been a fundamental change of method in dealing with the problems of dune movement resulting in a rush of small-scale dune dynamic studies, often of individual dunes, which have monitored sand movement, surface change, and wind flow patterns around the dune. Linear dunes have been particularly well represented (e.g. Lancaster 1985; Livingstone 1986, 1989; Tseo 1990, Tsoar 1983), with these studies beginning to provide information about the way in which linear dunes develop, and about the dynamics of dune surface sand transport.

Simultaneously, there has been an upsurge of interest in the role of vegetation in geomorphological processes (e.g. Viles 1988; Thornes 1990), and this has been extended into dune studies. The implicit assumption has frequently been that the main inhibition to dune movement is vegetation, and that vegetation cover is a response to rainfall. Recent work has shown both that rainfall totals are not the only controlling environmental parameter, and that sand

may move on dunes with some vegetation cover (Ash & Wasson 1983; Buckley 1987; Wasson & Nanninga 1986). The large range of reported values for rainfall limits for allegedly active and inactive dunes (see Thomas & Tsoar 1990, p. 475) is itself an indication that the controlling parameters are somewhat more complicated.

In Australia, Ash & Wasson (1983) concluded that there is 'no simple correlation between rainfall and dune mobility, modulated by vegetation cover' and further that 'no simple line can be drawn which separates "inactive" from "active" dunes' so that 'attempts to estimate climatic change from the difference between modern "active" dune zone and the relict zone are liable to large errors.' (Ash & Wasson 1983, p. 22–23). Ash & Wasson believed that any index of dune activity needed to take account of available wind energy, or windiness, as well as moisture, and they devised an index of mobility based on the balance between windiness and available moisture which Lancaster (1987) modified for the Kalahari. With sufficient wind energy, Ash & Wasson found that sand could be moved even where there was over 30% vegetation cover (also Wasson 1984).

It has become increasingly clear in recent years that vegetation cannot be considered some insignificant, dynamically-unimportant surface smear on landforms. Several authors have argued not only that vegetation could be present on active dunes but also that it may play a key role in the dynamics of some types of linear dune (e.g. Tsoar & Møller 1986; Thomas 1988; Thomas & Tsoar 1990). Thomas & Tsoar (1990) have suggested that vegetation might be important for dune dynamics in three main ways; as a surface stabilizer, as a focus for sand accretion and as a determinant of dune morphology.

Modes of dune activity

Much of the literature on stabilized, inactive or relict dunes inadvertently makes *a priori* assumptions about what dune activity actually is and assumes that activity and inactivity are discrete states. They are, however, merely end points on a continuum. Little is said about the intervening 'grey' area which in fact contains many complex 'dune states' that are responses to the total package of environmental conditions in the area concerned.

Some future study, therefore, needs to concentrate on defining what is meant by dune activity. As a preliminary, tentative classification, dunes might be divided into 'active', 'episodically active' and 'relict' states. **'Active'** dunes demonstrate contemporary sand movement which creates rippled surfaces and active avalanche faces. Some vegetation may be present, and levels of aeolian activity may vary seasonally, but surface sand movement occurs many times in a year. **'Relict'** dunes are dunes on which no sand has moved for some years, and they demonstrate some of the indicators of inactivity listed by Goudie (1983, p. 71) such as surface crusting, soil development, early stages of diagenesis, slope degradation and colluvial cover, deflation (lag) surfaces and woodland vegetation. Between 'active' and 'relict' dunes are **'episodically active'** dunes. These are dunes on which sand movement is at a low or negligible level and rippled surfaces and avalanche faces are rare or absent for long periods, even years, but 'episodically active' dunes do not demonstrate the indicators of lengthy inactivity listed for 'relict' dunes. During periods of activity the surface of 'episodically active' dunes may look similar to 'active' dunes, but for much of the time these dunes are essentially dormant.

There are clearly dunes in high energy wind regimes which support no vegetation and move rapidly, and are active. Equally, there are dunes, indisputably constructed from aeolian sediments that support dense vegetation communities and on which sand has not moved under aeolian transport for many years. There are also vegetationless dunes in arid areas that experience virtually no sand movement on their surfaces because they are currently in a low wind energy environment, such as the star dunes on the eastern fringe of the Namib Desert; and dunes with partial vegetation covers, including those in the southwest Kalahari, that experience surface sand movements for a few weeks a year under the influence of strong local winds. It would seem, therefore, that the issue surrounding what activity is and what defines it are rather more complex, and warrant greater consideration than previously given.

Elsewhere, Thomas (1992) has recently discussed the modes of activity of three fundamental desert sand dune types: **transverse**, **linear**, **star**. Each type results from a different wind direction regime (Fryberger 1979; Wasson & Hyde 1983), so that transverse dunes develop in unimodal wind regimes, linear dunes in bimodal regimes, and star dunes in complex, or multimodal, regimes. Consequently, transverse dunes progress downwind at reported rates between 0.08–80 m a^{-1}, while the plinths of linear and star dunes essentially remain fixed (notwithstanding reports from Hesp *et al.* (1989) and Rubin (1990) of a sideways shuffle by linear dunes) even though considerable sand movement may take place on the dune. There is, therefore, a fundamental difference between transverse dunes, which undergo net downwind migration, and linear and star dunes which appear to undergo negligible net lateral migration even though surface sand may be actively moving. Transverse dunes can, therefore, be typified as migratory forms, linear dunes as extending forms, and star dunes as sedentary forms (Table 1). Star dunes are often characterized as accreting forms (Lancaster, 1989*b*), although this is not an attribute exclusive to star dunes.

Table 1. *Modes of activity and associated wind regimes of fundamental desert sand dune types*

Dune type	Wind regime	Mode of activity
Transverse	Unimodal	Migrating
Linear	Bimodal	Extending
Star	Complex/ multimodal	Sedentary

The implication is that simply because the base of the dune does not migrate, it should not be assumed that the dune is inactive, and it may, therefore, be difficult to distinguish between active and inactive linear or star dunes. It is relatively easy to recognize an inactive transverse dune, because once sand is no longer transported over the dune it ceases to make downwind progress. On linear dunes, however, it may be that sand is transported along the dune with no net change of form apart from some (often very slow) extension of the downwind tip. There is also evidence that linear dune crests move seasonally back and forth although there is no **net** year-by-year change of cross-sectional form (Tsoar 1983; Livingstone 1989).

Indeed, the plinths, or lower slopes, of linear dunes may be particularly favourable sites for plant growth because of the combination of relatively immobile lower slopes with the moisture-retaining properties of the sand (Tsoar & Møller 1986). The nature of activity on active linear dunes may, therefore, render them liable to colonization by vegetation, whereas the constant reworking of windward slope sediments on transverse dunes, eventually resulting in the wholesale movement of all the sand in the dune body, means that vegetation usually fails to get a hold.

The notion of an active but non-migrating dune form is classic dynamic equilibrium theory as outlined by Hack (1960) whereby

'when in equilibrium a landscape may be considered as part of an open system in steady state of balance, in which every form is adjusted to every other. Changes in topographic form take place as equilibrium conditions change, but no particular cycle or succession of changes occurs through which the forms inevitably evolve ...'

Hack (1960).

At the scale of days, linear dune cross-profiles can be migrating (Livingstone 1989), and therefore represent an evolving system, and in seasonally bimodal wind regimes, linear dune cross-profiles demonstrate annual, cyclic development, but on a longer time scale of years, active linear dunes represent an equilibrium model so that the point in time at which the system is observed is immaterial.

Factors influencing dune activity

As a consequence of these issues, dune activity may result in the wholesale movement of the dune form, but it does not have to. This is a consequence of a series of factors which include wind regime and other environmental components such as sand supply, vegetation cover and surface crusting.

The nature of the overall wind regime determines the sedimentary and morphological response of a dune, as described above (Table 1) (Ash & Wasson 1983; Thomas 1992; Wasson & Hyde 1983). The frequency and duration of sand moving winds above the critical threshold for sand entrainment influences the duration of activity. If there are no other dune surface factors to be taken into consideration, this is a function of diurnal and seasonal wind regimes. Periodicity can, however, be considered at a range of scales. A dune can be considered to be experiencing activity (or inactivity) on a daily, seasonal or annual basis, or in the context of longer-term environmental change at a geological time scale when the dune may lapse into 'relict' status.

The role of wind in dune activity is complicated by other environmental factors. A supply of sediment which can be moved by the wind, generally sand-sized, is essential for continued dune activity. Indeed, for coastal dunes it is often the ready supply of sand which is the key to their activity, and some North American dunefields have become less active as a result of diminishing sand supply.

A total vegetation cover prevents aeolian action on the dune surface regardless of wind regime or the periodicity of potential sand moving winds. A partial vegetation cover, or one that is spatially discontinuous in a dunefield or varies across the dune profile, does not exclude sand movement from occurring in some places on the dune body, even though it may prevent the dune body as a whole from moving by, for example, anchoring the dune plinth. In the case of linear dunes, the morpho-dynamics of the dune can allow plants to gain a foothold on part of the dune surface especially when the periodicity of sand-moving winds is low.

In this context, it is important to recognize that vegetation is not simply a response to gross precipitation amounts, but that it too varies at a range of time scales, notably in response to the seasonality and interannual variability of dryland rainfall. Thus, when the activity of linear dunes is considered, the periodicity of both wind regime and vegetation communities needs to be taken into account.

Surface biological crusts are increasingly being identified as an important factor at the sediment–wind interface (Campbell *et al.* 1989; Forster 1990). Dune surface crusts formed of algae and in some cases lichens and mosses have been recognized in several desert dune fields such as the Negev (Yair 1990) and White Sands, New Mexico (Shields *et al.* 1957), but their full extent in desert dune systems is not known. Crusted surfaces on lower dune slopes in the Negev have been noted for their lack of evidence of aeolian activity, for their high silt content which may be a function of microorganisms trapping airborne dust, and for preparing the surface for colonization by higher order plants.

Key issues in understanding the dynamics of vegetated linear dunes

If it is possible or indeed probable that some linear dunes with partial vegetation cover might be active, this has major implications both for studies of dune dynamics and for palaeoenvironmental reconstructions, requiring an extension of the use of field techniques applied to vegetationless dunes and a reassessment of the interpretation of vegetated dunes as palaeomorphs from past climatic regimes. The preceding discussion permits the identification of a number of key questions to be addressed in future studies of partially vegetated linear dunes. Some are currently the subject of investigation in the Kalahari while others can only be approached when some of the more fundamental points have been considered.

Dune forming, dune modifying and dune maintaining activity

One of the most significant questions in the context of both the dynamics of linear dunes and the palaeoenvironmental interpretation of dunes concerns dune initiation. This is one area of linear dune study that has not been advanced in recent years; nor do there appear to be studies attempting to establish the conditions giving rise to linear dune genesis (see, however, Tseo 1990).

In the palaeoenvironmental context, an assumption of 'traditional' interpretations invoking precipitation thresholds is that vegetated dunes started life as unvegetated forms. If this line is pursued, but partially vegetated dunes are not seen as currently inactive, then it follows that a distinction can be made between dune-forming conditions, dune-maintaining conditions whereby the basic form is not altered, and dune-modifying conditions where dune morphology is changed. With respect to partially vegetated linear dunes, there is no normative model against which the concept of the type of activity can be tested, nor can the issue of separating conditions that are landform-forming from those that merely represent surface activity be readily investigated. An interesting related issue is that a nucleus for sand accumulation is an important criteria for dune initiation, and partially vegetated surfaces provide plenty of possible nucleii.

The periodicity of sand movement: a model of episodic linear dune activity

An important starting point in investigating the nature of partially vegetated linear dune activity is to collect field data on sand and windflow patterns over dunes. Given the nature of the environments that partially vegetated linear dunes are found in, this alone will not provide complete answers to questions concerning overall dune activity. Notwithstanding that there may have been changes in regional wind regime strengths in the Quaternary (Ash & Wasson 1983; Lancaster 1988), there are also variations in climatic parameters at shorter time scales that are pertinent to dune activity.

Interannual rainfall variability in the southwestern Kalahari is around 50% (Thomas & Shaw 1991) and is itself spatially variable, giving considerable variations in vegetation cover and the extent of bare ground. Temporal patterns in rainfall in southern Africa have been investigated by Tyson (1986) and Tyson et al. (1975)

who found a marked 18-year cycle of drought throughout the summer rainfall zone extending as far back as records go into the late nineteenth century. Given the way in which annual vegetation species respond to rainfall and moisture availability, it is probably most realistic to think of dunes in the 'grey' area as being episodic in activity in response to vegetation changes caused by sequences of drought years. Dunes are, thus, neither perennially active nor inactive. While such a model takes the inherent dynamism of dryland climates into account, it also means that field data collected over a few years may not necessarily be representative of overall dune processes (Fig. 5).

Superimposed on this level of episodic activity are likely to be variations at the intra-annual scale. This is a consequence of the contribution annual grasses make to dune vegetation in the southwestern Kalahari with biomass and ground cover decreasing during the course of the dry season, which extends from March to November. It is, therefore, interesting and probably highly significant from a dynamics perspective that while wind strengths are probably lower now than at some times in the past (e.g. Lancaster 1988; Newell et al. 1981), they display marked seasonal variations in the southwest Kalahari, with the greatest concentration of potential sand-moving winds in the period from August to December (Breed et al. 1979) when biomass and ground cover are likely to be least. The issue of activity needs to take these seasonal fluctuations into account.

Linear ridges and seif dunes

The sinuous, sharp-crested 'seif' linear dunes have sometimes been considered as a discrete morphological dune type compared with vegetated linear ridges (e.g. Tsoar 1989). Although Tsoar (1989, p. 508) has expressed the view that 'they differ in their genesis', it has also been observed (Thomas 1988; Tsoar & Møller 1986) that in some cases ridges can develop seif-like morphologies when vegetation cover is removed from the dune crest, even though the dune flanks may retain a plant cover. Understanding the relationship between seifs and ridges may be very important in identifying how a partial plant cover modifies wind and sand flow. The observations cited above also suggest that these differences in form may not represent discrete fundamental dune types, but simply points in the 'grey' area along the linear dune morphological continuum.

(a)

(b)

Fig. 5. High interannual rainfall variability has a significant influence on the cover of annual vegetation species on linear dunes in the southwestern Kalahari. This is demonstrated by dunes in the Twee Rivieren area of the Kalahari Gemsbok National Park in (**a**) 1990 and (**b**) 1991.

Prospectus

The intention here has been to broaden the debate about modes of dune activity in the hope that greater understanding of the dynamics of desert dunes will enable us to use them as more reliable palaeoenvironmental indicators than has been possible until now. In particular, a more explicit and necessarily more sophisticated view of what constitutes an active dune will need to be developed.

As working hypotheses for further research it is therefore proposed that:

- the nature of geomorphological activity varies from one dune type to another, and some dune types which appear inactive are in fact in dynamic equilibrium;

- the boundary between activity and inactivity is not controlled solely by rainfall totals but by a range of environmental parameters;
- the boundary between dune activity and inactivity is not a hard threshold but a soft gradation;
- dune activity may not respond to changes of environmental parameters in a simple stop-start fashion but in a rather more complex slow/fast manner;
- the presence of vegetation is not in itself necessarily an indicator of dune inactivity;
- in semi-arid environments where there is considerable rainfall variability, dune activity may be episodic on a number of time scales.

There is no doubt that some heavily vegetated dunes are truly inactive relicts which indicate a past environmental regime which was drier, windier or when sand supply was more plentiful. But it is also the case that some interpretations of vegetated dunes in the past have been over-simplistic. Even if further investigations do reveal that the dunes of the southwestern Kalahari are essentially fixed and inactive, this does not negate the need for further debate about the nature of dune activity and what constitutes a relict form.

This work is funded by a NERC grant. The paper was greatly improved by the comments of Drs Nick Lancaster and Haim Tsoar to whom we are most grateful. The views expressed remain our own.

References

ARAD, A. 1984. Relationship of salinity of groundwater to recharge in the southern Kalahari Desert. *Journal of Hydrology*, **71**, 225–238.

ASH, J. E. & WASSON, R. J. 1983. Vegetation and sand mobility in the Australian desert dunefield. *Zeitschrift für Geomorphologie Supplement Band*, **45**, 7–25.

BESLER, H. 1980. Die Dünen-Namib: Entstehung und Dynamik eines Ergs. *Stuttgarter Geographische Studien*, **96**.

BREED, C. S., FRYBERGER, S. G., ANDREWS, S., MCCAULEY, C., LENNARTZ, F., GEBEL, D. & HORSTMAN, K. 1979. *Regional studies of sand seas using LANDSAT (ERTS) imagery*. United States Geological Survey Professional Paper **1052**, 305–397.

BUCKLEY, R. 1987. The effect of sparse vegetation on the transport of dune sand by wind. *Nature*, **325**, 426–428.

CAMPBELL, S. E., SEELER, J. S. & GOLUBIC, S. 1989. Desert crust formation and soil stabilization. *Arid Soil Research and Rehabilitation*, **3**, 217–228.

FARR, J., PEART, R., NELISSE, C. & BUTTERWORTH, J. 1982. *Two Kalahari pans: a study of their*
morphometry and evolution. Botswana Geological Survey Report GS10/10.

FLINT, R. F. & BOND, G. 1968. Pleistocene sand ridges and pans in western Rhodesia. *Bulletin of the Geological Society of America*, **79**, 299–313.

FORSTER, S. M. 1990. The role of microorganisms in aggregate formation and soil stabilization: type of aggregation. *Arid Soil Research and Rehabilitation*, **4**, 85–98.

FRYBERGER, S. G. 1979. *Dune forms and wind regime*. United States Geological Survey Professional Paper **1052**, 137–169.

—— & GOUDIE, A. S. 1981. Arid geomorphology. *Progress in Physical Geography*, **5**, 420–428.

GLENNIE, K. W. 1983. Early Permian (Rotliegendes) palaeowinds of the North Sea. *Sedimentary Geology*, **34**, 245–265.

—— 1985. Early Permian (Rotliegendes) palaeowinds of the North Sea — reply. *Sedimentary Geology*, **54**, 297–313.

GOUDIE, A. S. 1970. Notes on some major dune types in southern Africa. *South African Geographical Journal*, **52**, 93–101.

—— 1983. *Environmental Change* (second edition). Oxford University Press, Oxford.

GROVE, A. T. 1969. Landforms and climatic change in the Kalahari and Ngamiland. *Geographical Journal*, **135**, 191–212.

HACK, J. T. 1960. Interpretation of erosional topography in humid temperate regions. *American Journal of Science*, **258A**, 80–97.

HEINE, K. 1982. The main stages of the Late Quaternary evolution of the Kalahari region, southern Africa. *Palaeoecology of Africa*, **15**, 53–76.

HESP, P. A., HYDE, R., HESP, V. & ZHENGYU, Q. 1989. Longitudinal dunes can move sideways. *Earth Surface Processes and Landforms*, **14**, 447–452.

KLEIN, R. G., CRUZE-URIBE, K. & BEAUMONT, P. B. 1991. Environmental, ecological and palaeo-anthropological implications of the late Pleistocene mammalian fauna from Equus Cave, northern Cape Province, South Africa. *Quaternary Research*, **36**, 94–119.

LANCASTER, N. 1980. Dune systems and palaeo-environments in southern Africa. *Palaeontologia Africana*, **23**, 185–189.

—— 1981. Palaeoenvironmental implications of fixed dune systems in southern Africa. *Palaeogeography, Palaeoclimatology, Palaeoecology*, **33**, 327–346.

—— 1985. Variations in wind velocity and sand transport on the windward flanks of desert sand dunes. *Sedimentology*, **32**, 581–593.

—— 1987. Formation and reactivation of dunes in the southwestern Kalahari: palaeoclimatic implications. *Palaeoecology of Africa*, **18**, 103–110.

—— 1988. Development of linear dunes in the south-western Kalahari, southern Africa. *Journal of Arid Environments*, **14**, 233–244.

—— 1989a. *The Namib Sand Sea*. Balkema, Rotterdam.

—— 1989b. Star Dunes. *Progress in Physical Geography*, **13**, 67–91.

LIVINGSTONE, I. 1986. Geomorphological significance of windflow patterns over a Namib linear dune. *In*: NICKLING, W. G. (ed.) *Aeolian Geomorphology*. Allen & Unwin, Boston, 97–112.

—— 1989. Monitoring surface change on a Namib linear dune. *Earth Surface Processes and Landforms*, **14**, 317–332.

NASH, D. J., THOMAS, D. S. G. & SHAW, P. A. 1993. Timescales, environmental change and dryland valley development. *In*: MILLINGTON, A. & PYE, K. (eds) *Effects of Environmental Change in Drylands*. Wiley, Chichester (in press).

NEWELL, R. E., GOULD-STEWART, S. & CHUNG, J. C. 1981. Possible interpretation of palaeoclimatic reconstruction for 18 000 BP in the region 60°N to 60°S, 60°W to 100°E. *Palaeoecology of Africa*, **13**, 1–19.

RUBIN, D. M. 1990. Lateral migration of linear dunes in the Strezelecki Desert, Australia. *Earth Surface Processes and Landforms*, **15**, 1–14.

SARNTHEIN, M. 1978. Sand deserts during the last glacial maximum and climatic optimum. *Nature*, **272**, 43–46.

SHAW, P. A. & COOKE, H. J. 1986. Geomorphic evidence for the late Quaternary palaeoclimates of the middle Kalahari of northern Botswana. *Catena*, **13**, 349–359.

——, —— & THOMAS, D. S. G. 1988. Recent advances in the study of Quaternary landforms in Botswana. *Palaeoecology of Africa*, **19**, 15–26.

—— & DE VRIES, J. J. 1988. Duricrust, groundwater and valley development in the Kalahari of southwestern Botswana. *Journal of Arid Environments*, **14**, 245–254.

SHIELDS, L. M., MITCHELL, C. & DROUET, F. 1957. Alga- and lichen-stabilized surface crusts as soil nitrogen sources. *American Journal of Botany*, **44**, 489–498.

THOMAS, D. S. G. 1984. Ancient ergs of the former arid zones of Zimbabwe, Zambia and Angola. *Transactions of the Institute of British Geographers*, **9**, 75–88.

—— 1988. The geomorphological role of vegetation in the dune systems of the Kalahari. *In*: DARDIS, G. F. & MOON, B. P. (eds) *Geomorphological Studies in Southern Africa*. Balkema, Rotterdam, 145–158.

—— 1992. Desert dune activity: concepts and significance. *Journal of Arid Environments*, **22**, 31–38.

—— & SHAW, P. A. 1991. *The Kalahari Environment*. Cambridge University Press, Cambridge.

—— & TSOAR, H. 1990. The geomorphological role of vegetation in desert dune systems. *In*: THORNES, J. B. (ed.) *Vegetation and Erosion*. Wiley, Chichester, 471–489.

THORNES, J. B. (ed.) 1990. *Vegetation and Erosion*. Wiley, Chichester.

TSEO, G. 1990. Reconnaissance of the dynamic characteristics of an active Strzelecki Desert longitudinal dune, south-central Australia. *Zeitschrift für Geomorphologie*, **34**, 19–35.

TSOAR, H. 1983. Dynamic processes acting on a longitudinal (seif) dune. *Sedimentology*, **30**, 567–578.

—— 1989. Linear dunes — forms and formation. *Progress in Physical Geography*, **13**, 507–528.

—— & MØLLER, J. T. 1986. The role of vegetation in the formation of linear sand dunes. *In*: NICKLING, W. G. (ed.) *Aeolian Geomorphology*. Allen & Unwin, Boston, 75–95.

TYSON, P. D. 1986. *Climatic Change and Variability in Southern Africa*. Oxford University Press, Oxford.

——, DYER, T. G. J. & MAMETSE, M. N. 1975. Secular changes in south African rainfall: 1880–1972. *Quarterly Journal of the Royal Meteorological Society*, **101**, 817–833.

VILES, H. A. (ed.) 1988. *Biogeomorphology*. Blackwell, Oxford.

WARD, J. D. 1984. *Aspects of the Cenozoic geology in the Kuiseb valley, central Namib Desert*. Unpublished PhD thesis, University of Natal.

WASSON, R. J. 1984. Late Quaternary palaeoenvironments in the desert dunefields of Australia. *In*: VOGEL, J. C. (ed.) *Late Cainozoic Palaeoclimates of the Southern Hemisphere*. Balkema, Rotterdam, 419–432.

—— & HYDE, R. 1983. Factors determining desert dune type. *Nature*, **304**, 337–339.

—— & NANNINGA, P. M. 1986. Estimating wind transport of sand on vegetated surfaces. *Earth Surface Processes and Landforms*, **11**, 505–514.

YAIR, A. 1990. Runoff generation in a sandy area — the Nizzana sands, western Negev, Israel. *Earth Surface Processes and Landforms*, **15**, 597–609.

Entrada Sandstone: an example of a wet aeolian system

MARY CRABAUGH & GARY KOCUREK

Department of Geological Sciences, University of Texas, Austin, Texas 78713, USA

Abstract: Wet aeolian systems are those in which the water table is shallow and the floors of the interdune flats are within the capillary fringe. Accumulation within a wet aeolian system occurs because of a relative rise in the water table, and these accumulations are characterized by the presence of both dune and interdune-flat deposits. The Jurassic Entrada Sandstone, studied along a 2.7 km traverse in NE Utah, is interpreted as having been formed in a wet aeolian system because of a preponderance of features indicating a shallow water table throughout accumulation of the unit, and regional correlation to marine-sabkha units that vertically stack, then onlap the aeolian-dominated strata. Features indicating a shallow water table and/or evaporites include subaqueous ripple deposits, contorted strata, breccias, collapse features, wavy bedding, foundered sets, loaded set bases, corrugated surfaces, polygonal fractures, ball-and-pillow structures, and trains of small dunes 'frozen' in place. Along the traverse, oriented roughly parallel to the predicted palaeoflow from the NNE, sets of cross-strata, bounding surfaces and 'flat'-bedded strata climb to the SW, while other surfaces and 'flat'-bedded strata are horizontal. Climb is interpreted to result from the migration of dunes and interdune areas during periods of a rising water table, whereas horizontal surfaces represent super-bounding surfaces formed during periods of static or falling water table, and horizontal 'flat' strata represent vertical sabkha accretion. The cross-stratified units consist of compound sets with two scales of cyclicity. The larger, most prominent cyclicity is formed of scalloped bounding surfaces systematically truncating foresets to the SW, and results from superimposed dunes migrating NW and SE along-slope larger bedforms that migrated toward the SW. The smaller scale of cyclicity consists of packages of wind-ripple laminae and grainflow strata truncated by bounding surfaces, and these cycles are thought to represent annual variations in wind direction. Some cross-stratified units are co-sets in which the foresets of the main set can be traced downward into bottomsets consisting of small sets separated by corrugated surfaces. The basal sets are thought to represent small satellite dunes moving along the base of the larger bedforms and across damp interdune floors. The recognition of super surfaces within the Entrada allow the identification of four genetic packages that comprise the formation. Within genetic packages the available sand supply can be estimated, and is considered as a function of the rates of sediment supply and water table fluctuations. These relative rates, coupled with close proximity of the study area to the Jurassic palaeoshoreline and the regional stratigraphic relationships, suggest that fluctuation in relative sea-level controlled the facies architecture seen in the Entrada Sandstone of NE Utah.

Recent work (e.g. Loope 1985; Talbot 1985; Kocurek 1988; Deynoux *et al.* 1989; Clemmensen & Blakey 1989; Kocurek *et al.* 1991; Havholm *et al.* 1992; Kocurek *et al.* 1992) demonstrates that aeolian systems, like other sedimentary environments, are dynamic and the ultimate facies architecture reflects internal and external controls on a broad span of processes. These processes, in the terminology of Kocurek & Havholm (1993), can be viewed as a hierarchy that proceeds from those instantaneous processes acting at the substrate level (deposition, bypass, erosion), to those that determine whether or not net deposition will occur (accumulation, super-surface formation), to those that determine whether or not any accumulation or surface will enter into the rock record (preservation, unconformity). Kocurek & Havholm (1993) propose that aeolian systems themselves can be viewed as a spectrum in which the end members are dry, wet, and stabilized aeolian systems. They contend that these are distinct systems because the dynamic processes that give rise to accumulations and their capping super-bounding surfaces are different in each system. Of specific interest here are wet aeolian systems, defined as those in which the water table or its capillary fringe is shallow and the moisture content at the interdune surfaces is at least sufficient to raise grain threshold values to the point where the surface is largely protected from deflation. Deposition, bypass or erosion at the substrate level is in part hydrodynamically controlled by rising, static, or falling capillary

From Pye, K. (ed.), 1993, *The Dynamics and Environmental Context of Aeolian Sedimentary Systems.*
Geological Society Special Publication No. 72, pp. 103–126.

fringe, respectively. The movement of the water table is measured with respect to the sediment column. Absolute changes in the water table can result from a climatic change to more humid conditions or the inland response to eustatic sea-level rise, whereas a relative change can result from subsidence of the sediment column through a static water table. Accumulation, in a wet aeolian system, occurs with a rise of the water table, and super surfaces form when the water table becomes static or falls. Wet aeolian systems differ from dry systems because in the latter, deposition, bypass and erosion at the substrate level are controlled by the local aerodynamic conditions only. Because of acceleration of the wind across interdune flats, accumulation is not favoured in dry systems until these flats have been largely eliminated by dune growth. Hence, the occurrence of interdune-flat deposits is characteristic of wet systems. The rise of the water table also raises the baseline of erosion, and preservation of the sequence can occur without, or in conjunction with, basin subsidence.

Previous work on the Middle Jurassic (Callovian) Entrada Sandstone of northeastern Utah by Kocurek (1980, 1981a, b) indicates that these well-exposed outcrops can be interpreted today

as representing a wet aeolian system (Fig. 1). The purpose of this study is to characterize the features of this formation and interpret its facies architecture in light of the new conceptual model of a wet aeolian system. The interpretation of the Entrada Sandstone as a wet system is based upon both the regional stratigraphic relationships and the features of the outcrops themselves. In addition to evaluating the Entrada in the new concept of wet aeolian system, three additional aspects of this paper allow a much more sophisticated treatment than in the original work by Kocurek. First, the methods used here to reconstruct the bedforms all postdate the original work. Second, the availability of better surveying techniques allows a more accurate definition of the attitude of bounding surfaces, so that in addition to the climbing surfaces originally recognized, candidates for super surfaces can also be recognized. Third, recent thinking on aeolian sequence stratigraphy allows the causes of the sequences recognized here to be approached.

Methods

With the study of Kocurek (1981a) it became evident that two-dimensional vertical sections

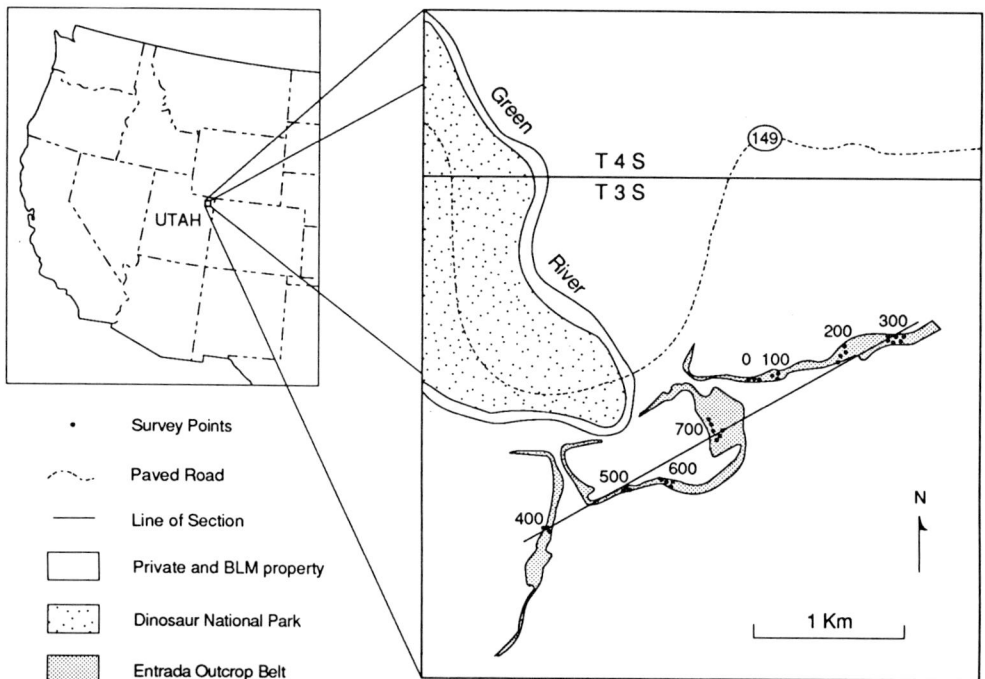

Fig. 1. Location map showing outcrop trend and location of 355 survey points measured in eight vertical sections. Line of traverse is N65E, and vertical sections are identified by number (0–700).

are inadequate to understand the facies architecture of aeolian systems. Instead it is necessary to study long lateral and three-dimensional traverses in order to correctly characterize the subtle three-dimensional relationships between sets of cross-strata, bounding surfaces and 'flat' beds. For this study, a virtually continuous 2.7 km outcrop of the Entrada was surveyed with a Zeiss Elta 4 electronic tacheometer surveying tool (Fig. 1). Correlation of surfaces and units between survey points was accomplished by walking the outcrop where possible or through visual correlations on the outcrop and in photomosaics. This method allowed a scaled reconstruction of the key elements of the outcrop (Fig. 2). Because of the three-dimensional nature of the outcrop, it is possible to mathematically define the correlated surfaces as planes utilizing the X, Y, Z coordinates of the survey (Table 1). The plane defining the contact between the Entrada and the underlying Carmel Formation (dipping 11.8° in a 327.4° direction) is used as datum. It is geologically reasonable to assume that this surface was originally horizontal because the contact is conformable and the Carmel in this area represents widespread sabkha facies (see Kocurek & Dott 1983). Moreover, the orientation of a shale drape horizon, occurring at approximately 24 m above the base of the section and traceable along the entire outcrop, is identical to the Entrada–Carmel contact. The shale drape, first documented in Kocurek (1981a), is now identified

as a bentonite layer and is almost certainly a time horizon. Using the attitude of these two planes as tectonic dip, the other correlated surfaces were rotated to their original depositional orientation using the computer program 'Stereonet 4.0-II' (Allmendinger 1989). Projecting the planes into the general line of section (N65E) allows a scaled diagram in which the angular relationships of the planes can be seen, as well as their apparent dip in the line of section (Fig. 3). Because the surveyed outcrop belt is relatively long and narrow, dip determination in the plane of the outcrop (N65E) is probably more accurate than perpendicular to the outcrop plane (N25W) (Fig. 1). For this reason those dip amounts of 0.1° and less, with a dip direction roughly perpendicular to the outcrop plane (i.e. 147°) (surfaces B, I, L, O in Table 1), are here considered horizontal.

Sets of cross-strata were interpreted using the relationships of bounding surfaces and cross-strata illustrated in the computer simulations of Rubin (1987). At key locations, the orientation of the cross-strata and internal bounding surfaces were measured with a Brunton Compass (c. 100 measurements) and supplemented with measurements from Kocurek (1980, 1981a). All measurements were rotated to remove the tectonic overprint using 'Stereonet 4.0-II' (Allmendinger 1989). Where sets are compound and result from the migration of superimposed dunes over a larger bedform, the stereonet method of Rubin & Hunter (1983) was

Table 1. *Orientation of mathematical planes determined from survey data. Apparent dip reflects dip-angle of plane as projected into line of section (N65E). True dip and dip direction is the calculated orientation after tectonic dip correction. Corr: corrugated surface; Polys: polygonal fractures*

Surface	Apparent dip in degrees	True dip and dip direction in degrees		Characteristics
A	0	0		Loaded
B	0	0.1	147	Loaded
C	0.1	0.2	126	Corrugated
D	0.2	0.4	5	Corrugated
E	0.17	0.2	57	Featureless
F	0.18	1.0	143	Corr. and polys
G	0.09	0.6	145	Corr. and polys
H	0.07	0.7	147	Featureless
I	0.02	0.1	147	Loaded
J	0	0		Featureless
K	Irregularly scoured bounding surface			
L	0.03	0.1	135	Corr. and polys
M	0.35	1.0	132	Corrugated
N	0.24	0.4	124	Featureless
O	0.04	0.1	147	Corrugated

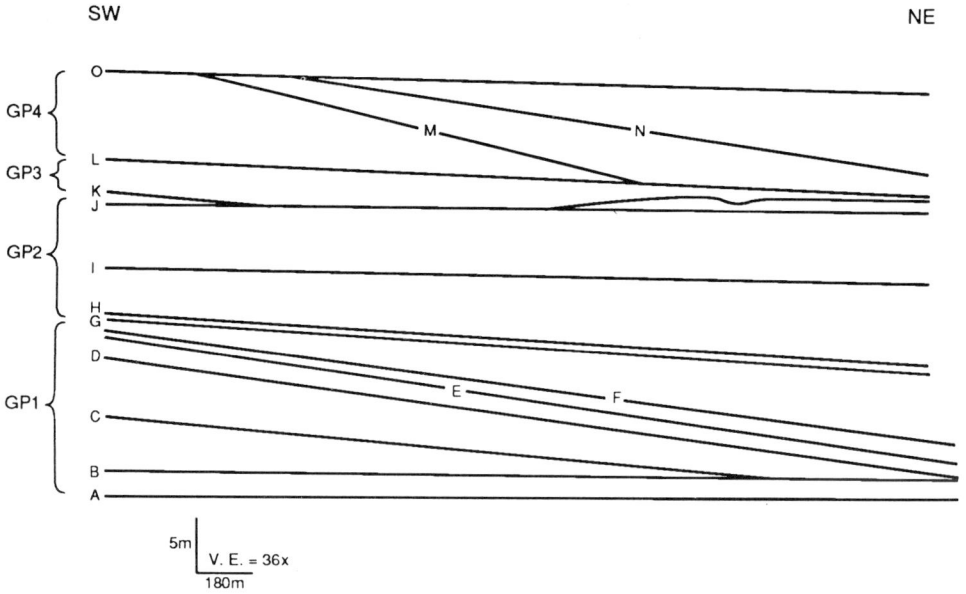

Fig. 3. Mathematical planes, for correlated surfaces with four or more X, Y, Z survey points, projected into line of section (N65E). Surfaces have been rotated to remove the tectonic dip (dip = 11.8^0; dip direction = 327.4^0) using Stereonet 4.0-II (Allmendinger 1989).

used to reconstruct the configuration of the superimposed and main bedforms. Then, utilizing all available outcrop evidence, the bedforms were simulated using the Rubin (1987) computer modelling program (Fig. 4).

General description

Palaeogeographic setting

The Entrada Formation is a very complex and extensive aeolian unit stretching from Utah to

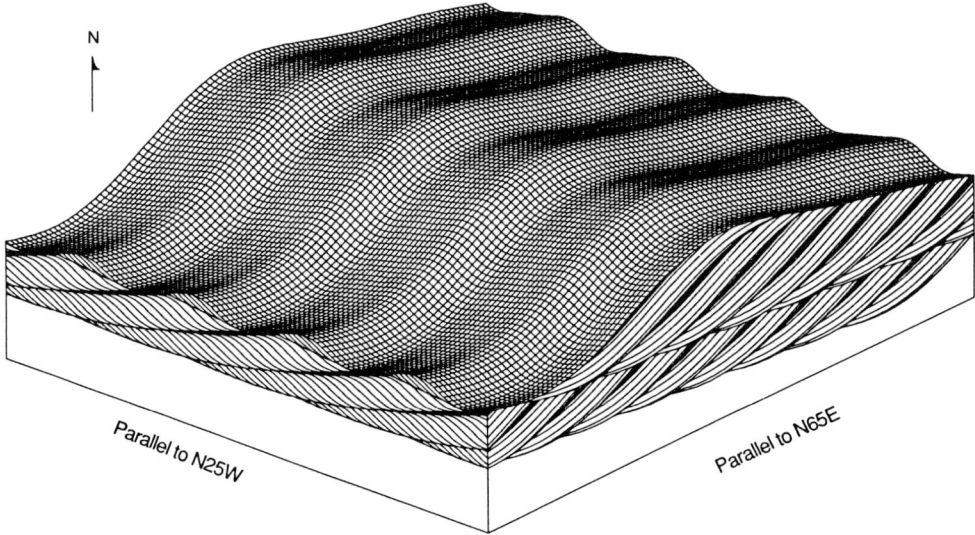

Fig. 4. Generalized Entrada Sandstone compound bedform as simulated using Rubin (1987). Examples of scalloped cross-strata are seen in panel parallel to N65E and climbing bedform structures are seen in N25W panel.

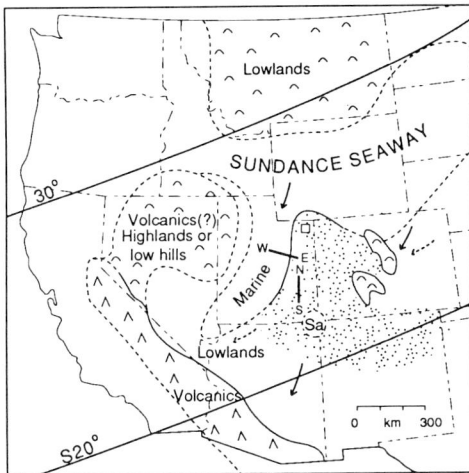

Fig. 5. Palaeogeographic map for the Middle Jurassic (Callovian) showing extent of Entrada erg (stippled pattern). Solid arrows indicate predicted summer wind patterns, and dashed arrows are winter wind patterns from Parrish & Peterson (1988). Study area indicated by small box. Section lines for Fig. 6 shown by W–E and N–S dark lines. Sa: sabkha within the Todilto Basin. Modified from Peterson (1988).

Outcrop description

The Entrada Sandstone consists of surfaces, 'flat'-bedded units and sets of cross-strata that are continuous over the length of the outcrop (Figs 2 & 3). The overall assemblage shows a distinct vertical tripartite division that can be recognized by differences in the architectural style (Figs 2 & 3). The term 'Division' is being used to informally subdivide the Entrada Sandstone for descriptive purposes and does not necessarily imply genetic associations. The Lower and Upper Divisions are dominated by compound sets of cross-strata separated by surfaces and relatively thin 'flat'-bedded units, all of which gently dip to the NE. In contrast, the Medial Division is dominated by 'flat'- to wavy-bedded units separated by thin, discontinuous sets of compound and simple cross-strata. In order to simplify description of the Entrada the major cross-stratified units have been numbered and the correlated surfaces have been lettered sequentially from the bottom of the sequence to the top (Fig. 2). The similarity of cross-strata throughout the Entrada makes it possible to describe them in detail in a separate section and not as part of the following architectural description.

the Texas panhandle and from Arizona and New Mexico north to southern Wyoming (Blakey *et al.* 1988) (Fig. 5). Palaeogeographic and stratigraphic reconstruction of the Entrada erg show that it was bounded on the north, west and partly on the south by units interpreted as representing marine and sabkha environments (Imlay 1980, Kocurek & Dott 1983; O'Sullivan & Pierce 1983; Blakey *et al.* 1988; Peterson 1988) (Fig. 5). Stratigraphically, after initial progradation of the Entrada, aeolian and marine-sabkha units vertically stack, followed by an overall onlap of marine-sabkha units onto the Entrada aeolian sand (Fig. 6).

Global circulation models for the Early and Middle Jurassic of the Western United States indicate prevailing wind directions fluctuated seasonally because of a strong monsoonal influence (Parrish & Curtis 1982; Parrish & Peterson 1988) (Fig. 5). These same studies also suggest the summer winds (from NNE) were greater than the winter winds (from NE) in strength. Between the Middle and Late Jurassic, palaeogeographic changes resulted in the breakdown of the monsoonal circulation. However, during the Middle Jurassic the monsoonal circulation was probably weakening and may have resulted in a degree of variability in the wind directions.

Lower Division

The Lower Division is bounded by Surfaces A and G and includes cross-stratified Units 1–5 (Figs 2, 3 and Table 1). Surface A is an overall horizontal surface but it does show low-amplitude undulations because of loading of the Entrada Sandstone into the underlying Carmel Formation (Fig. 7a, Table 1). Surface G is irregular, truncates sets of cross-strata and bounding surfaces (Fig. 2, Table 1), and shows both corrugation or microtopography and polygonal fractures (Fig. 7b). The Lower Division consists of sets of cross-strata separated by inclined bounding surfaces and 'flat'-bedded units that dip to the NE as well as a basal 'flat'-bedded unit (Fig. 2).

The 'flat'-bedded unit, between Surfaces A and B, is easily traced throughout the outcrop belt (Fig. 2). Surface B, like Surface A, is a horizontal suface with undulations caused by loading of dune cross-strata above into underlying 'flat' beds (Table 1). The unit is composed of wavy laminae and small sets of wind-ripple cross-strata with varying degrees of deformation.

Unit 1 (Fig. 2), resting on Surface B and bounded upward by a corrugated surface with a thin 'flat'-bedded unit and Surface C to the NE, is a series of thin sets of wind-ripple and thin

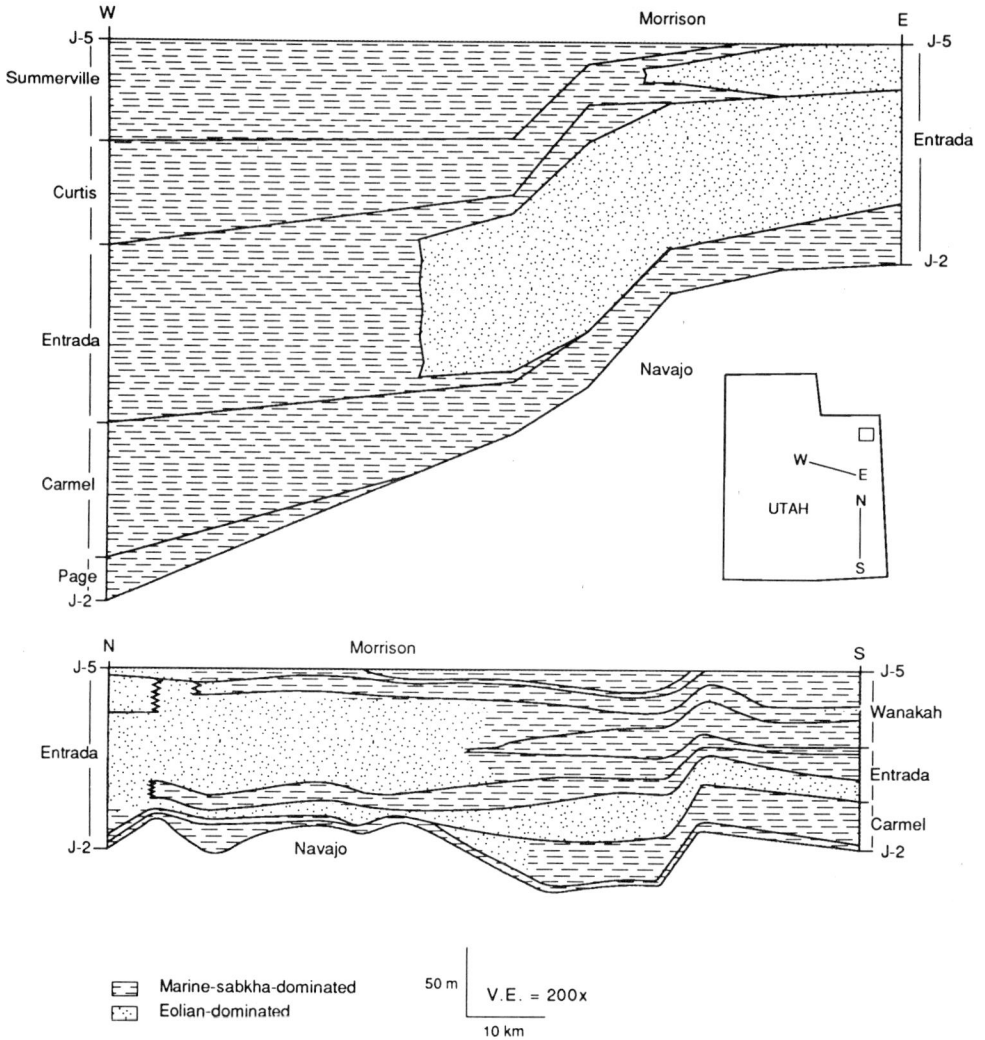

Fig. 6. General facies relationships for the Entrada and coeval formations of eastern Utah. Dotted pattern indicates predominantly aeolian strata within the Entrada. Horizontal line pattern indicates units coeval with the Entrada that are predominantly non-aeolian and largely shallow marine, and sabkha in origin. Figure shows transition from marine-sabkha-dominated deposition in the south and west to aeolian-dominated in the northeast. Refer to Fig. 5 for palaeogeographic setting. Because aeolian strata are contemporaneous with units of a marginal marine origin, and these vertically stack, then onlap onto the main body of the Entrada, accumulation of the aeolian strata is believed to have occurred during an overall sea-level rise. Modified from O'Sullivan and Pierce (1983) and O'Sullivan (1980).

grainflow cross-strata. The grain size of this unit is somewhat coarser overall than any of the other units. The surfaces separating the thin sets are corrugated and may show very thin (*c.* 5 cm) 'flat'-bed accumulations. Unit 1 can be traced from section 700 to 400 with good confidence; however, the unit cannot be traced from 700 to 0 because the outcrop has been removed by creek

erosion. The initiation of Unit 1 and its relation to Unit 3 cannot be determined with certainty, but Unit 1 probably originated between sections 700 and 0, and then progressively increases in thickness to the SW (Figs 1 & 2).

Resting on top of Unit 1 in sections 400 and 500 is Unit 2, which gradually expands from its initial thickness of a few centimetres between

(a)

(b)

(c)

(d)

Fig. 7. Features of the Lower Division. (**a**) Loading structure and deformation produced by emplacement of Entrada sands (light) onto underlying marine-sabkha sediments of the Carmel (dark). Field of view approximately 10 m. White in foreground is snow. (**b**) Horizontal view of Surface G in section 700, showing a well-developed polygonal fracture pattern. Lens cap for scale. Surface G caps the Lower Division and is the super surface of GP1. (**c**) Person's hand is resting on the corrugated upper surface of Unit 3 (Surface D) in section 300. Surface B is also labelled. Unit 3 is composed of a single set of compound cross-strata. (**d**) Looking obliquely across doubly compound cross-strata of Unit 4 in section 200. Surfaces E (loaded, featureless) and F (corrugated) indicated by letters. Arrows near kneeling person mark large-scale cyclicity (scallop bounding surface) produced by migration of superimposed bedforms. A second scalloped surface occurs beyond person. Smaller-scale cyclicity visible by truncating surfaces within scalloped sets.

sections 500 and 600 to 1.7 m in section 400 (Fig. 2). Internally it is composed of a single set of compound cross-strata, and it is capped by a corrugated to featureless surface (Fig. 2).

Occurring between inclined Surfaces C (loaded) and D (corrugated; Fig. 7c) (Table 1), Unit 3 grows from a 70 cm-thick accumulation of grainflow cross-strata in the NE into a 4 m-thick set to the SW (Figs 2–3). Internally Unit 3 is composed of a single set of doubly compound cross-strata that is easily correlated throughout the outcrop. The internal bounding surfaces are generally sharp and featureless. Between sections 500 and 400, Unit 3 appears to break down from one large set into one large set plus several smaller basal sets; however, this correlation is somewhat uncertain because a drainage separates the two outcrops.

The 'flat'-bedded strata that separate cross stratified Units 3 and 4 (Surfaces D to E) is relatively uniform in thickness throughout the outcrop belt, and gently dips to the NE (Fig. 2 and Table 1). Internally it is 'flat'- to wavy-bedded with scattered very thin, deformed sets of wind-ripple laminae.

Resting on Surface E (loaded to featureless) and capped by Surface F (corrugated and polygonally fractured), Unit 4 is composed of a single set of doubly compound cross-strata (Fig. 7d). In sections 700, 600, 500 and 400, thin 'flat'-bedded sands occur on Surface F. Unit 4 thins drastically from a maximum thickness of 7.0 m in section 200 to 0.6 m in section 400. Correlation from section 0 to 700 is not entirely certain because of the intervening creek bed. In section 300 a small set of compound cross-strata underlies and is separated by a corrugated surface from Unit 4. Laterally this set thins and pinches out before section 200.

Unit 5, bounded by Surfaces F and G (both corrugated and polygonally fractured), is composed of at least two sets of doubly compound cross-strata (Figs 2, 3 and Table 1). The bounding surface separating the sets is sharp and smooth. More precise correlation of the sets that make up Unit 5 (5a–5e, Fig. 2) is not possible because of a lack of outcrop between sections 0 and 700, and a large covered interval between sections 700 and 600 (Fig. 1). Unit 5 thins laterally from 1.5 m in the NE (section 300) to 0.6 m in the SW (section 400) (Fig. 2).

Medial Division

Bounded by Surfaces G (irregular, corrugated, polygonally fractured; Fig. 7b) and K (irregularly scoured, Fig. 2), the Medial Division can be subdivided into two sequences, each of which shows a progression upward from thin laterally discontinuous sets of cross-strata to 'flat' beds. The two sequences are divided by horizontal Surface I (Figs 2 & 3). Like the larger cross-stratified sets of the Lower Division, these sets generally have loaded bases where they overlie 'flat' beds, as well as having corrugated and polygonally fractured upper bounding surfaces. The sandbody geometry with respect to scale in this division is, however, distinctly different from the Lower Division. Individual sands cannot be traced over long distances, they are thin (a few centimeters to c. 1 m thick), discontinuous and commonly lensoid in appearance. Internally they may be simple or compound sets.

Occurring in the upper 5 m of this division are three bentonite layers identified by abundant unabraded, euhedral, hexagonal biotite, as well as apatite and zircon. The lowest bentonite layer, Surface J (horizontal, Table 1), can be

(a)

(b)

(c)

(d)

Fig. 8. Features of the Medial Division. (**a**) Arrows indicate span of two lensoid sets near section 500 showing complex internal stratification, corrugated upper surface, and adjacent 'flat' strata. Surface K, marked by shale drape, is indicated by letter. Lower lensoid set is approximately 5 m in width. (**b**) Internal stratification of small 'frozen' dune showing topsets and foresets. Pen for scale. Cross-strata indicate migration to WSW. (**c**) Arrows highlight fully preserved 'frozen' dune migrating to the WSW in section 100. Man is holding 1.5 m stick for scale. Surface G indicated by letter. (**d**) Wavy and contorted laminae seen in cross-section and interpreted to result from haloturbation (section 500). Field of view approximately 1 m.

(a)

(b)

(c)

(d)

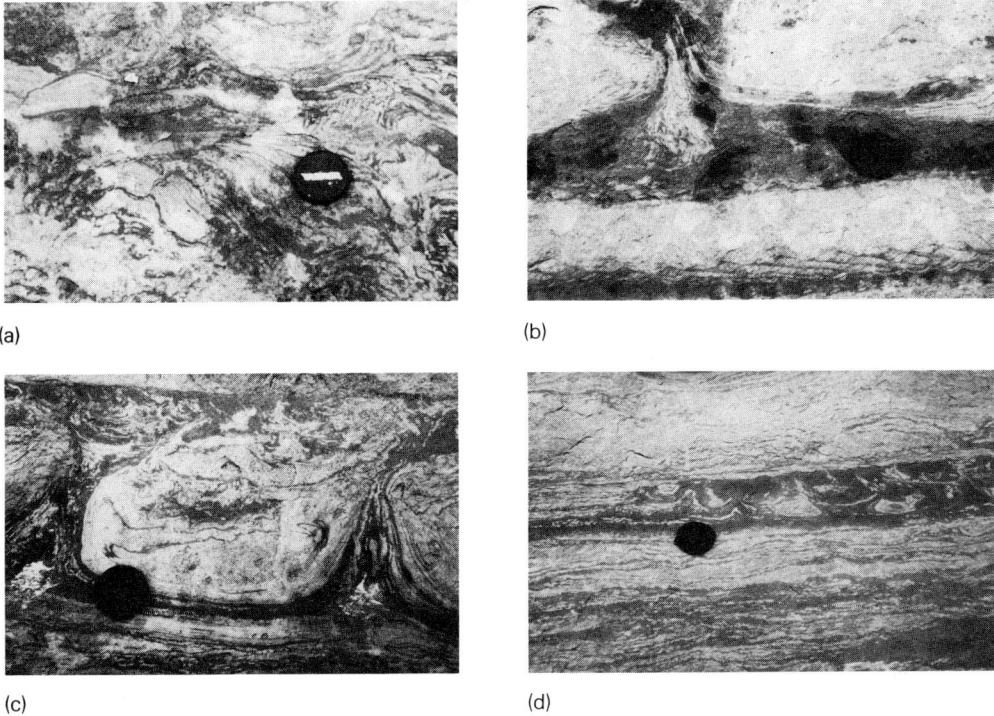

Fig. 9. Features of the Medial Division. (**a**) Brecciated laminae, in an oblique cross-section, caused by evaporite precipitation and dissolution within an aeolian sabkha. Lens cap for scale. (**b**) Base of photo shows climbing translatent strata and preserved subaqueous ripple forms. Fluid escape structures in upper portion of photo related to loading by overlying sands. Lens cap for scale. (**c**) Ball-and-pillow structure produced by loading of probable aeolian cross-strata into wet sabkha sediments (section 500). Lens cap for scale. (**d**) Ball-and-pillow structure and swirled bedding in the Medial Division near section 100, showing the lateral loss of bed integrity. Above and below failed bed are wavy and somewhat deformed laminae. Lens cap for scale.

traced over the entire outcrop and is, therefore, used as a time horizon. The two upper bentonites and several cross-stratified sets are clearly truncated by Surface K (Fig. 2, sections 0,100,200).

At a number of horizons within the Medial Division are distinctly lens-shaped cross-bedded units. Commonly they are approximately 1 m in height and tens of meters in length (Fig. 8a). Most of these lenses show a complex internal structure, but some show a single set of cross-beds, dipping to the SW. These latter lenses commonly have well-preserved topsets and foresets (Fig. 8b). In sections 0 and 100, at approximately 18 m above the base of the Entrada, there is a series of these small preserved dunes occurring on the same horizon migrating to the SW (Fig. 8c). Some of these lensoid features have associated thin accumulations (<20 cm) of cross-strata extending up to 100 m.

Predominating within the Medial Division are 'flat'-bedded strata, which are essentially the same in appearance as the thinner 'flat'-bedded units between sets of cross-strata in the Lower and Upper Divisions (see also Kocurek 1981a). Overall, the 'flat' strata consist of thin wavy to horizontal laminae with some degree of deformation being ubiquitous. Some very fine laminae are clearly wind-ripple strata, and deformed, thin (cm) sets of cross-strata occur, some of which can be traced laterally into thicker, more prominent sets. Most of the strata, however, consist of wavy and brecciated laminae characteristic of salt-ridge structures (Fryberger et al. 1983; Kocurek & Fielder 1982) (Figs 8d & 9a). Subaqueous strata are relatively common, especially symmetrical and asymmetrical ripple cross-strata and isolated ripple forms (Fig. 9b). Ripple cross-strata typically show critical to supercritical climb, and some horizons can be traced for tens of metres before being truncated or passing into wavy laminae. Some deformation features are distinct ball-and-pillow structures of failed beds (Fig. 9c), and many are mildly to dramatically contorted with tight, high-angle undulations or swirled bedding (Fig. 9d). Strongly contorted intervals generally occur immediately below sets of cross-strata. Polygonal fractures and microtopography are common within the Medial Division, especially along the upper surface of sets of cross-strata. Some distinct depressions, ranging from a few to tens of centimeters across, occur with an overall sagging of the strata or stair-step faults (Fig. 10a). A number of the depressions that occur in the 'flat' strata are directly aligned over polygonal fractures, which lie a few to several tens of centimeters below (Fig. 10b).

Upper Division

Bounded by Surface K (scoured) at the base and the very irregularly scoured contact with the overlying marine Curtis Formation, the Upper Division is very similar in character to the Lower Division in being dominated by compound sets of cross-strata (Fig. 2). Unit 6, the basal unit in the Upper Division, lies between Surfaces K and L (horizontal). Surface L is primarily corrugated in nature, but also shows polygonal fractures and a veneer of flat-beds capping it in sections 700 and 600. Unit 6 consists of a single set of doubly compound cross-strata that varies in thickness from 0.7 m in section 200 to 4.5 m in section 500 (Fig. 2, 10c).

Units 7a and b occur between Surfaces L and

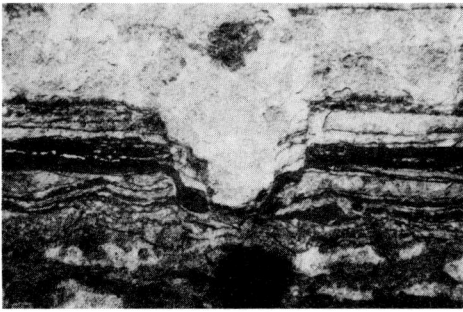

(a)

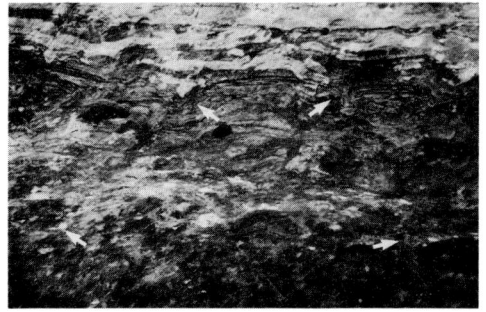

(b)

(c)

(d)

Fig. 10. Features of Medial and Upper Divisions. (**a**) Stair-stepped faults forming a depression thought to be related to evaporite dissolution from underlying strata. Lens cap for scale. (**b**) Upper arrows indicate depressions in the Medial Division aligned over underlying polygonal fractures (lower arrows). These depressions are believed to have formed with dissolution of evaporites within polygonal fractures. Lens cap for scale. (**c**) Unit 6, Surfaces K and L indicated by letters (section 700), showing climbing bedform structures migrating to the S. Arrows point to a climbing boundary surface. Face is oriented with N30W to the right. Distance from Surface K to L is 2 m. (**d**) Unit 7, composed of two sets of doubly compound cross-strata, measures 7 m from Surface L to M (indicated by letters). Lower set of arrows point to the corrugated surface separating 7a and 7b; upper set of arrows point to the inclined scallop bounding surface. Smaller-scale cycles can also be seen within the scalloped sets.

M, the latter of which is corrugated and has an apparent dip of 0.35° to the NE (Fig. 2; Table 1). Unit 7 is composed of two sets of compound cross-strata separated by a slightly corrugated surface (Figs 10d, 11a–c). Unit 7a begins between sections 200 and 100 where it is only a few centimeters thick, from there it thickens dramatically to the SW (7.3 m). Unit 7b begins to downcut into 7a in section 700, and progressively replaces it toward the SW end of the outcrop. Occurring between inclined Surfaces M and N is a 'flat'- to wavy-bedded unit that contains a few thin deformed cross-bedded sets.

Unit 8, between Surfaces N (featureless) and O (corrugated), clearly shows interfingering of cross-strata with the underlying 'flat' strata (Fig. 2). In section 300, Unit 8 is a 6.8 m-thick compound set of cross-strata that systematically thins to the SW. In section 200, the unit has thinned to 5.9 m and overlies a 'flat'-bedded unit (Fig. 11d). By section 0, Unit 8 is 3.5 m thick and consists of several small basal sets separated by thin 'flat'-bedded units, with an overlying thicker set (Fig. 2). The bounding surfaces separating the basal sets in section 0, traced laterally to the NE change from low-angle corrugated surfaces to much higher-angle featureless surfaces (Fig. 2). Between sections 100 and 200, the small basal sets can be clearly seen to merge into the main body of cross-strata. Given this interfingering of cross-strata with the

(a)

(b)

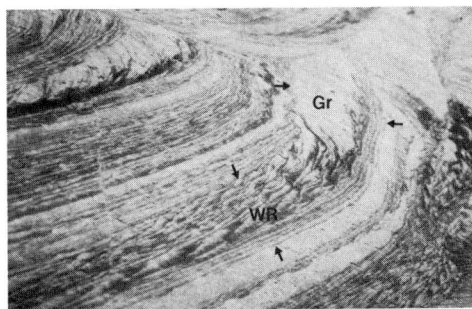

(c)

(d)

Fig. 11. Features of the Upper Division. (**a**) Unit 7, bounded by Surfaces L and M, showing large-scale cyclicity. Scallop bounding surfaces visible along trace of set, culminating in a more fully preserved trough marked by arrows on the left. Distance from surface L to M is approximately 3 m. (**b**) Unit 7 (section 500) illustrates the numerous small basal sets that occur in Unit 7a. These small sets can be traced laterally (left to right) into the cycles of the overlying main bedform of Unit 7a. Arrows show surface separating Units 7a and 7b. Unit 7 measures 7.3 m from Surface L to M. (**c**) Arrows indicate small-scale cyclicity in Unit 7 near section 600, which measures approximately 50 cm, consisting of bounding surface (lower arrows) conformably overlain by thin wind-ripple laminae, conformably overlain by grainflow strata (Gr) and thick wind-ripple laminae (WR), which are truncated by the next bounding surface (upper arrows). (**d**) Unit 8 in section 200 occurs between surfaces N and O (scoured) indicated by letters. Note wind-ripple–grainflow cycles composing this compound set. Unit 8 overlies and interfingers with 'flat'-bedded strata above surface N. Distance from Surface N to O is approximately 6 m.

'flat' beds in the NE segments of the traverse, Surface N is picked at the change from cross-stratified to 'flat'-bed dominated strata. Further to the SW in the section 500, Unit 8 is only 1.0 m thick and composed of a single set of compound cross-strata underlain by 'flat' beds (Fig. 2). The combination of poor exposure in section 700 and the lack of outcrop between sections 0 and 700 makes it very difficult to better resolve the facies relationships between sections 0 and 500.

Above Surface O lateral correlation of individual units is very difficult to impossible because of the deeply and irregularly scoured upper contact of the Entrada Sandstone. The stratigraphic and depositional relationships between the Entrada Sandstone and overlying Curtis Formation are thoroughly discussed in Eschner & Kocurek (1986).

Detailed description of cross-strata

All of the cross-stratified units within the traverse of the Entrada (Fig. 2), with the exception of Unit 1, are in part or totally composed of

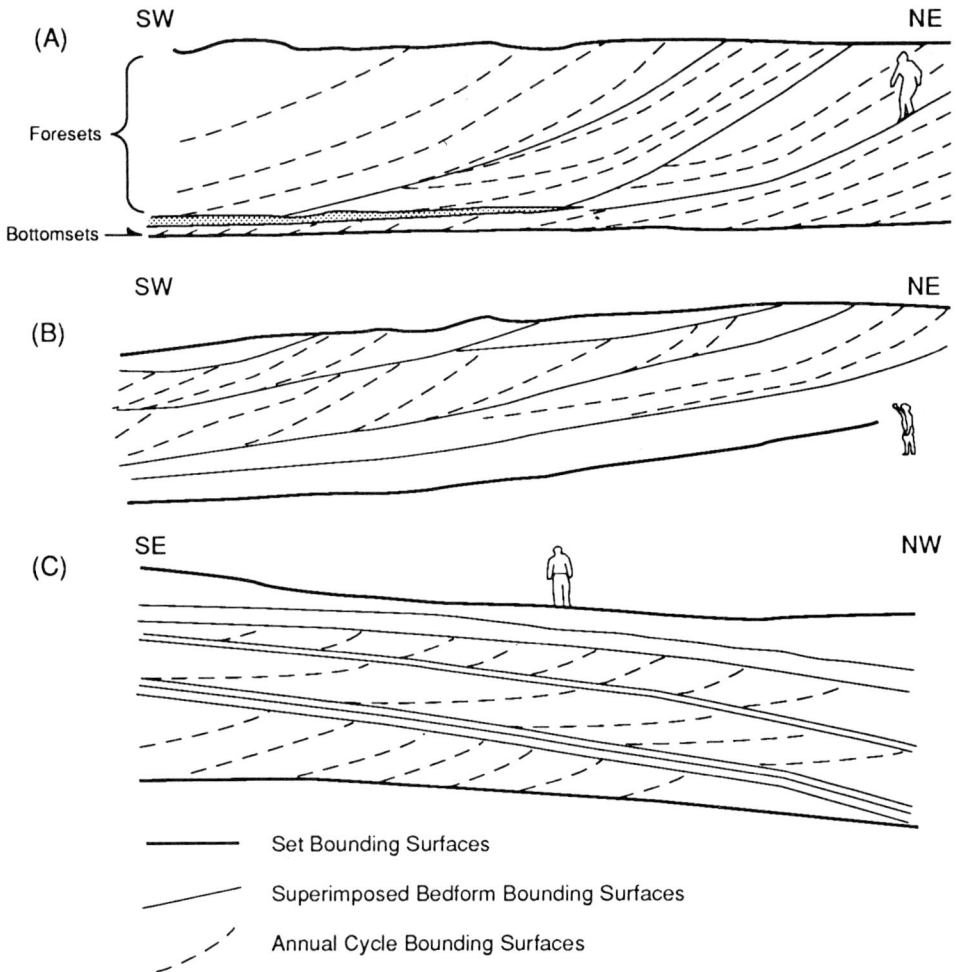

Fig. 12. Line drawings from photomosaics illustrating characteristics of internal bounding surfaces. (a) Co-set of cross-strata (Unit 8, section 200) illustrating that the small basal set when traced laterally becomes part of the overlying larger set. Patterned area above the basal set is 'flat'-bedded sabkha strata. Also note scalloped cross-strata showing truncation of annual cycles at the toes of the structures. (b) Low-angle scallop bounding surfaces (section 500, Unit 3) where the upper portions of the annual cycles are truncated. (c) Climbing bedform structures produced by the along-slope migration of a superimposed dune from right to left in this drawing (section 400, Unit 3). Note people for scale in all drawings.

sets of doubly compound cross-strata, so that a 'generic' style of cross-strata can be described (Figs 4 & 13). In addition, as noted above, some of the units are best described as co-sets showing a distinct set size bimodality, in which foresets of the main body of the set are replaced downward by bottomsets that consist of small sets with corrugated tops and, in some cases, intervening 'flat' beds (Figs 12a, 13).

Units composed of the compound co-sets include Units 3, 6, 7 and 8 (Fig. 2). These co-sets appear in vertical measured sections to consist of typically two, and up to five, sets of compound cross-strata separated by corrugated surfaces (Unit 8, section 200) (Fig. 2). For example Unit 8, between sections 0 and 300, can be described in a vertical section as a thin deformed set of cross-strata overlain by a thick compound set. However, if one traces the lower sets laterally to the NE they are seen to merge into the thick overlying set, thus indicating the sets should be considered as a whole and not as separate units (Figs 12a & 13). Unit 7 consists of a compound co-set (7a) overlain by a compound set (7b) (Figs 2 & 10d). Unit 7b downcuts into 7a such that in section 500 all that remains of the 7a co-set is its deformed amalgamated bottomset capped by a corrugated surface and immediately overlain by Unit 7b.

Internally the compound sets show two scales of cyclicity. The larger-scale and more prominent cyclicity is clearly seen in NE–SW-trending outcrops (i.e. sections 300, 200, 100, 0, 600, 500, Fig. 1), in which the foresets are systematically truncated to the SW by a bounding surface (Figs 7d, 10d, 11a, 12a–b). This cyclicity is the same in appearance as the 'scalloped cross-bedding' described by Rubin & Hunter (1983) in outcrops of the Navajo and Entrada sandstones. Scallop spacing ranges from one metre to tens of metres,

typically measuring 6–8 m in the thick sets. Within a single scallop the cross-strata dip directions deviate from the bounding surface dip directions; however, the deviation varies and is smallest immediately overlying the bounding surface, and greatest immediately below the next bounding surface (Figs 4 & 13). For example in Unit 4 (Fig. 2, section 200), a scallop bounding surface and the foresets that immediately overlie it have very similar dip directions and dip angles (bounding surface = 230.7° at 17.8°; foreset = 233.5° at 17.2°), whereas the foresets immediately underlying it in the previous scallop have a very different orientation (dip direction = 300.8°; dip angle = 20.1°) (Fig. 7d). This same cyclicity has a different appearance in outcrops oriented NW–SE, or perpendicular to the previous trend (sections 700 and 400; Fig. 1). In these NW–SE-trending walls, the scallops consist of sets showing distinct climb (Figs 4, 10c & 12c), and these features are here referred to as 'climbing bedform structures'. The apparent migration direction of these climbing bedform structures varies from NW to SE and is consistent within a set of cross-strata, with the exception of Unit 6. In Unit 6 climbing bedform structures climb both to the NW (between sections 100 and 200) and SE (section 700) (Fig. 10c).

The second scale of cyclicity is seen within the scalloped cycles. Internally the scallops are commonly composed of stacked sets of wind-ripple–grainflow cycles (Figs 11c–d, 13). These cycles, like the scalloped cross-bedding, have a concave-up wedge-shape geometry, and in NE–SW-oriented outcrops are systematically truncated to the SW (Fig. 13). Internally wind-ripple laminae overlie and parallel the basal bounding surface. These are immediately overlain by grainflow tongues that have a tangential

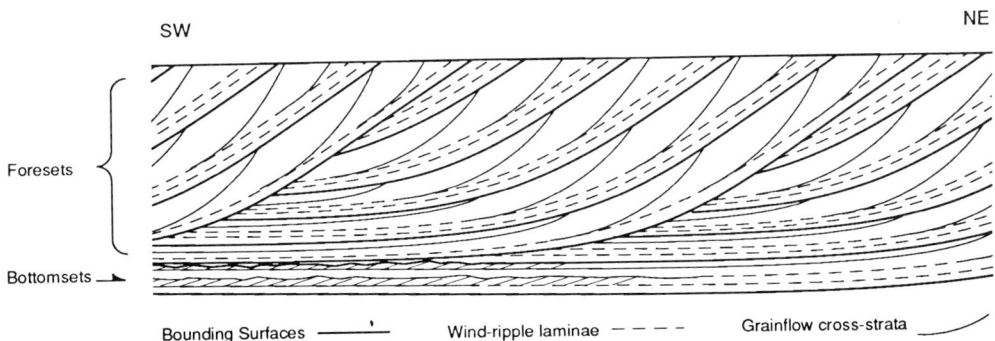

Fig. 13. Schematic diagram illustrating scalloped co-set of cross-strata with internal cyclicity. The basal small sets are separated from the overlying scalloped set by corrugated surfaces; however, traced laterally they become part of the previous scallop.

contact with the wind-ripple laminae. The grain-flow cross-strata are subsequently truncated by the upper bounding surface (Figs 11c–d). The bounding surfaces, wind-ripple laminae, and grainflow cross-strata all have very similar dip directions (e.g. Fig. 2, Unit 8 in section 300, wind-ripple and bounding surfaces = 228°; grainflow = 230°), but the grainflow cross-strata always have a higher dip angle than either the wind-ripple laminae or the bounding surfaces (e.g. Fig. 2, Unit 8 in section 300, grainflow = 24.9°; wind-ripple and bounding surfaces = 21.9°). This angular difference results in the systematic truncation of the grain-flow cross-strata by the upper bounding surface (Fig. 13). The cycles average 30–50 cm in thickness as seen in sections parallel to palaeoflow. These cycles are similar to 'compound cyclic crossbedding' of Hunter & Rubin (1983) in their variance of stratification type, the presence of an erosional bounding surface, and the similarity of dip direction between bounding surface and cross-strata.

In Unit 7 (section 600) (Fig. 2) wind-ripple–grainflow cycles show a distinct variation in wind-ripple laminae thickness (Fig. 11c). These particular exposures show wind-ripple laminae at the base of each cycle as well as associated with the toe of the grainflow packages. A smooth transition can be seen from thin basal wind-ripple laminae to the much thicker wind-ripple laminae associated with grainflow cross-strata, which are truncated by the next bounding surface and capped by the overlying thin wind-ripple laminae of the next cycle (Fig. 11c).

Interpretation

Entrada as a wet aeolian system

The Entrada is interpreted as a wet aeolian system based on both regional and outcrop evidence. At the regional scale, as shown in Figs 5 & 6, aeolian-dominated sections of the Entrada are coeval with marine-sabkha sediments to the north, west and partly to the south. Defining the exact location of a shoreline for any particular time during Entrada deposition is very difficult, because no high energy shoreface or beach facies have been identified in this area. The location of the shoreline can be approximated to the lateral transition from aeolian cross-strata-dominated facies to 'flat'-bed-dominated facies. This transition was never more than 50–75 km to the west and north based upon sections throughout the area (Fig. 5) (Kocurek 1981a). However, on this very low-relief coastal area this transition zone would be

expected to have fluctuated through time, and the predominance of 'flat'-bedded strata in the Medial Division suggests the shoreline for that time was located near and possibly inland of the study area. The stratigraphic architecture for the Entrada and coeval formations (Fig. 6) shows a vertical stacking of aeolian and marine-sabkha facies followed by a general onlapping pattern, indicating rising sea-level through Entrada time. Global sea-level curves of Haq et al. (1987) indicate two phases of sea-level rise during the Callovian, which together show an overall sea-level rise. At the outcrop scale, as already described, features of the sets, their bounding surfaces, and sedimentary structures within the intervening 'flat' strata show the presence of a shallow water table throughout Entrada accumulation. In addition, the presence of accumulations we interpret as interdune flat in origin, is in itself indicative of a wet aeolian system, according to the reasoning of Kocurek & Havholm (1993).

The implication of interpreting the Entrada in the study area as a wet aeolian system, and correlating this with coastal relative sea-level change, is to link the inland water table fluctuations with relative sea-level changes. The extent inland to which the water table can be controlled by sea-level at the coast, and any time lag between sea-level change and inland response, have yet to be determined theoretically and is certain to be influenced by numerous parameters. However, the relative close proximity of the shoreline to the study area, permeable substrate and the postulated arid climate, high evaporation rates and low topographic gradients for the Entrada make this hypothesis reasonable.

Interpretation of genetic packages

To reconstruct the sequence of events that resulted in the Entrada Sandstone it is essential to identify the vertical succession of genetic packages. Genetic package is used here for the aeolian units in the sense of Kocurek (1988) and Kocurek & Havholm (1993) as an aeolian accumulation capped by a super surface, thus forming the basic building blocks of aeolian sequences.

As a bedform migrates, its trough defines a bounding surface, which separates it from underlying strata. For a wet aeolian system where the water table controls accumulation (Kocurek & Havholm, 1993), and a series of bedforms migrate over a substrate where the water table is static and the capillary fringe has reached its maximum potential height, the amount of sediment deposited on the lee faces is balanced by the amount eroded from the stoss slopes, and no net deposition occurs (no accumulation). This scenario is one of bypassing in which all the bounding surfaces are coincident. However, if the relative water table is rising up through the bedforms as they migrate then the volume of sediment deposited on the lee face will be greater than the volume of sediment eroded from the stoss slope, resulting in net deposition (cross-stratified aeolian accumulation). The bedforms, their bounding surfaces, and interdune deposits move upward or climb (Allen 1963; Rubin & Hunter 1982), with the angle of climb determined by the rate of relative water table rise (Kocurek & Havholm, 1993) (Fig. 14a). The classic geometry of climbing bedforms given in Rubin & Hunter (1982) can be more complicated in nature because of spatial variation in the depth of scour over a field of migrating bedforms (Paola & Borgman 1991), and temporal variation in the depth of scour of migrating bedforms.

We suggest that larger features such as ergs, dunefields, and sabkha flats can also be considered to climb if they migrate while the water table rises (Fig. 14b). For example, large sand accumulations (ergs) separated by sabkha flats, and migrating under the influence of a relatively rising water table, would be expected to climb just as do individual dunes and interdune areas. In essence, any scale of feature that migrates while the depositional surface is rising can be expected to produce time-transgressive, climbing accumulations. It then becomes imperative to distinguish cross-stratified units

as representing either the product of simple bedforms or entire dunefields or ergs, and to recognize the origins of the surfaces. It should be noted that the term sabkha will be used in this discussion to refer to low relief areas of clay, silt or sand that are damp and evaporite encrusted (Glennie 1970), without implying areal extent or proximity to shoreline. By this definition we do not mean to imply any particular geomorphic setting, rather we see common depositional processes and resulting facies for a continuum from interdune sabkha areas to broad open sabkha flats.

As the cap for genetic packages, a super surface represents a shift in the sedimentary system from aeolian accumulation to non-aeolian accumulation (facies change), bypass, or erosion. Super surfaces can form in a wet aeolian system during a rising, static or falling water table (Kocurek & Havholm, 1993). A relatively rising water table that ultimately outpaces the aeolian sediment supply must yield a shift from aeolian strata to marine-sabkha units. Super bounding surfaces associated with a static water table are expected to mimic the water table level and be relatively horizontal and exhibit features such as corrugation and polygonal fractures. In the case of a falling water table, the resulting erosion and scouring that occurs can produce a very irregular super surface devoid of any damp sedimentary features; however, if deflation proceeds to the water table, features associated with deflation of a damp surface would be expected.

Following these criteria for identifying genetic packages in wet aeolian systems, four have been identified for the studied interval within the

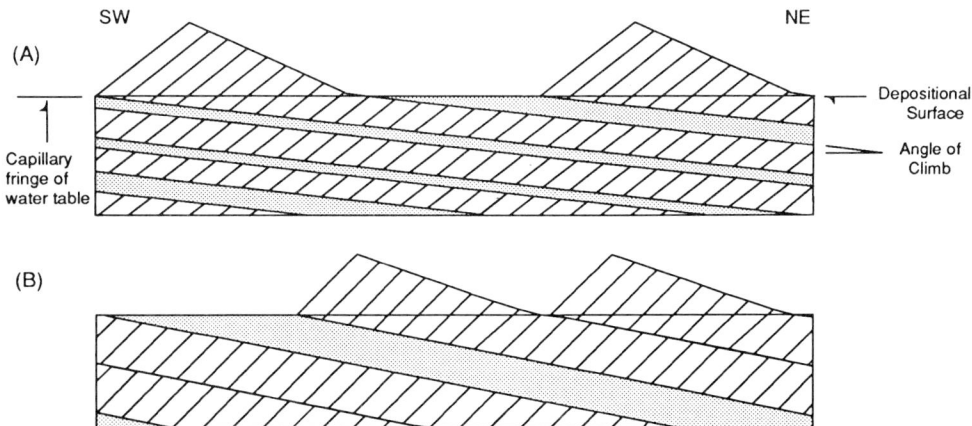

Fig. 14. (a) Wet aeolian system with dunes and interdunes climbing under the influence of a rising water table. (b) Dunefield and sabkha flat accumulations also climbing with rising water table.

Entrada Sandstone of Dinosaur National Monument. The entire Entrada idealized sequence is shown in Fig. 15. Identification of genetic packages here is based upon all the evidence at this extensive outcrop, but ideally a more regional basis for mapping the geometry of accumulations and super surfaces is desired, as in the case of the Page Sandstone in which formation-wide mapping was done (Havholm et al. 1993). In this sense, our interpretations are tentative until a more regional base is established.

Genetic Package 1 (GP1). Genetic Package 1 (Fig. 7), extends from Surface A to G, and consists of cross-stratified Units 1–5 and associated intervening 'flat'-bedded units, all of which are capped by Surface G. The 'flat'-bedded basal unit (Surfaces A to B; Figs 2 & 3) appears to be dominated by vertical accretion in a sandy supratidal or marine sabkha setting. Numerous features indicate this unit was deposited and accumulated under damp or wet conditions (e.g. wavy laminae, contorted bedding and loaded Surfaces A and B). The only features suggesting this package experienced dry conditions are thin wind-ripple laminae; no grainflow cross-strata are seen to indicate that bedforms with slipfaces formed. This basal unit is interpreted to represent the transition from the broad hypersaline intertidal to supratidal (marine sabkha) facies of the Carmel Formation, to the aeolian-

dominated accumulations of the Entrada Sandstone. The loaded nature of surfaces A and B suggests the contacts do not represent significant hiatuses.

The inclination of Units 1–5 and Surfaces C–F is interpreted to result from climb, and these climbing compound cross-strata and 'flat'-bedded units represent the accumulations of dunes and interdune sabkha flats migrating under the influence of a rising water table (Figs 14a & 15). Unit 1 may represent the initial organization of sand into bedforms. The unit is dominated by wind-ripple laminae and thin grainflow cross-strata suggestive of a sand-sheet environment with scattered small bedforms (Kocurek & Nielson 1986). Because Units 2–4 are composed of single sets of cross-strata, they are interpreted to represent the accumulations of individual compound bedforms (Figs 14a & 15). In contrast, Unit 5 is composed of at least two sets of compound cross-strata, and is thought to be the accumulation of at least two compound bedforms. The bounding surface separating these two sets in Unit 5 is distinct from the others in GP1 because it shows no evidence of 'flat'-bedded accumulations or formation under damp conditions. The down-cutting seen along this surface probably relates to temporal variations in the depth of scour of the individual bedforms, perhaps enhanced because the water table no longer served to limit the depth of scouring associated with dune

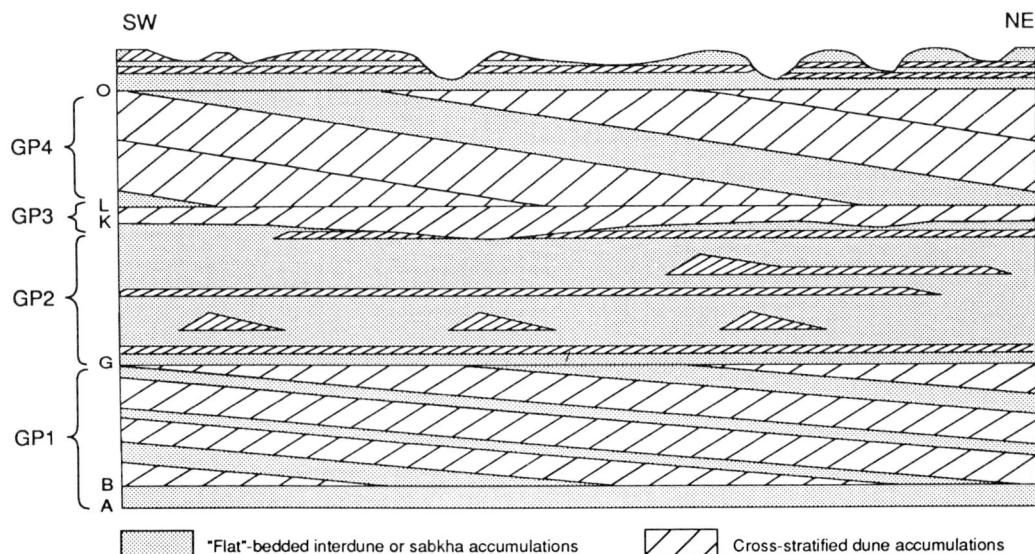

Fig. 15. Idealized vertical section of the Entrada Sandstone traverse. Major bounding surfaces indicated by letters.

migration. The accumulations of Unit 5 suggest a change in the style of deposition to a dry system. If this is correct, then this upward change in GP1 has implications for the dynamics of the system, as we discuss later. Surface G is interpreted as a super surface because it truncates the other inclined surfaces and units, and is overlain by the 'flat'-bed dominated Medial Division. The mathematically defined dip (Table 1) of Surface G is the artifact of drawing a straight line through this irregular surface.

Genetic Package 2 (GP2). Genetic Package 2, extending from Surfaces G to K, is composed of 'flat'-bedded accumulations and thin discontinuous simple and compound cross-strata (Figs 2 & 8). Sedimentary features recognized as typical of the 'flat'-bedded strata are the same as those seen in the thinner 'flat'-bedded units of the Lower and Upper Divisions. GP2 is interpreted to represent the deposits of a broad sabkha flat, which was marked by aeolian bedforms on an irregular basis (Fig. 15).

Evidence for the past presence of evaporites and a shallow water table is abundant in all the 'flat'-bedded strata. The overall fabric of the strata has long been associated with aeolian sabkhas (i.e. Glennie 1970), where the repeated growth and dissolution of the evaporite minerals, both on the surface and within the sediment column, gives rise to wavy to convoluted laminae. Modern sabkha flats are characterized by a depositional surface that is marked by salt polygons, salt ridges and varying amounts of algal mat, giving an inherent irregular waviness to the laminae (Glennie 1970; Fryberger *et al.* 1983, 1984, 1990; Kocurek & Nielson 1986; Schenk & Peterson 1991). Deformation is enhanced with collapse of salt ridges, dissolution of evaporite crystals, nodules and layers, and loading of the spongy substrate by migrating aeolian dunes (Figs 8d & 9a). In addition, polygonal fractures, interpreted elsewhere as fracturing of an evaporite-cemented sand (Kocurek & Hunter 1986), produce deformation within the fractures and in the adjacent host strata by repeated expansion and contraction. Corrugated surfaces are characteristic of deflation of damp or evaporite-cemented substrates (Cooper 1958; Sharp 1966; Carter 1978; Kocurek 1981*b*; Simpson & Loope 1985). Depressions along surfaces in the Entrada seem best interpreted as collapse features, probably resulting from evaporite dissolution (Fig. 10a). Hunter *et al.* (1992) show similar features produced in experiments simulating post-depositional evaporite dissolution. Cases in the Entrada where depressions overlie

polygonal fractures almost certainly originated with dissolution of evaporites within the fracture (Fig. 10b). Some apparently isolated depressions may actually overlie fractures but this relationship is less obvious because the outcrop slopes, whereas the concentric depressions are more vertical. Other isolated depressions may overlie previous evaporite nodules, or result from lateral flowage of underlying beds with evaporite dissolution. Clearly, some beds deformed in a brittle manner resulting in brecciation, and others show complete liquefaction with ball-and-pillow structures, contorted strata, fluid escape features, and flame structures (Figs 9c–d). The depositional surface was also periodically submerged, as evidenced by oscillation ripples, as well as marked by flowing water that gave rise to trains of subaqueous climbing ripples (Fig. 9b).

Emergence and the transport of dry aeolian sand is shown by the presence of wind-ripple laminae as well as the sets of cross-strata. Sets commonly have a loaded lower bounding surface caused by the bedforms migrating across water-saturated sabkha sediments and differentially loading down into the soft sediment. The upper bounding surfaces of the cross-stratified units are corrugated and polygonally fractured. The periodic trapping and burial by sabkha sediments of small aeolian dunes is indicated by the occurrence of the trains of fully preserved bedforms that were effectively 'frozen' in place (Figs 8a–c). Other aeolian sets can be seen to have foundered and become contorted. Because of the short extent of many of the sets, it was not determined whether they show climb.

GP2 is subdivided by Surface I into two wetting-upward cycles (Fig. 2). The two cycles grade from lensoid aeolian cross-strata at the base to 'flat'-bedded sabkha accumulations toward the top. Surface I is not a super surface because it does not mark the end of aeolian accumulation, but rather it appears to be a bedding plane across which sabkha sediments are replaced abruptly by aeolian strata. Similarly, Surface K caps sabkha sediments, but is distinctive in that it shows no features to suggest it formed under the influence of the water table, but rather shows marked deflation of the sabkha accumulations.

Genetic Package 3 (GP3). Between Surfaces K and L (Unit 6; Figs 2 & 3), Genetic Package 3 is unique in that it is composed of a single compound set of cross-strata, representing a large compound bedform that migrated across and

mantled the irregularly scoured Surface K. It is
not clear whether scouring of the substrate along
Surface K is associated with the migration of the
bedform or whether the set merely fills in a pre-
existing irregular scoured surface. It is also not
certain whether initial deposition and accumulation took place under wet or dry conditions.
The internal set surfaces are not corrugated, but
there are a few very thin, wavy-bedded laminae
at the base of Unit 6 in sections 0 and 100,
suggesting deposition under damp conditions
(Fig. 2). The rationale for defining a single set
as a genetic package (GP3) is that over the span
of the outcrop studied, Surface L is irregular, but
horizontal overall, and is interpreted as a super
surface (Figs 2 & 3). The surface itself, because
it is horizontal, corrugated and polygonally fractured is thought to have formed by deflation to a
static water table (Table 1; Fig. 2).

Genetic Package 4 (GP4). The final genetic
package addressed here (GP4; Surfaces L to O)
is composed of two thick cross-stratified units
(Units 7 and 8) separated by a 'flat'-bedded unit,
all of which are truncated by horizontal surface
O (Figs 2, 10d, 11a–d). The 'flat'-bedded unit
separating Units 7 and 8 is the result of vertical
accretion on a sabkha flat or interdune sabkha
that occasionally had small bedforms and wind-
ripples moving across it leaving thin deformed
accumulations. Because Units 7 and 8 climb
up from the same horizontal surface (Surface L)
they are probably contemporaneous features
(Fig. 2). At least two scenarios could explain the
relationship of Units 7 and 8 and the associated
'flat' bed. Because Unit 7 is composed of two
sets of compound cross-strata, it is possible to
consider it as the accumulation of at least two
separate bedforms representing a dunefield.
GP4 taken as a whole, therefore, could represent the accumulations of two climbing dunefields separated by a sabkha flat (Fig. 14b).
Alternatively, Units 7a, 7b and 8 could each
represent separate bedforms that were not
evenly spaced areally, and varied in their depth
of scour (Paola & Borgman 1991). This would
suggest that there was no interdune flat *per se*
between Units 7a and 7b, but a somewhat damp
interdune depression is indicated by the corrugated surface, while between Units 7b and 8 an
interdune sabkha flat existed. If these bedforms
were sinuous, variations in the size of the interdune flats separating the bedforms would be
expected along any single transect.

Bedform interpretation. Given the interpretations above and the nature of the cross-strata
described previously, a certain number of infer-

ences can be made regarding the bedforms,
including their size, nature of cyclicity, general
migration direction and overall geometry.
Assuming the deposits of GP1 result from
climbing dunes and interdune areas, the angles
created by the intersection of Surfaces C, D,
E and F with Surface B represent the angles
of climb; likewise for GP4 Surfaces M and N
reflect the angle of climb for those dunes and
interdune areas. For GP1 the angle of climb in
the line of section averages 0.16°, and for GP4
it averages 0.30° (Table 1). Orientation of the
mathematically defined planes from the horizontal in their maximum dip direction gives a
range of 0.2–1.0°, with a mean of 0.53° (Table 1).
Estimates of dune size are made based on dune
spacings and an assumed bedform index of 20 : 1
(ratio of dune spacing to dune height; Kocurek
et al. 1992). The dune spacing was estimated
for Units 3 and 4 by projecting mathematical
planes C, D, E and F to their intersection point
with Surface B (Fig. 3). The dune spacing averages 900 m resulting in a dune height estimate of
45 m using the assumed bedform index of 20 : 1.

The large-scale cyclicity, represented by the
scalloped cross-strata and climbing bedform
structures, is the result of the migration of
superimposed bedforms on the lee face of a
larger bedform. Superimposed bedforms and
scour pits migrating along the lee face of a larger
bedform are expected to scour into the face of
the larger bedform creating a bounding surface
(Rubin & Hunter 1983). In exposures that
roughly parallel the migration direction of the
main bedform this bounding surface is expected
to truncate the cross-strata of the previous
superimposed bedform cross-strata at a high
angle thus creating the scalloped boundary
(Rubin & Hunter 1983) (Fig. 4). The same
bounding surface in outcrops perpendicular
to migration direction delineates the climbing
bedform structures (Fig. 4). In outcrops of the
Entrada trending roughly E–W the scalloped
structure is clearly seen (Figs 7d, 11a, 12a & b),
and in outcrops trending roughly N–S the
climbing bedform structures occur (Figs 10c &
12c). Bedform trends, determined as outlined in
the methods section, indicate the main bedforms
were migrating generally to the WSW and the
superimposed bedforms were migrating along-
slope to the SE and to the NW (Table 2). Therefore, the traverse line of N65E (Fig. 1) roughly
parallels the migration direction of the main
bedforms. The migration direction of the main
bedforms is reasonable given the NE to NNE
predicted palaeowind directions of Parrish &
Peterson (1988) (Fig. 5). The apparent alternating migration directions of the superimposed

Table 2. *Bedform migration directions determined using the stereonet method of Rubin & Hunter (1983)*

Cross-stratified sets bounded by surfaces:	Main bedform migration direction in degrees	Superimposed bedform migration direction in degrees
Unit 3	W*	SW*
Unit 4	230	330
Unit 5	191	322
Unit 6	WSW*	SSE*
Unit 7	268	182
Unit 8	219	298

* Insufficient data to define more precisely migration direction.

bedforms may reflect wind variations over time, or the complex geometry of the main bedform (Table 2).

The angle of the scalloped bounding surface may vary as a function of the angle between the main bedform trend and the superimposed bedform trend. Rubin's (1987) bedform simulations indicate that superimposed bedforms migrating along the lee face parallel to the crest of the main bedform have high-angle bounding surfaces (i.e. Unit 8, *c.* 23°; Figs 2, 12a & 16a); whereas, those migrating at more oblique angles have much lower-angle bounding surfaces (i.e. Unit 3, *c.* 11°; Figs 2, 12b & 16b).

A second order of cyclicity recognized in these sets is represented by the wind-ripple–grainflow couplets and their truncating bounding surfaces (Figs 11c–d). These cycles are interpreted to represent periodic fluctuations in the wind regime with respect to the incidence angle of the wind to the superimposed dune crestline (see Kocurek 1991). The transition from one wind regime to the other can be seen as the gradual change through a cycle from thin wind-ripple laminae, to grainflow and thick wind-ripple laminae, followed by a truncating surface (Fig. 11c). Grainflow strata and downslope, thick wind-ripple laminae probably correspond to periods of a high incidence angle when a slip-face was developed, but where some along-slope transport still occurred. Thin wind-ripple laminae and the absence of grainflow strata correspond to times of more oblique incidence angles and pronounced along-slope transport. These wind-ripple–grainflow cycles are presumed to represent annual fluctuations in the wind directions because the amount of migration implied by each is reasonable for that scale of time. Similar features are described by Hunter & Rubin (1983) for annual cycles in the Navajo. Because these are interpreted as annual cycles, annual migration rates can be estimated. The cycles average 50 cm and show an approximately 3 : 1 ratio of grainflow to wind-ripple laminae. Annual fluctuations in wind directions would be consistent with monsoonal wind patterns postulated by Parrish & Peterson (1988) (Fig. 5). Parrish & Peterson (1988) also

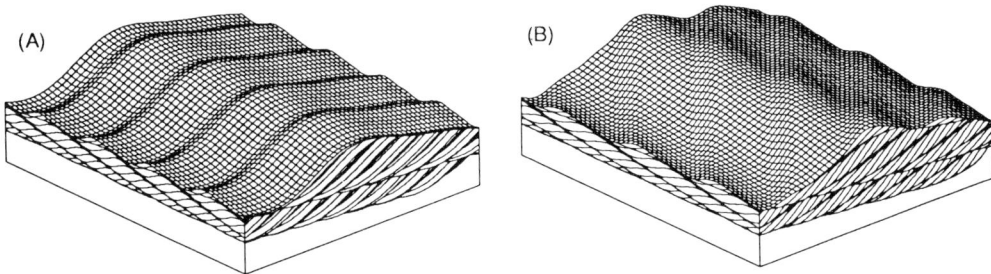

Fig. 16. The angle of the scalloped bounding surfaces (right side of block diagrams) can vary as a result of the angular difference between the main and superimposed bedform migration directions. (a) Higher-angle scalloped cross-strata with along-crest-migrating superimposed bedforms. (b) Lower-angle scalloped cross-strata with superimposed bedforms migrating 45° off the migration direction of the main bedform.

predicted stronger summer winds than winter winds. This would suggest that more sediment was transported during the summer, which would presumably correspond to the thicker grainflow cross-strata in these cycles. However, detailed reconstructions relating specific sedimentation patterns to predicted wind patterns is speculative, because all of the cross-strata studied in these outcrops represent sedimentation that took place on the superimposed bedforms and are, therefore, the result of the secondary airflow patterns on the lee face of the main bedform.

Several units are composed of compound co-sets (Units 3, 6, 7 and 8) (Figs 11b & 12a). The basal sets of these compound co-sets are the erosional remnants of smaller superimposed bedforms that moved along the toe or plinth of the main bedform. Similar features are described by Clemmensen & Blakey (1989) from outcrops of the Jurassic Wingate Sandstone. The contemporaneous nature of these bedforms with the larger superimposed bedforms is demonstrated by the lateral continuity of these sets with the overlying scalloped cross-strata (Fig. 12a). The corrugated surfaces or surfaces showing microtopography, which separate the basal sets vertically, result from differential wind erosion of damp or evaporite-cemented foresets. This suggests the capillary fringe of the water table was within the bedforms superimposed on the plinth of the main bedform, indicating the bedforms were migrating across a damp surface or interdune floor.

The overall picture for the Entrada bedforms is of relatively large bedforms with two scales of superimposed, roughly along-slope-migrating bedforms. Larger superimposed bedforms migrated along the lee, while smaller satellite bedforms occurred at the lee base and periodically extended into the damp interdune flats. Complex, seasonally varying winds are indicated. The data available allow modelling of the superimposed bedforms, and to a certain degree the orientation of the lee faces of the main bedforms upon which they are migrating. However, the overall shape of the main bedform is not known.

Discussion of controls on facies architecture

Recognition of genetic packages comprising the Entrada Sandstone, along with the theoretical inferences under which wet aeolian systems are believed to form (Kocurek & Havholm 1993), allow the controls on the facies architecture to be approached. Because wet aeolian systems reflect both hydrodynamic and aerodynamic controls, the available sand supply (dry, loose sand available for transport in bedforms) is considered to be a direct function of the rate of sediment supply (sand supplied to the area from the source area plus any sand liberated from the substrate) and the rate of water table change (relative change of water table with respect to the sediment column). For example, where the sediment supply is roughly balanced by the rate of the water table rise, there is little dry sand on the surface at any given time for dune building. Such a substrate would characterize a sabkha or a wet-damp sand sheet. Conversely, where the sediment supply exceeds the water table rise, then dry sand exists on the surface, and it is assumed from aerodynamic considerations that this sediment will be incorporated into bedforms to a large extent (see Kocurek & Havholm 1993). Hence, the surface coverage by dunes and interdune sabkhas or sabkha flats is a reflection of the available sand supply. From the modelling of Rubin (1987), the thickness of the dune and interdune accumulations approximates their surface coverage where the depositional surface is flat. From the Entrada traverse (Fig. 2), therefore, inferences of both dune/interdune surface coverage and available sediment supply can be made.

Because available sediment supply is the dependent variable controlled by the independent variables of rates of sediment supply and water table change, the overall relationship can be expressed in alternate ways (Figs 17a–b). Either sediment supply rate or the rate of water table change can be held constant, while the other independent variable fluctuates to reflect the available sand supply as determined from the rock record. In truth, neither can be assumed to be constant, and evidence from the outcrop can be used to better specify the conditions that occurred. It is interesting to note, however, that if the sediment supply is assumed constant (Fig. 17b), then the resulting rate of water table change becomes a relative sea-level curve. This is the consequence of considering the water table changes here to be the inland response to relative sea-level changes occurring on the coast.

Using the above approach, GP1 probably shows a gradual increase in the available sand supply through time, because sabkha accumulations at the base give way upward to wet system dune/interdune accumulations, (Surfaces A to B) then to the possible dry aeolian system in Unit 5 (Figs 2 & 17). The shift reflects increasing sand available and increasing areal coverage by dunes at the expense of sabkha and interdune flats. Surface G, because it is interpreted as a super surface, must reflect a change from a water table that was rising to one that is static or, more likely, falling, judging by the irregular nature of the surface. This change in the water table ends accumulation of GP1 and allows deflation of the surface. Downcutting of Unit 5a into 5b may reflect an already falling water table. Because no accumulation is preserved, there is no recorded information regarding the available sand supply during the time interval represented by Surface G. Deflation to Surface G also defines the baseline of erosion, hence, as with all wet systems, the rising water table both creates accumulation space and creates preservation space, in the terminology of Kocurek & Havholm (1993).

A significant change occurs in the depositional environment above Surface G, as indicated by the dominance of sabkha accumulations and the relative paucity of aeolian cross-strata in GP2. This genetic package overall reflects a significant decrease in the available sand supply from GP1, whereas fluctuations in the available sand supply are shown by the two wetting-upward cycles separated by Surface I (Fig. 17).

GP2 is capped by erosional Surface K, which is the only surface in the entire sequence where there is little evidence that deflation proceeded to the water table. Because the water table was near the surface when the underlying sabkha sediments formed, a drop of the water table of at least 3 m must have occurred during the formation of Surface K (Fig. 17), based on the depth of downcutting seen along the surface (Fig. 2). The sediment released by the deflation on Surface K may have served as an internal sediment supply for the accumulation that immediately overlies it (GP3). Super Surface L, showing damp-surface features, reflects a rise in the water table to a new base level (Fig. 17).

GP4 (Units 7 and 8) contains Unit 7, which shows an internal corrugated surface separating sets 7a and 7b, and sabkha sediments between Units 7 and 8, indicating that it formed under the influence of the capillary fringe of the water table. The preservation of GP4 occurs with the baseline of erosion defined by Surface O. Ultimately, Entrada accumulation ceases with onlapping of the marine Curtis Formation, the response to a sea-level rise that greatly outpaced the sediment supply (Fig. 17).

Lastly, because in wet aeolian systems the rate of accumulation is determined by the rate of the relative water table rise, where an estimate of time can be made for a vertical section, the accumulation rates and water table rise can be estimated. For example, for the Lower Division, the record of six bedforms of approximate spacing of 900 m are interpreted, and the average

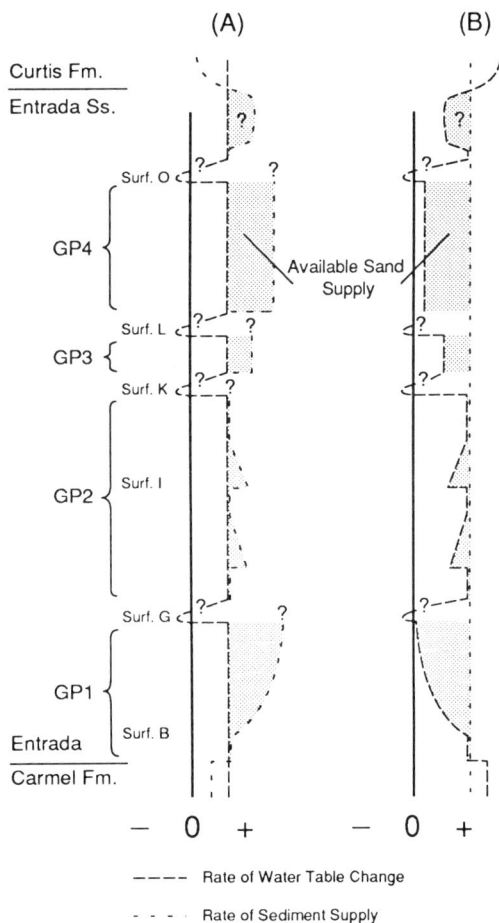

Fig. 17. Two models showing available sand supply as a function of water table and sediment supply throughout depositional history of Entrada. (a) The rate of the water table rise is constant, except at the super surfaces where the water table falls. (b) The rate of the sediment supply is constant. GP1–GP4, with major surfaces indicated. Vertical scale is stratigraphic thickness, except at the lettered bounding surfaces where scale is arbitrary. See text for discussion.

forward migration of the main bedforms determined from the annual cycles is 0.4 m a^{-1}. A length of time, therefore, of 13 500 years is implied. Because this assumes that all bedforms have been measured and none have been removed by spatial or temporal changes in the depth of scour, this time estimate is a minimum. The vertical section is about 15 m, yielding an average accumulation rate for the Lower Division of 1.1 mm a^{-1}. If the water table rise reflects the relative sea-level change on the coast, which in turn reflects eustatic sea-level rise plus subsidence, then the accumulation rate is also the relative rate of sea-level rise. From the small-scale interfingering of 'flat' beds and bottomsets seen at the base of some co-sets, accumulation rates can also be determined. For example, in Unit 8 (section 200, Fig. 11d) thin basal sets can be traced laterally back to the annual cycle to which they are related. The distance between the base of the basal set and the base of the main set represents a water table rise. Counting the number of cycles between the cycle related to the basal set, and the set directly overlying it, gives an estimate of time. By combining these two measurements it is found that for Unit 8 the water table rose approximately 8.0 mm a^{-1}. Higher rates are to be expected as reflecting more short-term changes than the longer-term average for the entire division.

Conclusions

1. Extensive surveying and correlation of well-exposed outcrops in northeastern Utah have made it possible to resolve the orientation of bounding surfaces, and to reconstruct very precisely the facies architecture of a wet aeolian system.
2. The Entrada is interpreted as a wet aeolian system because of the preservation of climbing dune and interdune sabkha accumulations, and a variety of damp and wet features in the 'flat' strata and the cross-strata.
3. The bedforms that formed these accumulations were large features with two scales of superimposed dunes. The larger superimposed dunes formed the prominent scalloped cross-strata as they migrated across the lee of the main bedform. The smaller satellite dunes migrated along the damp toe of the lee leaving thin, deformed sets of cross-strata with corrugated upper bounding surfaces. Internally the scalloped sets showed another level of cyclicity (wind-ripple–grainflow couplets) resulting from annual fluctuations in wind direction.
4. The facies architecture, as documented in this study of a wet aeolian system, reflects the available sand supply, or the dry sand available to be transported in bedforms. The available sand supply is a function of the interplay between relative water table level and sediment supply from the source area. The close proximity of the Entrada to the palaeoshoreline and regional stratigraphic

relationships argue for sea-level fluctuations as the underlying cause of the Entrada facies architecture.

We would like to thank British Petroleum for generously funding this project, and for Richard Steele's continued support and interest. Funding was also provided by the Geology Foundation of University of Texas at Austin. We would also like to thank Roberto Gutierrez as well as Andy Frank for the many hours of help in creating a computer program to define the mathematical planes. A large debt is owed to Jeff Crabaugh and especially Andy Frank for tolerating bugs and heat in Dinosaur National Monument. Finally we would like to acknowledge the assistance of Lars Clemmensen, David Loope and Karen Havholm in reviewing and improving the manuscript.

References

ALLEN, J. R. L. 1963. The classification of cross-stratified units with notes on their origin. *Sedimentology*, **2**, 93–114.

ALLMENDINGER, R. 1989. *Stereonet v. 4.0-II*. Computer program distributed by R. Allmendinger, Cornell University, Ithaca, New York.

BLAKEY, R. C., PETERSON, F. & KOCUREK, G. 1988. Synthesis of late Paleozoic and Mesozoic eolian deposits of the Western Interior of the United States. *In*: KOCUREK, G. (ed.) Late Paleozoic and Mesozoic Eolian Deposits of the Western Interior of the United States. *Sedimentary Geology*, **56**, 3–125.

CARTER, R. W. G. 1978. Ephemeral sedimentary structures formed during aeolian deflation of beaches. *Geological Magazine*, **115**, 379–382.

CLEMMENSEN, L. B. & BLAKEY, R. C. 1989. Erg deposits in the Lower Jurassic Wingate Sandstone, northeastern Arizona: oblique dune sedimentation. *Sedimentology*, **36**, 449–470.

COOPER, W. S. 1958. *Coastal sand dunes of Oregon and Washington*. Geological Society of America, Memoir **72**.

DEYNOUX, M., KOCUREK, G. & PROUST, J.-N. 1989. Late Proterozoic periglacial aeolian deposits on the West African Platform, Taoudeni Basin, western Mali. *Sedimentology*, **36**, 531–549.

ESCHNER, T. B. & KOCUREK, G. 1986. Marine destruction of eolian sands seas: origin of mass flows. *Journal of Sedimentary Petrology*, **56**, 401–411.

FRYBERGER, S. G., AL-SARI, A. M. & CLISHAM, T. J. 1983. Eolian dune, interdune, sand sheet, and siliciclastic sabkha sediments of an offshore prograding sand sea, Dhahran area, Saudi Arabia. *American Association of Petroleum Geologists Bulletin*, **67**, 280–312.

——, ——, ——, RIZVI, S. A. R. & AL-HINAI, K. G. 1984. Wind sedimentation in the Jafurah Sand Sea, Saudi Arabia. *Sedimentology*, **31**, 413–431.

——, KRYSTINIK, L. F. & SCHENK, C. J. 1990. Tidally flooded back-barrier dunefield, Guerrero Negro area, Baja California, Mexico. *Sedimentology*, **37**, 23–43.

GLENNIE, K. W. 1970. *Desert Sedimentary Environ-*

ments. Developments in Sedimentology, **16**, Elsevier, Amsterdam.

HAQ, B. U., HARDENBOL, J., VAIL, P. R. 1987. Chronology of fluctuating sea levels since the Triassic. *Science*, **235**, 1156–1165.

HAVHOLM, K. G., BLAKEY, R. C., CAPPS, M., JONES, L. S., KING, D. D. & KOCUREK, G. 1993. Eolian genetic stratigraphy: an example from the Middle Jurassic Page sandstone, Colorado Plateau. *In*: PYE, K. & LANCASTER, N. (eds) *Aeolian Sediments: Ancient and Modern*. International Association of Sedimentologists Special Publication, **16**.

HUNTER, R. E., GELFENBAUM, G. & RUBIN, S. M. 1992. *Clastic Pipes of Probable Solution-Collapse Origin in Jurassic Rocks of the Southern San Juan Basin, New Mexico*. United States Geological Society Bulletin **1808-L**.

—— & RUBIN, D. M. 1983. Interpreting cyclic crossbedding, with an example from the Navajo Sandstone. *In*: BROOKFIELD, M. E. & AHLBRANDT, T. S. (eds) *Eolian Sediments and Processes*. Developments in Sedimentology, **38**, Elsevier, Amsterdam, 429–454.

IMLAY, R. W. 1980. *Jurassic paleobiogeography of the conterminous United States in its continental setting*. United States Geological Survey Professional Paper **1062**.

KOCUREK, G. 1980. *Significance of bounding surfaces, interdune deposits, and dune stratification types in ancient erg reconstruction*. PhD Thesis, University of Wisconsin.

—— 1981*a*. Significance of interdune deposits and bounding surfaces in eolian dune sands. *Sedimentology*, **28**, 753–780.

—— 1981*b*. Erg reconstruction: the Entrada Sandstone (Jurassic) of northern Utah and Colorado. *Palaeogeology, Palaeoclimatology, Palaeoecology*, **38**, 125–153.

—— 1988. First-order and super bounding surfaces in eolian sequences—bounding surfaces revisited. *In* KOCUREK, G. (ed.) Late Paleozoic and Mesozoic Eolian Deposits of the Western Interior of the United States. *Sedimentary Geology*, **56**, 193–206.

—— 1991. Interpretation of ancient eolian sand dunes. *Annual Review of Earth and Planetary Sciences*, **19**, 43–75.

—— & DOTT, R. H. 1983. Jurassic paleogeography and paleoclimate of the central and southern Rocky Mountain region. *In*: REYNOLDS, M. W. & DOLLY, E. D. (eds) *Mesozoic Paleogeography of the West-Central United States*. Rocky Mountain Paleogeography Symposium **2**, Society of Economic Paleontologists and Mineralogists, 101–118.

—— & FIELDER, G. 1982. Adhesion structures. *Journal of Sedimentary Petrology*, **52**, 1229–1241.

— —& HAVHOLM, K. 1993. Eolian sequence stratigraphy—a conceptual framework. *In*: WEIMER, P. & POSAMENTIER, H. (eds) *Recent Advances in and Applications of Siliciclastic*

Sequence Stratigraphy. American Association of Petroleum Geologists Memoir (in press).

—— & HUNTER, R. E. 1986. Origin of polygonal fractures in sand, uppermost Navajo and Page Sandstone, Page, Arizona. *Journal of Sedimentary Petrology*, **56**, 895–904.

——, KNIGHT, J. & HAVHOLM, K. 1991. Outcrop and semi-regional three-dimensional architecture and reconstruction of a portion of the eolian Page Sandstone (Jurassic). *In*: MIALL, A. & TYLER, N. (eds) *Three-Dimensional Facies Architecture*. Society of Economic Paleontologists and Mineralogists Atlas, 25–43.

—— & NIELSON, J. 1986. Conditions favorable for the formation of warm-climate aeolian sand sheets. *Sedimentology*, **33**, 795–816.

——, TOWNSLEY, M., YEH, E., SWEET, M. & HAVHOLM, K. 1992. Dune and dunefield development stages on Padre Island, Texas: effects of lee airflow and sand saturation levels and implications for interdune deposition. *Journal of Sedimentary Petrology*, **62**, 622–635.

LOOPE, D. B. 1985. Episodic deposition and preservation of eolian sands: a late Paleozoic example from southeastern Utah. *Geology*, **13**, 73–76.

O'SULLIVAN, R. B. 1980. Stratigraphic sections of Middle Jurassic San Rafael Group and related rocks from the Green River to the Moab area in east-central Utah. *United States Geological Survey Miscellaneous Field Studies Map* MF-1247.

—— & PIERCE, F. W. 1983. *Stratigraphic diagram of Middle Jurassic San Rafael Group and Associated Formations from the San Rafael Swell to Bluff in southeastern Utah*. United States Geological Survey, Oil and Gas Investigation Chart OC-119.

PAOLA, C. & BORGMAN, L. 1991. Reconstructing random topography from preserved stratification. *Sedimentology*, **38**, 553–565.

PARRISH, J. T. & CURTIS, R. L. 1982. Atmospheric circulation, upwelling, and organic rich rocks in the Mesozoic and Cenozoic Eras. *Palaeogeography, Palaeoclimatology, Palaeoecology*, **40**, 31–66.

—— & PETERSON, F. 1988. Wind directions predicted from global circulation models and wind directions determined from eolian sandstones of the western United States—A comparison. *In*: KOCUREK, G. (ed.) Late Paleozoic and Mesozoic Eolian Deposits of the Western Interior of the United States. *Sedimentary Geology*, **56**, 261–282.

PETERSON, F. 1988. Pennsylvanian to Jurassic eolian transportation systems in the western United States. *In*: KOCUREK, G. (ed.) Late Paleozoic and Mesozoic Eolian Deposits of the Western Interior of the United States. *Sedimentary Geology*, **56**, 207–260.

RUBIN, D. M. 1987. *Cross-bedding, Bedforms and Paleocurrents*. Society of Economic Paleontologists and Mineralogists Concepts in Geology, **1**.

—— & HUNTER, R. E. 1982. Bedform climbing in theory and nature. *Sedimentology*, **29**, 121–138.

—— & —— 1983. Reconstructing bedform assemblages from compound cross-bedding. *In*: BROOKFIELD, M. E. & AHLBRANDT, T. S. (eds) *Eolian Sediments and Processes*. Developments in Sedimentology, **38**, Elsevier, Amsterdam, 407–427.

SCHENK, S. K. & PETERSON, F. 1991. Eolian sabkha sandstones in the Nugget Sandstone (Jurassic), Vernal Area, Utah. Rocky Mountain Section Meeting, Abstract. *American Association of Petroleum Geologists Bulletin*, **75**, 1139.

SHARP, R. P. 1966. Kelso Dunes, Mojave Desert, California. *Geological Society of America Bulletin*, **77**, 1045–1074.

SIMPSON, E. L. & LOOPE, D. B. 1985. Amalgamated interdune deposits, White Sands, New Mexico. *Journal of Sedimentary Petrology*, **55**, 361–365.

TALBOT, M. R. 1985. Major bounding surfaces in aeolian sandstones — a climatic model. *Sedimentology*, **32**, 257–265.

Fluvial–aeolian interactions in a Proterozoic alluvial plain: example from the Mancheral Quartzite, Sullavai Group, Pranhita-Godavari Valley, India

TAPAN CHAKRABORTY & A. K. CHAUDHURI

Geological Studies Unit, Indian Statistical Institute, 203 BT Road, Calcutta-700 035, India

Abstract: The coarse-grained alluvial succession of the Proterozoic Mancheral Quartzite encloses within it a few 2–6 m-thick, typically salmon red, fine-grained, very well-sorted sandstone units. Some 25–40% of the fine-grained sandstone units is aeolian — mostly adhesion laminae with subordinate adhesion cross-laminae and wind-ripple strata. The remainder of the sequence is aqueous; either massive or with faintly-developed trough cross-bedding. Based on their grain size, sorting and roundness, the bulk of the aqueous deposits appear to be reworked aeolian deposits. Thick adhesion laminated units within them show a number of superimposed drying-upwards sequences, represented by an upward decrease in laminae spacing within each of the sequences. Individual drying-upwards sequences, 5–15 cm thick, are in places bounded by disconformity surfaces marked by iron crusts.

Stratigraphic relationships with associated fluvial facies indicate that these aeolian sandstones formed in the distal part of the floodplain. An arid climate, vegetation-free landscape, quickly avulsing channel behaviour and rapid basin subsidence favoured development and preservation of these rather thick alluvial plain aeolian deposits. Fluvial dynamics, in contrast to the erg dynamics as interpreted from other interlayered aeolian–fluvial deposits, was the primary control on the depositional features and their internal organization within these aeolian sandstone units.

Much attention has been devoted in recent years to the study of the dynamics and products of interacting fluvial and aeolian systems (Sneh 1983; Langford 1989; Langford & Chan 1989; Clemmensen *et al.* 1989). However, all of these studies focus on the erg-margin situation, where an aeolian sand sea provides the background sediment and erg dynamics exert an overall control on the depositional facies and their organization. Exposed parts of the bar, sandflat or floodplain in the sand-dominant fluvial system are also commonly subjected to wind reworking during the periods of low flow (Rust 1972; Cant & Walker 1978; Bluck 1979; Tunbridge 1984; Shepherd 1987). Preservation of these aeolian deposits interlayered with the channel deposits is, however, rare and when preserved they tend to be very thin. In an arid climatic setting or in a pre-Silurian vegetation-free landscape, wind reworking of the exposed parts of the sandy, braided alluvial tracts may become significant and their products, if preserved, become a useful tool for environmental analysis.

This paper reports several well-sorted, fine-grained aeolian sandstones, 2 to more than 6 m thick, and interlayered with conglomeratic to pebbly, coarse-grained sandstones of fluvial origin, in the Proterozoic Mancheral Quartzite of the Pranhita-Godavari Valley, India. It also discusses the origin of the aeolian sandstones and attempts to focus on the dynamics of fluvial–aeolian interactions in an alluvial plain setting.

Geological setting

The Pranhita-Godavari Valley is one of the major Proterozoic basins of India which exposes two linear belts of Proterozoic rocks flanking an axial belt of Permo-Jurassic Gondwana rocks (Fig. 1). The Sullavai Group is the youngest group of the Proterozoic sequence exposed in this basin (Table 1). The Mancheral Quartzite, a constituent of the Sullavai Group, is very well exposed around Mancheral and Ramgundam in the southwestern Proterozoic belt (Fig. 1). The Quartzite unconformably overlies the rocks of the Middle Proterozoic Pakhal Group and is overlain by the Venkatpur Sandstone (Table 1). In and around Mancheral, the fluvial sequence of the Mancheral Quartzite gradationally passes upward through a transition zone of flat-bedded sabkha-playa sediments into the erg sequence of the Venkatpur Sandstone (Chakraborty 1991a).

From Pye, K. (ed.), 1993, *The Dynamics and Environmental Context of Aeolian Sedimentary Systems.*
Geological Society Special Publication No. 72, pp. 127–141.

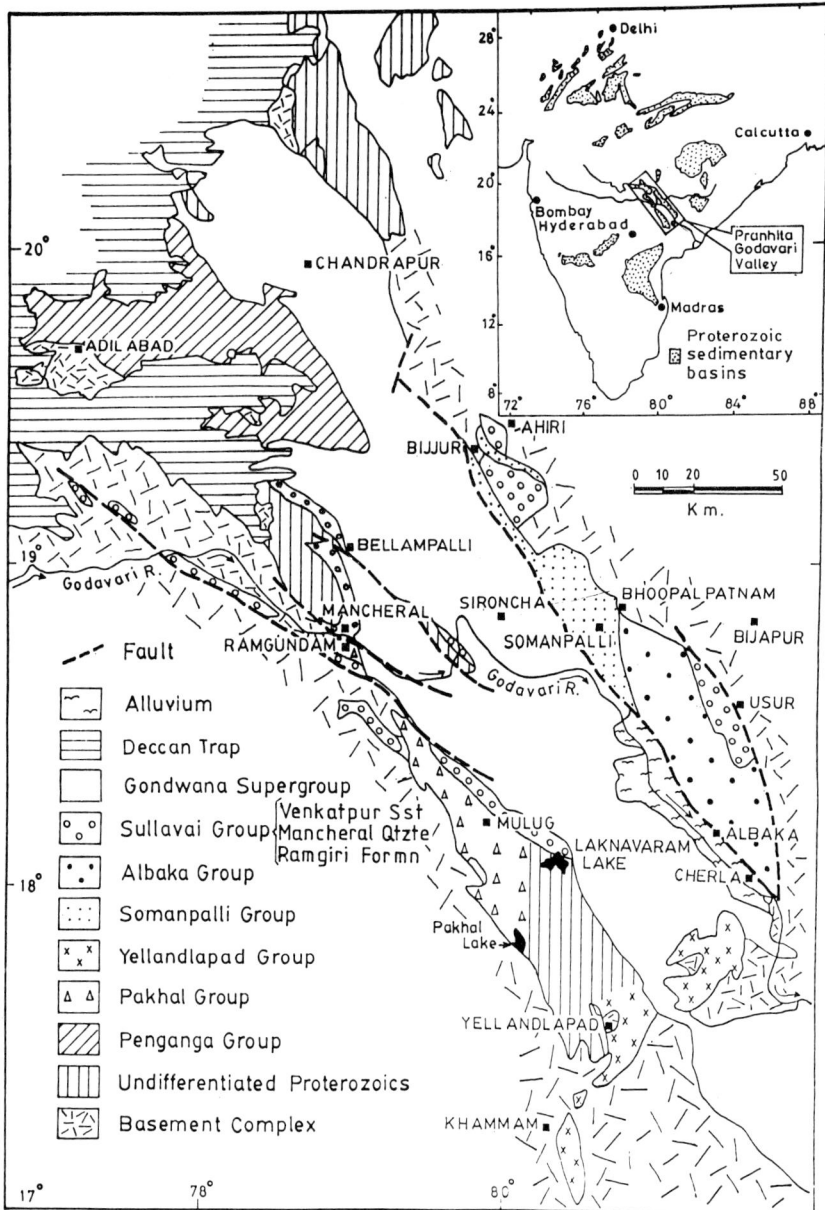

Fig. 1. Generalized geological map of the Pranhita-Godavari Valley. Inset map shows the major Proterozoic basins of India.

The Mancheral Quartzite comprises poorly-sorted, coarse-grained pebbly sandstones and minor conglomerates. Typically salmon red, fine- to very fine-grained sandstones occur at different stratigraphic levels and comprise a subordinate part of the Quartzite sequence. Several large-scale, concave-upward erosional surfaces have been recognized within the Mancheral Quartzite (Figs 2 & 3) which are overlain by fining-upwards sequences (Fig. 4). Some of these fining-upwards sequences are capped by 2–6 m thick salmon red, fine-grained sandstone units with abundant adhesion laminae. The sandstones and conglomerates of the Mancheral

Table 1. *Generalized stratigraphic sequence of the Proterozoic rocks (Sullavai Group) in the study area*

		Lithology	Depositional environment
Gondwana Supergroup (Permo-Jurassic)			
— Unconformity —			
	Venkatapur Sandstone (70 m)	Red, fine- to medium-grained, well-sorted sandstone with large planar cross-beds and flat beds; abundant aeolian strata.	Erg and erg-margin deposit
Sullavai Group (871 ± 14 Ma)	Mancheral Quartzite (50 m)	Mauve to deep red, coarse-grained to pebbly sandstone and conglomerate; F–U sequences; thin fine-grained sandstone units	Braided fluvial with inter-layered aeolian units
	Ramgiri Formation (456 m)	Red, conglomerate and pebbly arkose; conglomerates reverse graded; thin sheet-like sedimentation units.	Distal alluvial fan-braided river.
— Unconformity —			
Pakhal Group/Penganga Group (1330 ± 53 Ma)			
— Unconformity —			
Archaean (?) Granite			

P R O T E R O Z O I C

Fig. 2. Part of a large-scale concave-upward erosional surface in the Mancheral Quartzite. The surface sharply cuts into the thinly bedded fine-grained aeolian sandstone unit.

Quartzite are interpreted to have been deposited from high-gradient braided streams (Chakraborty 1988, 1991b). The broad facies recognized within the formation and their interpretation is shown in Table 2. The genesis of the salmon red, fine-grained sandstone units is discussed below.

Salmon red, fine-grained sandstones of the Mancheral Quartzite

The sandstone units consist of well-sorted and well-rounded, fine- to very fine-grained quartz sands (Fig. 5). The sandstones are generally 2–6 m thick but may attain a thickness of 12 m. The individual sandstone units are traceable along-strike for a distance of a few tens of metres to several hundreds of metres (Fig. 3). They are invariably overlain erosively by coarse-grained channel-lag conglomerates. The sandstone units either gradationally overlie or laterally inter-tongue with small-scale, trough cross-bedded coarse- to medium-grained sandstones (Facies IV in Table 2; Fig. 4) that are interpreted to have been deposited in the higher topographic levels of the channel complex or in the proximal part of the floodplain.

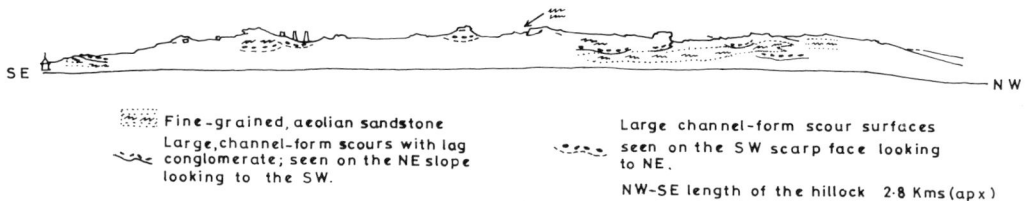

Fig. 3. Line drawing prepared from a photomosaic of the panoramic view of the Ramgundam Gutta (Hill), showing the position of several large-scale channel surfaces and the aeolian sandstone units.

Facies

The following facies have been recognized in the salmon red, fine-grained sandstones.

Fine-grained sandstone with adhesion laminae (F1)

The well-sorted, fine-grained sandstone of this facies is characterized by well-developed adhesion plane beds (Fig. 6). The facies units are 6–8 cm thick and grade rapidly laterally or vertically into massive or faintly developed trough cross-bedded sandstone units (F4, see below). Adhesion laminae are horizontal or in places gently inclined. Individual lamina thickness varies from less than 1 mm to 3 mm.

Adhesion laminated facies show internal, low-angle, bedding-parallel discordant surfaces which bound 5–15 cm-thick packages of adhesion plane beds (Fig. 7). Adhesion plane beds within such packages show a distinct upward-thinning of the laminae (Figs 7 & 8).

The adhesion-laminated facies comprise 20–35% of the fine-grained, salmon red sandstone units of the Mancheral Quartzite and overwhelmingly dominate other types of wind-laid strata in the sequence.

Interpretation. At lower (<80%) moisture contents of the substrate, adhesion plane beds develop over adhesion ripples (Kocurek & Fielder 1982). Parallel layers of adhesion plane beds accrete vertically until the moisture content of the substrate is lowered still further to a level where adhesion structures cease to develop altogether. Low-angle discordant surfaces within the sequence of adhesion plane beds indicate minor erosion along flat surfaces. The upward reduction of the adhesion laminae thickness within packages bounded by a pair of erosion surfaces reflects upward-decreasing capillary moisture. Lesser moisture results in a lesser amount of sand trapped by the substrate and, therefore, formation of thinner adhesion plane beds. The localized deflationary surfaces which repeatedly interrupt the drying-upwards adhesion laminae sequence were probably caused by periodic sand undersaturation of the wind.

Fine-grained sandstone with adhesion cross-laminae (F2)

Climbing adhesion ripple cross-laminae generally occur in this sequence as solitary, lenticular sets (Figs 9 & 10). Set thickness varies from 2–5 cm and sets are laterally traceable for

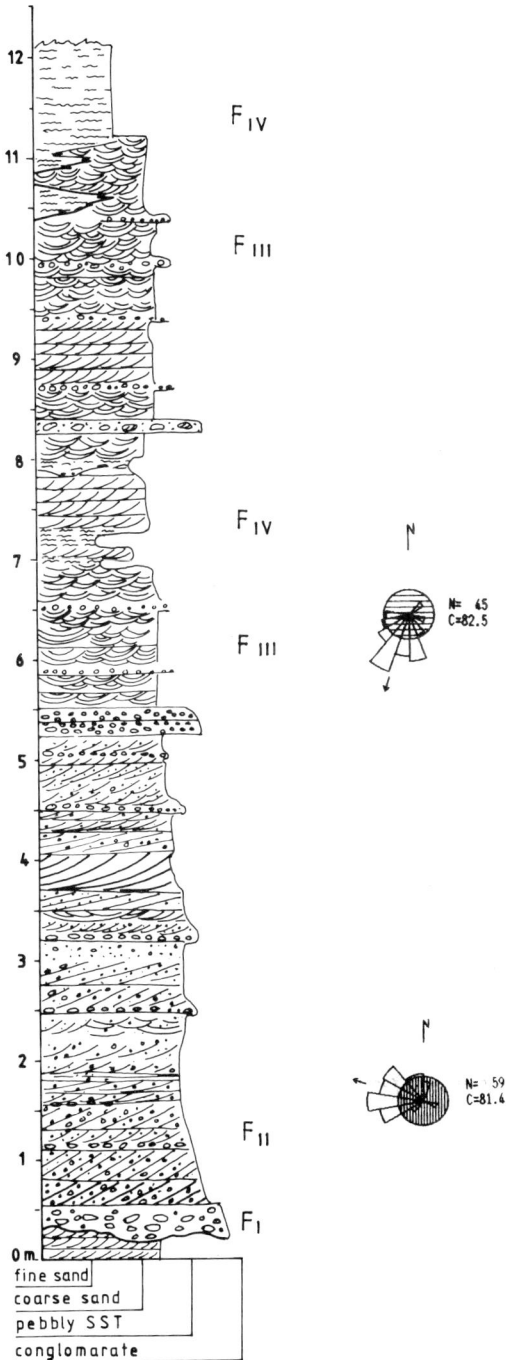

Fig. 4. Vertical log through the Mancheral Quartzite. The fining-upwards sequence is marked at the top by a fine-grained aeolian sandstone which also overlies and intertongues with a medium-scale trough cross-bedded unit (Facies IV, Table 2).

Table 2. *Broad facies recognized within the Mancheral Quartzite*

	Facies	Environment
I	Poorly-sorted conglomerate and conglomeratic sandstone overlying large-scale, concave-upward erosion surfaces (Fig. 2)	In-channel coarse lag deposits
II	Complex organized co-sets of planar cross-beds and a few trough cross-beds in the coarse-grained sandstone	Medial or lateral braid bars
III	Sheet-like units of medium- to coarse-grained sandstone made up of sets or co-sets of trough cross-beds; ripples, mud cracks in places	Proximal floodplain/higher topographic levels adjacent to the main channels
IV	Very fine- to fine-grained sandstones with adhesion structures; small aqueous cross-beds, shallow channel-fills	Aeolian/reworked aeolian deposits on the highest topographic levels in the floodplain

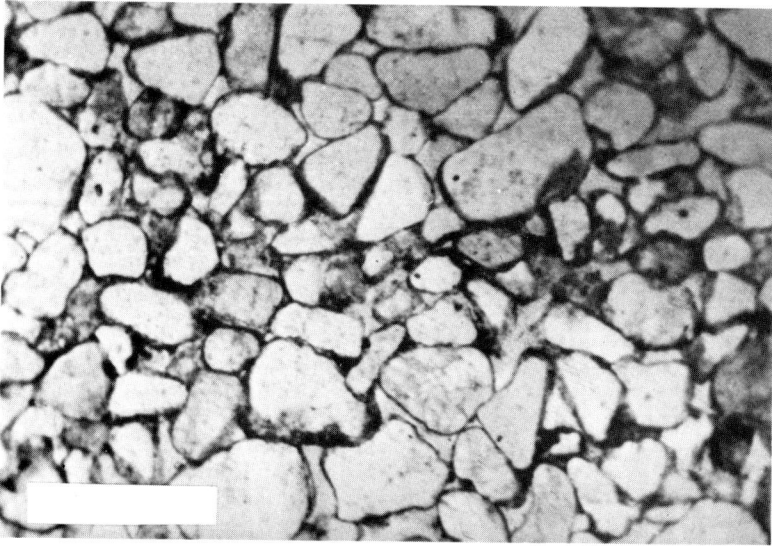

Fig. 5. Photomicrograph showing well-sorted, well-rounded sand grains of the aeolian units of the Mancheral Quartzite. Scale bar 0.25 mm.

Fig. 6. Adhesion plane beds in the fine-grained sandstone units. An aqueous cross-bed is in the central part of the photograph (A) and a water-injection feature in the lower left corner (I). Note a prominent rim of iron oxide around the water-injection feature.

Fig. 7. Sub-horizontal, parallel surfaces bounding packages of thinning-upwards adhesion plane beds. Note dark iron oxide layers mark the bounding surfaces (arrows).

Fig. 8. Close-up view of a thinning-upwards sequence of adhesion laminae. Pen for scale.

Fig. 9. Adhesion cross-laminae within aeolian units of the Mancheral Quartzite.

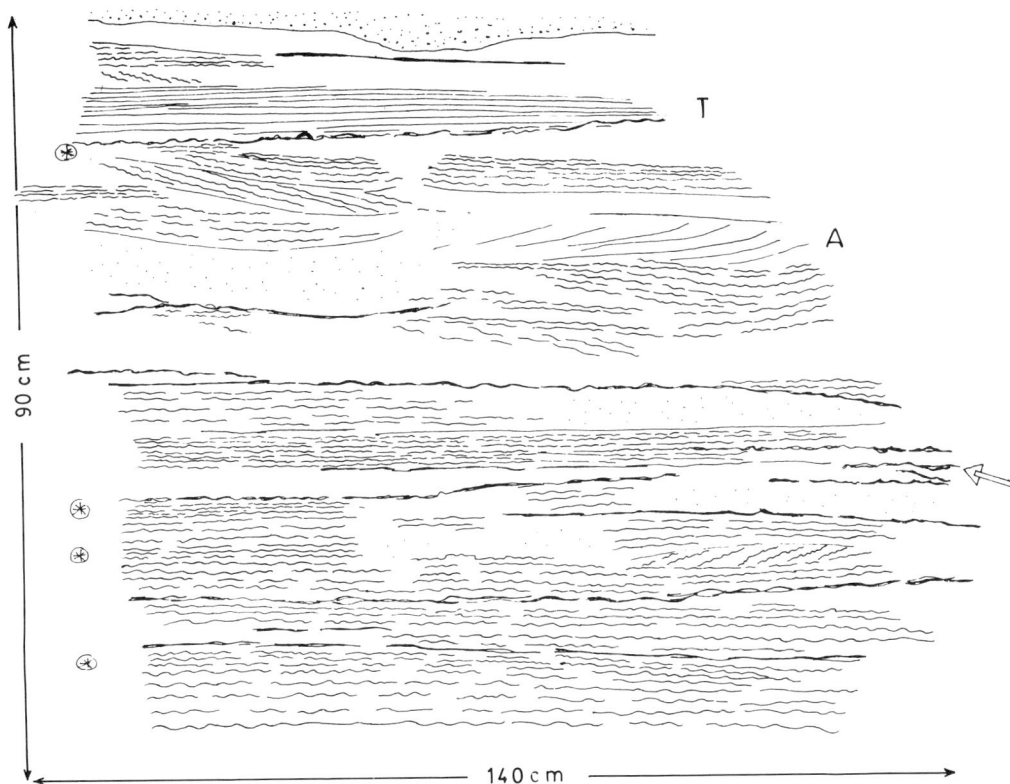

Fig. 10. Sketch showing a 90 cm-thick sequence dominated by adhesion laminae. The arrow marks the irregular iron-rich surfaces; asterisks mark thinning of adhesion laminae; T: translatent strata; A: erosively overlying aqueous beds.

15–40 cm. Sets of adhesion cross-laminae commonly grade both laterally and vertically into adhesion laminated strata. In places, however, adhesion cross-laminae overlie water-laid strata. Among adhesion structures the adhesion cross-laminae are volumetrically subordinate, although isolated sets are encountered quite often.

Interpretation. Adhesion cross-laminae form by the upwind migration and climb of the adhesion ripples on damp surfaces. Formation of adhesion cross-laminae requires both more than 80% water saturation of the substrate pore spaces, as well as adequate supply of dry wind-blown sand. Limited occurrence of the adhesion cross-laminae in this sequence and their rapid spatial and temporal gradation into adhesion plane beds may result from a paucity of any of the above two factors or a combination of the two (cf. Hummel & Kocurek 1984).

Wind-rippled fine-grained sandstone (F3)

Within these 2–6 m-thick aeolian sequences wind-ripple strata are conspicuous by their rarity. Sequences of wind-ripple strata, 10–30 cm thick, occur in places. Strata are flat-lying, laterally persistent, a few mm to 1 cm thick and show inverse grading (Fig. 10).

Interpretation. Thin units of F3 indicate the presence of localized wind-rippled dry sand cover. Thin inversely graded strata are typical of climbing wind ripples (Hunter 1977).

Massive or faintly cross-bedded fine- to medium-grained sandstone (F4)

These are fine- to medium-grained sandstone units with a lighter shade of red and the facies units vary in thickness from 5–25 cm, locally attaining thicknesses of up to 49 cm. The lower bounding surfaces of these beds are slightly irregular (Fig. 11), either concave-up or flat, giving rise to sheet-like or lensoid depositional units. The upper bounding surfaces of the beds are flat. The massive/cross-bedded units erosively cut into the underlying adhesion laminated units and grade upward into sandstones with adhesion structures. The F4 beds in places contain small clasts of fine sandstone/siltstone and in rare instances show a weakly developed fining-upwards grain size trend. The sediments have the same grain size and grain roundness but are relatively less well sorted than the adjacent adhesion laminated sandstones (Fig. 12). The facies units laterally as well as vertically grade within a short distance into the adhesion laminated or cross-laminated units (Fig. 10).

Interpretation. The channel-form shape of the sand beds, locally developed fining-upwards trends, presence of siltstone/fine-grained sandstone clasts and relatively poor sorting, collectively indicate an aqueous origin for these beds. It appears that these aqueous deposits inherited the characteristics of grain size and grain roundness from the aeolian deposits which were fluvially reworked in situ. The rapid lateral

and vertical transition into aeolian facies indicates the transient nature of the aqueous activity.

Iron-rich fine- to very fine-grained sandstone (F5)

The fine-grained sandstones of this facies are dark red in colour and are crossed by numerous irregular iron-rich surfaces (Fig. 10). Thin-section study reveals the presence of abundant iron oxide concentrated along solution planes (Fig. 13). The fine-grained iron-rich units mostly occur at the top of the drying-upwards sequences of adhesion laminae (Fig. 7) but may also occur within massive sandstone units or at the contact of the massive and adhesion laminated units (Fig. 10).

Interpretation. The iron-rich irregular surfaces occurring mostly at the top of the drying-upwards sequences of adhesion laminae are interpreted to represent iron-crusts formed on the exposed surfaces. Prolonged exposure at the damp surface–air interface would favour formation of an iron crust through evaporative con-

Fig. 11. Flat but very irregular erosive lower bounding surface of the massive sandstone units. Note underlying adhesion plane beds.

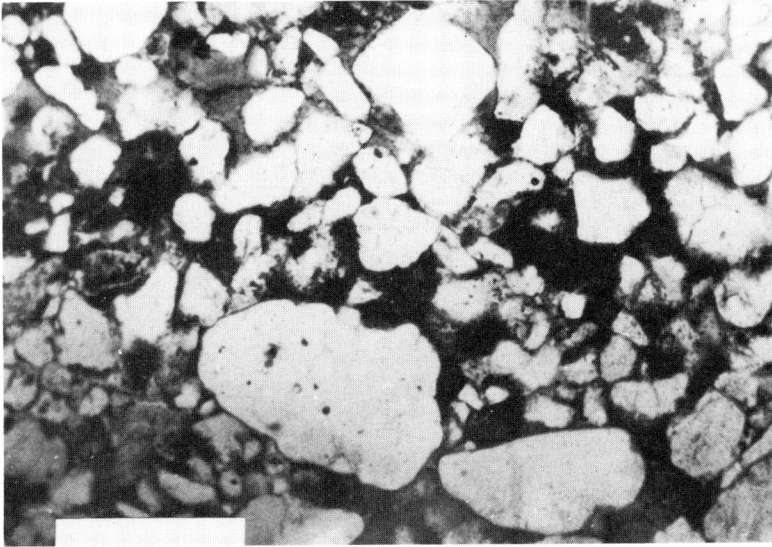

Fig. 12. Photomicrograph showing poor sorting of the aqueous units. Scale bar 0.25 mm.

centration and precipitation from capillary water. It is noteworthy that an iron oxide rim marks the margin of the sandstone dyke protruding through the adhesion laminated units shown in Fig. 6. The iron oxide rim presumably formed by similar processes at the interface of water saturated and damp/dry sediments.

Facies sequence

A typical sequence of salmon red, fine-grained sandstone exposed in the Ramgundam area is shown in Fig. 14 and the detail of a sequence dominated by adhesion structures is shown in Fig. 10. All of the three measured sections show the following common features.

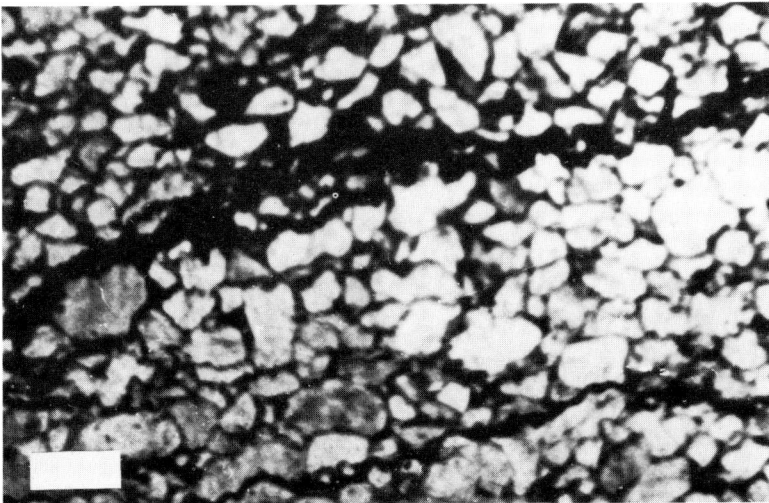

Fig. 13. Photomicrograph showing rocks with irregular iron-rich surfaces (F5); note the solution planes. Scale bar 0.25 mm.

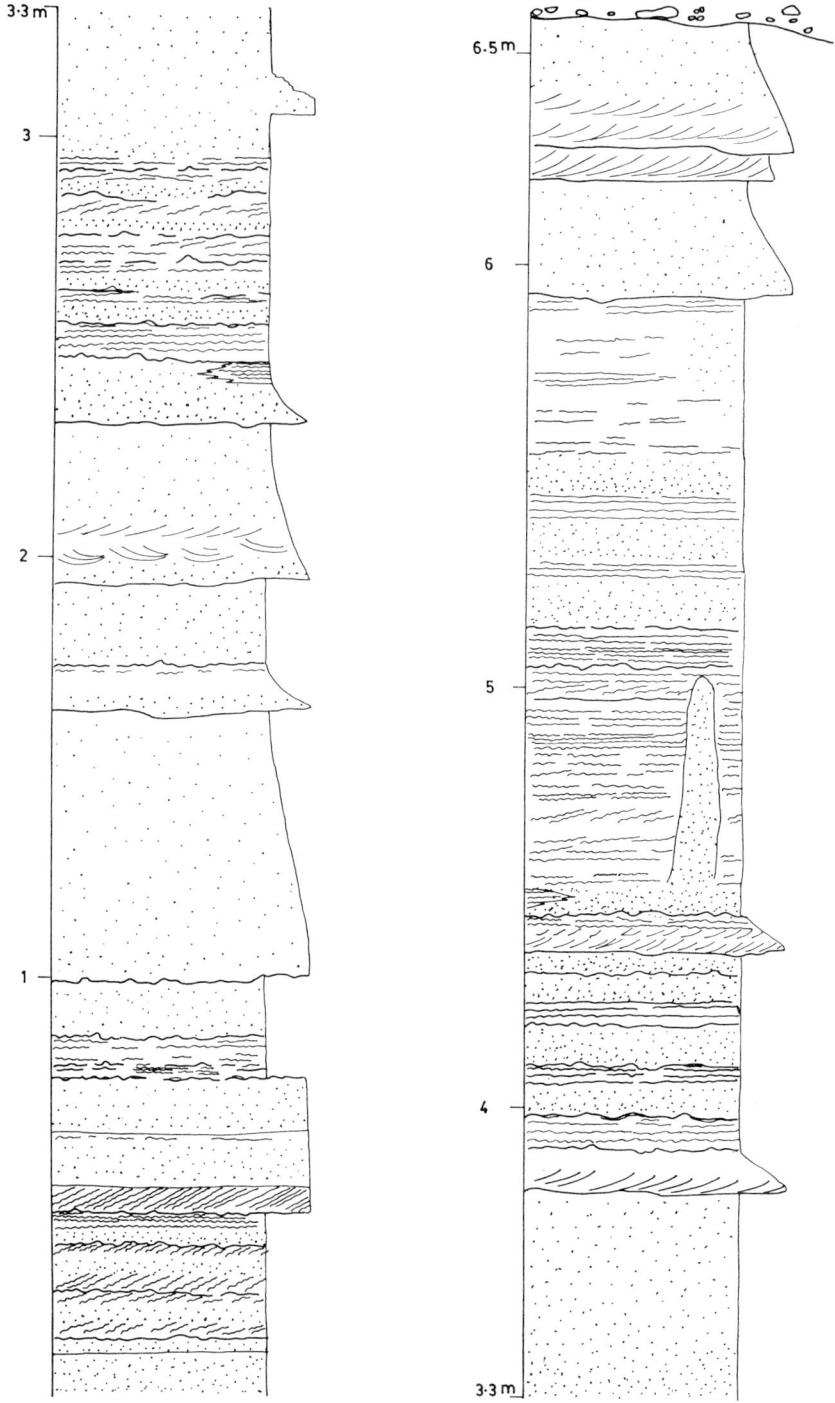

Fig. 14. A vertical log through the fine-grained sandstone unit of the Mancheral Quartzite. Note lack of grain-size differences in some of the massive units and the presence of a few floating adhesion laminae in some other massive sandstone units.

1. Some 25–40% of the total sequence is made up of aeolian strata and the rest consists of aqueous deposits. The aqueous beds have a very similar grain-size range to the aeolian strata. The aqueous beds are generally marked by sharp, erosional lower bounding surfaces and frequently show fining-upwards grain-size trends.
2. Vertical and lateral gradation from massive/cross-bedded units to adhesion laminated units is very common. Upward gradation of aqueous beds to adhesion plane beds through adhesion cross-laminae is rare. Adhesion cross-laminae are much less abundant than the adhesion laminae.
3. Drying-upwards sequences, exhibited by the upward decreasing spacing of adhesion laminae, are very common. These sequences are often bounded by horizontal deflation surfaces and/or irregular iron-rich surfaces (F5).
4. The sequences of sedimentary structures show very rapid lateral variation, and transitions between all the different facies are very common.

The sheet-like geometry of these aeolian or reworked aeolian sandstones, their thickness (which locally attains as much as 12 m) and its intertonguing relationship with the deposits of the higher topographic levels of the Mancheral channel complex or proximal part of the floodplain, indicate that these units developed in the distal floodplain or inter-channel area, where reworking by high-energy, active channels was virtually absent. Overall finer grain size, better sorting and presence of abundant well-rounded grains indicate little fluvial input. Channelized aqueous beds, therefore, indicate that during the floods only weak flows reached the depositional site, carrying with them the finest fraction of the fluvial load, and/or the flows largely reworked the wind-laid sands in situ. Similar low-energy shallow channels carrying fine-grained sediments are common in the higher topographic reaches/distal floodplain of many recent coarse-grained braided rivers (Williams & Rust 1969; Brierly 1991). However, aeolian sands on this scale have not previously been reported from any of them.

The nature and extent of the floodplain processes operating on the floodplain, magnitude of the flood events and the sediment load carried in the distal overbank area appear to have controlled the accumulation of these thick aeolian units in the Mancheral alluvial plain. This is unlike the other reported erg-margin aeolian–fluvial systems where erg dynamics exert the major control on the nature and internal organization of the interlayered aeolian–fluvial deposits.

Adhesion structures reported from many erg and erg-margin environments show well-developed drying-upwards sequences in which aqueous beds grade into adhesion cross-laminae and these in turn grade into adhesion plane beds (e.g. Kocurek (1981) amongst others). Such sequences are very rare in the aeolian sandstone units of the Mancheral Quartzite. At any one point profile, adhesion plane beds are the dominant structure and in most of the places aqueous deposits directly grade into adhesion plane beds without any intervening zone of adhesion cross-laminae. Compared to the modern day braided river floodplains, water seepage was much faster through the sandy overbank areas of the Mancheral fluvial system which was virtually mud free. As a result of the high moisture content required for the formation of the adhesion cross-laminae (100–80% of the substrate pore spaces filled with water) their occurrence was spatially and temporally very restricted. Lower levels of dampness, however, prevailed over wider areas for a longer time, allowing adhesion plane beds to form and dominate over other adhesion structures. On the other hand, capping of many of these adhesion plane bed sequences by erosion surfaces rather than by the wind-ripple strata reflect undersaturation of the air flowing over the surface. Given the surrounding wet environment and the presence of active channels in the vicinity, sand supply was poor and neither wind ripples nor any larger aeolian bedforms could develop. Even if formed, wind-ripple strata, being at the top of the drying-upwards sequences had the lowest preservation potential and might have been reworked by weak but erosive flood flows.

Stacking of a number of drying-upwards adhesion plane bed sequences without intervening aqueous deposits can probably be correlated with fluctuating ground water tables due to low-magnitude flood events. During these low-magnitude events the flood water did not actually reach these distal floodplain areas but they brought about a rise in the water table and thus renewed accretion of adhesion laminae in the floodplain.

Many of the massive sandstones were obviously laid down from shallow channelized flows. Many others, particularly those lacking obvious basal scours or a discernable fining-upwards trend (Fig. 14) might have formed by other processes. Several possibilities exist. They could be grainfall deposits accumulated in shallow ponds (cf. Weiner (1981) cited by Hummel & Kocurek 1984). Surface precipitation may locally emplace massive sandstones that lack erosive lower contacts. Alternatively, rising ground water tables may destroy aeolian stratification rendering the sands massive. Isolated pockets of adhesion structures floating in some of these beds (Fig. 14) support such an interpretation.

Conclusions

Previous work has indicated that in erg-margin aeolian–fluvial systems, the nature of dune–interdune morphology, their relationship with respect to the fluvial channels, and the relation between the direction of fluvial discharge and that of wind transport appear to be the main factors controlling the nature of the deposits (Langford & Chan 1989; Clemmensen *et al.* 1989). Flood surfaces and overbank–interdune sediments appear to be the most diagnostic deposits of this environment. However, such aeolian morphologies have been found to be absent in the Mancheral alluvial plain. Here, on the other hand, a delicate balance between the capillary moisture (which was again related to the fluvial discharge regime) and supply of wind-blown sand appears to have controlled the accumulation of fairly thick units of aeolian sand.

An arid climate and probable presence of a sand sea (Venkatpur Sandstone) in the interior part of the basin (Chakraborty 1988, 1991*a,b*) provided the general condition for the accumulation of the aeolian sands. Discharge fluctuation in the Mancheral braided rivers exposed wide alluvial tracts. The fluvial system being virtually mud-free, loose sand on the floodplains or higher topographic reaches was subjected to wind erosion. In the absence of land vegetation during pre-Silurian times, wind action was probably more vigorous than at present. On the other hand, coarse-grained fluvial deposits and a wet surrounding environment, restricted the availability of dry sand for wind transportation. It is in this setting that a unique aeolian sequence dominated by adhesion plane beds accreted. Probably, a significant part of the sequence was later reworked by flood flows and redeposited as massive sandstone or fine-grained sandstone with faintly developed cross-beds.

Stacking of large-scale channel-form surfaces throughout the sequence indicates that the channels probably did not migrate laterally for a considerable distance but were relocated time and again through avulsion. This channel behaviour was possibly related to pulsating tectonic activity in the rapidly subsiding basin. Pockets of large-scale soft sediment deformation present in the Mancheral Quartzite (for details see Chakraborty 1991*b*) probably attest to such periodic tectonic tremors. Rapid basin subsidence allowed a thick aeolian–aqueous sequence to develop in the floodplain areas and avulsive channel behaviour favoured their preservation in the rock record.

Since many of these features are likely to be present in other pre-Silurian fluvial systems of arid to semi-arid climatic regime, it is worth re-examining the finer-grained sediments of these systems for possible records of the alluvial plain aeolian deposits.

Field and other infra-structural facilities were provided by the Indian Statistical Institute. The authors are grateful to Ken Pye (University of Reading) for his critical comments on an earlier version of the manuscript.

References

BLUCK, B. J. 1979. Structure of coarse grained braided stream alluvium. *Transactions of the Royal Society of Edinburgh, Earth Sciences*, **70**, 181–221.

BRIERLY, G. J. 1991. Floodplain sedimentology of the Squamish River, British Columbia: relevance of element analysis. *Sedimentology*, **38**, 735–750.

CANT, D. J. & WALKER, R. G. 1978. Fluvial processes and facies sequences in the sandy braided South Saskatchewan River, Canada. *Sedimentology*, **25**, 625–648.

CHAKRABORTY, T. 1988. A preliminary study of the stratigraphy and sedimentation of the Late Proterozoic Sullavai Group in the southwestern belt of Pranhita-Godavari Valley (abstract). *In*: Workshop on Proterozoic Rocks of India (IGCP-217), Geological Survey of India, Calcutta, 20–21.

—— 1991*a*. Sedimentology of a Proterozoic erg: the Venkatpur Sandstone, Pranhita-Godavari Valley, south India. *Sedimentology*, **38**, 301–322.

—— 1991*b*. *Stratigraphy and sedimentation of the Proterozoic Sullavai Group in the south-central part of the Pranhita-Godavari Valley, Andhra Pradesh, India.* PhD thesis, Jadavpur University, Calcutta.

CLEMMENSEN, L. B., OLSEN, H. & BLACKEY, R. C. 1989. Erg-margin deposit in the Lower Jurassic Moenave Formation and Wingate Sandstone, southern Utah. *Geological Society of America Bulletin*, **101**, 759–773.

HUMMEL, G. & KOCUREK, G. 1984. Interdune areas of the back-island dune field, north Padre Island, Texas. *Sedimentary Geology*, **39**, 1–26.

HUNTER, R. E. 1977. Basic types of stratification in small aeolian dunes. *Sedimentology*, **24**, 361–387.

KOCUREK, G. 1981. Significance of interdune deposits and bounding surfaces in aeolian dune sands. *Sedimentology*, **28**, 753–780.

—— & FIELDER, G. 1982. Adhesion structures. *Journal of Sedimentary Petrology*, **52**, 1229–1241.

LANGFORD, R. P. 1989. Fluvial–aeolian interactions: Part I, modern systems. *Sedimentology*, **36**, 1023–1035.

—— & CHAN, M. A. 1989. Fluvial–aeolian interactions: Part II, ancient systems. *Sedimentology*, **36**, 1037–1051.

RUST, B. R. 1972. Structure and processes in a braided river. *Sedimentology*, **18**, 221–245.

SHEPHERD, R. G. 1987. Lateral accretion surfaces in ephemeral stream point bars, Rio Puerco, New Mexico. *In*: ETHRIDGE, F. G., FLORES, R. M. & HARVEY, M. D. (eds) *Recent Developments in Fluvial Sedimentology*. Society of Economic Paleontologists and Mineralogists, Special Publication No. **39**, SEPM, Tulsa, 93–98.

SNEH, A. 1983. Desert stream sequences in the Sinai Peninsula. *Journal of Sedimentary Petrology*, **53**, 1271–1280.

TUNBRIDGE, I. P. 1984. Facies model for a sandy ephemeral stream and clay playa complex; the Middle Devonian Trentishoe Formation of North Devon, UK. *Sedimentology*, **31**, 697–715.

WILLIAMS, G. E. & RUST, B. R. 1969. The sedimentology of a braided river. *Journal of Sedimentary Petrology*, **39**, 649–679.

Coastal dunefields

Foredune morphology and sediment budget, Perdido Key, Florida, USA

NORBERT P. PSUTY

Institute of Marine and Coastal Sciences, Rutgers — The State University of New Jersey, New Brunswick, New Jersey, USA 08903

Abstract: Along-shore foredune morphology on the eastern portion of Perdido Key is fitted to a model of foredune development related to sediment budget. The Perdido Key barrier island has a long history of shoreline erosion, up to 2 m a^{-1} for over a century. Sediment availability was altered in 1985 because of beach nourishment. Subsequent re-working of the beach fill has transferred sediment in the along-shore direction as well as cross-shore. Over three years of survey data, including passage of three hurricanes, and aerial photo coverage portray the morphological changes in the foredune system and the displacement of the shoreline. Measurements of the changes in a spatial context record sediment re-distribution from the nourishment area. Measurements of cross-sectional area changes support the model of foredune enhancement associated with negative sediment budget in the beach. Net along-shore changes of the foredune compare well with the barrier-island foredune development model.

There has been considerable attention given to sediment budget and exchange between beach and dune components of the dune/beach system (see Pye 1983; Goldsmith 1985; Carter 1988; Psuty 1988a, 1992; Nordstrom *et al.* 1990). Questions arise concerning temporal variability in the exchanges (Hughes & Chiu 1981; Vellinga 1982), as well as spatial variability in an along-shore direction (Short & Hesp 1982; Paskoff 1989). This paper will concentrate on sediment budget calculations representing exchanges between the upper and lower parts of the profile and portray exchanges over a 3.5-year period. It will report on observations of the variations in the profile response, and will describe the responses in an along-shore sequence.

Study site

Data are drawn from the eastern portion of Perdido Key, a barrier island located at the western margin of Florida, on the Gulf of Mexico (Fig. 1). This micro-tidal environment, mean tidal range of 0.4 m, is one of generally low wave energy primarily out of the southeast; mean significant wave height is 0.4–1.0 m and mean wave period is 3–6 s (Galvin & Seeling 1969; Hubertz & Brooks 1989). However, Perdido is also the site of frequent hurricane storm conditions that cause major redistribution of sediment on the profiles, incorporating super-elevation of the water level and waves that affect the coastal foredune.

The coastal foredune is of variable quality along the island, attaining elevations of 3–4 m in its well-developed areas and being entirely eroded in other locations. Inland of the shore-parallel foredune are a variety of parabolic dune forms that frequently extend the width of the barrier. Other locations are occupied by irregular dune forms that rise to about 7 m and form a loosely-connected inner dune belt.

Background

As reported by Balsillie (1986), the eastern portion of Perdido Key has a history of erosion for at least 135 years. There is an along-shore gradient in the erosion rates, reaching a maximum of 1.8 m a^{-1} near the eastern terminus and attaining a long-term average of no net change approximately 14–15 km to the west. Incorporated within this data set is a temporal gradient that suggests that the along-shore erosional trend has been shifting westward during this period. Thus, an interpretation of the data is that there is an increasing sediment deficit through time that is manifest in the loss of sand from the eastern portion of the island and that the enlarging deficit is causing the erosional shoreline to lengthen downdrift. Although the sediment budget may be decreasing naturally, dredging across the ebb tidal delta and within the inlet channel since the 1930s has exacerbated the trend. Much of the dredge spoil has historically been placed offshore on the downdrift side of the channel, creating and enlarging Caucus Shoal. The shoal has certainly affected the subsequent

From Pye, K. (ed.), 1993, *The Dynamics and Environmental Context of Aeolian Sedimentary Systems.* Geological Society Special Publication No. 72, pp. 145–157.

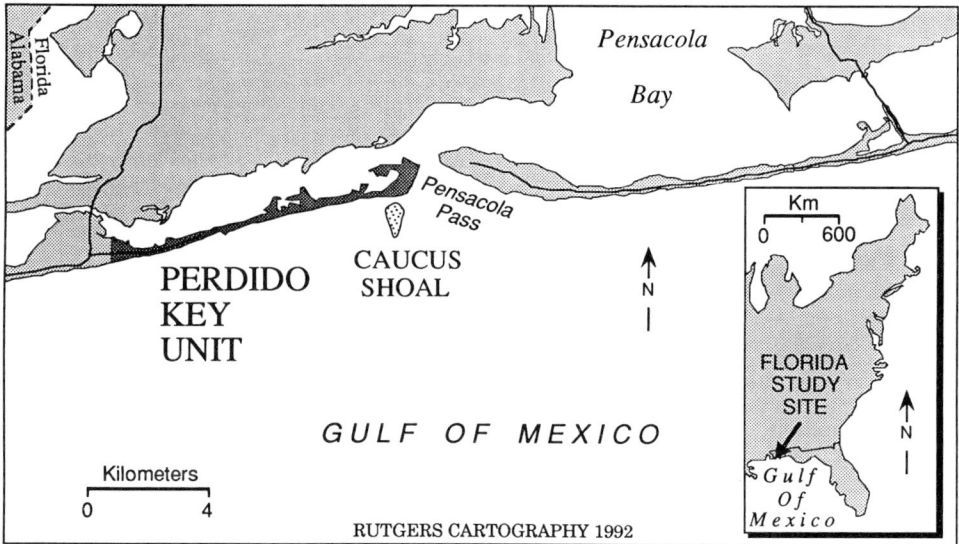

Fig. 1. Location map of the Perdido Key study site on the Gulf of Mexico.

sediment transport system by locally accumulating sediment and subtracting it from the downdrift transport, by refracting local wave systems and producing reorientation of the nearshore current direction, and by returning some of the sediment directly into the navigation channel where it is effectively removed from the along-shore transport system. The resulting sediment deficit has been estimated at 115 000–150 000 m^3 a^{-1} (Walton 1976), which affects the eastern portion of Perdido Key.

Eastern Perdido is presently a part of the Gulf Islands National Seashore. It is managed by the National Park Service. The General Management Plan for this portion of the Park calls for two different management strategies to be applied to Perdido (National Park Service 1978). The western portion of the Park has a developed infrastructure and it will continue to be used to accommodate as many people as can be served with the existing parking areas. The eastern portion of the Park is without any of the infrastructure and amenities and it will function as a 'natural area'. This means that there will be no construction, no managed trails, no attempts to control the natural processes at work in this portion of the Park.

The long history of shoreline erosion has led to a general inland migration of eastern Perdido as attested to by the ubiquitous washover features in the island's topography (Fig. 2). An aerial view of the eastern portion of the key in 1978 (Fig. 3) reveals several important aspects of

the geomorphology. This portion of Perdido has shifted inland about 200 m since the mid 1860s and it is apparent that most of the barrier has a well-developed, coherent foredune, is well-vegetated, and is backed by older sand dunes. The topography is criss-crossed with many tracks and trails, caused by vehicles driving in the dunes. This practice was halted in 1984 and the trails were allowed to re-vegetate naturally. Whereas most of the coastal foredune appears to be adequately developed and preserved, the immediate foreground in the 1978 photo shows that the coastal foredune is breached. The beach appears to be very narrow in this location and there is overwash penetrating into the back bay. Human disturbances may have contributed to the condition, but the narrow beach suggests that there is a negative sediment budget in the area and that the dune scarping and loss are part of the natural morphologies. Another characteristic is the presence of older parabolic forms in the topography, fronted by a foredune ridge in varying states of continuity. This suggests a complicated sequence of foredune stability and instability, periods of inland aeolian transfers, and periods of accumulation on the upper beach profile.

However, as erosion continued, this portion of the island narrowed to the point where the coastal foredune was severely reduced in dimension and washover was occurring (Fig. 4). Conditions in 1984 pointed to the loss of the consistent foredune and the replacement by

Fig. 2. Most of Perdido shows evidence of former washover fans which have penetrated across the island to form arcuate lobes on the lagoonward shoreline. A zone of recent overwash on the eastern portion of Perdido Key.

isolated foredune hummocks adjacent to washover fans. The situation was one of considerable inland transfer of sediment with very little higher portions of the profile on whch to accumulate sediment above the storm berm level. To counteract this problem, a decision was made to buffer the erosion and prevent an anticipated breach of the island which would isolate the end of the barrier from visitors and park management, and to accept material dredged from the inlet as beach nourishment even though the location was designated as a 'natural area' (Psuty 1988*b*).

Thus, in 1984, the shoreline and foredune situation at Perdido Key incorporated the long-term erosional trend caused by the negative sediment budget. Geomorphologically, the eastern portion of the barrier displayed a very narrow beach eroding back into high dunes associated

Fig. 3. Aerial view of the eastern end of Perdido in 1978. The dunes are cut by many off-road vehicle trails. The coastal foredune appears to be well-formed in the middle and far distance. The foredune in the near portion of the photograph is breached and there is some overwash. Photo courtesy of Dr R. Thackeray.

Fig. 4. Dissection of the foredune ridge results in a few vegetated hummocks amidst large washover features; April, 1985.

Fig. 5. A very narrow foreshore exists in front of the severely-scarped foredune at the eastern end of Perdido Key; April, 1985.

Fig. 6. The western portion of the study area has a more consistent coastal foredune, although occasionally breached by pedestrian pathways. The normal dune form is one of coalescing hummocks thinly covered by sea oats. The gap in the foreground is the only vehicular access to the beach and to the eastern terminus of Perdido Key.

with the inlet shoreline (Fig. 5). Immediately toward the west, this region graded into a dissected foredune area with a general scarping of the foredune remnants (Fig. 4), and gaps where there was evidence of inland transfers of sediment. The western portion of the Park had a more coherent foredune ridge (Fig. 6) which was irregular in elevation and plan view. It was situated atop a berm elevation of 2.0 m and extended to elevations of 4.5–5.0 m. It was relatively well vegetated by sea oats, *Uniola paniculata*.

Conceptual model of sediment budget and foredune development

Coastal foredune development can be related to the relative sediment budgets of the more active lower profile (wet beach related to wave and current processes) compared to the less active upper profile (dry, vegetated foredune related to wind and water processes) (Psuty 1992). In a four-part combination (Fig. 7), the budget scenarios can be linked to resulting geomorphological development, providing for maximum

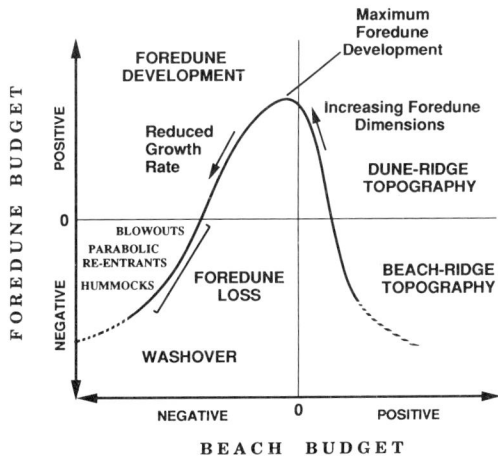

Fig. 7. Conceptual portrayal of dune and beach sediment budget combinations related to foredune morphologies. Foredune morphology passes through a sequential continuum which extends from little opportunity for development because of the abundant beach budget at one extreme to little opportunity for development because of the very negative beach budget at the other (after Psuty 1992).

accretionary topography at one end, and to attenuation and extinction at the other. Somewhere along this continuum, the conditions are favourable to maximize accumulation in the coastal foredune. The conceptual model portrays maximum foredune development in association with the threshold position of beach accretion/erosion because of the opportunity for continual accumulation at a slowly changing shoreline and because of the continued existence of large foredune topographies in association with modern erosional shorelines. The present study will apply the data from beach and foredune changes at Perdido to test the conceptual model.

Sediment enrichment

Because of the potential for island breaching and problems of access and of providing minimal safety measures on a detached island, the National Park Service agreed to the emplacement of 1.7×10^6 m^3 of dredged sand from the inlet at the eastern 1.2 km of the island. The purpose of the nourishment was to accommodate the need to dispose of the inlet dredge spoil in concert with a State of Florida mandate to make use of any dredged sediment suitable for beach fill, to buffer the shoreline erosional rate, to protect the zone of greatest washover occurrence, and to gain some time regarding a future management decision to accept an additional $4.0–4.5 \times 10^6$ m^3 of beach fill associated with enlargement of the navigation channel.

Data gathering

Interest in the performance of the beach fill and its interaction with downdrift morphologies led to a monitoring programme that was to begin prior to the emplacement of the fill in July and August 1985, and to extend for two years (Psuty & Jagger 1990). The first observations were made in the spring of 1985. Surveys were run from two sets of datums that were in place in the Park. One group of monuments was established by the State of Florida. They were spaced at intervals of 1000 feet (305 m) and numbered R-34 through R-65 (Fig. 8). The other group of monuments was established by the National Park Service and were spaced 300 m apart; augmented by several special purpose survey lines. Initially, there were 73 survey datums and profile lines that were occupied and run along 12 km of shoreline. The number decreased as storms and vandalism removed some of the datums. Although both the Florida datums and NPS datums were used during the course of this study, only the Florida data are presented here because of the overlap in the data sets.

This paper will focus on the subaerial portion of the beach, extending from mean sea-level inland, and, more specifically, on the upper portion of the profile, above the general level of wave and current processes on the profile, defined as above 2.0 m elevation. This threshold elevation is derived from an examination of the elevations of berm construction by near-shore waves and current processes. Elevations above 2.0 m, for purposes of this study, are considered to be the coastal foredune, whether created by aeolian or hydrodynamic processes. Profile data will be utilized for their elevational characterization, to calculate cross-sectional area, and by extension, to calculate volumes per metre of beach. Another measurement reported on is the distance to the mean sea-level intercept on the profile. This figure represents the distance from the profile datum to the 0.0 m elevation on the profile. All profile elevations are absolute values related to the National Geodetic Vertical Datum of 1929. A comparison of the distance measured during the different surveys provides a surrogate value for erosion or accretion on the beach profile, although there is some variation of the shoreline position associated with the development of 'caletas' in the beach/near-shore zone (Fig. 9). These features are associated with along-shore migrating circulation cells (Jagger et al. 1991) that provide a cusping of the beach in the form of Wright & Short's (1983) 'rhythmic bar and beach' intermediate morphodynamic state. These cusps occasionally penetrate sufficiently inland to scarp the foredune.

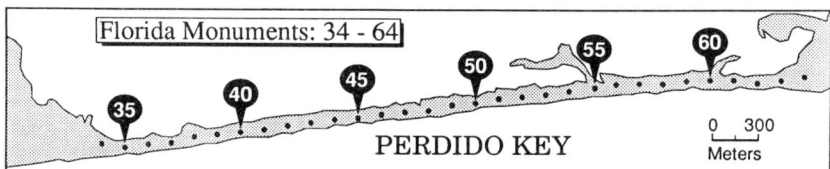

Fig. 8. Distribution of State of Florida survey datums along eastern Perdido Key, R-34 to R-65.

Fig. 9. The caleta embayments are the modal shoreline form along most of Perdido Key. The indentations translate along-shore in the direction of the prevailing drift.

Data sets

April 1985–September 1985

The pre-fill profiles demonstrated that the eastern portion of the island, R-59 to R-65, was essentially without a beach and the dunes were being scarped severely. The adjacent portion, R-52 to R-58, was undergoing destruction of the coastal foredune and washover was occurring through many breaches. The western portion, R-34 to R-51, generally had a well-developed and well-vegetated foredune with coherence of the ridge form (although broken occasionally by pedestrian paths from the road to the beach).

The island was hit by Hurricane Elena in late August–early September. The first profiles after the fill incorporated changes produced by the elevated water level and higher waves. Field observations indicated that the principal geomorphological effect of the storm was the building of a very high storm berm accumulation immediately in front of the foredune. If the foredune had been breached or eroded earlier, the site was a location of overwash. The common profile adjustment noted on most of the survey lines was a positive budget on the beach profile (Fig. 10). Only the exposed eastern margin of the island recorded a loss, as the foredune was scarped without storm berm accumulation. A spatial depiction of the several profile measurements (Fig. 11a) shows that the profiles accreted sand along most of the island during this period. The shoreline oscillated between small negative and positive values except for the beach fill area where it was displaced seaward as much as 170 m. There was a small inland displacement of the shoreline, about 20 m, a short distance downdrift of the fill, R-57, perhaps as a result of some local refraction. The very large foredune accumulation in the beach fill area was produced because the fill was elevated to about 2.5–3.0 m in a part of the zone, creating a surface higher than the foredune in the beach profile at Profiles R-62 and R-63.

September 1985–March 1986

An original objective was to conduct the surveys in the spring and fall of each year. In this way the effects of seasonal cycles could be identified as the higher-energy winter periods mobilized the beach portion of the profile, especially the protruding beach fill zone, followed by the accumulation phase of the low-energy summer periods. The immediate post-fill period was highly energetic as two more hurricanes were in the

Fig. 10. Sediment accumulation high on the profile associated with Hurricane Elena, September 1985, undergoing subsequent modification; December 1985.

Gulf and raised water levels and wave heights. Whereas Hurricane Elena tended to accumulate a high berm deposit, Hurricanes Juan and Kate had a variable effect. The beach fill was cut back on the order of 30–40 m (Fig. 11b). The material was apparently shifted downdrift a bit and the beach built out adjacent to the fill trapezoid following the model described by Dean (1983). West of R-57 the beach generally retreated, in some locations as much as 25 m. Profile R-65, at the inlet, recorded some deposition, otherwise the fill had a net loss during this period. The washover zone continued to display a lack of accumulation on the profile while the shoreline retreated a bit. The western portion of the study area gave mixed responses. The shoreline generally eroded but the high portion of the profiles continued to record accumulation in about half of the cases. Only two profiles had losses greater than 15 m². However, field observations and the profiles indicated that the storm berm accumulated during Hurricane Elena had been modified considerably by the succeeding two hurricane events, removing the seaward portion of the berm and transferring the accumulations higher up the profiles and into the foredunes.

September 1985–September 1986

One year after the first post-fill survey, the profiles offer some significant information regarding the sediment transfers (Fig. 11c). As expected, the beach fill is releasing sediment from the profile and the shoreline has receded as much as 60 m in this period. There is seaward displacement of the shoreline immediately to the west of the fill as the shoreline trend is straightened. After the several hurricanes, the washover zone has retreated slightly and there has been some loss of accumulation high on the profile. However, the western portion with its well-formed foredune has accumulated sediment high on the profile despite a general inland displacement of the shoreline. The quantities of sediment in the profile increase at first proceeding westerly from the overwash zone and then decrease toward the western limit of the study area.

September 1985–August 1987

Two years later (Fig. 11d), the data sets indicate that the surface area of the fill is being eroded,

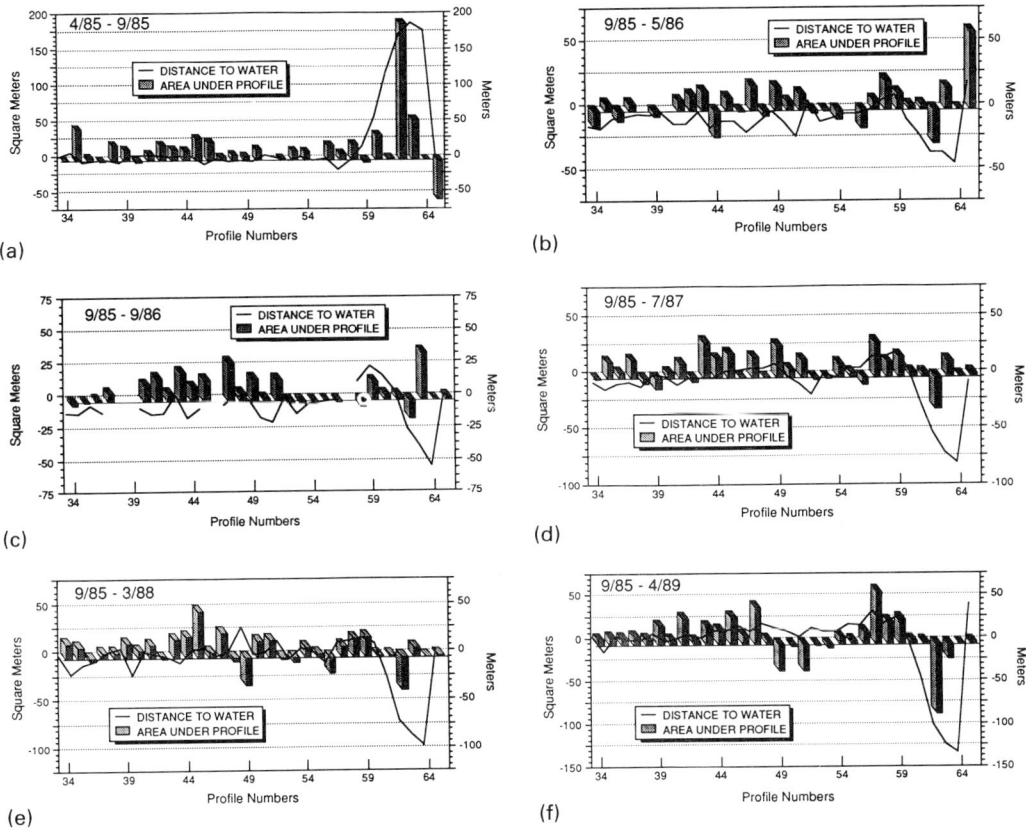

Fig. 11. Spatial distribution of cross-sectional area change and displacement of the shoreline. (**a**) April 1985 to September 1985; (**b**) September 1985 to March 1986; (**c**) September 1985 to September 1986; (**d**) September 1985 to August 1987; (**e**) September 1985 to March 1988; (**f**) September 1985 to April 1989.

the maximum retreat is greater than 80 m at R-63, and it is supporting the build-out of the beach in the washover zone, >30m at R-57. There is a type of cycling of the seaward displacement of the shoreline in which the downstream effect of sediment release from the fill decreases from R-59 to R-51, increases to a maximum at R-49, and then gradually decreases again to R-35. In the easternmost depositional cycle, about half of the stations record seaward displacements. In the western cycle, about two thirds of the displacements are landward. The washover zone has some loss as well as some gain in the upper profile, with a small net gain. The westerly foredune continues to record additional sediment accumulated high in the profile.

September 1985–March 1988

After 2.5 years (Fig. 11e), the end of the winter

season profiles suggest that the nearly 100 m of shoreline retreat at the fill zone, R-63, has contributed to a general recovery of the shoreline along most of the downdrift portion of the study area. A number of the profiles show an upper profile accumulation of >20 m². Except for the washover zone, most of the remaining profiles have a positive budget in the upper profile.

September 1985–April 1989

A severe winter storm at the beginning of 1989 raised water levels and lowered much of the upper beach as well as scarped the foredune along most of the study area. The washover zone was penetrated in many places, as were other locations where gaps existed in the foredune crestline. In April, the profiles had recovered and were going into their summer mode.

After 3.5 years (Fig. 11f), the beach fill zone

had lost about 75% of its surface area and had contributed to a general build-out of the shoreline that extended throughout most of the study area, R-34 to R-42 had recovered to a shoreline position similar to its original location. The portion of the study area immediately downdrift from the fill locations, R56 to R-59, recorded a seaward displacement of the shoreline and the greatest accumulation in the upper portion of the profile. The washover zone, originally from R-52 to R-58, had foredune recovery at the eastern margin but had a net loss throughout the remainder of the zone. The western portion of the coherent foredune had very high losses at R-49 and R-51, followed by a general increase in sediment accumulation high on the profile. The maximum tended to be to the east and gradually decreased towards the west.

Selected profiles

Beach fill

The super-elevation of the nourishment area effectively removed the pre-existing profile from interaction with most of the morphological processes (Fig. 12a). As expected, the fill was subjected to continuous erosion following emplacement and the record is one of inland displacement of the shoreline and a loss of sediment volume in the area. Contributions to the former foredune will not be initiated until the fill elevation is sufficiently eroded and reduced in elevation (<2 m) to allow the ambient processes to operate on the previous profile.

Downdrift adjacent

Erosion of the beach fill led to accretion in that portion of the barrier immediately to the west (Fig. 12b). Following an initial loss, the shoreline built-out 25 m and the upper portion of the profile showed a gradual accumulation of 11 m^2. Nearly all of the addition was in front of the existing foredune, causing it to increase in width slightly and to be accompanied by the seaward growth of a smaller foredune ridge.

Washover

Persistent erosion had narrowed this portion of the barrier to less than 200 m in width. The coastal foredune had been severely scarped and breached throughout much of the area, leaving only isolated hummocks separating broad washover fans. The profile elevations portray a low crest that was frequently topped by storm water levels and a displacement of sediment inland (Fig. 12c). The profiles from this region depict losses of the order of 8 m^2 and shoreline displacements of 20 m inland. Some recovery was observed on the eastern portion, where profiles R-50 and R-51 nearly attained a net cross-section equilibrium despite net inland shoreline displacement during the study period.

Coherent foredune

The western portion of the study area exhibited the greatest stability, in keeping with its historical pattern of a decreasing rate of change from east to west. However, some variations were observed and measured. Initially, the shoreline continued its recession under the influence of a negative sediment budget and the ambient wave and current conditions (Fig. 12d). It is likely that the effects of the eastern beach nourishment caused the sediment budget to become positive toward the end of the study period. The foredune zone recorded positive inputs to its volume throughout much of the period, with the greatest values at the eastern margin and decreasing, while remaining positive, toward the west. Most of the positive dune values occurred while the shoreline was being displaced inland. Towards the end of the survey period, the beach had returned to its initial position along much of this zone, whereas the dune portion of the profile had increased in volume.

Application to model

Although the record of dune/beach budget comparison is of limited duration, there does seem to be a spatial association of foredune development along the study area on Perdido Key. The fill area is an anomaly because this profile is one of continuous re-adjustment as erosion removes the high accumulation of sediment. However, once beyond the limits of the fill, the downdrift profiles are responding to the availability of sediment emplaced on the profile by the ambient processes. It is this sequence that can be related to the dune/beach sediment model.

According to the shoreline trends during the three years following the fill, the general sequence was one of positive sediment budget immediately downdrift from the fill, grading to an area that had a negative budget during most of the period, and then grading to an area that incorporated two different characteristics regarding its sediment budget. The first characteristic was of a decreasing negative budget from east to west, meaning that the deficit was larger at the eastern margin and became less towards

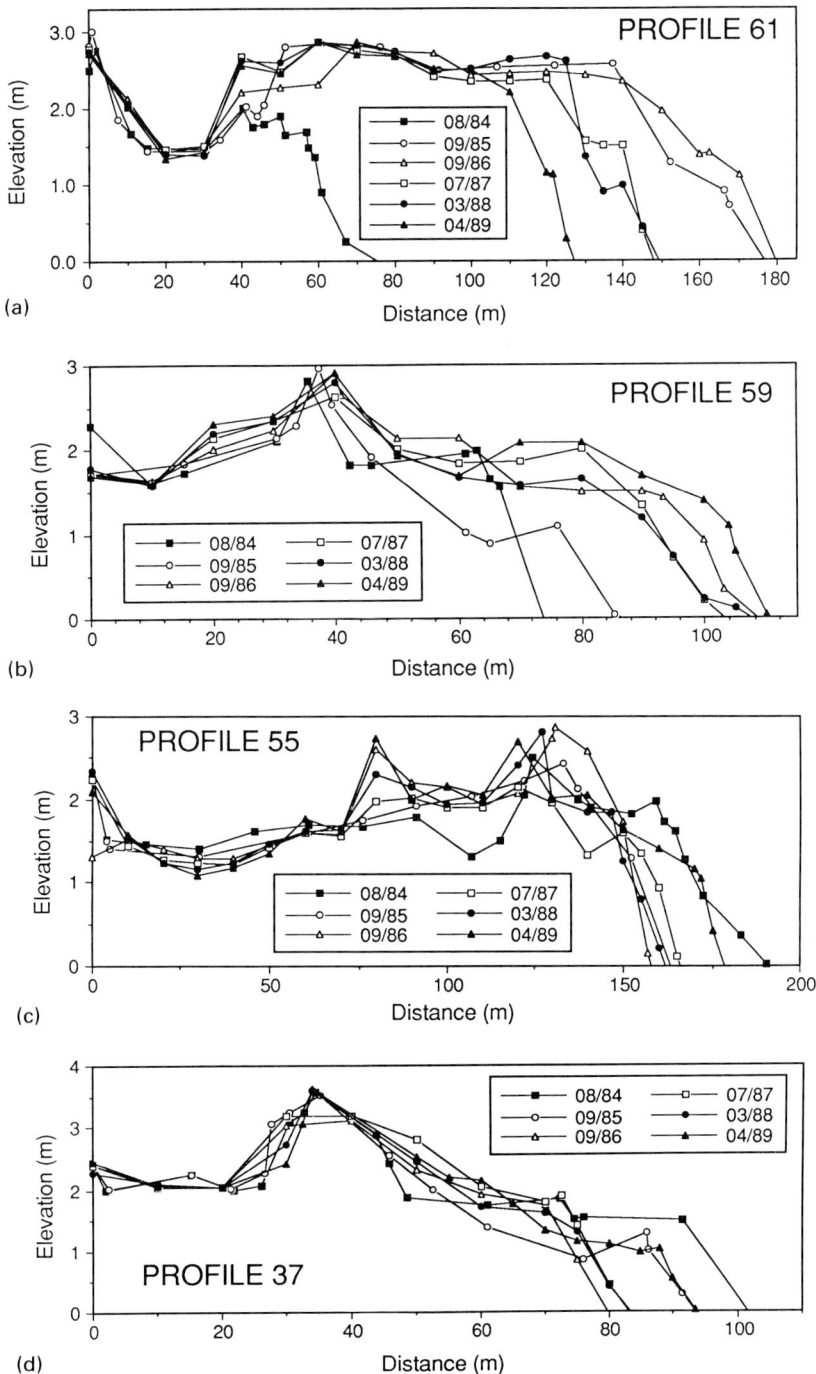

Fig. 12. Profiles. (**a**) R-61: beach fill completely covered the small foredune remnant that was located 35–40 m in front of the datum; (**b**) R-59: beach accretion widened the low portion of the profile as sediment was transferred from the nourishment zone into the immediate downdrift sector; foredune growth occurred along the dune face and at a seaward foredune ridge; (**c**) R-55: a lagoonward sloping washover fan with its crest adjacent to the active foreshore in the profile; isolated dune hummocks attain heights of about 50 cm above the washover feature; (**d**) R-37: accumulation high on the profile while the shoreline is being displaced landward.

the west. The second characteristic was that the deficit was reduced through time as the down-drift transport from the beach fill zone apparently began to enter this portion of the island, returning the beach position to its 1985 location. This latter characteristic is a reversal of the trend, albeit short-term, of the 135-year record.

The accompanying foredune budget shows a sequence that has large accumulations in the region immediately downdrift from the fill as the foredune both increased in width, height, and expanded seaward. The original washover zone shared in some of the foredune growth as sediment began to accumulate in the upper profile at its eastern margin. However, for most of the washover zone, there was no foredune accumulation. Toward the western, downdrift, margin of the washover zone, there was major erosion of some of the pre-existing foredune deposits. The end product of this sequence was a general downdrift shift of the washover topography. It is the next portion of the foredune system that may be the most instructive. The general foredune budget begins to increase as the topography shifts from the washover zone into the area of a coherent foredune ridge. The increase reaches a maximum a short distance from the washover zone and then decreases in the downdrift direction. Much of the increase in the foredune budget occurred during the time when the beach was being displaced inland under the influence of several hurricanes and winter storms. It is perhaps noteworthy that the positive foredune budget is decreasing in the downdrift direction while the negative beach budget is approaching a balance in the same direction.

There are many gaps in the record of foredune growth and development on Perdido Key. The data sets are assembled for a very short time and the sediment budget has been altered by a major fill episode that was an order of magnitude greater than the annual sediment budget for that portion of the island. There is no differentiation of the role of aeolian processes in foredune sedimentation from that of near-shore processes in removing or accumulating sediment in the upper portion of the profile. Additionally, there are no threshold values that tend to describe when foredunes maximize their growth and development. Yet, the 1978 aerial photographs indicate that foredunes continued to characterize the beach profile of the rapidly eroding portion of the barrier. Thus, there is evidence that the dune-forming processes were able to maintain a viable foredune, although possibly episodically, associated with a negative beach budget. Similar to situations repeated on many barrier islands, many of the eastern Perdido Key

foredunes are maintaining their mass despite a long-term negative budget that has seen the barrier shift inland nearly 200 m in the last century. It is possible that the foredune-forming process is linked with this inland migration that continues to transfer sediment to the upper portion of the dune/beach profile even while the beach is recording a net loss.

This study was supported by the Rutgers University Institute of Marine and Coastal Sciences and the National Park Service through the Cooperative Research Unit at Rutgers under Contract CA-1600-3-0005. Members of the Park Service contributed significantly to the completion of the field effort, especially Dr James R. Allen (North Atlantic Region, Boston) and Dr Robert Thackeray (Gulf Islands National Seashore, Pensacola). The Seashore provided ample and well-appreciated logistical support. Kathleen Jagger gave unstintingly in many aspects of this project. Jim Brosius and Anne Marie Newman assisted in the field surveying. Steve Namikas reduced the data and performed the calculations. Michael Siegel, Director of the IMCS Cartography Lab, prepared the illustrations. This is Contribution 92-19 of the Institute of Marine and Coastal Sciences.

References

BALSILLIE, J. H. 1986. *Longterm Shoreline Rates for Escambia County, Florida.* Technical and Design Memorandum No. 86-2, Florida Department of Natural Resources, Division of Beaches and Shores, Tallahassee, Florida.

CARTER, R. W. G. 1988. *Coastal Environments.* Academic Press, London.

DEAN, R. G. 1983. Principles of beach nourishment. *In:* KOMAR, P. D. (ed.) *Handbook of Coastal Processes and Erosion.* CRC Press, Boca Raton, Florida.

GALVIN, C. J. & SEELING, W. 1969. *Surf on the United States Coastline.* US Army Corps of Engineers, Coastal Engineering Research Center, Fort Belvoir, Virginia.

GOLDSMITH, V. 1985. Coastal dunes. *In:* DAVIS, R. A. (ed.) *Coastal Sedimentary Environments,* 2nd edition. Springer, New York, 171–236.

HUBERTZ, J. M. & BROOKS, R. M. 1989. *Gulf of Mexico Hindcast Wave Information.* Wave Information Studies Report 18. US Army Corps of Engineers, Coastal Engineering Research Center, Vicksburg, Mississippi.

HUGHES, S. & CHIU, T. Y. 1981. *Beach and Dune Erosion During Severe Storms.* University of Florida, Dept of Coastal and Oceanographic Engineering Report UFL/COEL-TR/043, Gainesville, Florida.

JAGGER, K. A., PSUTY, N. P. & ALLEN, J. R. 1991. Caleta morphodynamics, Perdido Key, Florida, U.S.A. *Zeitschrift für Geomorphologie, Supplement Band,* **81**, 99–113.

NATIONAL PARK SERVICE. 1978. *General Management Plan and Development Concert Plan, Gulf*

Islands National Seashore, Mississippi–Florida. US Department of the Interior, Denver Service Center, Colorado.

NORDSTROM, K. F., PSUTY, N. P. & CARTER, R. W. G. (eds) 1990. *Coastal Dunes: Form and Process.* John Wiley & Sons, Chichester.

PASKOFF, R. 1989. Les dunes du littoral. *La Recherche*, **212**, 888–895.

PSUTY, N. P. (ed.) 1988a. Dune/Beach Interaction. *Journal of Coastal Research*, Special Issue No. 3.

—— 1988b. Balancing recreation and environmental system in a 'natural area', Perdido Key, Florida, USA. *Ocean and Shoreline Management*, **11**, 395–408.

—— 1992. Spatial variation in coastal foredune development. *In*: CARTER, R. W. G., CURTIS, G. F. & SHEEHY-SKEFFINGTON, M. (eds) *Coastal Dunes: Geomorphology, Ecology and Management*. Balkema, The Hague, 3–13.

—— & JAGGER, K. A. 1990. *Final Report on the Shoreline Changes on Perdido Key, Florida, Gulf Islands National Seashore.* Center for Coastal and Environmental Studies, Rutgers – The State University of New Jersey, New Brunswick, New Jersey.

PYE, K. 1983. Coastal dunes. *Progress in Physical Geography*, **7**, 531–557.

SHORT, A. D. & HESP, P. A. 1982. Wave, beach, and dune interactions in southeast Australia. *Marine Geology*, **48**, 259–284.

VELLINGA, P. 1982. Beach and dune erosion during storm surges. *Coastal Engineering*, **6**, 361–387.

WALTON, T. L., Jr. 1976. *Littoral Drift Estimates along the Coastline of Florida.* Sea Grant Report 13, Coastal Oceanographic Engineering Laboratory, Gainesville, Florida.

WRIGHT, L. D. & SHORT, A. D. 1983. Morphodynamics of beaches and surf zones. *In*: KOMAR, P. D. (ed.) *Handbook of Coastal Processes and Erosion*. CRC Press, Boca Raton, Florida, 35–64.

Wind regime and sand transport on a coastal beach–dune complex, Tentsmuir, eastern Scotland

ABHILASHA WAL & JOHN McMANUS

Department of Geography & Geology, University of St Andrews, Fife KY16 9ST, UK

Abstract: Aeolian processes play a key role in the exchange of sediment between beaches and foredunes in coastal embayments. The present contribution examines the wind regimes characterizing one stretch of temperate coast (Tentsmuir, Scotland), the routing and volumes of the resultant aeolian sediment transport on the beaches and foredunes and presents a summary of wind-induced aeolian bedforms along the coast.

The wind regimes identified in the Tentsmuir area include high-energy seasonal 'unimodal' (offshore or onshore) and 'bimodal' (both offshore and onshore) patterns. Landward aeolian sand transport results in foredune accretion, whereas seaward and longshore components contribute to beach growth despite reworking by waves at high water. Under the influence of the prevalent wind regimes and the resultant aeolian sediment transport, foredune accretion at Tentsmuir leads to the burial of the landward margin of the beach face.

Many previous studies of coastal dunes have been limited to descriptive accounts of dune form and evolution (Cooper 1958; Jennings 1957; Olson 1958; Smith & Messenger 1959), their ecological importance (Ranwell 1972; Woodhouse 1978) and physical development in relation to near-shore processes (Davies 1980). The wide diversity of approaches used hitherto has led to the recognition of a need for the adoption of a generally accepted technique to permit the characterization and quantification of the various factors controlling coastal aeolian activity and dune development.

During the last fifty years, research on the rate of aeolian sand transport within the beach and dune environments has not been very extensive. Most studies have been confined to wind tunnels in the laboratory except for a few sporadic field measurement contributions by Svasek & Terwindt (1974), Howard et al. (1978), Hotta et al. (1985), Sarre (1988), Illenberger & Rust (1988), Gares (1987), Bauer et al. (1990) and Goldsmith et al. (1990). Most of the measurements of sand transport rate on beaches have been aimed at the comparison of predictive models (Bagnold 1941; Kawamura 1951; Kadib 1965) with measured transport rates (Sarre 1988; Hotta et al. 1985), or at defining the limits (moisture, shear velocity) of the transport rate expressions (Svasek & Terwindt 1974). Little has been achieved in the way of defining *detailed* aeolian sediment transfers and patterns of deposition along the coast, for example the variability of transport rate across the beach (lower, middle and upper beach face) and dune (foredune and primary dune: for definitions refer to Psuty &

Millar 1989; Hesp 1988) sub-environments, except for the recent works of Gares (1987), Bauer et al. (1990) and Goldsmith et al. (1990). In consequence, a programme to determine sand transport rates was undertaken on the Tentsmuir beach through emplacement of Leatherman (1978) vertical sand traps in a range of beach and dune environments and a Cassella cup rotation anemometer array along the middle beach face.

There are essentially three areas of study discussed in this paper: potential drift rates and direction from wind records; measured sand transport rates; and a classification of sedimentary structures observed on the beach and the dunes under different transport conditions.

Study site

The 12 km-long Tentsmuir beach–dune complex is situated on the macro-tidal (tidal range 4–6 m) coast of northeast Fife, Scotland (Fig. 1). It is a site of long-term net accretion, with over 3.5 km of shoreline advance in 5000 years (Ferentinos & McManus 1981). Foredune accretion follows the coastal orientation, with north–south-trending parallel dune ridges in the central portion of the shoreline, and less well-defined broken dune ridges aligned northwest–southeast in the extreme northeast at Tentsmuir Point.

At low water the low gradient beaches are 85–400 m wide. The beaches are wider towards the northern and southern extremities of the Tentsmuir platform, towards the mouths of the Tay and Eden estuaries, respectively (Fig. 1). There is no evidence to suggest that the dunes

From Pye, K. (ed.), 1993, *The Dynamics and Environmental Context of Aeolian Sedimentary Systems.* Geological Society Special Publication No. 72, pp. 159–171.

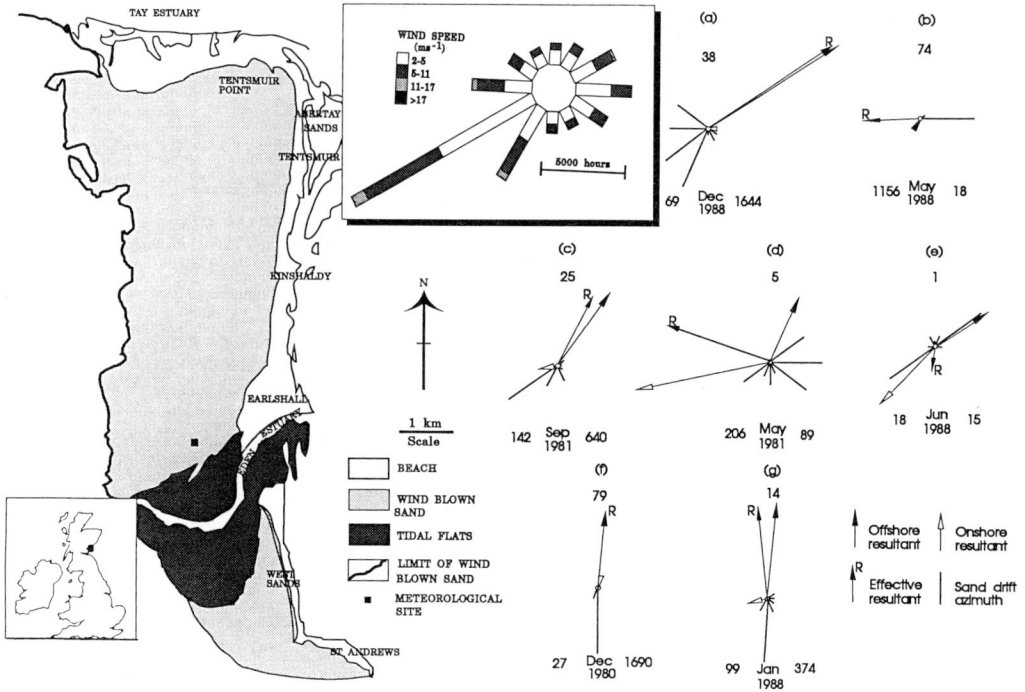

Fig. 1. Study area showing the location of the meteorological site and the average wind rose for 1980–90. Also shown are the monthly sand roses showing (**a**) unimodal offshore; (**b**) unimodal onshore; (**c**) bimodal offshore; (**d**) bimodal onshore; (**e**) bimodal equal; (**f**) unimodal longshore; and (**g**) bimodal longshore wind regimes. Lines represent the sand-drift potential from that particular azimuth and arrows point in the direction of the resultant drift potential. Numbers on the left, right and top centre of each sand rose are the onshore drift potential, offshore drift potential and reduction factor, respectively.

are 'coupled' to the estuary-mouth circulations or bedforms. Backshore dune accumulations occur at Tentsmuir Point (2–3 m high), Kinshaldy (1–2 m high) and West Sands (4 m high).

Grain-size analysis of coastal beach surface sediments from West Sands to Tentsmuir Point (45 samples) revealed that the mean grain size of the sediments was 0.25 mm. In general, there was a northward coarsening of the sands from 0.17 mm in the south to 0.28 mm in the north. Beach gradients also increased gradually northwards. At any single locality the sands became finer up the beach so that beach face slopes decreased landward at most sites. In general, the sorting of the sands, as expressed by the standard deviation of the grain-size distribution, decreased northwards and down-beach.

The prevailing winds of this region are south-westerlies, which long-term Meteorological Office records reveal to blow for 23.5% of the year at Leuchars air station (11 years of record)

on the northern shore of the Eden estuary (see wind rose in Fig. 1), and for 35% of the year at the Bell Rock lighthouse 20 km to the northeast (11 years of record). However, the dominant wind, defined as that which induces most coastal erosion, blows from the southeast for no more than 10% of the average year (Ferentinos & McManus 1981).

Wind regimes and sand-drift potentials

Sand-drift potentials for the Tentsmuir area have been calculated from the long-term meteorological wind records using the method of Fryberger & Dean (1979) who estimated the potential sand drift (*Q*), expressed numerically in vector units (VU), as being proportional to the product of a weighting factor and the percentage of total wind recorded in time (*t*), as:

$$Q \; \alpha \; [V^2 - (V - V_t)]^1 \; \times t \quad (1)$$
$$[\text{weighting factor}]^1$$

where, V is the wind velocity at 10 m height for the time period t, and V_t is the threshold velocity (10 m height) at which the sand (average md. 0.25–0.30 mm) moves (5.9 m s^{-1}). Sand roses have been used to represent the potential sand drift from the twelve compass directions. The arm lengths of the rose are proportional to the potential sand drift from a given direction expressed in vector units. Vector unit totals from the different directions have been vectorially resolved to three single resultants: representing onshore (for 0 to 180° azimuth); offshore (180° to 360° azimuth); and the effective (0° to 360°) resultant vectors. The resultants indicate the direction in which the wind blows, unlike the arms of the sand rose which represent the direction from which the wind is derived.

Wind regime classification

Analysis of monthly sand roses from 1980–90 has permitted the classification of wind regimes in terms of the offshore, longshore and onshore components of wind velocity, and also provided an estimate of the potential aeolian sand-drift rates at Tentsmuir.

Wind regimes are classed as 'unimodal' (Figs 1a & b) when either the offshore or onshore component is the dominant wind resultant (i.e. the ratio of the resultant drift potentials (RDP) of the two components (offshore and onshore) is <9) and 'bimodal' (Figs 1c, d & e) when both the offshore and onshore resultants are present in significant magnitudes (ratio of RDP of the two components is between 2 and 8). 'Bimodal offshore' (Fig. 1d) and 'bimodal onshore' (Fig. 1e), respectively, can be considered as variants of the bimodal regime. When the offshore and onshore components are nearly equal (ratio of the RDP is between 1 and 1.5) the wind regime is classed as 'bimodal equal' (Fig. 1e). 'Longshore' wind regimes (both unimodal and bimodal), whether with northward or southward direction of motion, are approximately parallel to the coastline (Figs 1f & g).

Description of wind regimes

The complete scenario of drift potential between 1980 and 1990 is shown in Fig. 2. The months October to January are dominated by unimodal offshore southwesterly winds. Extremely high sand-drift potentials (700 VU to 3000 VU) are recorded during this season. The month of March also is not uncommonly subject to these wind conditions (1986, 1989, 1990).

The unimodal onshore wind regime is characteristic of the months of February to June, when the onshore sand-drift potentials range from 197 VU in June to 855 VU in March. Still higher values of onshore drift potentials (1158 VU) are identified occasionally, as, for example, during May 1988 (Fig. 1b).

Dominantly offshore bimodal winds are much more abundant than the bimodal onshore winds and are experienced throughout the year, although particularly from September to December, occasionally in March and rarely in April. During some of the months the onshore component of the bimodal offshore wind regime is substantial (e.g. February 1981: 718 VU; January 1984: 616 VU; October 1980: 423 VU; December 1983: 416 VU).

Bimodal onshore winds occur during spring (April and May), late winter (January and February) and rarely in the months of November and December. The onshore sand-drift potentials during the bimodal onshore winds range from 104 VU to 841 VU.

Bimodal equal wind regimes, with nearly equal magnitudes of the onshore and offshore drift potential resultants, operate during the summer months of June, July, August (Low RDP: 25– 74 VU), occasionally in October and November (High RDP: 241–471 VU) and also in April (intermediate RDP: up to 188 VU). Exceptionally high sand-drift potentials have been observed during February 1981 when the onshore and offshore drift potential were 718 and 1062 VU respectively.

Longshore winds of the winter months (September to February) often blow to the north but rarely towards the south with drift potentials ranging from 115–1690 VU.

From the above data it is evident that:

1. sand transport vector resultants are related to the seasonal component of the wind regime: strong offshore winds during the peak of winter; strong onshore winds during spring; bimodal dominantly onshore winds during late winter (January and February), autumn and spring (April and May); cyclic sea breeze and weak onshore winds during the summer months of June, July and August;

2. unimodal to bimodal wind regimes are prevalent at Tentsmuir. The magnitude of wind variability (ratio of the resultant drift potential and the drift potential) is between 0.55–0.85 indicating moderate to low variability of the prevalent winds. In general the intermediate (200–400 VU) to high drift potentials (> 400 VU) of Fryberger & Dean (1979) are present in the area.

Resultant Drift Potential (vector units)

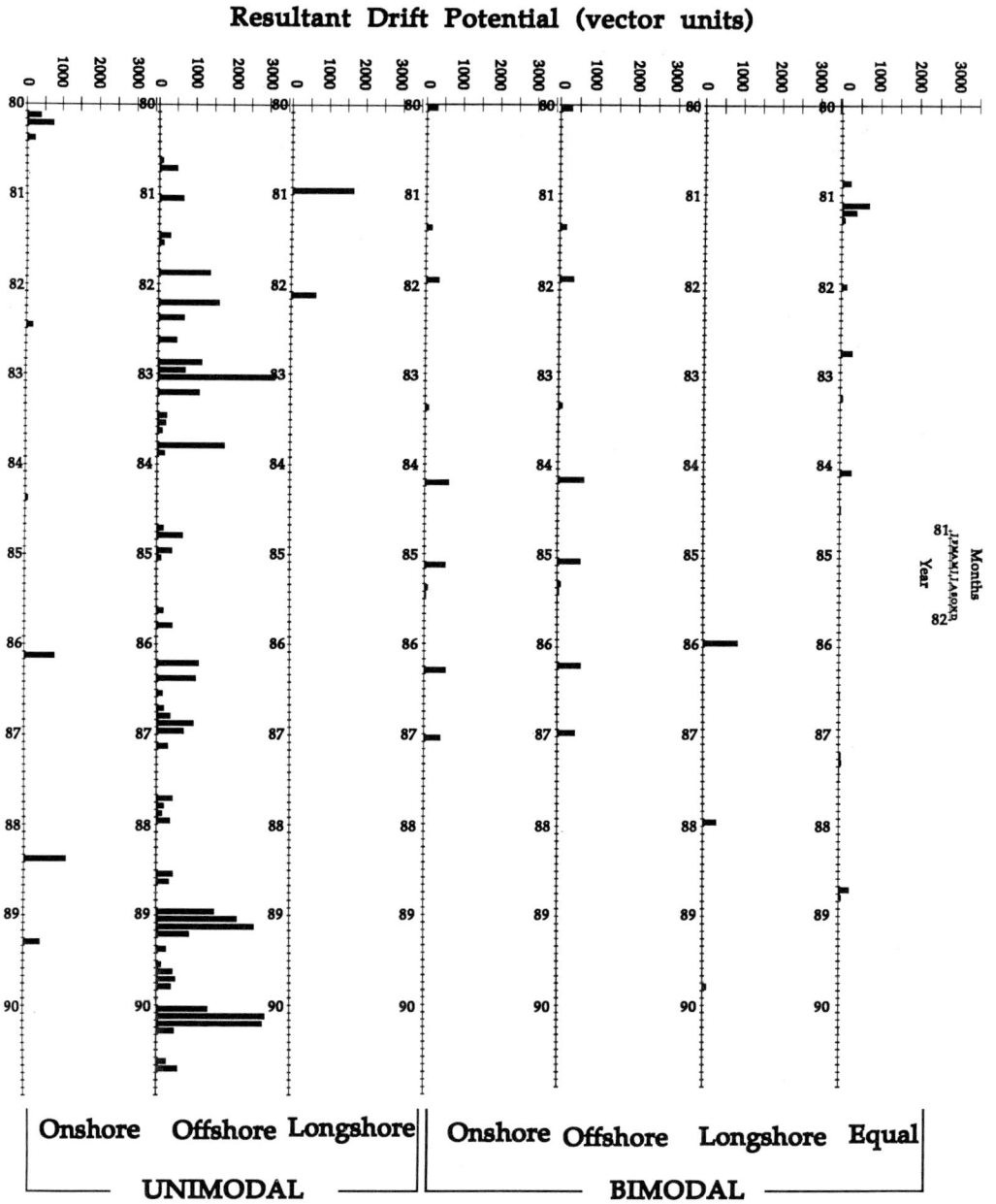

Fig. 2. Summary of wind regimes based on drift potential variation between 1980–90.

Short-term sand transport rate

Although it is possible to compute the potential movement of sand under any given set of conditions, it is important that verification be attempted. On the Tentsmuir beach, rates of sand transport have been determined on 15 occasions under a range of wind conditions principally through the emplacement of vertical sand traps in a range of beach and dune environments.

Technique

Transects were established across the beach at several sites along the coastline, namely Tentsmuir, Kinshaldy and Earlshall (Fig. 1) for repeated trapping measurements to enable computation of aeolian rates of sand transport. The traps were emplaced in an array of 5 to 8 installations in various beach–dune sub-environments during conditions of dominantly offshore, onshore or longshore winds. Sub-environments examined were the primary dune, foredune, upper (UBF), middle (MBF) and lower (LBF) beach face areas, into which traps were inserted by digging a hole in the ground with a hand trowel and positioning the entry slit base flush with the ground surface. If more than a single trap installation was made in a particular sub-environment, the landward trap was denoted by a suffix 'I' e.g. MBF I. It was impracticable to place the traps very close to the water's edge due to the problem of groundwater seepage into the base of the trap. However, reduction of the length (length embedded into the ground surface) of the trap enabled the traps to be placed within 10 m of the water's edge in order to explain the limiting role of surface moisture on the rate of sand transport. The traps were exposed to winds of 4–20 m s^{-1} for 15 minutes. Once exposed, the traps were monitored very carefully to ensure that they were not clogged by the moist sand and that the slit faced the dominant wind direction. The traps were exposed for 15 minutes and several experimental runs on fifteen different days between March 1990 to September 1991 were made, yielding 56 trap results. After the period of exposure, the sand in the trap was emptied into air-tight polythene bags, brought to the laboratory and weighed immediately to ensure minimum moisture loss. The samples were then oven dried at 110°C, re-weighed and the moisture percentage determined. Sieving and grain-size analysis of the samples was then carried out to establish the mean grain size of the trapped sand. In all cases the direction of the wind was determined indirectly from measurement of the axes of aeolian bedforms produced during sand migration across the beach surface.

Wind velocity measurements

For measuring velocity profiles (to derive u_*) it is necessary to deploy vertical arrays of anemometers and employ some form of recording device. The anemometer array was emplaced in the middle of the beach and wind velocity determinations made at 0.5 m, 1.5 m and 2.0 m above the ground surface. Each anemometer produces one output pulse per revolution which is recorded into an integrated circuit pulse counter, the potential being supplied by a 5.6 V d.c. dry-cell battery. A precision timer gates the pulses to the counter for a pre-set period of 60 s and total count is displayed on an LCD screen. The wind direction was determined by the axes of the aeolian bedforms produced on the beach. However, for taking multiple simultaneous measurements the three anemometers are connected to a data logger, which records the number of revolutions of the cups during a minute. This number is then converted to the speed in m s^{-1} by a manufacturer's calibration graph supplied with the anemometer. The wind speed measurements were then plotted versus height on semi-logarithmic graph paper. Lines were drawn across the points and the gradient or the slope of the line, which is also a measure of the shear velocity, was calculated using equation 2 (Horikawa 1987).

$$u_* = (u_{10} - u_1)/5.75 \qquad (2)$$

The wind speeds ranged from 4–20 m s^{-1} at 2 m height. Wind velocity profiles during sand movement on the Tentsmuir beach, showed that the shear velocity ranged from 0.185 m s^{-1} (at the threshold of sand movement) to 0.52 m s^{-1} (Fig. 3) during active sand transport.

The projected focal point (u', z', $\pm 5\%$ instrumental error) on the Tentsmuir beach corresponds to a wind speed of 1.75 m s^{-1} at a height of 0.3 mm (Fig. 3). Numerous workers have reported their inability to successfully compute the empirical determination of the focal point (Zingg 1953; Howard et al. 1978; Sarre 1988) and they have either resorted to the use of formulas for the determination of the z' as a function of the grain diameter or have used the values determined empirically by other workers. A number of workers have given different values for the focal point u', z', e.g. Bagnold (1941, p. 60) (2.5 m s^{-1} at 0.3 cm), Belly (1964) (2.7 m s^{-1} at 0.3 cm), Horikawa et al. (1986) (2.5 m s^{-1} at 0.6 cm), Chepil & Woodruff (1963) (2.14 m s^{-1} at 0.6 cm), Hsu (1973) (2.75 m s^{-1}). The importance of the focal point has been emphasized by Bauer et al. (1990), as it links the shear velocity to the wind speed at any arbitrary height.

The wind direction records from the adjacent Leuchars meteorological station during the simultaneous measurement of wind flows in the field with three anemometers indicate that the flow field was fairly constant in approach to the

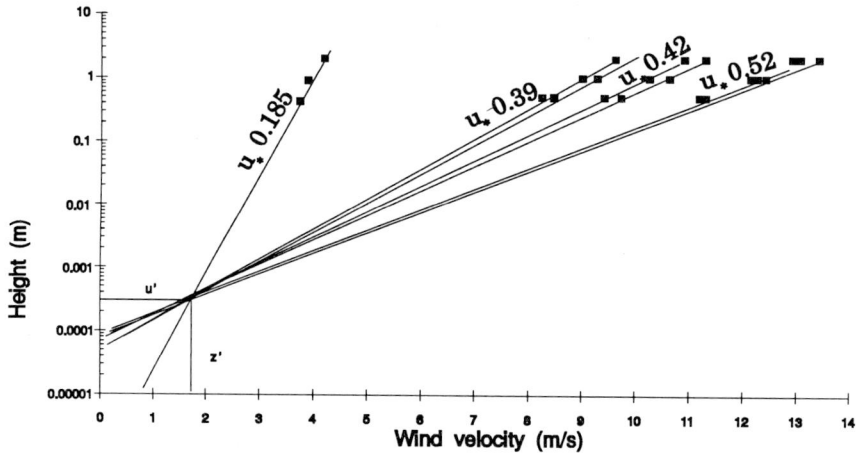

Fig. 3. Wind velocity distributions on Tentsmuir beach during sand movement. u_* is the shear velocity (m s^{-1}). Focal point $(u'\, z')$ corresponds to $u' = 1.75$ m s^{-1}, $z' = 0.3$ mm.

beaches. Thus, any differences in the measured sediment transport cannot be attributed to a spatially varying directional flow field. Since, a single anemometer array was available it was difficult to measure the *absolute* spatial variations in the magnitude of flow velocity across the beach–dune profile. While conducting wind velocity measurements across a beach and dune at Castroville, California, Bauer *et al.* (1990) found that the temporal fluctuations in wind speed and direction across the beach and dune system were similar. However, they also observed the spatial variation in the absolute magnitude of the wind velocity noting that the fastest flow velocities occurred at the dune crests and smallest in the lee of the dunes.

Measured sand transport rates

Before embarking on the description of the rates of sand transport on the beach, it will be necessary to comment on the source of sand that is available for entrainment by the wind. Any wind (whether offshore, onshore or longshore) causes the drier exposed parts of the beach to become deflated. The offshore winds, in particular, deflate the exposed dune face and also both the upper and middle beach face areas. The longshore winds move the sand along the exposed dune face and along the drier upper beach face zone. The onshore winds, on the other hand, move the dry sand available in the middle beach face only (during rising tide) and also the lower beach face during low tide. Thus, at any one time the source of sand, for entrainment by offshore or onshore wind, is compartmentalized and

emanates from a well-defined strand line beyond which the sand is relatively dry.

Figure 4 shows a plot of measured transport rates based on 56 trap results against the corresponding shear velocities measured during field observations. The rates of sand transport vary from sub-environment to sub-environment and no clear relationship between sand transport rate and shear velocity is discernible. However, sand transport rates in the different sub-environments plot in distinct envelopes (Fig. 3). Maximum sand transport rates are observed in the middle beach face sub-environment ranging from 0.01–0.3 kg s^{-1} m^{-1} for shear velocities ranging from 0.2–0.46 m s^{-1}. The lowest sand transport rates, for a similar shear velocity spread, occur in the forcdune (Fig. 3). Lower beach face and upper beach face I envelopes fall between the middle beach face and foredune envelopes. The sand transport rate envelope of the upper beach face sub-environment lies in a very broad area and envelopes both dune and lower beach envelopes (Fig. 4).

Strong offshore winds (> 9 m s^{-1}) create deflation surfaces in the upper beach area because all the sand is being transported towards the sea (Fig. 5a). Hence, low sand transport rates ranging from 0.02–0.03 kg s^{-1} m^{-1} are observed along the upper beach face once this zone has been stripped of the mobile sand. The sand is moved from the upper beach face towards the middle beach face sub-environment, and in consequence this zone becomes characterized by the presence of large volumes of mobile sand yielding high local sand transport rates ranging from 0.08–0.16 kg s^{-1} m^{-1}. As a result of the bonding effects near the water

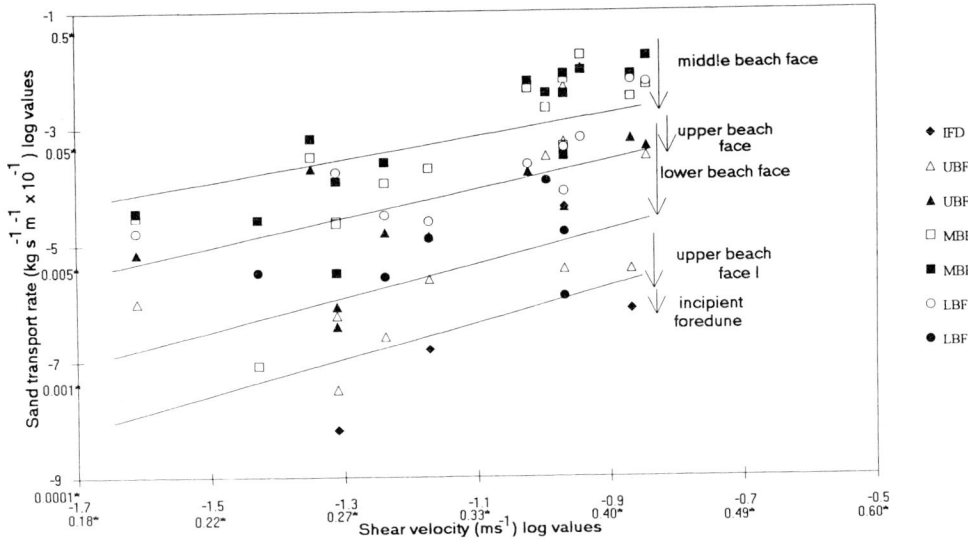

Fig. 4. Shear velocity versus sand transport rate plot. Lines represent envelopes of sand transport variation in different beach–dune sub-environments. LBF: lower beach face; MBF: middle beach face; UBF: upper beach face; IFD: incipient foredune, '*' represents non-log values.

margin, the rate of sand transport decreases (0.019–0.03 kg s^{-1} m^{-1}) across the lower beach face. The transport rates determined during medium velocity winds (< 9 m s^{-1}) followed the same trend of maximum transport rates along the middle beach face ($\simeq$ thrice that of the UBF and LBF values) with reductions along the upper and lower beach face, though the rates were smaller in magnitude. Across the landward face of the primary dune ridge, offshore sand transport rates were very low in comparison with those on the open beach. However, the seaward-facing dune face composed of unvegetated sand blowouts yielded high transport rates (0.04 kg s^{-1} m^{-1}), comparable to those of the upper beach face area, during similar wind conditions.

Field measurements of sand transport rates during onshore winds (> 9 m s^{-1}) (Fig. 5b) on the beach showed increasing rates of sand transport landward of the low-water mark and a subsequent decrease of sand transport beyond the incipient foredune, especially when the area is densely covered with the long dune grass (e.g. *Ammophila arenaria*). An abrupt gradient in surface shear stress between the wind and vegetation causes sand to be deposited quickly in this zone. This results in the accumulation of large volumes of aeolian sand in the backshore and incipient foredune areas. Maximum sand transport rates are found in the middle beach face (0.12–0.13 kg s^{-1} m^{-1}), whereas net deposition occurs along the incipient foredune areas (rates

of approximately 0.01 kg s^{-1} m^{-1} are encountered in this zone) as no sediment transport takes place in the primary dune area beyong the foredune zone. The transport rates along the lower beach face range from 0.01–0.04 kg s^{-1} m^{-1}.

The longshore winds (>9 m s^{-1}) are able to mobilize sand much more efficiently (possibly due to a larger erosional fetch) along the upper and middle beach face sub-environments than the offshore winds (Fig. 5c). Very high upper and middle beach face longshore transport rates (0.1–0.13 kg s^{-1} m^{-1}) have been found at Tentsmuir Point.

Lower wind velocities (<7 m s^{-1}) result in low sand transport rates on the beach. Beach width exposure did not appear to have a limiting effect on the sand transport, especially during high wind velocities. Storm winds (>10 m s^{-1}) were found capable of producing high sand transport even during limited beach exposures. This is in agreement with the findings of Borowka (1990). Our trapping experiments have shown that sand movement by wind was not restricted to totally dry conditions and even moist sand (moisture content 1.2–1.5%) was moved on many occasions. However, the threshold velocity required was greater than for dry sand as revealed by Belly (1964), Johnson (1965) and Sarre (1988).

In general, variation in wind velocity, presence or absence of vegetation, ground surface moisture and the sand source limitation accounts

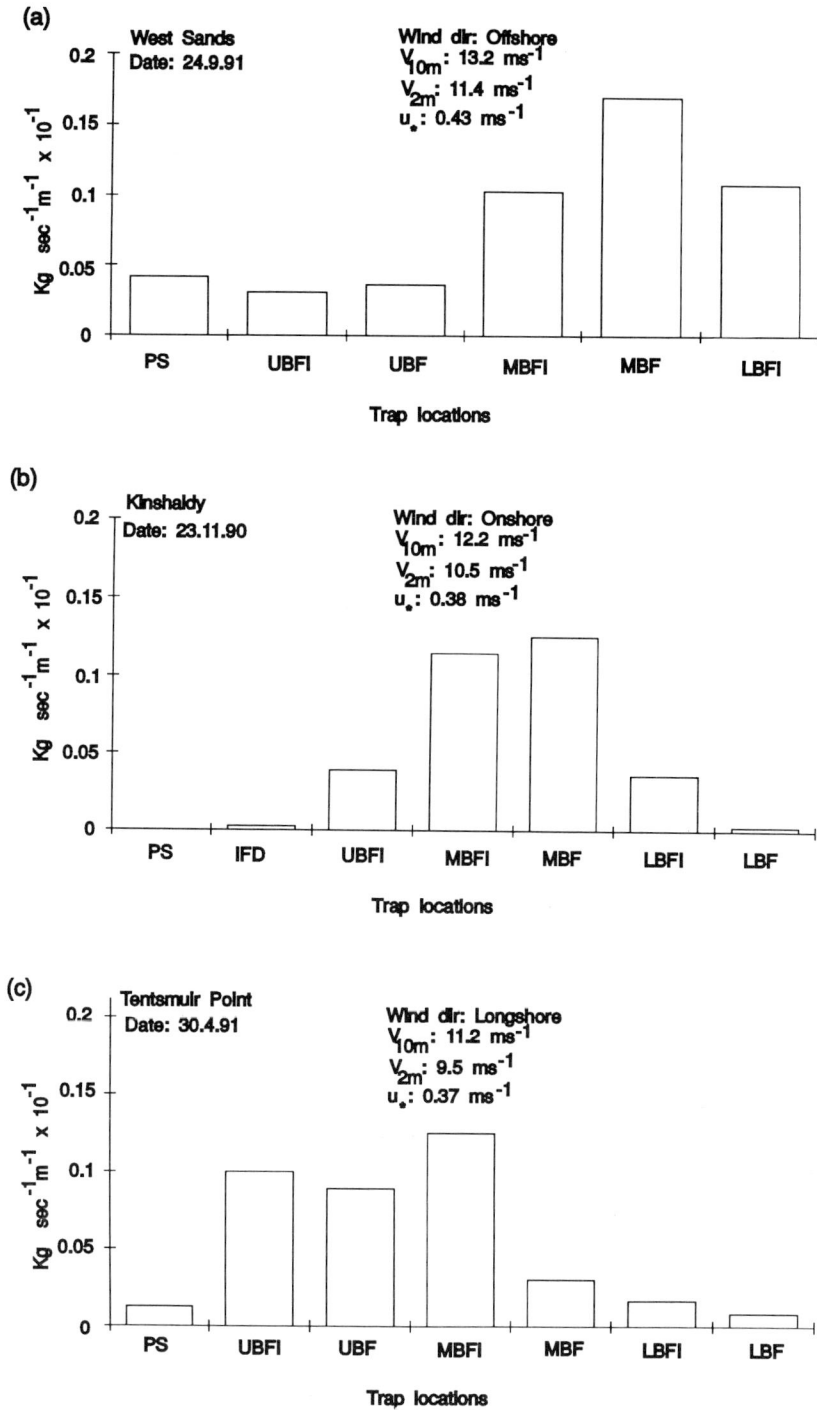

Fig. 5. Measured aeolian sand transport rates in various beach–dune sub-environments during (**a**) offshore; (**b**) onshore; (**c**) longshore winds. LBF: lower beach face; MBF: middle beach face; UBF: upper beach face; IFD: incipient fordune; PS: primary dune seaward slope. V_{2m} and V_{10m} is wind velocity at 2 m and 10 m height, respectively, and u_* is shear velocity.

for the variability of the short-term rates of aeolian sand transport on the Tentsmuir beach and dune sub-environments.

Offshore/onshore transport and the sediment budget

Rates of sand transport are greatest in the middle beach face sub-environment during both offshore and onshore wind regimes. The offshore/onshore net sediment exchange results in a positive sediment budget along the incipient foredune (semi-vegetated) area and the backshore, as there is a high sand transport/accumulation rate in this area during an onshore wind and very low sand transport rate during offshore winds. The middle beach face has a high sand transport rate during both onshore and offshore winds, denoting an excessive amount of mobile sand in this area, and this zone is also the potential source of sand accumulated in the backshore or incipient foredune area by onshore winds. An increase in the frequency of the onshore winds, with a consequent growth of the foredune, causes the sediment budget of the beach to become increasingly negative.

However, the role of the offshore and longshore winds in creating a positive beach sediment budget, due to spatial compartmentalization in response to the limiting effect of the tidal margin, should not be ignored. The general paradigm of relating dune development and growth to the onshore wind frequency *only* is seen to be inappropriate in the light of the present study. This work emphasizes the importance of the middle and upper beach face sediment budget spatial compartmentalization by offshore and longshore winds. Such winds result in aeolian sand accumulation often in the form of small-scale barchans and wind shadows on the foreshore. The chaotic assemblage of these structures often leads to the formation of dune-ridge loci. A number of dunes round the globe (Cooper 1967; Psuty 1969; Leatherman 1976; Rosen 1979) have formed due to offshore and longshore aeolian sediment transport (Sherman & Hotta 1990).

Summary of the wind conditions under which aeolian bedforms are generated

On the basis of field observations of aeolian sand transport and the resultant bedforms produced under offshore, longshore and onshore winds on the beaches of Fife, a synthesis or model interlinking winds and the resultant structures is presented in Table 1. This model will aid in studying the aeolian process–response relationships at the mesoscale in order to establish sediment exchange mechanisms between the beach and the dune. Each sub-environment is characterized by the presence of a particular sedimentary structure/structures formed under the influence of a particular wind regime. The

Table 1. *Model of wind-induced bedforms and structures formed during offshore/onshore sand transport*

		4.3–6 m s⁻¹	6–9 m s⁻¹	>9 m s⁻¹	
S A N D T R A N S P O R T	**Onshore**	Intermittent sand movement on the drier parts of the foreshore	*	Aeolian sand accumulation over and across the crest[3] Aeolian sand accumulation[2]	Dune backshore
			Lee[1] dune (landward pointing)	Sand strips	H.W.M.
			Deflation surface	Deflation surface	L.W.M.
	Offshore	Intermittent sand movement on the drier parts of the foreshore	Rippled duneface	Pyramidal dunes	Dune backshore
			Deflation surface Rippled sand lobes	Deflation surface Sand strips Barchans	
			Lee[1] dune		H.W.M.
		Adhesion structures	Adhesion plane bed Adhesion ripples	Adhesion plane bed Adhesion ripples	L.W.M.
WIND VELOCITY		4.3–6 m s⁻¹	6–9 m s⁻¹	>9 m s⁻¹	

[1], Refers to lee dunes (wind shadow) formed downwind of tidal debris or any other obstacle found on the foreshore. *, same as [2] and [3] but smaller volumes of sands will be entrained and deposited here. 'Rippled' refers to ballistically (wind-induced) rippled deposits. L.W.M. is low-water mark (tide); H.W.M. is high-water mark (tide).

variables within the model are wind speed, foredune height, and ground surface roughness (as a consequence of vegetation and moisture). Three wind velocity categorizations (4.3–6 m s^{-1}; 6–9 m s^{-1}; and > 9 m s^{-1} at 2 m height) have been made as a result of field observations. The structures/bedforms form in response to a decreasing wind velocity from > 9 m s^{-1}. However, due to fluctuations in wind flow, the bedforms produced under the three wind velocity categories may be seen to coexist.

Low wind velocities, whether offshore or onshore, but slightly above the threshold (> 4.3 m s^{-1}) when accompanied by occasional gusts, cause intermittent sand movement across the beach.

During very high wind velocities (> 9 m s^{-1}, both offshore and onshore) sand strips move across the beach and the foreshore forms an almost flat surface, as described by Bagnold (1941), Wilcoxon (1962) and Carter (1988). The surface of the damp beach is deflated leaving behind scattered ridges of sand leading to the formation of scour remnant ridges.

Offshore wind-induced structures

During high velocity (> 9 m s^{-1}) offshore winds the sand in the lee of the primary dune ridge (1.5 –2 m high) is moulded into a pyramidal wind-shadow structure pointing downwind (Hoyt 1966; Hesp 1981; Tsoar 1983). Seaward from the pyramidal dunes, an offshore wind creates a shelly deflation surface (Carter 1976) in the upper beach face zone. Beyond this deflation surface, ballistically rippled foreshore sand lobes are formed in the zone of mobile sand in the middle beach face sub-environment. The rest of the sand flux is deposited in the form of ballistically rippled wind shadows (Allen 1984) in the lee of tidal weeds, brought ashore by the tides along the high-water mark. Sand strips formed during wind velocities in excess of 9 m s^{-1} will transform into barchans if continued high velocity winds (in areas of abundant sediment supply and space) persist over a period of days (4 to 10). If such high velocity winds persist only for a couple of days the sand strips transform into crescent shaped foreshore sand lobes. The probability of the formation of barchans in response to offshore or longshore winds is greater than in response to onshore winds due to (1) presence of sediment source on the windward side; and (2) higher erosional fetch in the case of longshore winds. Adhesion ripples (Glennie 1970) characteristically develop along the moist/wet tidal margin during high

velocity offshore and long-shore winds. Away (1–1.5 m) from the tidal margin at Tentsmuir, an adhesion plane bed (Kocureck & Fielder 1982) is formed.

Longshore wind-induced structures

Longshore winds mobilize the sediments towards the northern and southern extremities of the beach (i.e. the mouths of the Eden and Tay estuaries). Along West Sands beach, the Eden estuary outlet and the Swilken burn limit the longshore sand movement. The sediments in the drier backshore and middle beach face sub-environments are entrained by the longshore winds, leading to the formation of shore-parallel sand accumulation (sand strips, wind shadows, current crescents, barchans). The potential of the longshore winds to transport sediments is high as they have a much wider fetch than the onshore or offshore winds. Adhesion structures have also been observed at the land–water interface.

Onshore wind-induced structures

Onshore winds deposit sand on the windward dune slope and over the crest and landward of the dune ridge (height > 1.5 m). Moderate to high wind velocities (6–10 m s^{-1}) result in the development of ballistic ripples on the dry foreshore sand, landward-pointing shadow dunes in the lee of the tidal weeds and sand accumulation along the upper beach face or potential foredune area. During high velocity onshore winds, sand strips move across the beach and a very high rate of sand accumulation is encountered in the backshore and incipient forcdune area. Adhesion structures can form on a beach with onshore winds if there is a sand source on its seaward side, but they would not be common (Glennie, pers. comm.).

The onshore winds deflate the lower beach face area and deposit sand along the middle and upper beach face. The area of the beach to be deflated will depend on the width of beach exposed to the wind at a given time.

Mode of foredune growth

At Tentsmuir (along areas of foredune accumulation, e.g. Tentsmuir Point, Kinshaldy), onshore wind regimes from January to May (onshore RDP: 800–1200 VU) and July to December (onshore RDP: up to 600 VU) result in significant foredune development in the manner described above. Examples of excessively high rates of foredune aeolian sand accumulation have been recorded in February 1966 and March 1981, during which the vertical growth of vegetation was unable to keep up with the high rate of sand transport (warden reports, Nature Conservancy Council, pers. comm.). Longshore winds (from S and SE) during November 1968 led to the development of approximately one metre-high barchan dunes across the beach. A number of workers (Short & Hesp 1982; Pye 1983; Davidson-Arnott & Law 1990; Chapman 1990) have reported that onshore winds transport sand landwards resulting in a positive foredune budget, while others (Olson 1958; Chepil & Woodruff 1963; Bressolier & Thomas 1977; Goldsmith 1973) have laid emphasis on the importance of pioneer vegetation (e.g. *Elymus farctus*) flourishing on the backshore to act as a suitable trap for aeolian sand. The onshore winds at Tentsmuir play a significant role in foredune development as they transport considerable volumes of sand beyond the upper beach face sub-environment. Once the sand has been transported to the potential foredune zone by these winds it is stabilized by vegetation and so there is little net transfer of the sediments from this area back to the beach during an offshore wind event.

However, the greater frequency of the offshore winds, results in the accumulation of large quantities of aeolian sand in the form of pyramidal sand accumulations across a major dune ridge, and also barchans or crescent-shaped foreshore sand lobes. These accumulations along the middle beach face and backshore, in the course of time, if not destroyed by spring tides or under high aeolian transport conditions, can accrete into a foredune ridge.

Conclusions

At Tentsmuir the potential offshore rates of sand transport are much higher than the onshore transport rates; the former contribute to the development of a positive beach sediment budget due to the limiting effect of the tide line. This inhibits further aeolian sand movement beyond this physical barrier. However, the net beach sediment budget is the result of a complex interplay of aeolian, wave and tidal processes. A positive beach sediment budget is the much needed sand supply for a progradational shoreline to develop, otherwise a negative-feedback cycle is initiated through the narrowing of the beach and erosion of the dune by wave attack (Psuty 1988). The offshore component of the wind regime in an area creates a positive beach face sediment budget, from where (the beach) the onshore winds potentially

can carry the sand landwards resulting in the development of the foredune behind the beach. In fact the offshore winds alone are capable of generating foredune growth along the margins of the primary dune ridge and the beach in the form of pyramidal dunes and wind-shadow structures in the lee of obstacles (tidal weeds or pioneer dune grass e.g. *Elymus farctus* or *Cakile maritima*). These wind-shadow structures eventually accrete and coalesce to form a foredune ridge. The role of the offshore component of the wind regime is related to foredune formation and growth both directly (formation of seaward-dipping wind-shadow structures) and indirectly (by producing a positive beach budget). This differs from the generalization made by other workers of relating the foredune growth to the onshore wind frequency only.

Foredune accretion at Tentsmuir, in the global context, can be considered as one of dune development under the present erosional conditions found worldwide, where the dune budget is positive and the beach budget is negative, a concept advocated by Psuty (1986, 1988).

The potential rates of aeolian sand transport may differ from the actual rates due to the limiting effect of a whole host of factors — e.g. wet or crusted beaches, vegetation, tidal limits, precipitation (Sarre 1989; Pye & Tsoar 1990). However, the Fryberger & Dean (1979) method of analysing meteorological information is useful in giving an idea of the magnitude and direction of potential sand transport, although the values are over-estimations as a mobile dry sand surface is assumed. This technique is extremely useful for analysing the long-term wind velocity distribution in an area and relating it to dune growth, and may also be significant in decision making for the management of coastal land use.

The analysis of wind velocity data has resulted in the understanding of the prevalent wind regime in terms of the directional and magnitudinal variability within the temporal and spatial frames of reference. Both unimodal and bimodal winds of intermediate energy are the characteristic wind types at Tentsmuir. The wind types, as in many desert locations (Glennie 1970; Havholm & Kocurek 1988), in turn are controlled largely by the seasonal component of the wind regime. The study of wind regimes and the understanding of aeolian sediment erosion, transport and depositional processes are fundamental to the development of potential beach–dune transport models (Bauer *et al.* 1990).

This work forms part of a PhD project for AW funded by the Foreign & Commonwealth Office, London. We are extremely grateful to Mr P. Kinnear, Scottish

Natural Heritage, Scotland and the Meteorological Office, Edinburgh for providing useful information about the area. Instrument facilities have been provided by the Department of Civil Engineering, Aberdeen and Tay Estuary Research Centre, Newport. The authors have greatly benefitted from numerous discussions with K. Glennie and R. Duck. Thanks are extended to M. Khanna for help during fieldwork. Lastly we appreciate the diligent efforts of R. Sarre and an anonymous reviewer for critical review of the manuscript and for their valuable comments which have greatly improved the present script.

References

ALLEN, J. R. L. 1984. *Sedimentary Structures*. Elsevier, Amsterdam.

BAGNOLD, R. A. 1941. *The Physics of Blown Sand and Desert Dunes*. Methuen, London.

BAUER, B. O., SHERMAN, D. J., NORDSTROM, K. F. & GARES, P. A. 1990. Aeolian transport measurement and prediction across a beach and dune at Castroville, California. *In*: NORDSTROM, K. F., PSUTY, N. P. & CARTER, R. W. G. (eds) *Coastal Dunes: Form and Process*. John Wiley & Sons Ltd, Chichester, 39–56.

BELLY, P. Y. 1964. *Sand Movement by Wind*. US Army Corps of Engineers. Coastal Engineering Research Center. Technical Memorandum, No. 1.

BOROWKA, R. K. 1990. The Holocene development and present morphology of the Leba Dunes, Baltic coast of Poland. *In*: NORDSTROM, K. F., PSUTY, N. P. & CARTER, R. W. G. (eds), *Coastal Dunes: Form and Process*. John Wiley & Sons Ltd, Chichester, 289–311.

BRESSOLIER, C. & THOMAS, Y. F. 1977. Studies on wind and plant interactions on the French Atlantic coastal dunes. *Journal of Sedimentary Petrology*, **47**, 331–338.

CARTER, R. W. G. 1976. Formation, maintenance and geomorphological significance of an eolian shell pavement. *Journal of Sedimentary Petrology*, **46**, 418–29.

—— 1988. *Coastal Environments*. Academic Press, London.

CHAPMAN, D. M. 1990. Aeolian sand transport — an optimised model. *Earth Surface Processes & Landforms*, **15**, 751–760.

CHEPIL, W. S. & WOODRUFF, N. P. 1963. The physics of wind erosion and its control. *Advances in Agronomy*, **15**, 211–302.

COOPER, W. S. 1958. *Coastal Sand Dunes of Oregon and Washington*. Geological Society of America Memoir **72**.

—— 1967. *Coastal Dunes of California*. Geological Society of America Memoir **104**.

DAVIDSON-ARNOTT, R. G. D. & LAW, M. N. 1990. Seasonal patterns and controls on sediment supply to coastal foredunes, Long Point, Lake Erie. *In*: NORDSTROM, K. F., PSUTY, N. P. & CARTER, R. W. G. (eds), *Coastal Dunes: Form and Process*. John Wiley & Sons Ltd, Chichester, 177–200.

DAVIES, J. L. 1980. *Geographical Variation in Coastal Development*, 2nd edition. Longman, London.

FERENTINOS, G. & McMANUS, J. 1981. Nearshore processes and shoreline development in St Andrews Bay, Scotland, U. K. *In*: NIO, S. D., SHIITTENHELM, R. T. E. & VAN WEERING, T. C. E. (eds) *Holocene Marine Sedimentation in the North Sea Basin*. Special Publication International Association of Sedimentologists, **5**, 161–174.

FRYBERGER, S. H. & DEAN, G. 1979. Dune forms and wind regime. *In*: McKEE (ed.) *A Study of Global Sand Seas*. US Geological Survey Professional Paper No. 1052. 137–169.

GARES, P. 1987. *Eolian Sediment Transport and Dune Formation on Undeveloped and Developed Shorelines*. PhD thesis, The State University of New Jersey, Rutgers.

GLENNIE, K. W. 1970. *Desert Sedimentary Environments*. Developments in Sedimentology **14**, Elsevier, Amsterdam.

GOLDSMITH, V. 1973. Internal geometry and origin of vegetated coastal sand dunes. *Journal of Sedimentary Petrology*, **43**, 1128–1142.

——, ROSEN, P. & GERTNER, Y. 1990 Eolian transport measurements, winds, and comparison with theoretical transport in Israeli coastal dunes. *In*: NORDSTROM, K. F., PSUTY, N. P. & CARTER, R. W. G. (eds) *Coastal Dunes: Form and Process*. John Wiley & Sons, Chichester, 79–102.

HAVHOLM, K. J. & KOCUREK, G. 1988. A preliminary study of the dynamics of a modern draa, Algodones, southeastern California, USA. *Sedimentology*, **35**, 649–669.

HESP, P. 1981. The formation of shadow dunes. *Journal of Sedimentary Petrology*, **51**, 101–111.

—— 1988. Morphology, dynamics and internal stratification of some established foredunes in southeast Australia. *Sedimentary Geology*, **55**, 17–41.

HORIKAWA, K. (ed.) 1987. *Nearshore Dynamics & Coastal Processes—Theory, Measurement and Predictive Models*. University of Tokyo Press, Japan.

——, HOTTA, S. & KRAUS, N. C. 1986. Literature review of sand transport by wind on a dry sand surface. *Coastal Engineering*, **9**, 503–26.

HOTTA, S. KUBOTA, S. KATORI, S. & HORIKAWA, K. 1985. Sand transport by wind on a wet sand surface. *In*: *Proceedings of the 19th Coastal Engineering Conference. Houston*. 1265–1281.

HOWARD, A. D., MORTON, J. B., GAD-EL-HAK, M. & PIERCE, D. B. 1978. Sand transport model of barchan dune equilibrium. *Sedimentology*, **25**, 307–38.

HOYT, J. H. 1966. Air and sand movements to the lee of dunes. *Sedimentology*, **7**, 137–143.

HSU, S. A. (1973) Computing eolian sand transport from shear velocity measurements. *Journal of Geology*, **81**, 739–743.

ILLENBERGER, W. K. & RUST, I. C. 1988. A sand budget for the Alexandria coastal dunefield, South Africa. *Sedimentology*, **35**, 513–21.

JENNINGS, J. N. 1957. On the orientation of parabolic or U-dunes. *Geographical Journal*, **123**, 473–480.

JOHNSON, J. W. 1965. Sand movement on coastal dunes. *In*: *Proceedings of the Federal Inter-agency Sedimentation Conference*. United States Department of Agriculture Miscellaneous Publication, No. 970, 747–755.

KADIB, A. A. 1965. *A Function of Sand Movement by Wind*. Hydraulic Engineering Laboratory Technical Report HEL-2-12, University of California, Berkeley.

KAWAMURA, R. 1951. *Study of Sand Movement by Wind*. Report Institute of Science & Technology, University of Tokyo 5.

KOCUREK, G. & FIELDER, G. 1982. Adhesion structures. *Journal of Sedimentary Petrology*, **52**, 1129–1241.

LEATHERMAN, S. P. 1976. Barrier island dynamics: overwash processes and aeolian transport. *In*: *Proceedings of the 15th Coastal Engineering Conference. New York*, 1958–1974.

—— 1978. A new aeolian sand trap design. *Sedimentology*, **25**, 303–306.

OLSON, J. S. 1958. Lake Michigan dune development 1. Wind velocity profiles. *Journal of Geology*, **66**, 254–63.

PSUTY, N. P. 1969. Beach nourishment by aeolian processes, Paracas, Peru. *In*: *Proceedings of the 10th Meeting, New York–New Jersey Division, Association of American Geographers*, **3**, 117–123.

—— 1986. Principles of dune beach interaction related to coastal management. *Thalassas*, **4**, 11–15.

—— 1988. Sediment budget and beach/dune interaction. *Journal of Coastal Research, Special Issue No. 3*, 1–4.

—— & MILLAR, S. W. S. 1989. Coastal dunes: A concept of zonality. *Essener Geogr. Arbeiten*, **18**, 149–169.

PYE, K. 1983. Coastal dunes. *Progress in Physical Geography*, 7, 531–557.

—— & TSOAR, H. 1990. *Aeolian Sand and Sand Dunes*. Unwin Hyman, London.

RANWELL, D. S. 1972. *Ecology of Salt Marshes and Sand Dunes*. Chapman & Hall, London.

ROSEN, P. 1979. Aeolian dynamics of a barrier island system. *In* LEATHERMAN, S. P. (ed.) *Barrier Islands*. Academic Press, New York, 81–98.

SARRE, R. D. 1988. Evaluation of aeolian sand transport equations using intertidal zone measurements, Saunton Sands, England. *Sedimentology*, **35**, 671–679.

—— 1989. Aeolian sand drift from the intertidal zone on a temperate beach: potential and actual rates. *Earth Surface Processes & Landforms*, **14**, 247–58.

SHERMAN, D. J. & HOTTA, S. 1990. Aeolian sediment transport: theory and measurement. *In*: NORD-STORM, K. F., PSUTY, N. P. & CARTER, R. W. G. (eds) *Coastal Dunes: Form and Process*. John Wiley & Sons Ltd, Chichester, 17–39,

SHORT, A. D. & HESP, P. A. 1982. Wave, beach and dune interactions in southeastern Australia. *Marine Geology*, **48**, 259–284.

SMITH, H. T. U. & MESSENGER, C. 1959. *Geomorphic Studies of the Provincetown Dunes, Cape Cod*. Massachusetts Technical Report No. 1, Office of Naval Research Geography Branch University of Massachusetts, Amherst.

SVASEK, J. N. & TERWINDT, J. H. J. 1974. Measurements of sand transport by wind on a natural beach. *Sedimentology*. **21**, 311–22.

TSOAR, H. 1983. Wind tunnel modelling of echo and climbing dunes. *In*: BROOKFIELD, M. E. & AHLBRANDT, T. S. (eds) *Eolian Sediments and Processes*. Elsevier, Amsterdam, 247–259.

WILCOXON, J. J. 1962. Relationship between sand ripples and wind velocity in a dune area. *Compass*, **39**, 65–76.

WOODHOUSE, W. W. 1978. *Dune Building and Stabilisation with Vegetation*. SR-3, CERC, US Army Corps of Engineers, Fort Belvoir, VA.

ZINGG, A. W. 1953. Wind tunnel studies of the movement of sedimentary material. *In: Proceedings of the 5th Hydraulics Conference Bulletin* **34**, 111–35. Institute of Hydraulics, Iowa City.

Aeolian processes and deposits in northwest Ireland

R. W. G. CARTER & PETER WILSON

*Department of Environmental Studies, University of Ulster at Coleraine,
Coleraine, County Londonderry BT52 1SA, UK*

Abstract: The exposed Atlantic coast of northwest Ireland supports numerous dunefields derived largely from glacigenic deposits in the area of the present continental shelf. The present-day dunes began to form about 5000 to 6000 years BP at, or just after, the mid-Holocene sea-level peak which reached a maximum of + 3 m OD on parts of the County Antrim coast. As the sea-level fell so sediment was transferred onshore accumulating as beach ridges and dunes. Some dunes are perched on gravel ridges of marine origin that were exposed long enough to aeolian processes to allow the formation of ventifacts. Since about 4000 years BP the primary sediment budget has become strongly negative, allowing morphological reworking of many dune systems. The processes of erosion may well have been triggered by climatic changes or human impacts or a mixture of both. Periods of dune instability are marked by the widespread engulfment of former soil surfaces by large-scale mobile bedforms, the transport of sand inland and the formation of extensive sand sheets. At Maghera, County Donegal and Downhill, County Londonderry there are ramp dunes against former sea cliffs, and between Portstewart, County Londonderry and Portrush, County Antrim extensive cliff top dunes formed in the late Holocene which are now isolated from their original sediment source. Non-coastal aeolian sands are of limited extent in northwest Ireland, but some local examples are described.

Northwest Ireland is one of the most exposed and windiest parts of Europe (Troen & Petersen 1989). Within Ireland, the highest average annual wind speed (8 m s^{-1}) is recorded at Malin Head in north County Donegal (Rohan 1986). In addition, the area has an abundance of fine clastic sediment, mostly derived from glacial or paraglacial (areas affected by glacial processes) sources, so that it is not surprising that aeolian dunes are a common landform throughout the area. Dunes fall into two categories; first there are extensive coastal dunes, and second there are small, but nonetheless significant, inland dunes and sand sheets.

The aim of this paper is to describe the evolution of the dune environments of northwest Ireland and to examine aspects of their present day sedimentology, structure and morphology. One justification for such an approach is that there have been few regional descriptions and syntheses of dune systems in the British Isles. In view of the anticipated vulnerability of the British dunes to near-future climate change (Carter 1991), it is important to recognize the diversity of types and to understand something of their origins.

Study area: geology and geomorphology

The northwest of Ireland (Fig. 1) comprises the county of Donegal, plus small parts of Counties Londonderry, Antrim, Fermanagh, Tyrone, Leitrim and Sligo. Much of the west of this area falls within the NE–SW-trending Dalradian metasediments of Precambrian age and the surface rocks include quartzites, pelites and granites (Pitcher & Berger 1972). To the east of the major fault-line along the Foyle Valley, the geology is dominated by Oligocene basalts, overlying Cretaceous chalk.

The basement rocks have been heavily scoured by Pleistocene glaciations, to the extent that much of central and west County Donegal presents a largely ice-eroded surface. While the general pattern of glaciation in Ireland is not disputed (Edwards & Warren 1985), much of the detail remains contentious (Warren 1985; McCabe 1987; Eyles & McCabe 1989). For the purposes of this paper it is sufficient to note that northwest Ireland probably acted as both a centre for small ice-caps (mainly in the Donegal Highlands) and for small basinal ice accumulations in the Erne, Foyle and, to the east, Lough Neagh lowlands. Much of the present landscape owes its origins to the activity of local ice towards the culmination of the Midlandian glaciation between 22 000 and 16 000 years BP. Considerable deposition of glacial debris appears to have been into water; possibly the sea, as Eyles & McCabe (1989) argue, or alternatively lakes. What is clear is that large amounts of clastic sediments, in the form of glacial diamictons,

From Pye, K. (ed.), 1993, *The Dynamics and Environmental Context of Aeolian Sedimentary Systems.*
Geological Society Special Publication No. 72, pp. 173–190.

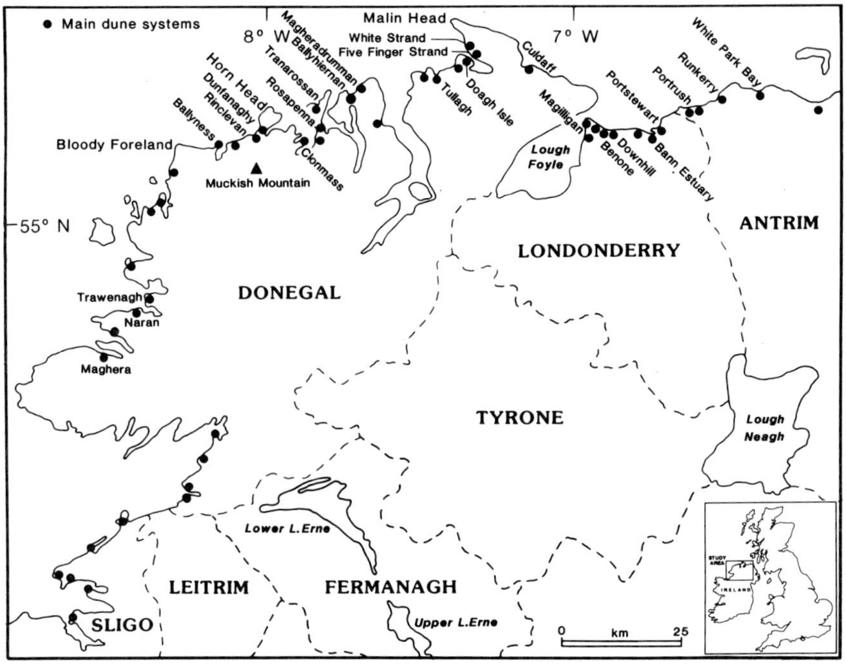

Fig. 1. Locations of dune systems in northwest Ireland.

glaciofluvial sands and gravels were deposited across the present-day coastline or onto the shelf (Evans *et al.* 1980).

The geomorphology of northwest Ireland is dominated by features of glacial erosion and deposition and in places, there are extensive periglacial landforms (Lewis 1985; Wilson 1990*a*). During and immediately after the final glacial phases there was a massive volume

(several orders of magnitude above present) of mobile sediment available to create landforms (Fig. 2). This period of almost over-supply was marked by aggradation as well as a tendency for downslope and onshore sediment transport, resulting in the shoreline zone becoming a locus for deposition. As time proceeded so the sediment availability fell, to be replaced by a general scarcity, interrupted only spasmodically and

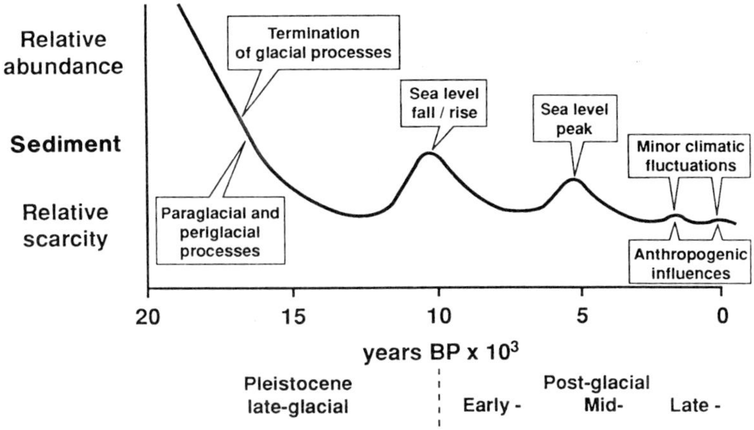

Fig. 2. Coastal sediment economy for northwest Ireland over the last 20 000 years.

locally when environmental conditions changed. In effect, the sediment economy has shifted from one of abundance to one of scarcity over the last 20 000 years. The history and development of the aeolian dunes needs to be placed in this context.

The coastal dunes

Sea-level control and the Holocene development of dunes

The course of relative sea-levels in northwest Ireland during the late- and post-glacial stages is fundamental to the origins and evolution of the coastal landscape.

At the end of the Midlandian, sea-level probably stood 10 to 15 m above the present shoreline. There are numerous late-glacial raised shoreline remnants, particularly in north Donegal (Stephens & Synge 1965), although such features may also be associated with extreme events, which even today can exceed 10 to 12 m above the mean surface (Carter 1983; McKenna 1990). What is indisputable is that the late-glacial sea-level fell rapidly, so that by 12 000 years BP it was probably around −15 to −30 m (Carter 1982a; Devoy 1983), at which time much of the Malin Shelf would have been either exposed or covered with a much shallower water body. The low stand sea-level period lasted for about 4000 years (until 8500 years BP), when the sea-level rose again and flooded the present coastline to a depth of 2 or 3 m (Fig. 3). During the period of low sea-level, terrestrial sediment would have been transported onto what is now the inner shelf shoreface.

The rising, relative sea-level after about 11 000 years BP resulted from a declining degree of isostatic adjustment (Carter 1982a) and a worldwide eustatic increase of oceanic volume. Under the impetus of the rising sea, the glacial debris was eroded, sorted and later incorporated into coastal sedimentary deposits. The general pattern of this movement can be associated with the landward migration of an erosional front (Orford et al. 1991) across which sediment is reworked and partitioned into various textural components. The speed of such changes is related to the rate of sea-level rise (Carter et al. 1989), but may be controlled locally by other factors, including the development and subsequent breakdown of gravel barriers (Carter et al. 1990b).

The evolution of the coast of northwest Ireland over the critical period from 12 000 to 5000 years BP may be divided into three distinct phases (Fig. 4). The first, (Fig. 4I) represents the period when glacigenic sediment accumulated on the shelf. The second phase (Fig. 4II) approximates to the transgressive shoreline moving under a rapidly rising sea-level and the third (Fig. 4III) to the time when sea-levels were stationary or falling slightly, as material was moved onshore.

Stratigraphic and structural evidence for the first phase does not exist. Results from offshore exploration in the North Channel and Malin Sea (Caston 1976; Evans et al. 1980) indicate a current-swept shelf comprising rock outcrops, shell- and gravel-lags, sand ribbons and sand wave fields. Even where present the late Pleistocene/Holocene shelf sequences tend to be greatly attenuated (Keary & Keegan 1975). The relatively dynamic tidal streams together with the long period (up to 16 s) swell wave climate

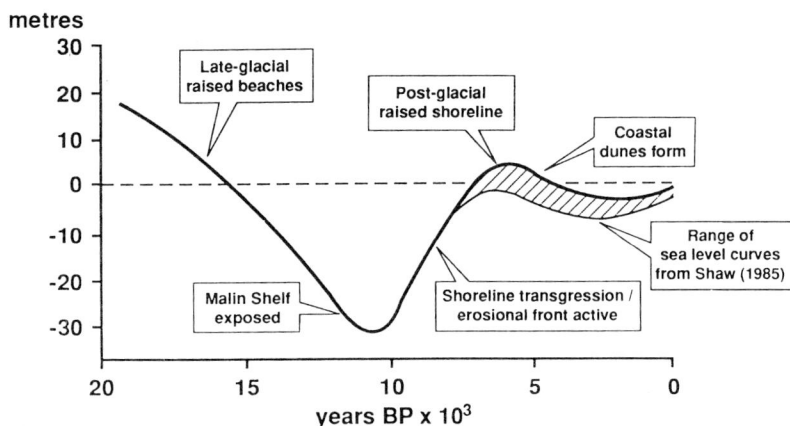

Fig. 3. Sea-level curve for northwest Ireland over the last 20 000 years.

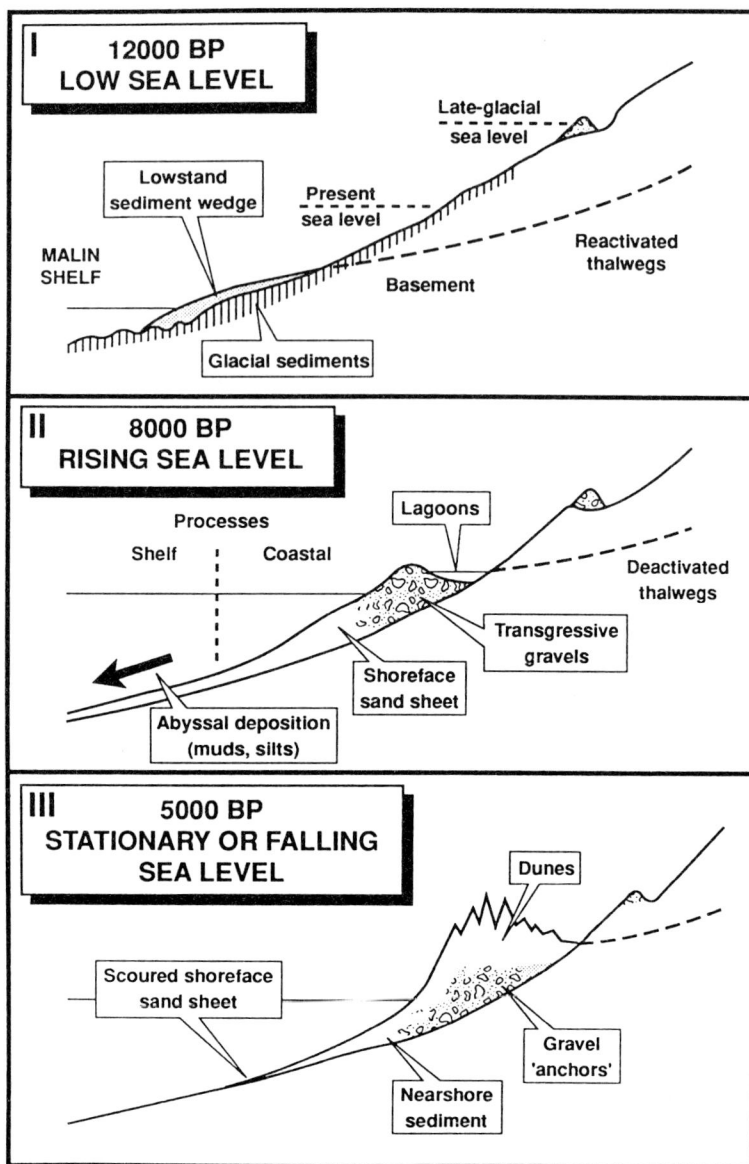

Fig. 4. Phases of coastal evolution in northwest Ireland. Each phase is explained in the text.

(producing significant bottom stresses in 50 to 100 m water depths) have conspired to remove much of the evidence for former low stand strandlines. Much of the exposed seabed is current-swept and extensive shoreface and near-shore sand occurs only in the lee of major head-lands or within estuaries (Evans *et al.* 1980). Directional structures — mainly asymmetric sand waves — indicate dominant easterly trans-port into and through the North Channel.

The impact of the subsequent sea-level rise is preserved at various sites. In the Bann Estuary there was a major freshwater lagoon which was overwhelmed by rising sea-level sometime after 8960 ± 110 years BP (Beta-34315). (All ^{14}C dates quoted in this paper are in uncalibrated ^{14}C years BP.) A similar situation appears to have existed at Portrush, where Jessen (1949) described an extensive lagoon/wetland, although this outcrop has now been destroyed.

However, ^{14}C dates from a nearby exposure of peat lying beneath dune sand (see Wilson & Carter 1990) suggest that the lagoon was forming well before 7310 ± 100 years BP (Beta-36944). The sea-level rose rapidly, peaking about 2 to 3 m above present OD around 6500–5500 years BP in north County Antrim (Carter 1982a) but further west, this peak declines in elevation until it is below present sea level in west County Donegal (Shaw 1985). Shaw also believes the peak sea-level may have been as late as 4500 years BP in the west.

The transgressive phase before the deceleration of sea-level rise around 6000 years BP is important as it is during this period that the shoreline barriers developed. These barriers, mainly formed of gravel, provide the anchor points for later aeolian deposition. On a relatively dissipative coast formed of a dominant sand population, but with a subsidiary gravel mode, wave action will isolate the larger gravel clasts and move them onshore. The asymmetric flow structure and shear stress distribution of shoaling waves favour the preferential onshore transport of gravel (Carter 1988) so that it accumulates at the shoreline. Each gravel clast is ejected from the sand bed and then overpassed until it reaches the shoreline. Gravel ridges, either as solitary structures or in sets, underlie most of the dune systems in northwest Ireland. At Runkerry, Portrush, Portstewart, Rosapenna and Ballyness, solitary gravel beach ridges crop out beneath aeolian sand. At Horn Head and White Strand, there are numerous gravel ridge sets orientated sub-parallel to the present shoreline. The difference in ridge form probably reflects sediment supply during development. The presence of gravel is important geomorphologically as it provides a deflation-limit above the water table, forming dry-core dunes.

At Portstewart, Wilson (1991a) has described ventifacts from the surface of the gravel ridge. One hypothesis would be that the ventifacts formed in the period between that of the ridge construction by waves and burial by wind blown sand, while a second might suggest development within transport corridors that remained active as the surrounding dunes built-up. The ventifacts indicate a phase (or phases) of intense winds, but marked by only a limited supply of fine sediment, permitting wind abrasion of the gravel ridge surfaces. The length of time required for ventifact formation is probably in the order of hundreds of years.

At the time of the maximum mid-Holocene sea-level, it would appear that the inner shelf and shoreface were covered with a relatively thick sheet of well-sorted marine sands. Some of this material may well have been moving into the numerous estuaries and bays along the coast, but by and large it was residing offshore. As the sea-level stabilized, then began to fall away, ideal conditions would have been created to feed sediment onshore, especially from the inner shoreface (above wave base), perhaps to a depth of 15 to 20 m. The gradual withdrawal of the sea would facilitate a number of processes. First, it would remove the reflective element (the gravel ridge) from the backshore, removing the likelihood of standing waves and multiple nearshore bars. Second, it would optimize conditions for the construction of sand beach ridges, and third it would progressively lower wave base, allowing renewal of the offshore sand supply.

The formation of beach ridges is perhaps the key to development of dunes. Beach ridges may form in a number of ways (Carter 1986), but in northwest Ireland two modes are common. The first involves the slow in situ accumulation of sediment around the swash mark on the upper beach, while the second sees the rapid stranding (or welding) of a nearshore bar onto the beach.

At Magilligan, the falling sea-level resulted in the development of a major beach ridge plain, some 3200 ha in extent. The ridges, which fall in four distinctly-orientated sets, may be examined in cross-section along the Lough Foyle (west) shore where they have been eroded. Wilson & Farrington (1989) have dated the Magilligan beach ridge plain to between 4500 and 1500 years BP, indicating that it developed during the mid- to late Holocene. Beach ridge plains, although much smaller, are also found at White Strand, Five Finger Strand, Rosapenna and Rinclevan.

The beach ridge forms one primary source of sediment for the dunes. The deflation of these ridges leads to the development of foredunes, in time, these lead to established dunes. Most dunes in northwest Ireland appear to have formed sometime after 6000 years BP. There is some circumstantial evidence that the site examined by Jessen (1949) at Portrush included 'dune' sands *below* early Holocene (Atlantic substage) peats although this site has been destroyed and it is not possible to examine the sands. Unequivocally early dunes are found at Grangemore on the south side of the Bann Estuary. These dunes have been dated to 5315 ± 135 years BP (UB-937), by the presence of a thin organic layer intercalated with aeolian sands (Hamilton & Carter 1983).

Dunes forming from beach ridges may take several forms depending on (i) the entrainment potential of the sediment, (ii) the supply of sand

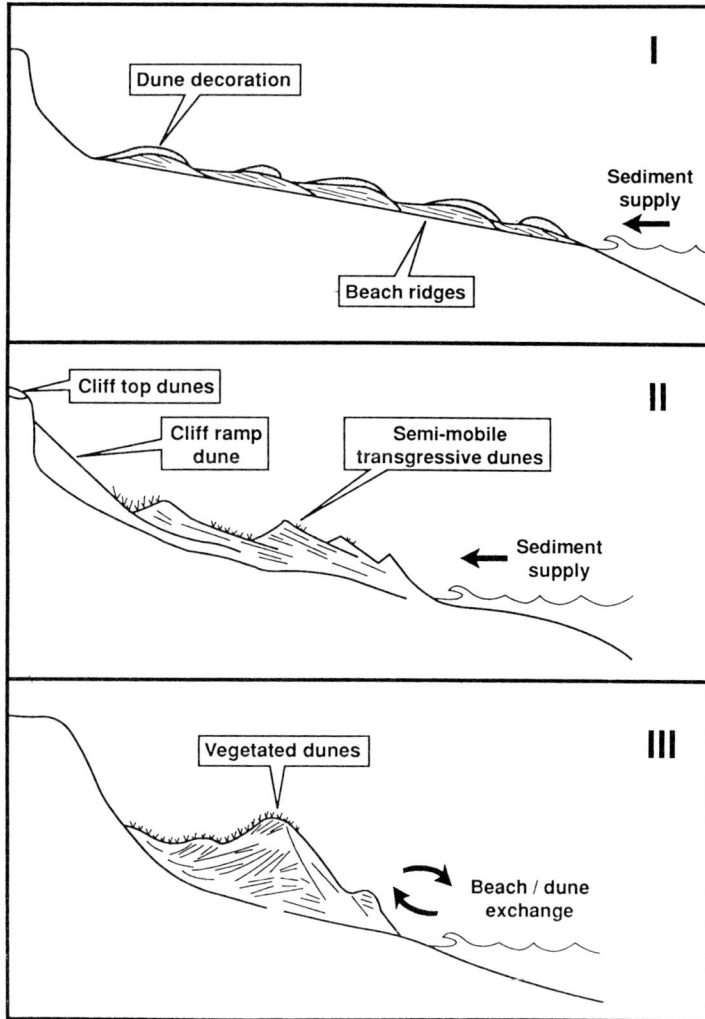

Fig. 5. Variations in dune types formed from beach ridge deflation. See text for explanation.

(especially manifest as the rate of accretion), and (iii) the presence or absence of vegetation. Figure 5 shows three forms. In Fig. 5I the beach ridges remain relatively undisturbed by aeolian activity. This is the case at Magilligan, where the aeolian component on many ridges is limited to decoration. There are several possible reasons for this, including the beach ridge plain being relatively wet. At Magilligan it would appear that the water table, augmented by a number of allogenic streams, remained high enough to suppress deflation. Many of the lower, inter-ridge hollows at Magilligan supported fresh-water lagoons or mires, which served simply to disrupt further the permeability of the ridge plain. Also, the widespread occurrence of shells

in the Magilligan beach ridges led to the development of deflation pavements which served to cut-off the sand supply from the dunes (Carter 1976).

Figure 5II shows the evolution of a more complex dune system in which primary supply to the foredunes is limited by onward sediment transfer and dispersion into the established dunes. Here the key role is played by vegetation. If vegetation is sparse then there is a reasonable chance that the dunes will not be fixed, and become mobile forms. Alternatively, the presence of a reasonably dense sward will lead to development of fixed ridges. There is a clear feedback here, as the vigour of the dune building vegetation is *inter alia* a product of the rate of sand supply. Incoming sand supplies nutrients, especially nitrogen as well as reducing interspecific competition, so that one or two plants dominate. Research at Magilligan (Carter & Wilson 1990) suggests that if the sand supply does not exceed the equivalent of 0.6 to 0.8 m of vertical accretion per year, then the vegetation grows vigorously, and the sand dune is fixed. Above this rate of accretion the dune is unstable. However, there is some evidence (Fay & Jeffrey 1992), that the sediment accumulation rate may itself be dependent on the amount of nitrogen influx so that dunes in sites where there is, for example, extensive sub-tidal algal production to supply nutrients, the vegetation might grow more vigorously and trap more incoming sand.

In the mid-Holocene, semi-mobile dunes may have been quite common. Large-scale cross-bedding indicative of mobile dune bedforms often underlies buried soils of Neolithic or early Bronze Age, as at Maghera and Trawenagh in west County Donegal. Such forms were probably due to the large volumes of available sediment moving onshore and inland. In places this sediment became trapped against the raised cliffs as ramp dunes, or even overtopping them, to form cliff-top dunes. Much of the west, and parts of the north coast include examples of these types of dune. At both Downhill and Maghera very large ramp dunes are banked against bedrock cliffs. Occasional rockfalls and landslips have modified the Downhill dunes, creating complex intercalations of the two deposits. Cliff-top dunes are found on Doagh Isle and east of Portstewart.

The third type of dunes (Fig. 5III) occur where and when sand supply is relatively limited and the beach ridge can be reworked, over a period of time, into the dune. This development includes repeated dune/beach exchange so that on occasions material is transferred from the dune to the beach and then returned to the dune. In this manner the coastal dune grows as a complex structure, with numerous landward- and seaward-dipping, often tabular cross-sets. The slow formation of this type of dune system means that it develops at the shore, and landward migration is necessarily muted.

It would appear that once the coastal dunes began to form, the relatively abundant sediment supply ensured they grew rapidly. This influence is supported by evidence of massive cross-bedded structures towards the base of many dunes. By 4000 years BP many of the Irish dunes had stabilized indicating a declining rate of sediment supply, tightening vegetation control (probably accompanied by organic-rich soils, carbonate leaching and enhanced water reten-

tion) and morphodynamic change. The latter would be associated with the depletion of the shelf sand bodies, the adjustment of the shoreface and the probably slow switch towards dissipative–intermediate nearshore conditions, allowing the development of a balanced beach/dune exchange.

The initial building phase of coastal sand dunes in northwest Ireland ended around 4000 years BP. There are numerous ^{14}C dates (see Wilson 1990b) as well as archaeological evidence (summarized in Mallory & McNeill 1991) pointing to the fact that stabilized dunes were commonplace by the Neolithic (6000 to 4000 years BP). Indeed, much of the early destabilization may be ascribed to early man, using the sandhills for hunting and fishing camps.

The establishment of balanced beach/ dune conditions

In a two-dimensional sense the culmination of the coastal dune evolution was marked by the emergence of balanced beach/dune exchange systems in which the shoreline remained relatively fixed, despite continuing sand fluxes. A balanced beach/dune exchange is a rarity, as in almost all examples there is likely to be a leakage, if not in a cross-shore direction, then alongshore. Psuty (1986) notes that four conditions may exist related to whether or not the beach or dune sediment budget is positive or negative. The most rapid dune growth occurs when both are positive although dune development can take place under all combinations, at least over short periods. However, at many sites in northwest Ireland — White Park Bay, Portstewart, Benone, Culdaff, Trawenagh and Maghera — it is difficult to discover any evidence of long-term shoreline changes, yet there is a clear seasonal flux of material from the dune front to the beach/nearshore and back again in the order of 10^4 m^3 km^{-1} of dune front.

Establishment of reasonably balanced conditions around 4000 years ago probably led to the widespread stabilization by vegetation of the primary dune systems, with well-vegetated dunes extending to the shoreline. Under these conditions, the supply of sand will accumulate either on the dune front, just over the dune crest or as a dispersed apron on the landward slopes (Fig. 6). Inevitably there will be a slow retreat as some sediment is transported landward. The dune develops as a complex structure, converging with the example shown in Fig. 5III, composed of massive slump and chute structures, avalanche bedding, small back beach

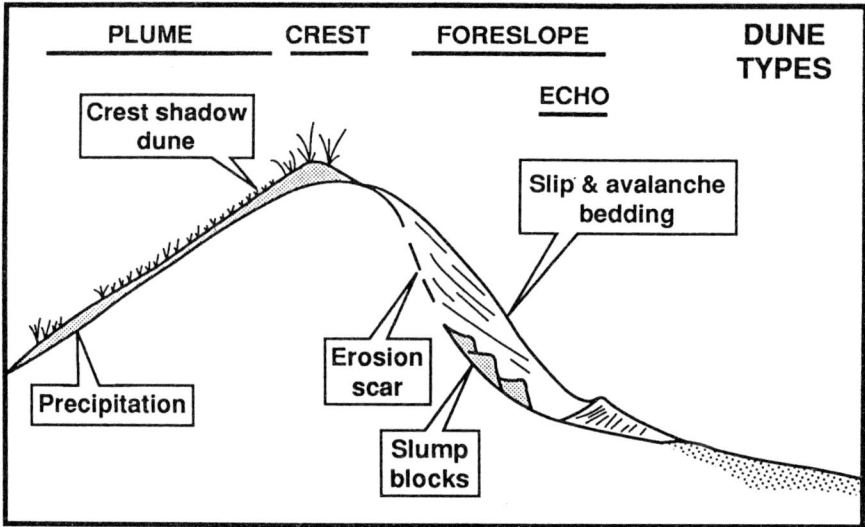

Fig. 6. Complex dune form resulting from slumping and avalanching on retreating seaward slopes and landward dispersion of sand.

echo dunes, crestal shadows and landward dispersal plumes (Carter *et al.* 1990*a*). It also seems likely some sediment is lost onto the lower shoreface, below wave base, so that shore-normal profiles under low or negligible sediment supplies always have a propensity towards retreat, regardless of sea-level tendency.

Figure 7 summarizes the Holocene development of the coastal sand dunes of northwest Ireland, in relation to the sediment balance. What has not been discussed is the impact of man on the coastal dunes which has been substantial, both directly from the point of view of sand surface instability (Carter 1987) and, more

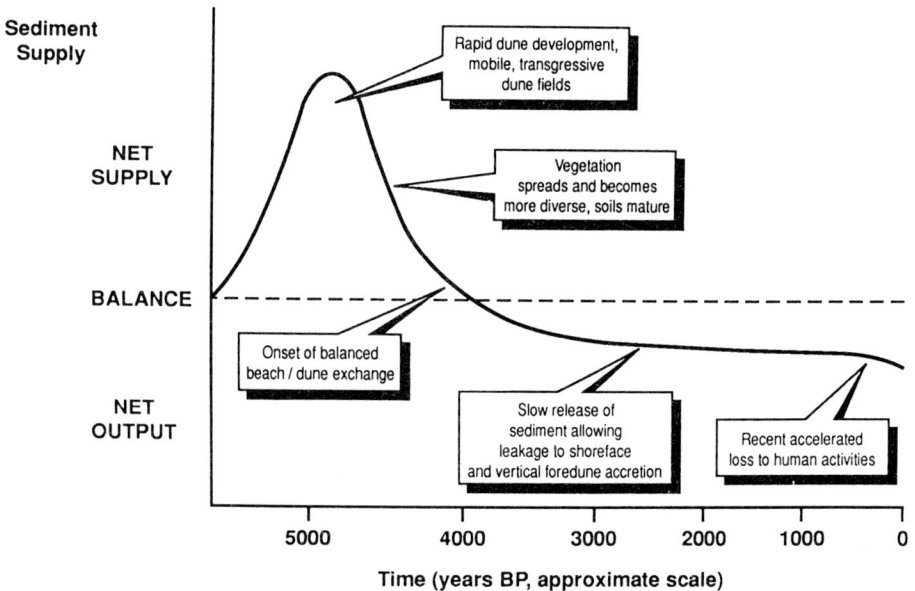

Fig. 7. Holocene development of coastal dunes in northwest Ireland in relation to the sediment balance.

indirectly, through the reclamation of estuaries which has led to closure of tidal inlets, and the adjustment of sediment budgets. In several cases, at Magilligan, Rosapenna and Five Finger Strand, this has appeared to herald the creation of new dunefields (Carter & Wilson 1990). This relationship is discussed further below.

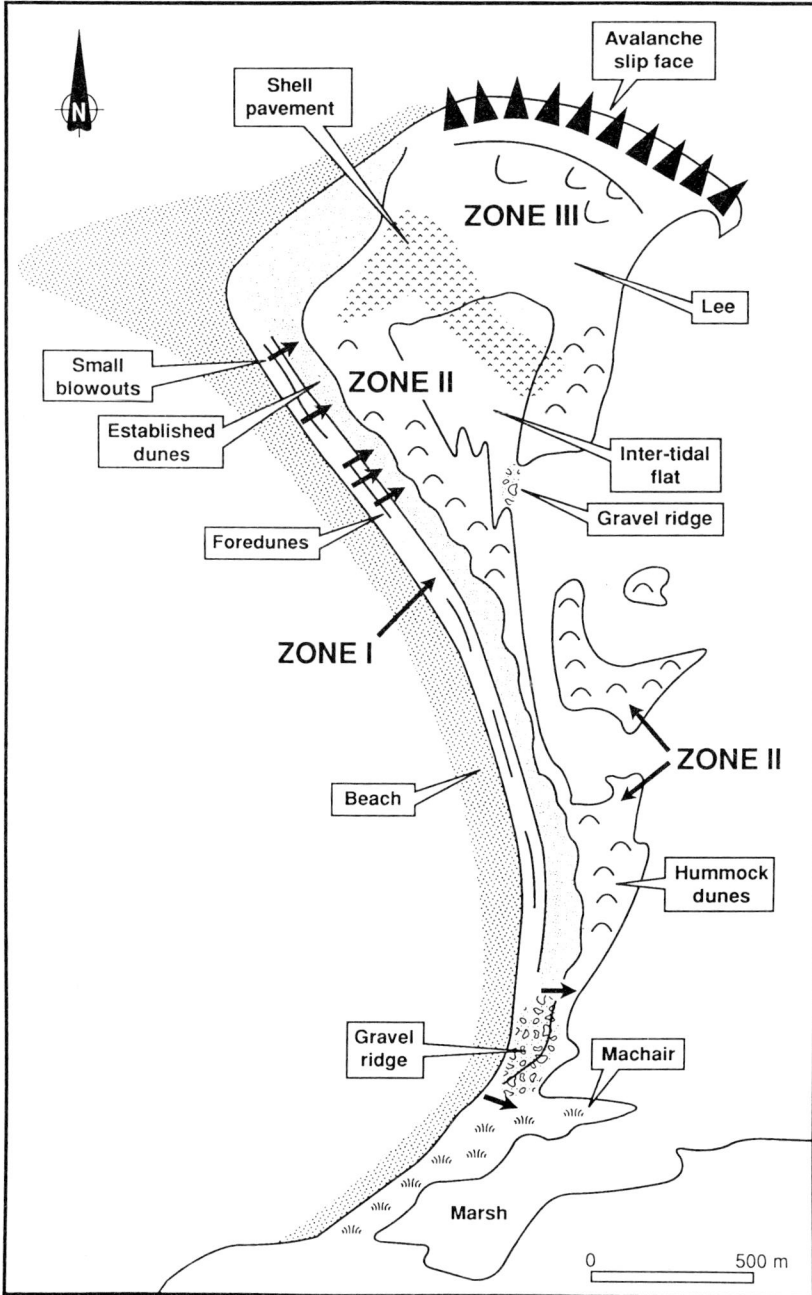

Fig. 8. Morphology and structure of the Ballyness dune system.

Geomorphology of coastal dunefields

(i) Progradational or depositional forms. The dunefields of northwest Ireland tend to be confined to discrete swash-aligned bays, and are usually bounded by rock cliffs. Consequently, there is very little longshore transport of material. The establishment and maintenance of cross-shore drainage is very important in dune system dynamics, as river mouth/estuaries provide the only major alongshore transport-initiating processes.

Where dune systems abut river mouths or estuary mouths, beach/dune/inlet circulations tend to develop. Sediment is moved between the sub-aerial and sub-marine environments, with the inlet jet acting as a pump (Carter 1990). Dunes forming in this manner receive high volumes of sediment, and at Ballyness the amount is sufficient to preclude vegetation stabilization and create a massive free-form dune structure about 2000 m^2 and 30 m high. This dune both receives and supplies material from the nearshore bedforms. The dunes at Magilligan, Tra-na-Rossan, Five Finger Strand, Dunfanaghy and Maghera all owe their origins to similar pumping mechanisms although the typical sand fluxes today are nowhere as high as at Ballyness.

The morphology and structure of the Ballyness site is of considerable interest (Rutherford 1979; Carter 1990; Fig. 8) as it includes many of the quintessential attributes of the northwest

dunes. Shaw (pers. comm. 1991) has dated organic marsh sediments underlying dune sand to 2370 ± 90 years BP (Beta-22237), although it is likely that parts of the dune system formed before this date. Ballyness must be viewed as a composite dunefield, evolving within both a wave shadow (created by an island, 1.5 km offshore) and an estuary mouth. The basement of the dune system would appear to be a series of gravel ridges, outcropping today both at the proximal end of the system and in several other locations within it. These gravel ridges were fed from the cliffs of the Bloody Foreland to the west and probably formed around 5000 years ago.

Ballyness is composed of mixed calcareous/quartz sand, with mean grain sizes around 1.2 to 2.0 ϕ and sorting between 0.3 ϕ (good) and 0.7 ϕ (moderate). There is clear environmental discrimination between the sedimentary environments (Fig. 9), particularly in terms of mean grain size, and skewness. The beach and tidal flat sediments are relatively coarser and positively skewed.

The dunes divide into three types (Fig. 8). First there are foredunes (Zone I) running parallel to the oceanside beach. These dunes extend into established, well-vegetated dune ridges and are connected by 16 transverse blow-out gullies, orientated between 270° and 30°N with a mean of 321°N. The second type Zone II is an area of lower hummock dunes which occupy the inner parts of the dune complex grading into the higher salt marsh. This morphologically

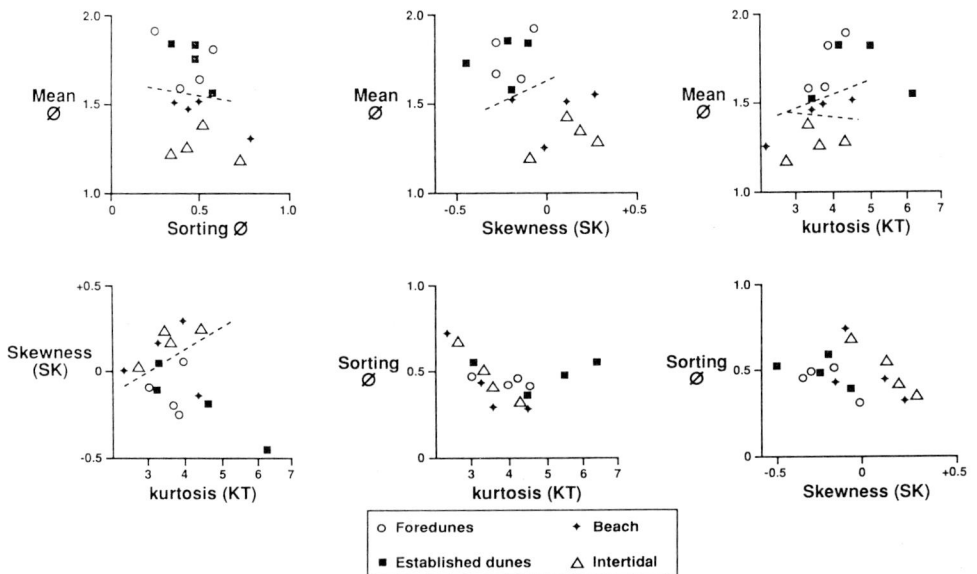

Fig. 9. Textural parameters for sedimentary environments at Ballyness.

sheet-like deposit may represent an earlier phase of dune formation, but there is no clear evidence that this is so. Zone III is perhaps the most interesting as it comprises a massive dune ridge, almost 30 m high, with almost no vegetation cover. Approaching from the west, the dune rises from a compacted shell and gravel pavement, reaching a precipitous slip face and avalanching into the estuary channel to the east. The major dune is covered with numerous small-scale barchan or transverse dunes, often 2–4 m in height, which are migrating across it from west to east. This dune is seen as the sub-aerial component of the estuary mouth tidal pump, which is powered by inlet currents and acting to maintain a constant sediment flux through the dune. The transport rate is unknown, but it appears to be sufficiently high to preclude the stabilisation of the dune by vegetation. Tidal pump maintained dunes are common in Ireland (Carter 1990) and probably elswhere where there is an excess of sediment within an inlet-driven circulation.

As well as morphological contrasts there are structural differences between the dune zones at Ballyness. Exposures in blowout sidewalls and storm scarps suggest that the foredunes (Zone I) are composed at low angle, concave and tangential, cross-beds (> 50% less than 10° from horizontal), while the older dunes (Zone II), have a higher proportion of steeper beds, suggesting they formed under a more dynamic wind regime (Fig. 10). Zone I cross-beds dip offshore, mainly due to slippage along the scarpline. Zone II shows sub-dominance of off- and onshore bedding, while Zone III (from a sample of only 25 observations) shows no clear pattern, although there is a higher proportion here of steeply inclined units (> 40% exceeding 20°).

Prograding dunes are also found at Magilligan Point and are associated either with the reduction in the tidal prism of Lough Foyle during the

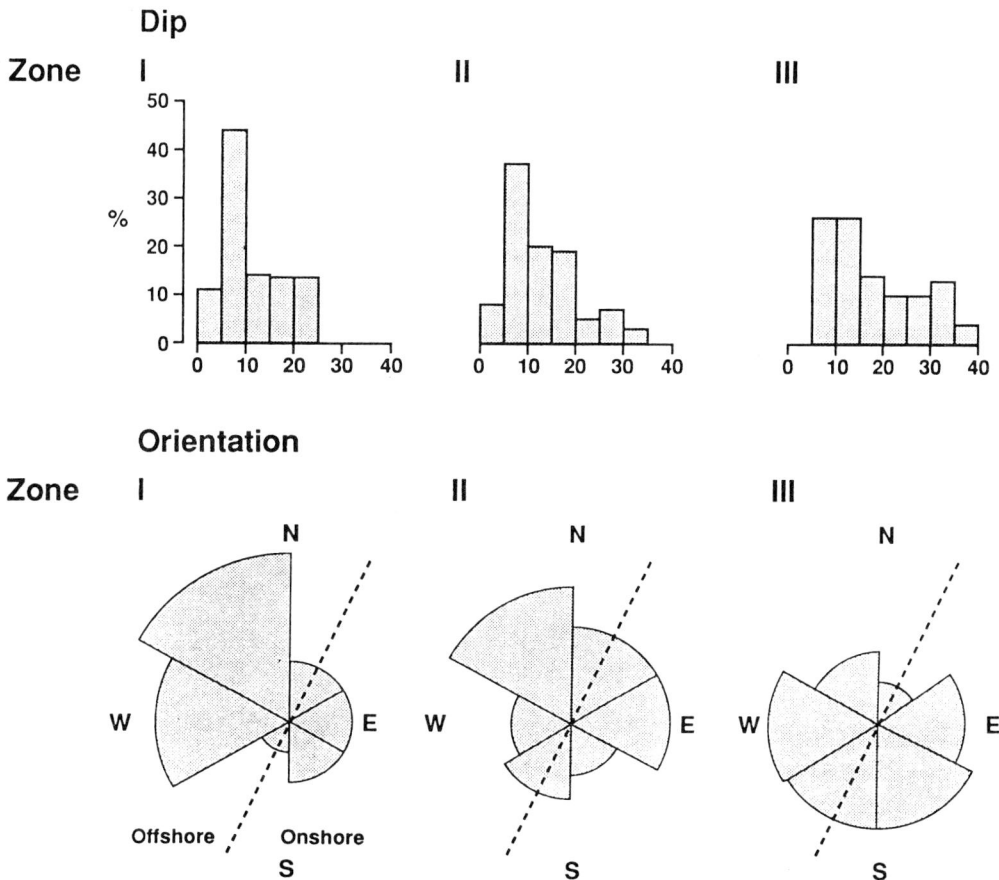

Fig. 10. Dip and orientation of dune bedding in dune zones I, II and III, Ballyness.

mid-nineteenth century (due to land reclamation) and/or a cyclic transfer of material — driven by secular changes in the climate and wave fields — between the offshore banks and the shoreline (Carter et al. 1982). The Point is fed by sediment derived from the erosion of the adjacent ocean- and Lough-side dune cliffs, at a rate of around 50 000 to 70 000 m^3a^{-1} (Carter 1986). The prograding dunes develop on the nuclei of beach ridges developed after storms. Sediment is blown from the ridges, accumulating in the foredunes. Foredune growth is most rapid (rates of 3 to 4 $m^3 m^{-1}$ per week have been recorded) immediately after bars are stranded, but the controls over sedimentation (mainly the development of shell pavements) make such rapid accumulation unsustainable (Carter & Wilson 1990).

The foredune ridge tends to grow rapidly to a height of 3 to 4 m above HWM, after which it stabilizes. To a large extent the height of a ridge reflects the length of time it has been exposed to the sediment supply, so that during the most recent phase of rapid progradation (1965 to 1980) a number of small ridges, 1–2 m high formed. The cessation of progradation, after 1983, has led to the re-activation of the 1976 and 1980 ridges, which have now developed as composite forms, with later sediment engulfing previously vegetated surfaces. The nature of the prograding ridges is such that gaps, present in the original duneline, often remain as areas of non-deposition as the dune grows up around them. These 'wind-gaps' are often used as footpaths, and may eventually develop as small blow-outs. Magilligan Point is typical of prograding dunes in northwest Ireland, other examples occur at Rosapenna and Five Finger Strand (Carter 1982b). At Clonmass similar foredunes developed after the estuary channel migrated northwest in the 1950s (Shaw 1985). Seaward advancement only occurs due to a reliable supply of material from nearby eroded dunes.

(ii) Recessional and eroding dunes. Throughout northwest Ireland most of coastal dunes are retreating. The dunefields show abundant evidence of morphosedimentary reactivation, with numerous secondary forms and deposits. Retreat is clearly linked to pervasive negative sediment budgets. At first these are probably manifest as beach erosion which may stimulate foredune growth and lead to the accumulation of high barrier dunes along the shoreline as at Portrush, Magilligan and Trawenagh. The seaward edges of these dunes are often cliffed, with frequent wave-scarping of the base causing numerous sub-aerial slope failures (Carter & Stone 1989). As recession proceeds, the eroding dune line often becomes irregular and blowouts develop. Initiation of blow-outs may arise from a number of processes (Carter et al. 1990a), but in northwest Ireland may seem to arise from selective erosion of the dune scarp allowing wind-funnelling through the crest and ultimately breaching. Figure 11 shows the dunes at Trawenagh, where blow-outs in various stages of evolution may be observed. The shoreline is marked by a series of small shore-normal gullies (15 in number) at irregular intervals. Of these, three have developed sufficiently to breach the crest, allowing sand to be moved inland. In addition there are several major blowouts. One, in the centre of the bay, is active, inasmuch as it is allowing transfer of material from the beach to the established rear dunes. A second, to the south is now cut off from the beach by a new foredune ridge, but is still deflating with a large sand plume engulfing the dune to the west. The landward edge of this blowout has developed an encircling rim dune 45 m above the beach. Transverse sections through the rim dune indicate a very complex structure of small-scale, reversing cross- and avalanche-beds, forming a massive unit over 10 m in thickness. Elsewhere there are several saucer blowouts within the dunefield some active and some healed. Figure 11 reveals a very complex pattern of morphology, typical of northwest Irish dunefields. The chaotic topography evidences a long history of instability with examples of the engulfment of agricultural land along the eastern border of the dunes. To the north, blown sand has covered the rocky headland, giving rise to hummocky topography characterized by depauperate *Ammophila*.

The dunes at Trawenagh are being dissected by blowouts, initiated at the shoreline by gullies, allowing slow landward encroachment of the sand. A survey of the entire northwest coast (Fig. 12) shows 98 major blowouts, of which 60 (62%) are shore-normal. At those sites where the coast is not orientated to the west or southwest (into the prevailing wind), it is evident that the blowouts will reorientate themselves to face the wind. At Portstewart, where the beach faces north, the dominant blowout slip faces are shore-parallel.

Moisture and vegetation act together to stabilize the dune systems, so that fully-mobile sands are uncommon today, although they may have existed in the past. Bare sand in blowouts is often deflated to either the water table or until an immobile surface lag forms. Within blowouts, much of the activity is internal with sand re-

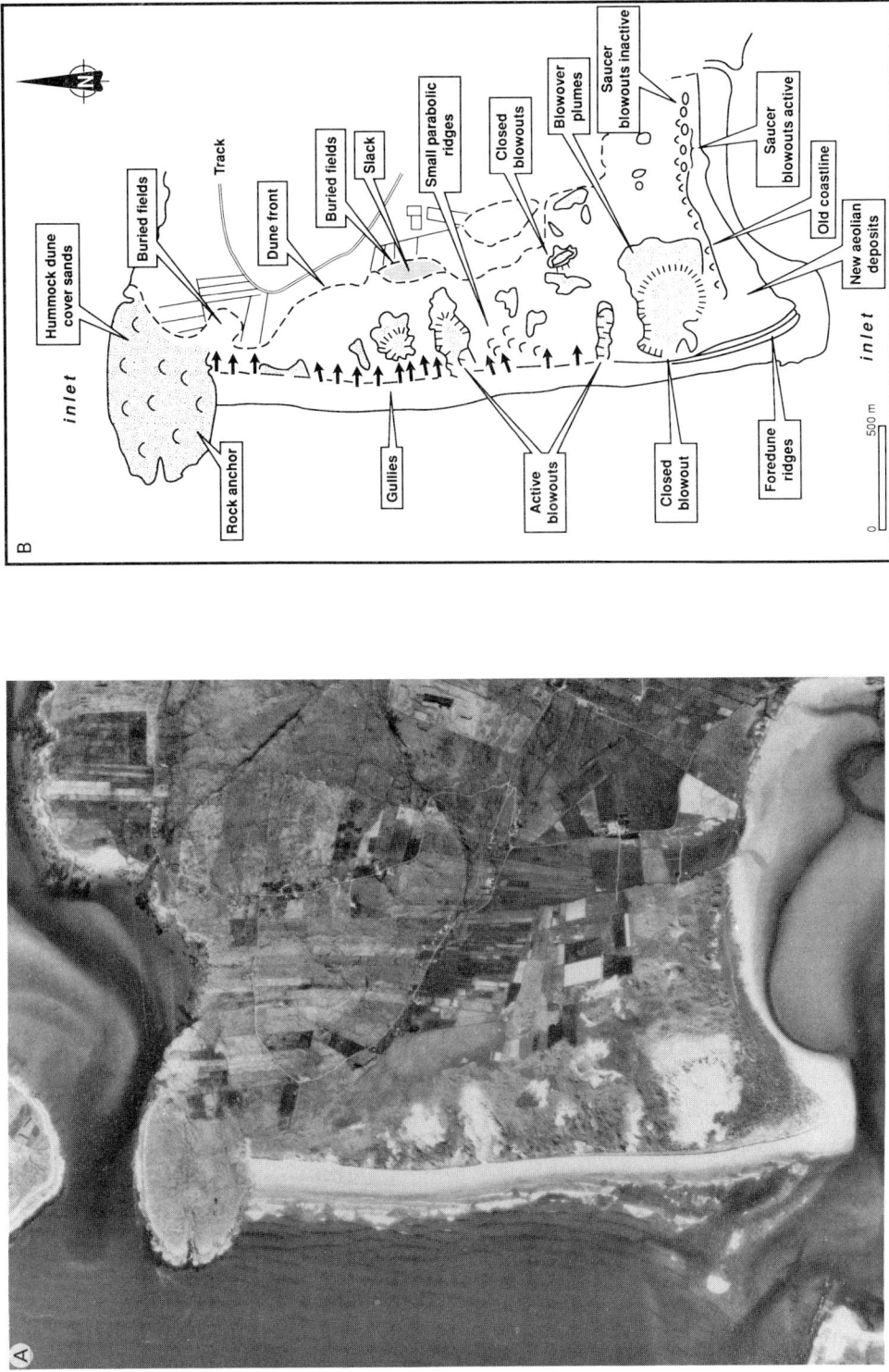

Fig. 11. Dune system at Trawenagh showing active and inactive blowouts and areas of sand encroachment onto agricultural land. (**A**) 1976 Aerial photograph and (**B**) morphological interpretation.

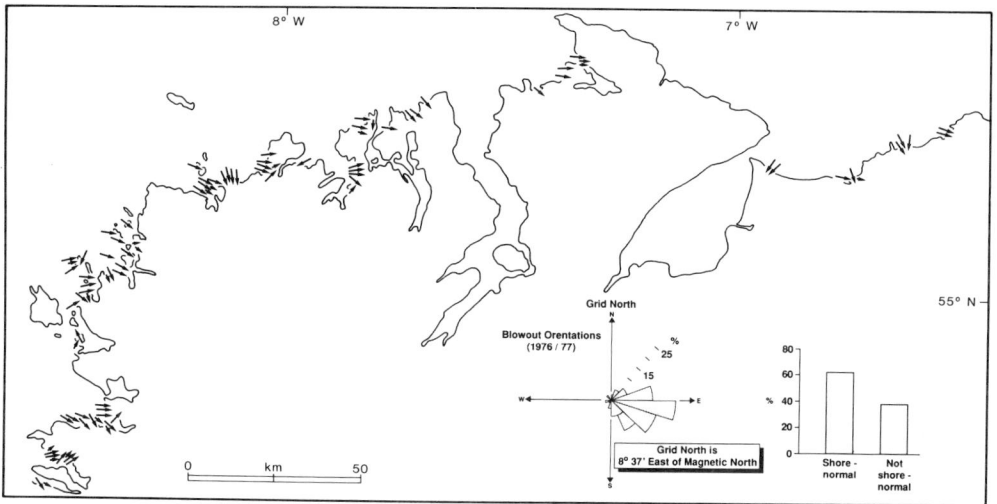

Fig. 12. Orientations of major blowouts in dune systems of northwest Ireland. Data compiled from 1976/1977 air photographs.

circulating between the side walls and the floor (Carter *et al.* 1990*a*). Where sand escapes from the blowouts it usually forms a dispersed blowover plume rather than a coherent slip face, perhaps marked by vigorous growth of *Ammophila*. However at Runkerry, Trawenagh and Naran, migrating slip faces do occur at the landward edges of some blowouts.

Massive deflation of dunefields may lead to the formation of machair (Bassett & Curtis 1985; Quigley 1991). Machair is common in north and west Ireland, and may be formed as a consequence of strong onshore winds and slow variations in groundwater regime (under the influence of sea-levels and precipitation). At Ballyhiernan, Magheradrumman and Doagh Isle, machair surfaces show abundant evidence of slow accretion as sand is released from the shoreline. This forms a gently undulating but landward-dipping plain, often terminating in a seepage lake. Cultural influences on the development of machair are also likely, in the late-eighteenth century the dunes at Tullagh were disturbed by ploughing, resulting in the formation of a deflation plain 2 km wide (Carter *et al.* 1990*a*).

The influence of deflation can be seen at Maghera, a dunefield of about 100 ha at the mouth of the Loughros Beg estuary, a narrow tidal estuary some 4.5 km in length. The south side of the estuary is cut into glacial deposits, and there are several drumlin islands along the estuary shore. One of these, opposite Carrickshanbally is covered with a 1 to 2 m veneer of aeolian deposits. Maghera dunes themselves are most notable for the fact that they are fronted by one of the widest 'beaches' in Ireland, approximately 0.8 km from the dune front to the mean HWM. This 'beach' is in fact an aeolian deflation surface, lowered to the water table, but often covered by mega-ripples, sand waves and small barchan dunes up to 1.5 m high.

Cliff-top aeolian sands

Although ramp or under-cliff dunes occur at Downhill and Maghera, they do not reach to the cliff-top and associated cliff-top dunes are lacking. Between Portstewart and Portrush a low (30 m OD) north-facing basalt cliff is capped by a thin and discontinuous spread of blown sand (Wilson & Manning 1978), but the cliff is fronted by wave-cut platforms and boulder beaches, and sand is absent (McKenna 1990). However, at both east and west ends of the cliff the sand passes into large dune accumulations that rise > 20 m above their source beaches. Given the prevailing westerly winds, it is not necessary to hypothesize why present-day under-cliff dunes are absent from the site (cf. Jennings 1967); the cliff-top sand could be explained as a downwind extension of the Portstewart dune system to the west. Recent examination of exposures in the sand indicates that this explanation is too simplistic. Available evidence now suggests that some of the sand may be derived from dunes to the east and/or that it has been reworked from earlier cliff-top accumulations.

Wilcock (1976) and Wilson & Manning (1978) considered the sand to be mid-Holocene in age and related to a higher stand of sea-level. Carter *et al*. (1989) suggested that the sand could represent reworked dune sand associated with the increased frequency of storms between the eleventh and thirteenth centuries. These age estimates are conjectural; ^{14}C dates now reveal a more complex (multi-phase) history of sand deposition.

A section at the Strand Hotel, Portstewart, described by Wilson (1991*b*) revealed two depositional units within the sand, the lower of which contained a 1.4 m thick humus podzol. A ^{14}C date of 4780 ± 45 years BP (SRR-3766) was obtained from the organic-rich horizon of a buried stagnogley soil developed in a basalt-rich diamicton beneath the sand. The date provides a maximum age estimate for deposition of the podzolized sand unit. A ^{14}C date of 525 ± 45 years BP (SRR-3765) from the organic-rich horizon of the podzol gives a maximum age estimate for burial of the podzol by deposition of the upper sand unit. The earlier of these dates indicates that the sand did not accumulate while sea level was as its mid-Holocene peak (cf. Wilcock 1976; Wilson & Manning 1978) which, in the Portstewart–Portrush area, is constrained by ^{14}C dating to between *c*. 5400–5900 years BP (Hamilton & Carter 1983; Wilson & Carter 1990). The younger ^{14}C date places deposition of the upper sand unit in the Late Middle Ages (cf. Carter *et al*. 1989), but whether this was associated with climatic deterioration and increased storminess (cf. Lamb 1982) or simply represents localized sand redistribution in response to surface disturbance by human activities, remains to be established (Wilson 1991*b*). The proximity of this site to the Portstewart dune system favours derivation of the lower sand unit from the dunes.

A temporary exposure in the cliff-top sand 2 km west of Portrush in 1991 showed 1.3 m of well-stratified, carbonate-deficient sand, containing a sand-ranker soil, resting on a sequence of thin organic-rich sands and sandy peat that were in turn underlain by a basalt-rich diamicton. A ^{14}C date of 1360 ± 80 years BP (Beta-45228) was obtained on the upper part of the peat. The stratigraphy, particle-size characteristics and pedology of this section, along with the age estimage, are significantly different to those at the Strand Hotel, suggesting a different source for the sand.

At Portrush, 6 m of structureless and carbonate-deficient sand is banked against the eastern end of the basalt cliff (Wilson & Bateman 1990). Three weakly-expressed buried soil organic horizons are present within the sand but no ^{14}C dates are yet available. The sand is most likely derived from the adjacent beach to the east, which possesses a relatively low shell carbonate content at its southern end and may also be the source of the sand that extends at least 2 km to the west.

Although depicted on the geological drift map as Holocene sands, it is becoming clear that even over a relatively limited spatial extent the history of cliff-top sand accumulation cannot be ascribed to a single source at a point in time.

Non-coastal aeolian sands

Non-coastal aeolian sands are of limited occurrence in Ireland, although this may be attributed in part to the lack of research directed towards their identification and description. Lewis (1985) commented on the marked absence of periglacial wind-blown sediments from an otherwise rich and varied suite of periglacial features. Since then, little more has been done to rectify this situation. The extensive loess and coversand deposits found across much of lowland Britain and mainland Europe have no apparent equivalents in Ireland. Loess-like material comprising the infill of periglacial wedge casts has been briefly described by Lewis (1979), and Dardis (1986) identified small-scale, low-relief aeolian dunes on glaciolacustrine deltas. However, these reports form minor components of papers dealing with the broader issue of periglacial wedge structure development; consequently little is known about the sedimentology of these materials although a cold and arid environment at the time of deposition was inferred.

In the uplands of Ireland aeolian sediments are known from only one site. At 650 m OD on Muckish Mountain, flow-aligned sand shadows (shadow dunes) and a dissected sand sheet occur (Fig. 13; Wilson 1988, 1989). The shadow dunes comprise tapering accumulations of sand, with minor quantities of fine gravel, silt and clay, developed leeward (up-slope) of small cairns that were probably constructed as navigational aids for quarry workers in the late nineteenth century. Dune orientation indicates sand transport by northwesterly winds; the source of the sand is a series of friable quartzite beds that crop out immediately below the northern edge of the summit plateau. The location, morphology and sedimentological characteristics of the dissected sand sheet lend strong support to the notion that sand has been transferred from the steep northern slopes of the mountain to the plateau. Maximum sand thickness occurs at the plateau edge and declines with increasing distance from the edge, and the lateral extent of the sand sheet along the plateau edge corresponds closely to

Fig. 13. Dissected sand sheet on summit plateau of Muckish Mountain.

the outcrop of friable quartzite along the escarpment. The sand has similar particle size characteristics to the friable quartzite and possesses an identical suite of quartz grain surface textures, reflecting short distance and duration of aeolian transport. Radiocarbon dates obtained from thin peat horizons below and within the sands indicate two phases of sand accumulation; one between c. 5300 years BP and c. 2650 years BP, the other after c. 1910 + 1760 years BP. The upslope aeolian transport of sand from escarpment to plateau is by no means unusual. The nature of airflow over windward facing escarpments was outlined by Bowen & Lindley (1977), and Marsh & Marsh (1987) have used their model to explain the pattern of sand deposition along Lake Superior bluffs. The Muckish sand described above also conforms to this model.

Conclusion

The dunefields of northwest Ireland are complex landforms that have formed as a result of many interacting processes over a period of several thousand years. It is possible to identify a phase of coastal dune growth in the mid-Holocene, which has been followed by constant modifications, some ascribable to climate change, others to man and some to both. While the present overall sediment budget is strongly negative, there are locations where sand is recycled between beach, nearshore and dune, which allows development of new foredunes. Many of the dunescapes, however, are heavily dissected with sediment transport active through blowouts.

Many thanks to Mary McCamphill and Kilian McDaid for technical assistance. Ken Park abstracted the data on which Fig. 12 is based.

References

BASSETT, J. A. & CURTIS, T. G. F. 1985. The nature and occurrence of sand-dune machair in Ireland. *Proceedings of the Royal Irish Academy*, **85B**, 1–20.

BOWEN, A. J. & LINDLEY, D. 1977. A wind-tunnel investigation of the wind speed and turbulence characteristics close to the ground over various escarpment shapes. *Boundary-Layer Meteorology*, **12**, 259–271.

CARTER, R. W. G. 1976. Formation, maintenance and geomorphological significance of an aeolian shell pavement. *Journal of Sedimentary Petrology*, **46**, 418–429.

—— 1982a. Sea-level changes in Northern Ireland. *Proceedings of the Geologists' Association*, **93**, 7–23.

—— 1982b. Recent variations in sea-level on the north and east coasts of Ireland and associated shoreline response. *Proceedings of the Royal Irish Academy*, **82B**, 177–187.

—— 1983. Raised coastal landforms as products of modern process variations, and their relevance in eustatic sea-level studies: examples from eastern Ireland. *Boreas*, **12**, 167–182.

—— 1986. The morphodynamics of beach-ridge formation at Magilligan, Northern Ireland. *Marine Geology*, **73**, 191–214.

—— 1987. Man's response to change in the coastal zone of Ireland. *Resource Management and Optimization*, **5**, 127–164.

—— 1988. *Coastal Environments*. Academic Press, London.

—— 1990. The geomorphology of coastal dunes in Ireland. *In*: BAKKER TH. W. M., JUNGERIUS, P. D. & KLIJN, J. A. (eds) *Dunes of the European Coasts*. Catena Supplement, **18**, 31–39.

—— 1991. The impact of near-future sea-level rise on coastal dunes. *Landscape Ecology*, **6**, 29–39.

—— & STONE, G. W. 1989. Mechanisms associated with the erosion of sand dune cliffs, Magilligan, Northern Ireland. *Earth Surface Processes and Landforms*, **14**, 1–10.

—— & WILSON, P. 1990. The geomorphological, ecological and pedological development of coastal foredunes at Magilligan Point, Northern Ireland. *In*: NORDSTROM, K. F., PSUTY, N. P. & CARTER, R. W. G. (eds) *Coastal Dunes: Form and Process*. John Wiley, Chichester, 129–157.

——, DEVOY, R. J. N. & SHAW, J. 1989. Late Holocene sea levels in Ireland. *Journal of Quaternary Science*, **4**, 7–24.

——, HESP, P. A. & NORDSTROM, K. F. 1990a. Erosional landforms in coastal dunes. *In*: NORDSTROM, K. F., PSUTY, N. P. & CARTER, R. W. G. (eds) *Coastal Dunes: Form and Process*. John Wiley, Chichester, 217–250.

——, LOWRY, P. & STONE, G. W. 1982. Sub-tidal ebb-shoal control of shoreline erosion via wave refraction, Magilligan Foreland, Northern Ireland. *Marine Geology*, **48**, M17–M25.

——, ORFORD, J. D. & JENNINGS, S. C. 1990b. The recent transgressive evolution of a paraglacial estuary as a consequence of coastal barrier breakdown: Lower Chezzetcook Inlet, Nova Scotia. *Journal of Coastal Research, Special Issue*, **9**, 564–590.

CASTON, G. G. 1976. *The floor of the North Channel, Irish Sea; a side-scan sonar survey*. Institute of Geological Sciences (NERC) Report **76/7**.

DARDIS, G. F. 1986. Fossil ice and sand wedges in south-central Ulster, Northern Ireland. *Irish Geography*, **19**, 51–57.

DEVOY, R. J. N. 1983. Late-Quaternary shorelines in Ireland: an assessment of their implications for isostatic land movement and relative sea-level changes. *In*: SMITH, D. E. & DAWSON, A. G. (eds) *Shorelines and Isostasy*. Academic Press, London. 227–254.

EDWARDS, K. J. & WARREN, W. P. (eds) 1985. *The Quaternary History of Ireland*. Academic Press, London.

EVANS, D., KENALT, N., DOBSON, M. R. & WHITTINGTON, R. J. 1980. *The Geology of the Malin Sea*. Institute of Geological Sciences Report **79/15**.

EYLES, N. & McCABE, A. M. 1989. The late Devensian (< 22000 BP) Irish Sea basin: the sedimentary record of a collapsed ice sheet margin. *Quaternary Science Reviews*, **8**, 307–351.

FAY, P. & JEFFREY, D. W. 1992. The foreshore as a nitrogen source for marram grass *Ammophila*

arenaria (L) Link. *In*: CARTER, R. W. G., CURTIS, T. G. F. & SHEEHY-SKEFFINGTON, M. J. (eds) *Coastal Dunes: Geomorphology, Ecology and Management for Conservation*. Balkema, Rotterdam, 177–188.

HAMILTON, A. C. & CARTER, R. W. G. 1983. A mid-Holocene moss bed from eolian dune sands near Articlave, County Londonderry. *Irish Naturalists' Journal*, **21**, 73–75.

JENNINGS, J. N. 1967. Cliff-top dunes. *Australian Geographical Studies*, **5**, 40–49.

JESSEN, K. 1949. Studies in late-Quaternary deposits and flora-history of Ireland. *Proceedings of the Royal Irish Academy*, **52B**, 85–290.

KEARY, R. & KEEGAN, B. F. 1975. Stratification by in-fauna debris, structure, a mechanism and a comment. *Journal of Sedimentary Petrology*, **45**, 128–131.

LAMB, H. H. 1982. *Climate, History and the Modern World*. Methuen, London.

LEWIS, C. A. 1979. Periglacial wedge-casts and patterned ground in the Midlands of Ireland. *Irish Geography*, **12**, 10–24.

—— 1985. Periglacial features. *In*: EDWARDS, K. J. & WARREN, W. P. (eds) *The Quaternary History of Ireland*. Academic Press, London, 95–113.

MALLORY, J. P. & McNEILL, T. E. 1991. *The Archaeology of Ulster*. Institute of Irish Studies, Belfast.

MARSH, W. M. & MARSH, B. D. 1987. Wind erosion and sand dune formation on high Lake Superior bluffs. *Geografiska Annaler*, **69A**, 379–391.

McCABE, A. M. 1987. Quaternary deposits and glacial stratigraphy in Ireland. *Quaternary Science Reviews*, **6**, 259–299.

McKENNA, J. 1990. *The Morphodynamics and Sediments of Basalt Shore Platforms*. DPhil Thesis, University of Ulster.

ORFORD, J. D., CARTER, R. W. G. & JENNINGS, S. C. 1991. The evolution of gravel barriers under rising sea-level. *Quaternary International*, **9**, 87–104.

PITCHER, W. S. & BERGER, A. S. 1972. *The Geology of Donegal*. Wiley-Interscience, New York.

PSUTY, N. P. 1986. A beach/dune interaction mode-land dune management *Thalassas*, **4**, 11–15.

QUIGLEY, M. (ed.) 1991. *A Guide to the Sand Dunes of Ireland*. EUDC, Leiden.

ROHAN, P. K. 1986. *The Climate of Ireland* (2nd ed.). Stationery Office, Dublin.

RUTHERFORD, J. H. 1979. *A Study of Ballyness Dunes, County Donegal*, MSc Thesis, The New University of Ulster.

SHAW, J. 1985. *Holocene Coastal Evolution in County Donegal, Ireland*. DPhil Thesis, University of Ulster.

STEPHENS, N. & SYNGE, F. 1965. Late-Pleistocene shorelines and drift limits in County Donegal. *Proceedings of the Royal Irish Academy*, **64B**, 131–153.

TROEN, I. & PETERSON, E. L. 1989. *European Wind Atlas*. Roskilde, Denmark.

WARREN, W. P. 1985. Stratigraphy. *In*: EDWARDS, K. J. & WARREN, W. P. (eds) *The Quaternary History of Ireland*. Academic Press, London, 39–65.

WILCOCK, F. A. 1976. *Dune Physiography and the Impact of Recreation on the North Coast of Ireland*. DPhil Thesis, The New University of Ulster.

WILSON, H. E. & MANNING, P. I. 1978. *Geology of the Causeway Coast, Vol. 1*. Memoirs of the Geological Survey of Northern Ireland, HMSO, Belfast.

WILSON, P. 1988. Recent sand shadow development on Muckish Mountain, County Donegal. *Irish Naturalists' Journal*, **22**, 529–531.

—— 1989. Nature, origin and age of Holocene aeolian sand on Muckish Mountain, County Donegal, Ireland. *Boreas,* **18**, 159–168.

—— 1990a. Characteristics and significance of protalus ramparts and fossil rock glaciers on Errigal Mountain, County Donegal. *Proceedings of the Royal Irish Academy*, **90B**, 1–21.

—— 1990b. Coastal dune chronology in the north of Ireland. *In*: BAKKER, TH. W. M., JUNGERIUS, P. D. & KLIJN, J. A. (eds) *Dunes of the European Coasts*. Catena Supplement, **18**, 71–79.

—— 1991a. Sediment clasts and ventifacts from the North Coast of Northern Island. *Irish Naturalists' Journal*, **23**, 442–446.

—— 1991b. Buried soils and coastal aeolian sands at Portstewart, County Londonderry, Northern Ireland. *Scottish Geographical Magazine*, **107**, 198–202.

—— & BATEMAN, R. M. 1990. Portrush-Dhu Varren. *In*: WILSON, P. (ed) *North Antrim and Londonderry*. Irish Association for Quaternary Studies, Field Guide **13**, 39–45.

—— & CARTER, R. W. G. 1990. Portrush-Mill Strand. *In*: WILSON, P. (ed.) *North Antrim and Londonderry*. Irish Association for Quaternary Studies, Field Guide **13**, 35–39.

—— & FARRINGTON, O. 1989. Radiocarbon dating of the Holocene evolution of Magilligan Foreland County Londonderry. *Proceedings of the Royal Irish Academy*, **89B**, 1–23.

A complex dune system in Baix Empordà (Catalonia, Spain)

LLUISA CROS[1] & JORDI SERRA[2]

[1]Institut de Ciències del Mar, CSIC, Passeig Nacional s/n, Barcelona, 08039 Spain
[2]Departament de Geologia Dinàmica, Universitat de Barcelona, Zona de Pedralbes,
Barcelona, 08028 Spain

Abstract: The dune system of Baix Empordà, Catalonia, Spain is composed of sediments supplied by the Ter and Daró rivers, transported by the effective north-northwesterly local wind called the *Tramuntana*.

Dune formation is mostly related to the vegetation on the plains area and to topographic obstacles in the Begur Massif zone. Morphologically, these dunes are mainly parabolics and blowouts. Climbing, falling and 'passadis' dunes are present in the mountainous zone, the latter being a newly proposed name for a type of lee dune.

The sands are differentiated into distinct groups according to their sedimentological features and marine faunal content. Their evolution and age have been determined by mapping, and by geological and historical information, respectively.

Important Quaternary aeolian sand deposits occur in the Baix Empordà, NE Catalonia, Spain (Martinez Gil 1972; Palli & Bach 1987). These deposits, occupying an area of 14.5 km^2 (Fig. 1) and a volume of $190 \times 10^6 \text{m}^3$ (Cros 1987), correspond to dune formations composed of well sorted sands (IGME 1981). Early this century pine trees were planted on the dunes in order to stabilize them.

Many of the dunes are located in the Pals Bay area, but some extend towards and over the mountains of Begur (Fig. 1). Pals Bay is located along the open coast of the Neogene Empordà basin and is bordered to the north and south by the Montgri and Begur mountains, respectively. Both the Ter, a river originating in the Pyrenees, and the Daró, a torrential river draining a very small basin, flow into the bay.

Coastal dynamics show a predominant sediment drift direction to the south, reflecting dominant waves from the E–NE. The headlands create a closed sedimentary compartment (cf. Davies 1974).

In this part of Catalonia there exists a type of wind that is very characteristic to the area called the *Tramuntana* — a cold, strong, dry wind that blows from the NNW and has over the years transported sand from the southern part of the Pals plain and bay towards the Begur massif. From October to April, the *Tramuntana* — the strongest wind of this area — blows at speeds stronger than 45 km per hour (6 on the Beaufort scale) at least twice a month (Pascual & Flos 1984).

Dune types and their geographic distribution

The dune zone was mapped with the help of air photographs but only a few morphological types were distinguished due to the vegetation and the size of the sand bodies. Later, the forms observed in the photographs were verified in the field, as well as those unobserved because of their size. Measurements were also taken to determine orientation, dimensions and shape.

In the Baix Empordà a large variety of dune types can be found but there are few examples of what might be called typical dunes because they are often deformed or interrelated with each other. There are many compound and complex dune combinations (terminology of McKee 1979). Most dunes have an elongate shape with a NNW–SSE orientation.

The morphology of the majority of these dunes is related to obstacles. These obstacles can be vegetation, as is true of the blowouts, parabolic dunes and, to a lesser degree, the dome dunes, sand shadows and foredunes. They can also be topographic obstacles, as in the case of the climbing, falling and 'passadis' (corridor) dunes.

In Baix Empordà there exist two well-differentiated geographic areas of dune morphology (Fig. 1): the plains area with dunes mostly influenced by vegetation and the mountainous area (Begur massif) with dunes mostly related to topographic obstacles.

From Pye, K. (ed.), 1993, *The Dynamics and Environmental Context of Aeolian Sedimentary Systems.*
Geological Society Special Publication No. 72, pp. 191–199.

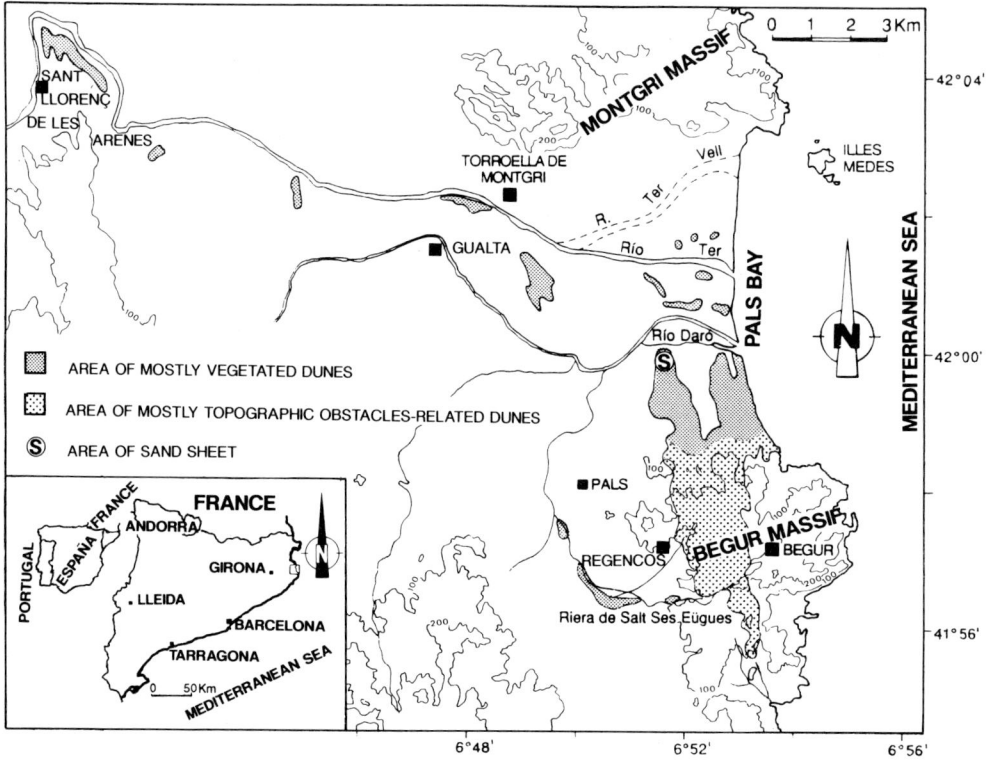

Fig. 1. Geographical location and distribution of morphological dune types.

Dune types of the plains area

The most common dune forms found in the plains area are blowouts and parabolic dunes.

The blowouts are often small and elliptical in shape with their main axis not longer than 20 to 30 m and lying in a N–S direction (345° to 350° axis orientations are not unusual). The north side of these dunes is often not fully developed and can even be non-existent in some cases. Some transitional forms between blowouts and small parabolic dunes have been observed. Superimposed blowouts on other dune types are not uncommon.

The parabolic dunes have an elongate or hairpin shape with their trailing points facing north. They have an average long axis orientation of 350°, parallel to the direction of the wind that formed them (cf. Seppälä 1972), and can reach up to 300 m in length with a mean of about 30 m. Nested and compound dunes are not uncommon. They are often deformed, with parts (arms) separated from the main sand mass (nose), as described by Landsberg (1956). These arms appear as vegetated linear dunes.

The dome-shaped dunes are generally small and elliptical in plan morphology.

In the coastal plain zone there are foredunes and shadow dunes. The shadow dunes are practically the only active dunes that exist in the Baix Empordà area. On the days that the *Tramuntana* blows it is possible to observe the formation of small sand shadows along the Pals beach with axes orientated between 345° and 355°. When two or more of these sand shadows merge and become colonized by *Ammophila arenaria*, they can actually begin a dune-forming process which, if continued, can form a foredune. Today, due to a decrease of fluvial sediment contributions and construction work going on along the beach, this process has practically come to a standstill.

Along the coast foredunes are found parallel to the beach. Blowouts and parabolic dunes superimposed on the foredunes can be found.

A small area of sand sheets has been detected in the northern internal dune zone of the Pals plain. Based on earlier studies by Mainguet *et al.* (1983) and Kocurek & Nielson (1986), we believe that this formation is very likely to be a result of adverse conditions for the formation of dunes: i.e., lack of sufficient sand.

A few small barchans, with theirs horns facing south, are also been found, though generally on river floodplains (especially SW of Regencos, Gualta and Sant Llorenç de les Arenes). Normally they are grouped together or form dune complexes with blowouts or small parabolic dunes superimposed on the windward side of compound barchans. They generally do not measure more than 20 to 30 m in width.

Dune types of the mountainous area

There are also a number of dunes controlled by topographic obstacles in the Baix Empordà area. These are found on the Begur massif. Their formation can be related both to the relief of the mountain and the changes in the direction of the wind as it blows through the mountains.

On the northern slopes of the mountains are climbing dunes which become thinner as they climb higher up the mountain. It is interesting to note that towards the lower part of the mountains, in the north–south orientated gullies cut in the northern slope, there are elongate dunes ascending the mountain towards the south.

The falling dunes found on the southern slopes accumulate greater quantities of sand than the climbing dunes due to deceleration of the wind as it passes over the crest of the mountains.

The greatest accumulations of sand in the Begur area, however, correspond to the 'passadis' dunes (Cros 1987) which form behind passes or open gaps between the mountains in much the same way as the 'sand drifts between obstacles' described by Bagnold (1941). 'Passadis' (corridor) dunes are formed by a funnelling effect produced as the wind passes between two obstacles, in this case two mountains. Gaylord & Dawson (1987) found this type of accumulation in the Ferris dune field behind Windy Gap, Wyoming. Pye & Tsoar (1990) discuss these accumulations and group them together with lee dunes. The exact limits of these dune formations in the Begur massif are not always clear because of their lateral transition to the falling dunes and the following climbing dunes.

This assemblage of dunes subdues and buries the previous relief by filling in hollows and blocking up valleys. Only the most vigorous rivers manage to prevent the sands choking their channels.

On the most subdued zones, superimposed blowouts and parabolic dunes are not uncommon.

Sedimentology of the sand deposits

Characterization of the sands was carried out on the basis of a specific granulometric index and a statistical analysis of the principal components of the sands and the presence and distribution of marine microfauna (Cros 1987; Cros & Serra 1991).

Granulometric distribution, $I_{i(160)}$ index

Using classical grain-size parameters, it can be stated that these sands are mostly of medium grain size and are well sorted (Cros 1987).

Most of them have a mode of 250 μm, though samples taken near the Ter river and along the coast have modes of up to 500 μm. Superimposed on this modal distribution we find another mode at approximately 160 μm. This bimodal character becomes quite significant in samples taken in the Pals inland area and the Begur massif, but is absent in the coastal area.

In order to clearly reflect the importance of the 160 μm fraction, the $I_{i(160)}$ index was devised. This index is obtained using the following formula:

$$I_{i(160)} = P_{160} (P_{250} + P_{500})^{-1} \qquad (1)$$

where P is the weight of the corresponding granulometric fraction.

A graphic representation of the values obtained (Fig. 2) shows a long coastal strip along which the values are minimal. Likewise, those taken along the Ter river are also very low, except in the inner and southern La Fonollera zone that have $I_{i(160)}$ values between 0.2 and 0.4. Pals inland area, Begur Massif and the area southwest of Regencós have the highest values with samples where the $I_{i(160)}$ index becomes higher than 0.5 (Fig. 2).

Quartz grain morphology

No great differences were observed in the sphericity and roundness of the grains from the different regions (Cros 1987). Nevertheless, there are definite indications of greater roundness in the southern samples. These can even reach a C-D level in the tables compiled by Shepard & Young (1961). These differences are consistent with slight selective transport of more rounded grains, or with progressive rounding due to abrasion during limited aeolian transport.

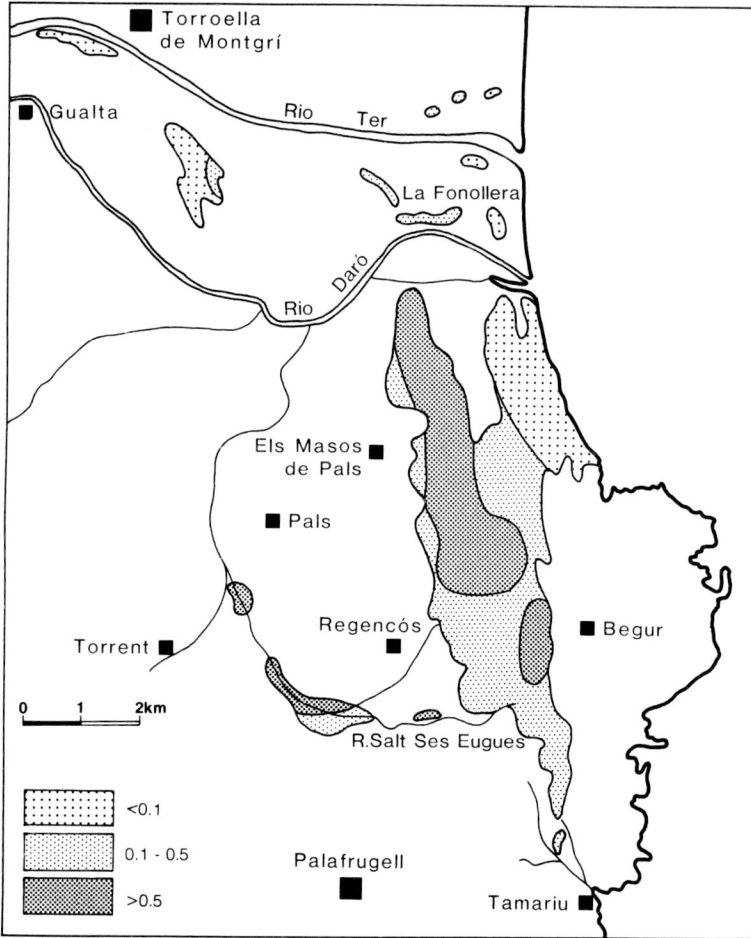

Fig. 2. Geographical distribution of the $I_{i(160)}$ Index values.

Observations of the surface texture of quartz grains using scanning electron microscopy (Cros 1987; Cros & Serra 1991) also showed that development of characteristic aeolian grain textures is limited.

Composition and faunal content of the sands

The sands are basically composed of quartz, feldspars, fragments of limestone, slate and volcanic rock. Near streams, rare gravels and small quantities of fine sediments are present. The percentage of heavy minerals is generally lower in the Pals inland sands and the Begur mountains compared with the coastal sands (Cros 1987).

The calcium carbonate content ranges from 5 to 7%, with extreme values of 1.8 and 11% (Cros 1987).

The faunal content of these aeolian sands consists of terrestrial mollusc fragments and detrital fossil remains probably derived from the surrounding Eocene rocks and transported mainly by the Ter river. Likewise, the remains of recent marine foraminifera and molluscs can be identified.

A study of the spatial variation in abundance (Fig. 3), shows that the sands near the river have no marine bioclasts, while the coastal sands contain small quantities of fragments, as well as whole specimens of molluscs — especially adult bivalves. Marine organisms, especially euryhaline microfauna (*Elphidium crispum*, *Ammonia beccari ammoniformis*, and the *Quinqueloculina* genus), are common in the

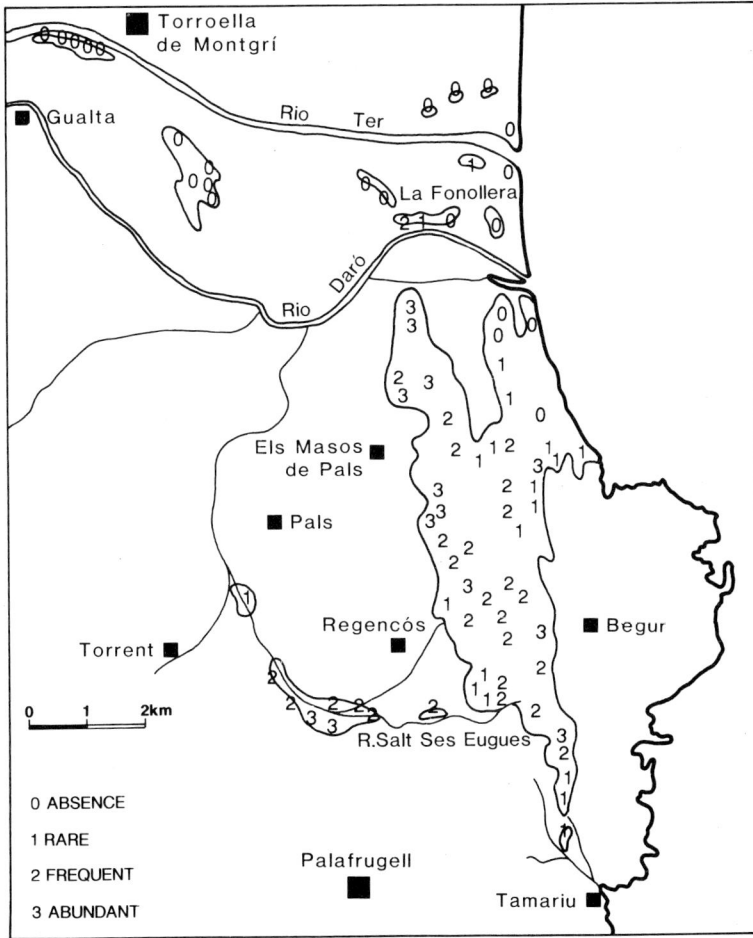

Fig. 3. Abundance distribution of marine organisms.

inland sands of Pals, Begur massif and SW of Regencós. The quantity and state of preservation, especially of the *Quinqueloculina* genus, diminish considerably towards the south of the Begur massif, suggesting southerly transport. Specimens of young gastropods and bivalves — especially *Glycemeris* — are also found in this area.

The important faunal content of euryhaline species in the inland sands of Pals, the Begur massif and to the SW of Regencós suggests that the source area of these sands could be a sedimentary environment with frequent and substantial variations in salinity.

Statistical analysis

Principal component analysis (PCA) was carried out on the data relating to calcium carbonate content, the different granulometric fractions and the silt and clay aggregates in the sands (Cros 1987; Cros & Serra 1991). A specially-written program was used (LAWI, J. Lleonart, I.C.M., Barcelona) after smoothing of the data by logarithmic transformation (Legendre & Legendre 1983).

The three primary components represent 78.39% of the total. The first component has an explanatory value of 42.28% and the second 24.40%.

The first component separates the coarse and very coarse sands (positive correlation) from the fine sand (negative correlation). Thus, aeolian sands near the Ter river and the coast are distinguished from the aeolian inland sands which contain an important 160 μm fraction.

The second component is negatively correlated with gravel size, very fine sands, silt and clays but positively correlated with the carbonates and the medium-size sands. This component has fresh water influences. It separates the alluvial sands (Ter river zone) from the coastal sands that are richer in carbonates and medium-size grain sands (Fig. 4). Within the inland sand group, the samples showing stream influence (with small incorporations of silt and clay, and gravels) are in a different quadrant from the those samples that, because of their geographic situation, have no stream influence.

Within this bidimensional space created by the two components, three regional groups of sand-types can be identified: one near the Ter river, another near the coast, and another called the 'inland sands' zone which includes Pals inland sands, Begur massif sands, SW Regencós sands and southern inland La Fonollera sands.

Origin and evolution of the sand groups

Based on the above results, three types of aeolian sediments can be distinguished: 'alluvial

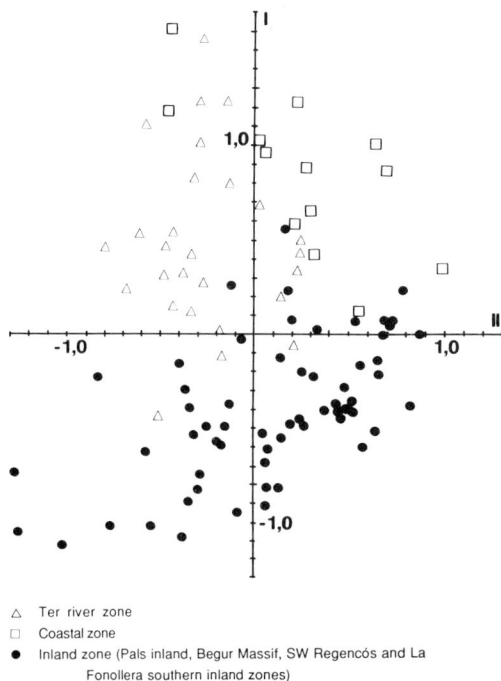

△ Ter river zone
□ Coastal zone
● Inland zone (Pals inland, Begur Massif, SW Regencós and La
        Fonollera southern inland zones)

Fig. 4. Results of principal component analysis (PCA) showing the distribution of three aeolian sand groups in the bidimensional space created by the first and second components.

sands', 'coastal sands' and 'inland sands' — each according to its particular origin and/or geographic location.

Alluvial sands

These aeolian sands are the most immature sands of the three in terms of granulometric composition and grain morphology. They have a high content of coarse and very coarse sands and a small percentage of silt, all of which points to a fluvial influence. These sands possess low $I_{i(160)}$ values and there is an absence of marine fauna. They are derived from fluvial sediments and form the alluvial plain dunes. The most important are those of the area around Sant Llorenç de les Arenes and those of Gualta (Fig. 5a).

Coastal sands

These are mostly medium-grained sands, though they do also include a significant fraction of coarse sand. They are typically unimodal and they show very low $I_{i(160)}$ values. On the whole they do not contain marine microfauna. Nevertheless, some bivalve and gastropod fragments are present. These sands differ from the former in their marine reworking. They show important aeolian characteristics, especially the more southern sands. They are derived directly from the Pals beach and indirectly from the Ter river. Geographically, they are located on the coastal margin of the Pals plain and ascend the coastal side of the Begur massif (Fig. 5b).

Inland sands

What differentiates this group of sands from the two former groups is the presence of a significant quantity, though no great diversity, of marine microfauna (benthic foraminifera of euryhaline species, as well as numerous gastropods and molluscs, both young and in their larval stage). This would suggest that these sands originated in a salt-water lagoon, restricted bay or marsh with important, and irregular, contributions of fresh water.

Mineralogically they contain lower percentages of heavy minerals than the other two types of sands and they are more rounded than the other two groups.

Granulometrically these sands are finer than those of the other two groups. They show bimodality and high $I_{i(160)}$ values. The bimodality suggests that the sedimentary environment in which they originated was supplied with sediment from at least two different sources, prob-

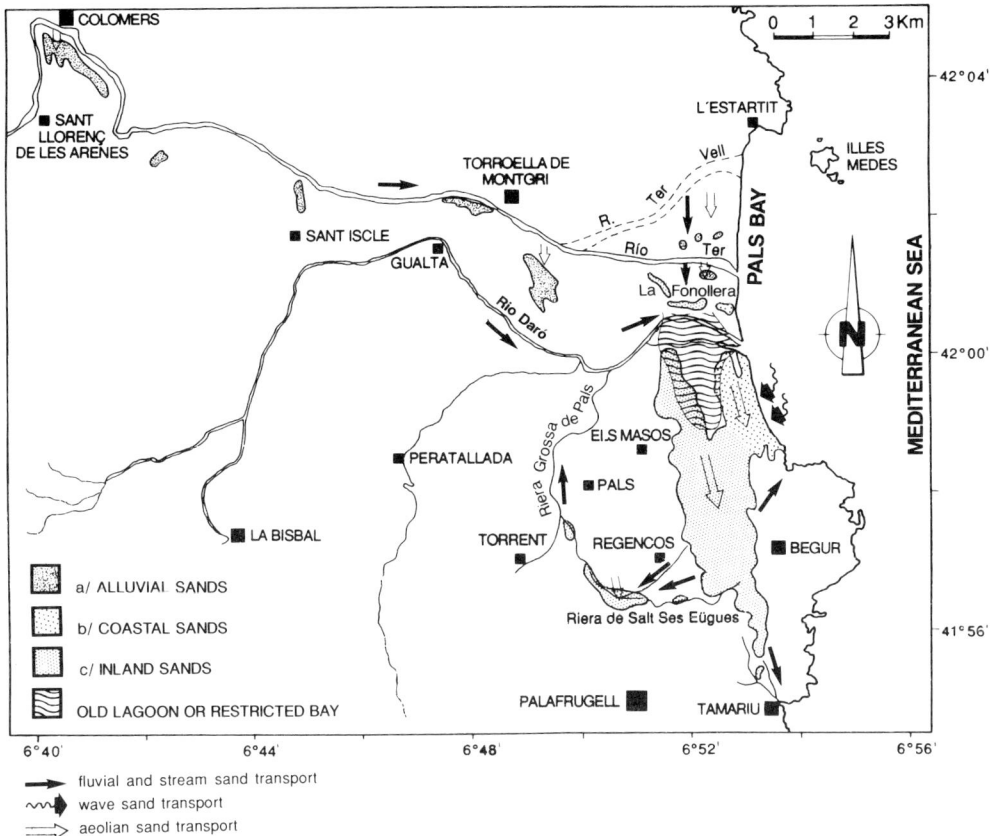

Fig. 5. Distribution of sand types and evolutionary diagram of the Baix Empordà aeolian deposits.

ably the Ter and Daró rivers (in the past the Daró drained the Pals and Ullastret ponds).

This group is the largest of the three. Geographically, it is located on the inland plain area of Pals, the Begur massif and the area SW of Regencós. The inner southern aeolian sands of the La Fonollera zone, also belong to this group (Fig. 5c).

Evolution of aeolian deposits

Taking into account the general characteristics and the geographic location of these three genetically different groups of sands, a map has been prepared showing the dynamic cycle of the aeolian deposits of Baix Empordà (Fig. 5). This map illustrates the source areas of sands, their different transport paths and the feedback linkages between the main processes: *Tramuntana* wind transport, coastal drift and fluvial transport.

The sediments arrive at the Empordà plain principally by way of the Ter river. The *Tramuntana* reworks the sediments from the floodplains and forms the alluvial plain sand dunes.

The Ter river also transports sediments to the sea, which in turn reworks them to form the Pals beach sands. Later, wind deflation of the beaches forms the coastal sand dunes.

The inland sands come from a restricted marine sedimentation system which no longer exists — presumed to be a restricted-bay or lagoon that had variable contributions of river-water from at least two different sources. Deflation of sandy deposits in this environment by the *Tramuntana* formed the inland sand dunes that made their way up the Begur massif. The eastern Begur massif streams reworked them and in so doing carried some sands back to the sea. The western Begur massif streams fed the hypothesized restricted bay or lagoon once again and redeposited the sands on the flood-plain formed at the junction of the Regencós and Salt Ses Eügues streams. Later, the *Tramuntana* formed the SW Regencós dunes by deflation of these sands.

Timing of dune formation

Historical records, both written and verbal, testify that the dunes were active in the last century and even early this century.

In the Begur massif the aeolian sands cover red clay deposits dated by Martí & Villalta (1974) as being Mindelian.

Soundings made at the Riera Grossa de Pals stream reveal a well established ancient hydrographic network fossilized by more recent Quaternary terrain among which exist lens-shaped patches of sands of equal grain size to those of the Begur massif dunes (Martinez Gil 1972).

Archaeological excavations carried out in the Fonollera area, between the present-day mouths of the Ter and Daró rivers, revealed the existence of a layer of sands burying a Roman village (Pons 1984).

In Sant Llorenç de les Arenes the dunes partially fill an eleventh century church, and completely cover the whole village north of the church. Today, the area is being excavated for the extraction of industrial and building sands. The sands that cover the area belong to the alluvial aeolian group.

In the Pals bay area there exists a sixteenth century watch-tower, called the Torre de Pals, which shows evidence of aeolian erosion on its northern side and is partly filled up with sand, especially affecting the adjacent buildings. These coastal aeolian sands have been fixed as a result of pine groves having been planted on them.

Two eighteenth century maps, (Cartotheque of the Collegi d'Arquitectes, Girona), also offer clear evidence of the sedimentary conditions needed for the formation of the inland sands as described above.

Conclusions

Sedimentological, microfaunal and statistical criteria, especially the $I_{i(160)}$ granulometric index, have proved useful in distinguishing three sand groups in the complex Baix Empordà dune system. The three groups were derived mainly from alluvial, coastal and inland sand sources, respectively.

The distribution of aeolian sand groups reflects the sand sources, the direction of the dominant wind (the *Tramuntana*), and recent geomorphological changes in the coastal area.

The inland sand group is the oldest. Alluvial and coastal aeolian sands are the most recent sand groups and had their development after the maximum Holocene transgression.

This system was partially active as recently as the last century.

References

BAGNOLD, R. A. 1941. *The Physics of Blown Sand and Desert Dunes*. Methuen, London.
CROS, L. 1987. *Estudi Sedimentològic dels Dipòsits Eòlics del Baix Empordà*. Tesi de Llicenciatura, Universitat de Barcelona.
—— & SERRA, J. 1991. Origen y evolución dinámica de los depositos eólicos del Baix Empordà (Abstract). *VIII Reunion Nacional sobre el Cuaternario*, Valencia.
DAVIES, J. L. 1974. The coastal sediment compartment. *Australian Geographical Studies*. **12**, 139–151.
GAYLORD, D. R. & DAWSON, P. J. 1987. Airflow-terrain interactions through a mountain gap, with an example of eolian activity beneath an atmospheric hydraulic jump. *Geology* **15**, 789–792.
IGME 1981, *Mapa Geologico de España* 1 : 200.000, n° 35. Instituto Geologico y Minero de España, Madrid.
KOCUREK, G. & NIELSON, J. 1986. Conditions favourable for the formation of warm climate aeolian sand sheets. *Sedimentology* **33**, 795–816.
LANDSBERG, S. Y. 1956. The orientation of dunes in Britain and Denmark in relation to wind. *Geographical Journal,* **122**, 176–189.
LEGENDRE, L. & LEGENDRE, P. 1983. *Numerical Ecology*. Elsevier. Amsterdam.
MAINGUET, M. & CHEMIN, M. C. 1983. Sand seas of the Sahara and Sahel: an explanation of their thickness and sand dune type by the sand budget principle. *In*: BROOKFIELD, M. E. & AHLBRANDT, T. S. (eds) *Eolian Sediments and Processes*. Elsevier, Amsterdam, 353–363.
MARTI, C. S. & DE VILLALTA, J. F. 1974. Un yacimiento Mindeliense en las cercanias de Bagur (provincia de Gerona). *Acta Geológica Hispánica*, **9**, 4–9.
MARTINEZ GIL, F. J. 1972. *Estudio hidrogeológico del Bajo Ampurdán (Gerona). Contribución a la metodología de los estudios hidrogeológicos regionales*. Memorias, IGME, **84**. Madrid.
McKEE, E. D. 1979. Introduction to a study of global sand seas. *In*: McKEE, E. D. (ed.) *A Study of Global Sand Seas*. Geological Survey Professional Paper, **1052**, 1–19.
PALLI, L. & BACH, J. 1987. *Itinerari Geològic pel Baix i Alt Empordà*, I.C.E. Universitat Autonoma de Barcelona.
PASCUAL, J. & FLOS, J. 1984. Meteorologia i oceanografia. *In*: ROS, J., OLIVELLA, I. & GILI, J. M. (eds) *Els Sistemes Naturals de les Illes Medes*. Institut d'Estudis Catalans. Barcelona, 75–114.

PONS, E. 1984. *L'Empordà: de l'edat del bronce a l'edat del ferro*. Centre d'investigacions Arqueologiques de Girona, Serie 4. Diputació de Girona.

PYE, K. & TSOAR, H. 1990. *Aeolian Sand and Sand Dunes*. Unwin Hyman, London.

SEPPÄLÄ, M. 1972. Location, morphology and orientation of inland dunes in northern Sweden. *Meddelander Från Uppsala Universitets Geografiska Institutioner*, Series A, **253**, 85–104.

SHEPHARD, F. P. & YOUNG, R. 1961. Distinguishing between beach and dune sands. *Journal of Sedimentary Petrology*. **31**, 196–214.

Late Holocene dune formation on the Sefton coast, northwest England

KENNETH PYE & ADRIAN NEAL

Postgraduate Research Institute for Sedimentology, University of Reading,
Whiteknights, Reading RG6 2AB, UK

Abstract: Coastal dunes fringe the coast between Southport and Liverpool in northwest England, forming a natural barrier which prevents marine flooding of low-lying agricultural land in West Lancashire and north Merseyside. The dunes are best developed at Formby Point, where they reach 25 m above sea-level and blown sand extends some 4 km inland. This paper presents the results of an initial programme of field drilling, sediment analysis and radiocarbon dating which suggest that the dune complex developed from a large offshore sandbank, which was in existence by 6800 years ago. Behind the bank lay a large area of intertidal mudflats and sandflats with fringing saltmarsh. Dunes may have first become established on the emergent sand bank around 6000 years ago, but the oldest dunes for which there is direct dating evidence formed between 5700 and 5800 years ago. A period of dune stability, possibly associated with seaward progradation of beach and foredune ridges, appears to have occurred between 2500 years ago and the early Middle Ages. During the earlier part of this period oak and alder woodland covered the dunes and the sands were deeply podsolized. Historical and documentary evidence indicates that extensive sandblowing occurred all along the coast during the early Middle Ages, which was a period of exceptional storminess and rapid coastal erosion. Several settlements and farms were lost to the sea or were overwhelmed by blown sand as the coast eroded. Active sand tongues extended across the full width of the barrier and buried the backbarrier peats and silts along the western margin of Downholland Moss. Dune instability has continued intermittently until the present, although on a smaller scale, and consequently most of the present dune forms are relatively recent. Sand stabilization and foreshore reclamation measures undertaken in the second half of the nineteenth century were highly effective and contributed to a net seaward progradation of the entire Formby coastline between 1850 and 1900. Abandonment of these measures around the time of the First World War contributed to, but did not ultimately cause, a return to shoreline erosion and a new phase of frontal dune instability which is still in progress.

There is increasing worldwide interest in the possible effects which changes in sea-level and wind/wave climate may have on coastal environments during the next century. Low-lying sandy and muddy coasts in northwest Europe are already experiencing erosion and submergence in many places, and these problems may be expected to become more serious if there is a significant increase in the rate of sea-level rise or an increase in storminess associated with global warming. In order to be able to predict the nature and magnitude of the likely effects, there is a need to obtain a deeper understanding of the morphodynamics and age structure of these coastal sedimentary systems.

The Sefton coast in northwest England (Fig. 1) was chosen for detailed investigation for a number of reasons:

(1) it includes one of the largest areas of coastal windblown sand in the UK;
(2) previous work in the area (Tooley 1970, 1978; Pye 1990) has suggested that coastal change and dune formation has been epi-

sodic in the last few thousand years, with episodes of transgressive dune activity separated by periods of relative stability and extensive soil formation;

(3) part of the dune frontage at Formby Point is rapidly eroding at present and the local authorities are increasingly concerned about the resulting implications for coastal zone management (Houston & Jones 1987; Houston 1989; Pye & Smith 1988).

The dunes form an important natural defence against marine flooding for a large area of high-grade agricultural land and settlements in West Lancashire and north Merseyside. The dune belt and adjoining beaches are also of high conservation and recreational value, with much of the land being managed as nature reserves by English Nature, the National Trust, and Sefton Borough Council. Dune restoration works have been employed since the late 1970s, but have not succeeded in stopping the coastal erosion between Formby Point and Ainsdale.

In order to provide more information about

From Pye, K. (ed.), 1993, *The Dynamics and Environmental Context of Aeolian Sedimentary Systems.*
Geological Society Special Publication No. 72, pp. 201–217.

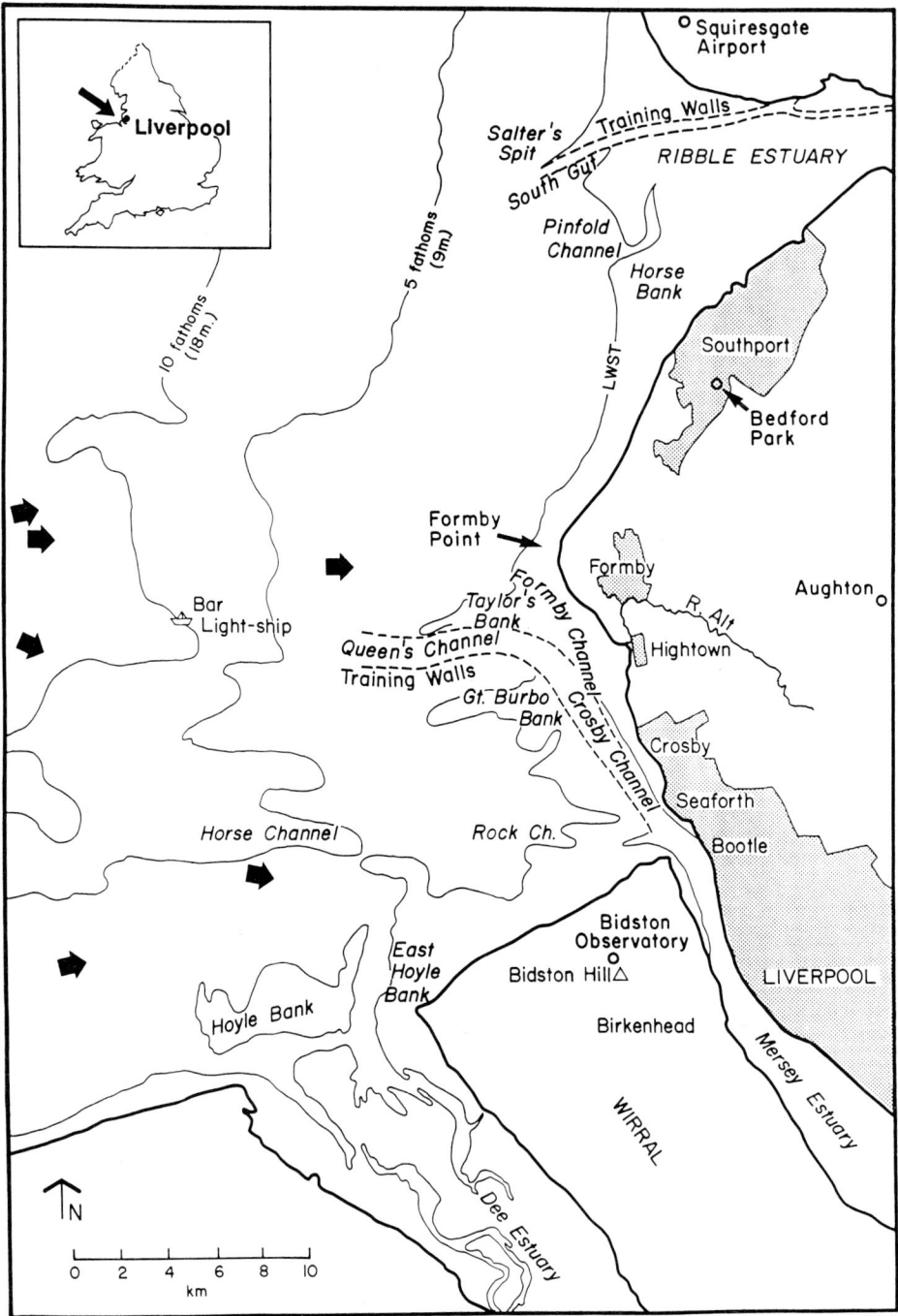

Fig. 1. Location of the study area in northwest England. Large arrows indicate net sediment transport direction.

the sedimentological character of the coast and its historical development during the later Flandrian Period, a programme of drilling and sediment sampling was begun in October 1990. Existing information from boreholes has been re-evaluated, and additional stratigraphic information gathered by examination of coastal sections and trial pits. This paper presents the results obtained so far.

Geological and environmental background

The present-day Sefton coast (Fig. 1) is transitional between open coast and estuarine regimes, being influenced by processes both in the eastern Irish Sea and in the Ribble and Mersey estuaries (Smith 1982; Pye 1990). Sediment transport and depositional processes are dominated by strong tidal currents and moderate wave energy conditions in the eastern Irish Sea. Mean spring tidal range at Formby Point exceeds 8.5 m and the maximum wave fetch to the west-northwest is more than 200 km. As a result, most of the coast is sandy and is backed by coastal dunes. Present-day mud accumulation is mainly limited to the higher intertidal flats in the more sheltered Ribble estuary and near the mouth of the River Alt.

Behind the dune belt lies an extensive area of peat mossland which overlies several metres of blue-grey estuarine clays interbedded with freshwater alluvium, peat and marls. These sediments, in turn, partially overlie aeolian coversands (Shirdley Hill Sands), Late Devensian glacial till and Triassic bedrock (de Rance 1872; Reade 1871; Wray *et al.* 1948; Tooley & Kear 1977; Wilson *et al.* 1981). The stratigraphy, palaeoecology, depositional history and evidence of sea-level change provided by the sediments on Downholland Moss have been investigated in detail by Tooley (1969, 1970, 1974, 1976, 1978, 1982, 1985a,b), but relatively little previous work has been undertaken on the morphostratigraphy and age structure of the dune belt.

The main source of sediments in the eastern Irish Sea is provided by wave and tidal current reworking of glacial deposits on the sea bed (Wright *et al.* 1971; McQuillin *et al.* 1969). During the later Holocene the Rivers Dee, Mersey and Ribble have transported mainly fine-grained sediments to the coastal zone, and for much of the last 7–8000 years their estuaries have acted as sinks for sediment brought both from the land and from the adjacent offshore areas (O'Connor 1987). Boomer profiling and vibrocoring have shown that the sands and gravelly sands offshore from Formby overlie former intertidal flat deposits, gravels and glacial till at relatively shallow depths (BGS 1984). However, the relationship of the submerged intertidal flat sediments to those on Downholland Moss has not been proven.

The tides in the eastern Irish Sea are semi-diurnal with a period of 12.5–13 hours. The predicted mean neap tidal range at Formby is about 4.5 m, increasing to 8.5 m at springs. Actual tidal heights can vary from predicted values by as much as 2 m due to the effects of storm surges (Lennon 1963a,b; Graff 1978; Pye 1991). Flood tidal current velocities in Liverpool Bay are higher than the ebb tidal velocities, resulting in a net landward drift of sediment near the bed. This net drift direction is confirmed by the alignment of sand waves on the sea floor and the results of sea bed drifter studies (Sly 1966; Best *et al.* 1973; Halliwell 1973). The flood tidal streams diverge off Formby Point and flow towards the Ribble and Mersey estuaries. The lower velocity ebb tidal streams diverge offshore from Birkdale and eventually merge with those of the Mersey and Ribble. Landward transfer of sediment into the Mersey and Ribble estuaries is also enhanced by near-bed residual currents, generated by density differences, wind stress and wave effects, which also have a net landward component (Price & Kendrick 1963; Bowden & Sharaf-El Din 1966; Ramster & Hill 1969; Halliwell 1973.

The dominant regional wind direction is southwesterly, but since the direction of greatest fetch (200 km) lies to the west and northwest the biggest waves recorded at the Mersey Bar approach from this direction (Sly 1966; HRS 1969a). Offshore wave energy is moderate, although no detailed records of wave conditions inshore at Formby are available. Wave recordings at the Mersey Bar light vessel (in 17.5 m of water, 16 km offshore), recorded over a twelve-month period in the 1950s, showed that the most common conditions were those with a significant height of 0.6–0.9 m and a zero crossing period of 4.0–4.5 s (Draper & Blakey 1969). Waves higher than 2 m in height occur only about 3% of the time, although their incidence increases to 10–15% of the time during the winter months. The maximum wave height recorded was approximately 8.0 m and the maximum recorded wave period 9.0 s.

Wave energy is presently focused on Formby Point due to refraction over the offshore banks and due to wave reflection off the north side of Taylors Bank. Westerly wave crests approach approximately normal to the shore at Formby Point, but diverge at an oblique angle both to the north and south (Gresswell 1953; Sly 1966). An upper foreshore littoral drift divide is located near Victoria Road, Freshfield (HRS 1969a,b; Parker 1974, 1975).

The large tidal range results in a wide multi-barred foreshore (Gresswell 1953; Parker 1971, 1975; Wright 1976, 1984). The morphology of the foreshore displays seasonal and longer-term variation, with the ridges and runnels being more pronounced in the summer months than in the winter when the foreshore adopts a more planar characteristic. The foreshore is narrowest and the ridges most closely spaced at Formby Point, where six or seven ridges are usually visible above low-water mark. The ridges are typically breached at intervals of 100–200 m by rip channels. In general, the foreshore becomes flatter and the ridges and runnels less pronounced, towards Southport and towards the River Alt mouth.

The sand comprising the beaches and dunes is generally fine grained and well-sorted (Pye 1977, 1991; Vincent 1986), but lower foreshore samples locally contain a few percent of mud and shell debris. The sand is predominantly quartzose, although the carbonate content increases towards the north and exceeds 10% near Southport.

The adjacent Irish Sea is relatively shallow, with the 18 m depth contour located about 15 km west of Formby Point. Examination of marine charts of different dates indicates rapid shoaling in Liverpool Bay during the last 150 years, with major sediment accumulation offshore and to the north and south of Formby Point (HRS 1958; Sly 1966; Pye 1977). Some of this sediment has been supplied by erosion at Formby and by dumping of dredged material from the Mersey approaches, but the majority is believed to have been moved eastwards and southeastwards into Liverpool Bay by currents and waves. Sediment accumulation within the Mersey and Ribble estuaries has also undoubtedly been enhanced by the construction of training walls and dredging in the main

Fig. 2. Oblique air photograph looking towards Ainsdale and Birkdale, taken c. 1963. The truncation by recent erosion of the parallel foredune ridges formed in the period 1880–1925 can clearly be seen.

approach channels to the ports of Liverpool and Preston (Barron 1938; Cashin 1949). These activities are believed to have concentrated the ebb flow in the trained channel, leading to enhanced residual flood transport over the sand banks on either side (HRS 1958, 1965, 1968, 1975, 1980; Price & Kendrick 1963; Sly 1966).

Evidence from Ordnance Survey maps indicates that the entire shoreline around Formby Point prograded seawards between 1845 and 1906. However, after this date erosion of the frontal dunes, preceded by a narrowing and steepening of the upper beach, began between Victoria Road and Lifeboat Road (Gresswell 1937, 1953). The northern limit or erosion has continued to move northwards towards Southport, and now lies approximately 1 km north of Fisherman's Path (Fig. 2). The change to erosion involved the establishment of a negative beach sediment budget and gradual reduction in backshore width, thereby increasing the susceptibility of the dunes to storm wave attack (Gresswell 1937; Parker 1975; Smith 1982). Since 1906 the average rate of erosion at Formby Point has been 3–4 m a^{-1}, although most of the retreat has occurred during a limited number of severe storms which recur with an average frequency of 5–6 years. During one such storm, in February 1990, 12–14 m of frontal dune recession occurred between Victoria Road and Fisherman's Path (Pye 1991). The shoreline at Lifeboat Road, Formby, is presently in the same position which it occupied at the beginning of the nineteenth century when the lifeboat station was built (the foundations of this structure can still be seen on the upper beach). Further north, around Wick's Lane and Victoria Road, Freshfield, the present shoreline lies inland of its position at the beginning of the nineteenth century (Pye & Smith 1988).

Most of the inland dunes are now largely stabilized by vegetation, including conifer plantations which were established in the later nineteenth and early twentieth centuries. However, a number of blowouts and active transgressive dunes occur in the dunes between Ainsdale and Southport and on the south side of Formby Point. The eroding section of shore is fronted by partially vegetated hummock dunes which are growing upward and moving landward as coastal erosion proceeds (Fig. 3). A major active blowout and transgressive sand sheet, orientated southwest–northeast, extends inland from a breach in the foredune ridge at Massam's Slack, Ainsdale. The morphology and dynamics of the dunes are discussed more fully in Pye (1990) and Pye & Neal (in press).

Fig. 3. Recently formed hummock dunes, partially vegetated by *Ammophila arenaria*, located on the eroding dune frontage near Victoria Road, Freshfield.

Investigative methods

Three main methods were used in the first phase of this investigation: (1) examination of sediment sequences exposed on the foreshore and frontal dune cliffs following storms; (2) shell and auger drilling, supplemented by hand augering and excavation of trial pits; and (3) re-examination of borehole records held by the British Geological Survey, Sefton Metropolitan Borough Council and the National Rivers Authority. During April–May 1991, 13 shell and auger holes were drilled to a maximum depth of 14 m along three transects between Ainsdale and Formby Point (Figs 4 & 5). Borehole logs were compiled and samples collected for sedimentological analysis and radiocarbon dating. The height of all boreholes relative to OD was determined by levelling with EDM equipment. Sediment samples from the shell and auger holes, hand-auger holes and inspection pits were analysed to determine the sediment grain size, carbonate content, mineral composition, and chemical composition. Where necessary, the sand and mud fractions were first separated by wet sieving through a 63 μm sieve and the size distribution of the sand-size fraction determined by dry-sieving. The silt and clay fractions were analysed by Coulter Counter. All data were processed using the computer programme GRANNY and the sediments classified according to the terminology of Folk (1968). Calcium carbonate content was determined using a calcimeter, and organic carbon content by titration. The mineralogy of representative bulk samples was determined by X-ray powder diffraction. Suitable materials were submitted for radiocarbon dating at Beta Analytic Inc., Coral Gables, Florida.

Fig. 4. The shell and auger rig used in the investigation. Behind the drilling site is one of the conifer stands planted in the early part of this century.

Results

In the eroding dune cliff sections along the shore north of Lifeboat Road four distinct dune morphostratigraphic units can be recognized lying directly above the intertidal muds and sands. Each unit is defined by a peat or humic horizon and has a characteristic soil type or types developed on it. The units have variable extent and a full sequence cannot be identified in any single section.

The lowermost unit, designated Unit 1, is a podsolized dune sand which outcrops at the base of the eroding foredunes at Lifeboat Road, Formby Point. The unit lies directly on an intertidal and supratidal mud unit which is described below. This unit is 1.5–1.25 m thick and can be traced northwards along the shore from Lifeboat Road for several hundred metres. The A1 horizon of the soil (elevation approximately 4.9 m OD) is 20 cm thick and forms a gently undulating indurated sandy peat which outcrops on the backshore (Fig. 6). The peat contains in situ roots and stumps of *Quercus*, one of which previously gave a radiocarbon age of 2510 ± 120 ^{14}C years BP. (Pye 1990). A similar date of 2335 ± 120 ^{14}C years BP (Hv 4709) was reported from this unit by Tooley (1970), who also demonstrated that the pollen spectrum of the A_0 horizon is dominated by oak and alder. The soil

is very similar in appearance to that described from Magilligan Point in Northern Ireland by Wilson (1991), but appears to have formed in a shorter time interval (*c.* 700 years).

The second dune unit directly overlies the podzol of dune unit 1. It has a distinctly mottled orange colour, is 0.7–2.0 m thick, and is capped by a thin (10–15 cm) humic soil horizon (Fig. 7). It can be traced along the eroding foredunes with a typically flat relief. Internally the sands show sub-horizontal stratification at their base, dipping gently (2–3°) in a landwards direction. Towards the top of the unit large-scale, low-angle, concave-up cross-stratification is developed. This suggests only low dune development during the deposition of the sands. The typical soil profile observed is that of a peaty gley, as defined in the modern dunes by James & Wharfe (1989). In a number of places the upper part of the unit has been removed by erosion associated with the deposition of the younger dune sand unit above. The lack of relief shown by this unit, and the peaty gley developed upon it, suggests soil development in a dune slack environment with undulating topography under conditions of a high water table.

Dune Unit 3 is very variable in thickness, lateral extent and degree of soil development. The unit is capped by a thin humic horizon and

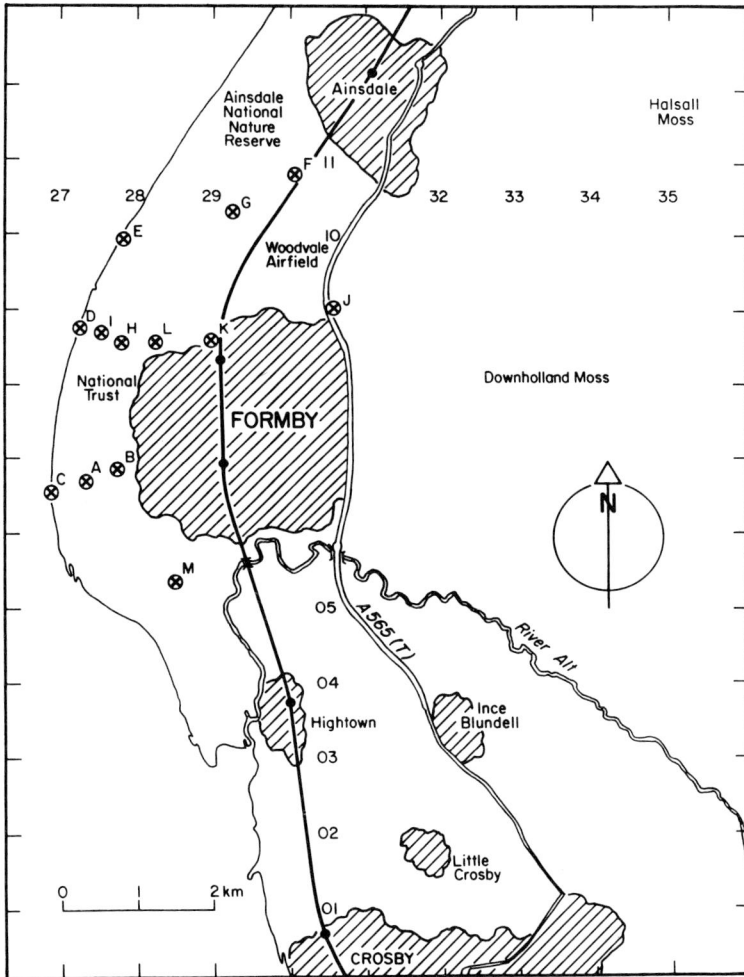

Fig. 5. Location of boreholes A to M.

Fig. 6. Organic-rich soil. A horizon containing roots and fallen trunks of oak exposed on the foreshore north of Lifeboat Road after a storm in February 1990. The organic horizon overlies a bleached A2 horizon and dark-brown Bh horizon developed in beach and dune platform sands.

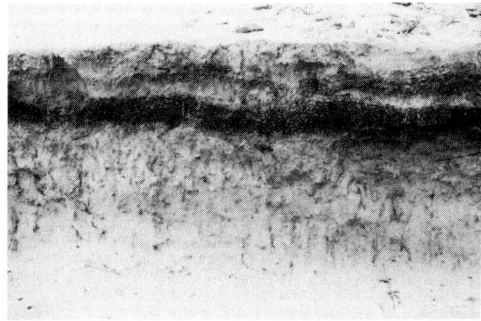

Fig. 7. Organic-rich sandy peat (dark) and underlying orange- mottled sands of a peaty gley soil developed in a former dune slack, exposed by foredune erosion near Victoria Road, Freshfield. The soil has been intermittently buried by more recent blown sand.

displays dune relief of up to 5 m. Internally, large-scale, high-angle, concave-up cross-stratification is common, indicating the development of dune topography at least several metres high. In topographic lows the humic horizon is seen to coalesce with that of the underlying unit. In these lows a typical peaty gley soil is developed. On topographic highs, pine stumps, roots and pine needles can be observed, and the remnants of a poorly developed *acid sand* soil type (James & Wharfe 1989) can be identified. This acid sand soil type is presently associated with pine woodland on the modern dunes. The stumps noted above are the legacy of an extensive pine planting programme begun in the late nineteenth century by local landowners (Gresswell 1953). This effectively gives a maximum age of 100 years to this dune surface, although the underlying sands themselves may be older. Where pines are not present, a *sand pararendzina* soil (James & Wharfe 1989) is developed. This is typical of the modern stabilized dunes with a continuous grass or heath vegetation cover.

The youngest dune unit (Unit 4) is up to 5 m thick and locally rests on all three of the previously described units. Internally the sands display a structure similar to that of Unit 3. Once again this is due to the development of high dune topography. Vegetation cover, mainly marram grass, is generally sparse with little or no soil development. The sands of this unit represent the active frontal dune ridge which is slowly migrating landwards over the older sand terrain behind. The formation of this unit is related to the onset of coastal erosion around the Point after 1900. Along most of the frontage, the dune ridge is only a few years to a few decades in age.

A complex and spatially variable sequence of intertidal muds and muddy sands is periodically exposed along the eroding foreshore northwards from Formby Point. The deposits are best exposed during periods when moderate storm events coincide with neap or average tides, when sand is eroded from the mid and upper foreshore (Fig. 8). The exposures rarely extend more than a few tens of metres in a shore normal direction at any one time, but at different times the deposits may be seen to extend more than 300 m towards low-water mark. Drilling and trenching has indicated that the deposits lie below the beach between a point approximately 200 m south of the old Lifeboat Station and Fisherman's Path, Ainsdale. The surface of the deposits dips seawards at a low angle (<5°). Along the line of the present upper foreshore the deposits are relatively thin (maximum thickness 0.7 m), and they wedge out in a landwards

Fig. 8. Extensive outcrop of late Holocene silts exposed on the foreshore near Dale Slack Gutter, Freshfield North, in 1977. Erosion has since removed much of the silt.

direction, terminating just landwards of the frontal dunes between Victoria Road and Lifeboat Road. North of Victoria Road the muds wedge out on the upper foreshore or backshore. The sediments are generally muddy with varying amounts of fine sand. Fresh exposures are typically black in colour, but extensive oxidation imparts a brown to reddish colour. Sub-horizontal lamination is well-developed in the deposits, particularly in the northern half of the exposure, leading to a step-like outcrop. On the seaward side of the exposures the contact with the underlying shelly sands is gradational, but on the landward margin there is a relatively sharp basal contact with the underlying orange-coloured shelly sands. The shell assemblage in these sands is similar to that in the modern beach sands.

In the north, around Dale Slack Gutter and Fisherman's Path, numerous dead *Scrobicularia plana* in life position and occasional *Cerastoderma* shells are observed in the top 0.2 m of the deposits. A radiocarbon date obtained from a number of the *Scrobicularia* shells (Beta-47681) indicated a recent age of 310 ± 70 ^{14}C years BP (Table 1). Since the *Scrobicularia* only occur in the uppermost 0.2 m of the deposit, and particularly in the top 5 cm, it suggests that they burrowed down into the top of the beds at a later date rather than having lived in the sediment as it was being deposited.

Several layers within the mud deposits contain hoof prints (Fig. 9), mainly of deer and a large bovine animal which is interpreted as domestic ox. Human footprints and horse tracks have also been observed, although at least some of these are modern features formed since re-exposure of the beds by erosion of the overlying beach sand.

Table 1. *Summary of radiocarbon dates obtained in this study. Reported radiocarbon ages are those supplied by the radiocarbon dating laboratory without correction for isotopic fractionation or calibration. Dates for shell samples have not been corrected for the local carbon 'reservoir' effect in British coastal waters since this is unnecessary when ages have not been normalized relative to $\delta^{13}C = -25‰$ (after Harkness 1983). Calibrated ages take account of variations in atmospheric ^{14}C concentration over time, based on dated tree ring series (after Stuiver & Pearson 1986; Pearson & Stuiver 1986; Pearson et al. 1986). Calibrations were performed using the computer program presented by Stuiver & Reimer 1986)*

Sample	Description	Reported radiocarbon age (^{14}C years BP)	Calibrated age calendar years BC/AD (range of 1σ uncertainty)
Beta-47679	Whole, disarticulated *Scrobicularia* shells	5960 ± 110	4997–4770 BC
Beta-47680	Wood from a dune slack peat	5110 ± 70	3994–3814 BC
Beta-47681	*In-situ*, whole, articulated *Scrobicularia* shells	310 ± 70	AD 1474–1654
Beta-47682	*Alnus* (alder) growing within desiccation cracks	3230 ± 80	1615–1427 BC
Beta-51922	Woody material from the top of a peat, with evidence of in-blowing sand	3380 ± 60	1747–1620 BC
Beta-51923	Disseminated organic material from a dune slack peat	2260 ± 60	396–234 BC

To the south of Wick's Lane the sediment characteristics are slightly different, with millimetre- to centimetre-scale alternations of brown clay and shelly sands. The fauna consists of both in situ and disarticulated, often broken *Scrobicularia* and *Pholas* shells, indicating a slightly higher-energy environment of deposition. Just north of the beach access point at Lifeboat Road the most elevated exposure of these muddy units is found on the upper foreshore just in front of the eroding frontal dunes. The upper surface of the beds is charac-

Fig. 9. Hoofprints of red deer on the surface of the foreshore silts exposed near Victoria Road, Freshfield.

Fig. 10. The remains of alder stems and roots which grew within desiccation polygons at the top of the intertidal mud unit, north of Lifeboat Road, Formby.

terized by the development of large desiccation polygons which contain the woody stems and roots of *Alnus* (Fig. 10). *Phragmites* stems are also present in places. The desiccation polygons clearly indicate subaerial exposure of this surface, and the growth of alder within the polygons indicates complete removal of marine influence for a period of at least several years. A sample of in situ alder root collected from this site at an elevation of +3.51 m OD, gave a radiocarbon age of 3230 ± 80 ^{14}C years BP (Beta-47682). This date gives a minimum age for the underlying intertidal flat deposits and a maximum age for the overlying podzolized dune Unit 1.

Shell and auger drilling at this site (borehole C, Figs 5 & 11) demonstrated that below the muddy sand unit, sands extend to at least −2 m OD. Drilling on the upper beach north of Victoria Road (borehole D) indicated only the presence of sands to a depth of −6 m OD. A similar sequence was encountered in a hole drilled on the upper beach at Fisherman's Path (borehole E). Previous drilling work on the foreshore near Pontin's holiday camp at Ainsdale has also demonstrated the existence of sands with occasional thin silty lenses to a depth of at least −7.5 m OD (Surveying Services Ltd, 1982).

Drilling behind the modern foredune ridge at Lifeboat Road showed that the muddy sands and muds extend approximately 100 m inland of the present shoreline. In boreholes A and B (Figs 5 & 11) they show a downward transition into clayey, muddy and silty sands and then clean sands containing numerous intact but disarticulated valves of juvenile *Cerastoderma edule* and *Scrobicularia plana*. A radiocarbon date obtained from a number of *Scrobicularia* shells from a depth of −5.5 m OD in borehole A gave an uncorrected radiocarbon age of 5960 ± 110 ^{14}C years BP (Beta-47679).

Drilling along a transect north of Victoria Road showed that the foreshore mud unit does not extend inland beyond the frontal dunes as a discrete unit although the sands at the corresponding elevation in borehole H were found to be slightly muddy. Borehole L, on the landward side of the main dune ridge, revealed a sequence of dune sands with interbedded soils and dune slack peats down to a depth of +1.25 m OD, overlying a 2 m-thick sequence of muddy sands which grade down into clean shell-rich intertidal or sub-tidal sands. A bulk organic sample from the intradune peaty soil at +4.5 m OD gave a radiocarbon age of 2260 ± 60 ^{14}C years BP (Beta-51923).

Further to the east, near Freshfield station, the dune sands rest on a well-developed peat, interpreted as a former salt or brackish marsh deposit, at an elevation of +3.8 m OD. This peat caps a 0.5 m-thick intertidal clay unit which in turn overlies at least 5 m of muddy sands grading downwards into clean shelly sands. The intertidal mud unit thickens eastwards towards borehole J, located at the landward end of this transect, on the eastern side of the Formby by-pass. At this point a relatively thin layer of blown sand (c. 3 m thick) overlies a well-developed freshwater peat layer which marks the top of a sequence of at least 9 m of Downholland Silt. Woody material from the top of this peat layer gave a radiocarbon age of 3380 ±60 ^{14}C years BP (Beta-51922), indicating a maximum age for the onset of deposition of the overlying aeolian sand. This compares with the date of 4090 ± 175 ^{14}C years BP (Hv 4705) previously reported by Tooley (1978) for the topmost peat beneath blown sand at the western margin of Down-holland Moss. However, the sand overlying the peat is mostly brown in colour and incompletely decalcified, suggesting that it may have been deposited within the last few hundred years.

Borehole G, located in the middle of the Ainsdale National Nature Reserve, showed that the dune sands rest on a foundation of slightly muddy, shell-rich marine sands with a number of thin mud horizons. A thick sequence of sandy silt and sandy mud was encountered below the dune sands at the base of borehole F, located near the eastern margin of the reserve. An intra-dune peat layer encountered in this borehole at 2.7–2.8 m OD yielded a date of 5110 ± 70 ^{14}C years BP, indicating that dunes existed in this area at least by 5800 calendar years ago (Table 1).

A peat bed and the remains of an ancient forest are exposed on the foreshore at Blundell-sands Sailing Club, near Hightown, and extends southwards towards Hall Road, Blundellsands (Reade 1871; Travis 1926; Fig. 12). The surface of the peat lies at an altitude of approximately +3.4 m OD and contains the branches and trunks of *Alnus*, *Quercus* and *Betula*. The bed has been exposed by landward movement of the River Alt channel across the foreshore during the past century. It is underlain by *Phragmites* peat, blue esturarine clay, sand, more blue clay and shell-rich sand containing fragments of *Cerastoderma edule*, *Macoma balthica*, *Barnea candida* and *Chlamys opercularis* (Tooley & Kear 1977). Landwards and stratigraphically above the peat bed lie 4–6 m of dune sands. A radiocarbon date from the base of the organic deposit at 3.10 m OD yielded a date of 4545 ± 90 ^{14}C years BP (Hv 2679; Tooley 1970). Palaeo-ecological data from the site have been inter-preted as indicating a progressive change from high-energy marine conditions, through low-energy saltmarsh to freshwater reedswamp and fen conditions, possibly aided by a slight fall in relative sea-level (Tooley 1970, 1978). The age of the overlying dune sand at this site has not been determined directly, but the date of 4545 ±

90 ^{14}C years BP gives a maximum age for the onset of its accumulation.

Grain-size analysis showed that the dune sedi-ments between the River Alt and Ainsdale are predominantly composed of fine, very well-sorted sand (mean size 2.25 ϕ to 2.38 ϕ; sorting 0.28–0.36 ϕ). Beach and sub-tidal sand samples obtained from the boreholes are predominantly fine to very fine (mean 2.17–3.32 ϕ), with sorting values which range from moderately to very well sorted (standard deviation range 2.62–0.69 ϕ). Intertidal sediment samples from the landward side of the barrier were also found to be composed of fine to very fine sands (mean size 2.24–3.79 ϕ) which is moderately well to very well sorted (0.21 to 0.74 ϕ).

The mineralogical composition of the dune sands is dominated by quartz (75–90%), with 5–20% K-feldspar, 3–6% plagioclase feldspar, 0–2% calcite, 0–2% dolomite and 0.5–2% heavy minerals. The beach, intertidal flat and sub-tidal sands, some of which contain fines, from beneath the dunes contain 60–70% quartz, 5–28% K-feldspar, 2–10% plagioclase, 1–3% calcite, 1–3% dolomite, 0.5–2% heavy minerals and 1–6% clay minerals. The backbarrier inter-tidal sediments, which are mostly muddy sands and sandy muds, contain 60–65% quartz, 12–15% K-feldspar, 10–15% plagioclase, 0–2% calcite, 3–5% dolomite, 0.5–2% heavy minerals and 8–9% clay minerals.

Discussion and interpretation

The sequence of estuarine silts and peats present on Downholland Moss does not extend con-tinuously beneath the dune belt at Formby Point. The western and central sections of the dune belt rest on a foundation of shelly marine sands, locally muddy, which extend to a depth of at least −7.5 m OD. Although the bottom of this unit was not reached during drilling, earlier boreholes have shown that the maximum thick-ness of drift overlying the Triassic bedrock in the Formby Point area is about 20 m (BGS Geological Sheet 83, Drift Edition). The Formby Number One Deep Borehole, located near the western edge of Downholland Moss, proved a thickness of approximately 30 m of estuarine alluvium, peat and marl overlying glacial till and Trias. Between Ainsdale and Southport, earlier boreholes have shown that the blown sands rest on a foundation of peat and estuarine alluvium at approximately 2.5 m OD and these deposits, which extend to at least −15 m OD, overlie glacial till. Similarly, south of the River Alt as far as Seaforth, the late Holocene dune sands overlie peat and blue

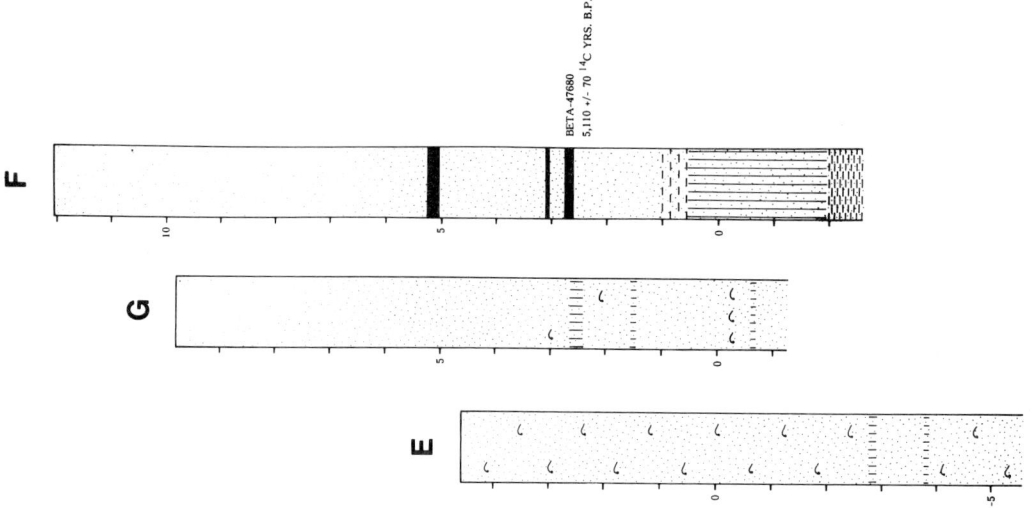

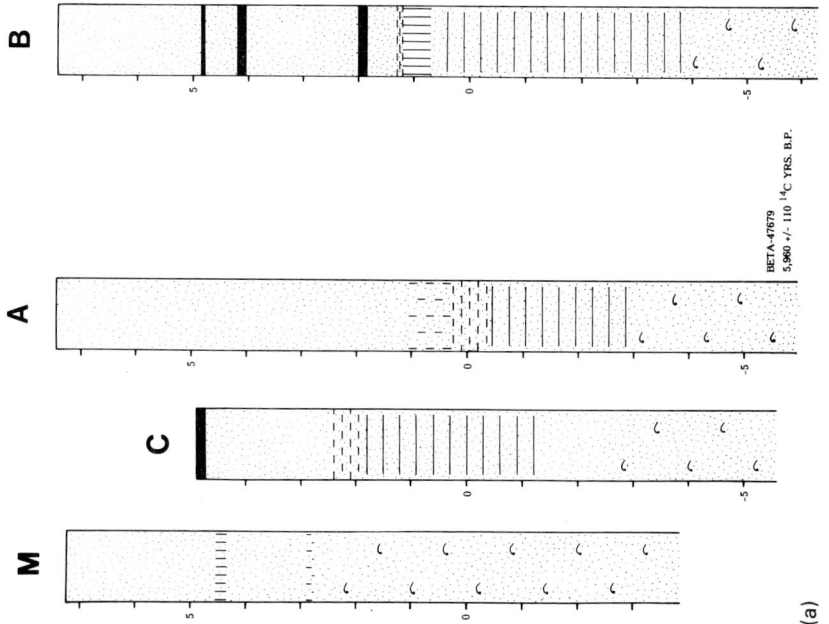

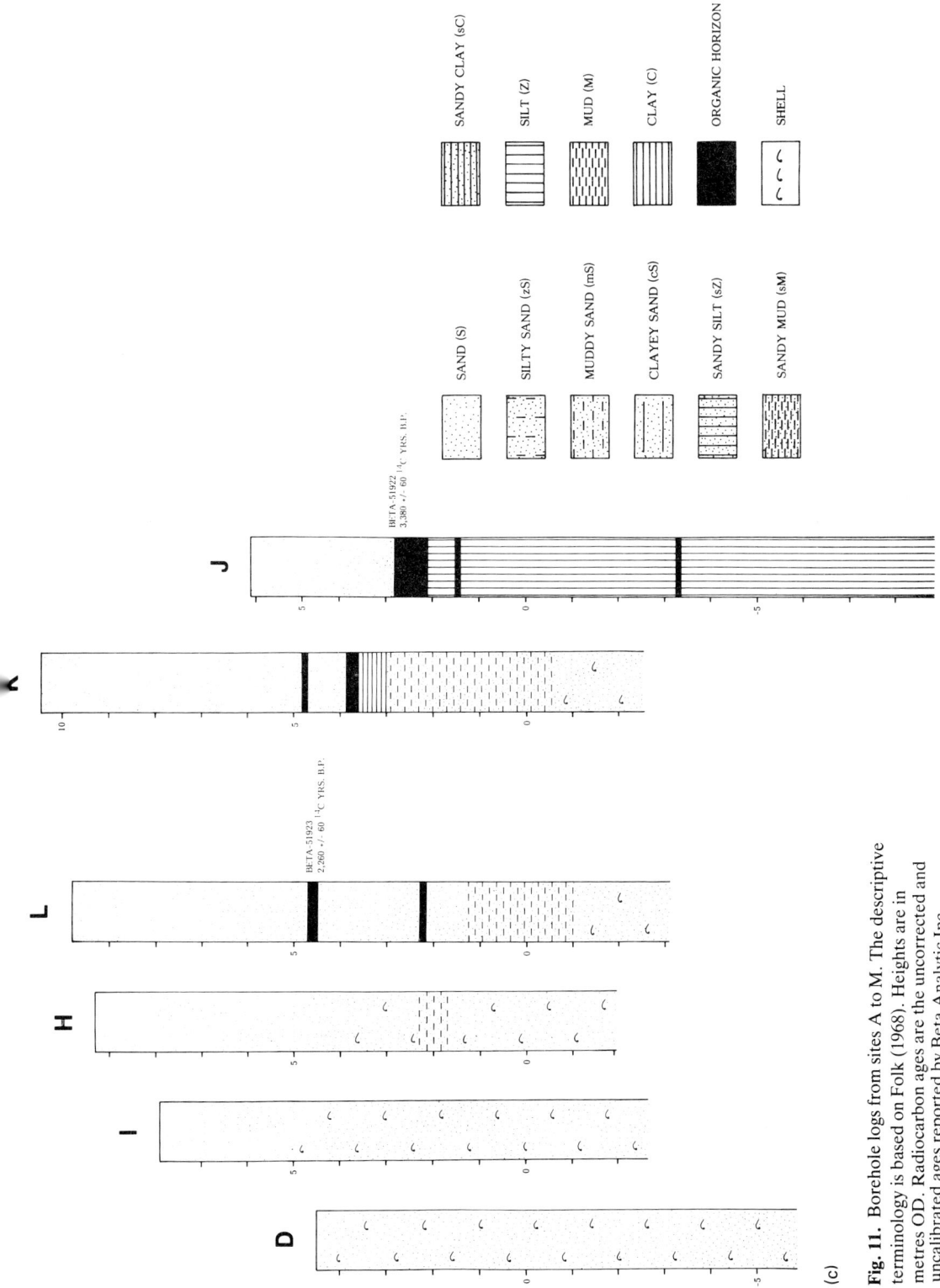

Fig. 11. Borehole logs from sites A to M. The descriptive terminology is based on Folk (1968). Heights are in metres OD. Radiocarbon ages are the uncorrected and uncalibrated ages reported by Beta Analytic Inc.

(c)

Fig. 12. The intertidal peat bed at Hightown being examined by R. K. Gresswell during the 1930s.

estuarine clays which extend to a depth of −15 to −20 m OD (de Rance 1872).

Tooley (1969, 1974, 1978) demonstrated that the predominantly estuarine sequence of fine-grained sediments on Downholland Moss began to accumulate about 8000 years ago and ended approximately 4500 years ago. He suggested that the rate and extent of estuarine sedimentation varied over time in response to fluctuations in sea-level. Our data indicate that subtidal and intertidal sands had started to accumulate in the Formby area at least by 6800 years BP, probably in the form of an offshore bar or bank which extended between Ainsdale and just north of the River Alt mouth. There is direct evidence that low dunes existed in the area of the Ainsdale National Nature Reserve around 5800 years ago and in the Formby Point area by at least 3000 years ago.

Formation of the muddy sediments presently exposed on the foreshore at Formby appears to have begun some time after 4500 years BP and ended by about 3500 years ago. They formed as intertidal flat deposits which onlapped a pre-existing sandy barrier beach. The accumulation of intertidal mud indicates a reduction in wave energy, due either to the growth of a north–south-trending sand bank which offered shelter, or a possible regional reduction in wave energy during the late Atlantic and early Sub-Boreal Periods, when the atmospheric circulation was

generally less vigorous and westerlies took a more northerly track over Europe than at present (Lamb 1977; Musk 1985). The trend towards intertidal flat progradation and seaward movement of high-water mark could also have been encouraged by a slight fall in sea-level around this time as suggested by evidence from Downholland Moss and adjacent areas (Tooley 1974, 1978). Landward of the developing mudflat, aeolian activity on the original barrier system would have been very restricted.

An increase in wave energy led to a change from mud deposition to sand deposition along the Formby coastline sometime after 3300 BP. Erosion caused the shoreline to move landwards, and beach ridges with low dune cappings were formed on top of the intertidal mudflat deposits in the vicinity of the present shoreline at Formby Point. Sand was blown inland at least as far as a line through the middle of the Ainsdale NNR and the western edge of Formby village, forming low dunes and sand sheets. Radiocarbon evidence from the Lifeboat Road exposures and from borehole L at Ainsdale indicates that this major phase of sandblowing had ceased by 2350 years BP, leading to the formation of extensive peaty soils beneath a mature oak woodland vegetation cover.

Historical and archaeological evidence suggests that much of the sand belt at Formby remained relatively stable until the early Middle

Ages (Ashton 1909, 1920). Documentary accounts suggest that high dunes did not exist in the area until this time, and that the low dunes were extensively used for cultivation and grazing. Several agricultural settlements existed within the sand belt during early medieval times (Kelly 1973; Harrop 1985). During the thirteenth century, a major phase of coastal erosion and widespread sandblowing was initiated. The period AD 1200–1400 was one of great climatic instability, with both severe droughts and severe storms. Cooler, slightly wetter and more stormy conditions continued during much of the Little Ice Age (AD 1430–1850). The scale of sandblowing and dune mobility during the earlier Middle Ages was such that several settlements, including Argameols and Ravenmeols, were abandoned (Ashton 1909; Kelly 1973; Harrop 1985). This interpretation is consistent with our stratigraphic and sedimentological data, which indicate that by far the largest part of the Formby dune complex is very young in age.

Although radiocarbon dates of between 4000 and 3300 years BP have been obtained from the topmost peats beneath blown sand at the western margin of Downholland Moss, they provide only maximum ages for the onset of sand deposition. The sands themselves display pedogenic and sedimentological characteristics which suggest it is more likely that the major phase of sand encroachment occurred during the Middle Ages. Independent support for this conclusion is provided by preliminary optical luminescence dating of sand from just above the peat contact (Stokes & Pye, unpublished data).

Drilling showed that most of the dunes at Formby and Ainsdale are composed of uniform sand with only weak pedogenetic differentiation, implying that the present forms are at most only a few hundred years old. The possibility that large dunes formerly existed but were largely reworked during the Middle Ages or later times is considered unlikely on the basis of documentary evidence. The period 2500–600 BP is, therefore, interpreted to be one in which there was slow progradation of beach ridges and low foredunes around the whole coast.

Sandblowing and localized dune activity appears to have continued from the Middle Ages until the nineteenth century, when the local landowners took stronger control measures, including more extensive planting of marram grass and conifers, and erection of brushwood fences on the beach. Large areas of the dunes were successfully stabilized in the late nineteenth century and early twentieth century, although in areas subject to heavy visitor pressure active sandblowing continued. The stabilization measures reinforced a trend towards rapid progradation of the shoreline all around Formby Point during the later nineteenth century (Gresswell 1937, 1953). However, control measures effectively ceased during the First World War and were not renewed until the late 1970s (Houston & Jones 1987). This abandonment undoubtedly contributed significantly to the return of conditions of rapid coastal erosion and active sandblowing.

Conclusions

The Formby dune complex probably originated on an offshore sand bank which was in existence at least by 6400 years ago. The area landward of the sand bank filled up with intertidal silts, saltmarsh peats, river alluvium and freshwater peats beginning around 7000 years ago. Intertidal muds also accumulated on the seaward side of the barrier between about 4500 and 3500 years ago, possibly in response to a reduction in wind/wave climate and a slight fall in sea-level. After 3500 years BP there was a return to higher wave energy conditions which may have been associated either with stronger regional winds or a slight rise in sea-level. The shoreline retreated landwards and there was a change from mud to sand deposition, including the formation of low dunes between 3500 and 2500 years BP. This was followed by a period of widespread sand stabilization and woodland development after about 2500 years BP. The dune belt appears to have remained mainly stable, without high dune development, until the early Middle Ages, when increased storminess caused rapid coastal erosion and instigated sandblowing on a massive scale. Sand tongues and sheets extended right across the barrier, burying the late Holocene peats along the western margin of Downholland Moss. Most of the high dune forms seen today are relatively recent, post-dating the Middle Ages and possibly related to the introduction of marram grass in the seventeenth century.

Assistance provided by Tony Smith and John Houston of Sefton Metropolitan Borough Council is gratefully acknowledged. Permission to drill on private land was granted by the National Trust, English Nature, Formby Golf Club and the Formby Land Company. We are particularly grateful to Martin Garbett, David Wheeler and the Sefton Ranger Service for logistic support, discussion and provision of unpublished information. The research was supported by a grant from the Nuffield Foundation and by a NERC Research Studentship held by Adrian Neal. University of Reading PRIS Contribution No. 209.

References

ASHTON, W. 1909. *The Battle of Land and Sea on the Lancashire, Cheshire and North Wales and the Origin of the Lancashire Sandhills.* W. Ashton and Sons Ltd, Southport.

—— 1920. *The evolution of a Coastline.* E. Stanford Ltd, London, and Ashton & Sons Ltd, Southport.

BARRON, J. 1938. *A History of the Preston Navigation from Preston to the Sea.* Preston Corporation & Guardian Press, Manchester.

BEST, R., AINSWORTH, G., WOOD, P. C. & JAMES, J. E. 1973. Effects of sewage sludge on the marine environment, a case study in Liverpool Bay. *Proceedings of the Institution of Civil Engineers,* **55**, 43–66; Discussion **55**, 755–765.

BOWDEN, K. F. & SHARAF-EI DIN, S. H. 1966. Circulation and mixing processes in the Liverpool Bay area of the Irish Sea. *Geophysical Journal of the Royal Astronomical Society,* **11**, 279–292.

BGS 1984. *Liverpool Bay. Sheet 53N 04W Sea Bed Sediments & Quaternary. Scale 1 : 250,000.* British Geological Survey, Keyworth.

CASHIN, J. A. 1949. Engineering works for the improvement of the estuary of the Mersey. *Journal of the Institution of Civil Enginers,* **32**, 396–455.

DRAPER, L. & BLAKEY, A. 1969. *Waves at the Mersey Bar light vessel.* Institute of Oceanographic Sciences, Report.

FOLK, R. L. 1968. *Petrology of Sedimentary Rocks.* Hemphill Publishing Co., Austin, Texas.

GRAFF, J. 1978. Abnormal sea levels in the northwest of England. *Dock & Harbour Authority,* **58**, 366–368.

GRESSWELL, R. K. 1937. The geomorphology of the southwest Lancashire coastline. *Geographical Journal,* **90**, 35–349.

—— 1953. *Sandy Shores in South Lancashire.* Liverpool University Press, Liverpool.

HALLIWELL, A. R. 1973. Residual drift near the sea bed in Liverpool Bay; an observational study. *Geophysical Journal of the Royal Astronomical Society,* **32**, 439–458.

HARKNESS, D. D. 1983. The extent of natural ^{14}C deficiency in the coastal environment of the United Kingdom. *In*: MOOK, W. G. & WATERBOLK, H. T. (eds) *Proceedings of the First International Symposium on [14]C Dating and Archaeology.* PACT Journal Volume 8, Council of Europe Publication Series, Strasbourg, 351–364.

HARROP, S. 1985. *Old Birkdale and Ainsdale; Life on the Southwest Lancashire Coast 1600–1851.* The Birkdale and Ainsdale Historical Research Society, Southport.

HOUSTON, J. 1989. The Sefton Coast Management Scheme in Northwest England. *In*: VAN DER MEULEN, F., JUNGERIUS, P. D. & VISSER, J. H. (eds) *Perspectives in Coastal Dune Management.* SPB Academic Publishing, The Hague, 249–253.

—— & JONES, C. R. 1987. The Sefton Coast Management Scheme: project and process. *Coastal Management,* **15**, 267–297.

HRS 1958. *Radioactive Tracers for the Study of Sand Movement: A Report on an Experiment Carried out in Liverpool Bay in 1958.* Hydraulics Research Station, Wallingford.

—— 1965. *Investigation of Siltation in the Estuary of the River Ribble, July 1965.* Hydraulics Research Station, Wallingford, Report Ex 281.

—— 1968. *Notes on Engineering Works to Reduce Dredging in the Ribble Estuary.* Hydraulics Research Station, Wallingford Report Ex 391.

—— 1969a. *The Southwest Lancashire Coastline, a Report of the Sea Defences.* Hydraulics Research Station, Wallingford, Report Ex 450.

—— 1969b. *Computation of Littoral Drift.* Hydraulics Research Station, Wallingford, Report Ex 449.

—— 1975. *Sand Winning at Southport.* Hydraulics Research Station, Wallingford, Report Ex 754.

—— 1980. *River Ribble Cessation of Dredging.* Hydraulics Research Station, Wallingford, Report Ex 948.

JAMES, P. A. & WHARFE, A. J. 1989. Timescales of soil development in a coastal sand dune system, Ainsdale, northwest England. *In*: VAN DER MEULEN, F., JUNGERIUS, P. D. & VISSER, J. (eds) *Perspectives in Coastal Dune Management.* SPB Academic Publishing, The Hague, 287–295.

KELLY, E. (ed.) 1973. *Viking Village, The Story of Formby.* The Formby Society, Liverpool.

LAMB, H. H. 1977. *Climatic Change,* Volume 2. Methuen, London.

LENNON, G. W. 1963a. Identification of weather conditions associated with the generation of major storm surges along the west coast of the British Isles. *Quarterly Journal of the Royal Meteorological Society,* **89**, 381–394.

—— 1963b. A frequency investigation of abnormally high tide levels at certain west coast ports. *Proceedings of the Institution of Civil Engineers,* **25**, 451–484.

McQUILLIN, R., WRIGHT, J. E., OWEN, B. & LISTER, T. R. 1969. Recent geological investigations in the Irish Sea. *Nature,* **222**, 365–366.

MUSK, L. F. 1985. Glacial and post-glacial climatic conditions in North-west England. *In*: JOHNSON, R. H. (ed.) *The Geomorphology of Northwest England.* Manchester University Press, Manchester, 59–79.

O'CONNOR, B. A. 1987. Short and long-term changes in estuary capacity. *Journal of the Geological Society, London,* **144**, 187–195.

PARKER, W. R. 1971. *Aspects of the Marine Environment at Formby Point, Lancashire.* PhD thesis, University of Liverpool.

—— 1974. Sand transport and coastal stability, Lancashire, U.K. *Proceedings of the 14th Conference on Coastal Engineering,* 828–850.

—— 1975. Sediment mobility and erosion on a multibarred foreshore (Southwest Lancashire, U.K.) *In*: HAILS, J. R. & CARR, A. P. *Nearshore Sediment Dynamics and Sedimentation.* Wiley, London, 151–177.

PEARSON, G. W. & STUIVER, M. 1986. High precision calibration of the radiocarbon timescale, 500–2500 BC. *Radiocarbon,* **28** (2B), 839–862.

——, PILCHER, J. R., BAILLIE, M. G. L., CORBETT, D. M. & QUA, F. 1986. High precision ¹⁴C measurement of Irish oaks to show the natural ¹⁴C variations from AD 1840 to 5210 BC. *Radiocarbon*, **28** (2B), 911–934.

PRICE, W. A. & KENDRICK, M. P. 1963. Field and model investigations into the reasons for siltation in the Mersey estuary. *Journal of the Institution of Civil Engineers*, **24**, 473–517.

PYE, K. 1977. *An Analysis of Coastal Dynamics Between the Ribble and Mersey Estuaries With Particular Reference to Erosion at Formby Point*. Unpublished BA Dissertation, University of Oxford.

—— 1990. Physical and human influences on coastal dune development between the Ribble and Mersey estuaries, northwest England. *In:* NORDSTROM, K. F., PSUTY, N. P. & CARTER, R. W. G. (eds) *Coastal Dunes: Form and Process*. Wiley, Chichester, 339–359.

—— 1991. Beach deflation and backshore dune formation following erosion under storm surge conditions: an example from northwest England. *Acta Mechanica Supplementum*, **2**, 171–181. Springer Verlag, Berlin.

—— & NEAL, A. (in press) Coastal dune erosion at Formby Point, north Merseyside: causes and consequences. *Earth Surface Processes and Landforms*.

—— & SMITH, A. J. 1988. Beach and dune erosion and accretion on the Sefton coast, northwest England. *Journal of Coastal Research Special Issue 3*, 33–36.

RAMSTER, J. W. & HILL, H. W. 1969. Current systems in the northern Irish Sea. *Nature*, **224**, 59–61.

RANCE, T. M. DE 1872. Explanation of the Quarter-Sheet 90 N.E. Of the One-Inch Geological Survey Map of England. *Memoirs of the Geological Society of England and Wales*. Longman's Green & Co., London.

READE, T. M. 1871. The geology and physics of the Post-Glacial Period, as shown in the deposits and organic remains in Lancashire and Cheshire. *Proceedings of the Liverpool Geological Society*, **2**, 36–88.

SLY, P. G. 1966. *Marine Geological Studies in the Eastern Irish Sea and Adjacent Estuaries, with Special Reference to Sedimentation in Liverpool Bay and the River Mersey*. PhD thesis, University of Liverpool.

SMITH, A. J. 1982. *A Guide to the Sefton Coast Data Base*. Metropolitan Borough of Sefton (unpublished).

STUIVER, M. & PEARSON, G. W. 1986. High precision calibration of the radiocarbon time scale. AD 1950–500 BC. *Radiocarbon*, **28** (2B), 805–838.

—— & REIMER, P. J. 1986. A computer program for radiocarbon age calibration. *Radiocarbon*, **28** (2B), 1022–1030.

SURVEYING SERVICES LTD 1982. *Report No. 8217 Ainsdale Foreshore*. Unpublished report to Sefton Metropolitan Borough Council.

TOOLEY, M. J. 1969. *Sea-level Changes and the Development of Coastal Plant Communities During the Flandrian in Lancashire and Adjacent Areas*. PhD thesis, University of Lancaster.

—— 1970. The peat beds of the southwest Lancashire coast. *Nature in Lancashire*, **1**, 19–26.

—— 1974. Sea-level changes in the last 9000 years in northwest England. *Geographical Journal*, **140**, 18–42.

—— 1976. Flandrian sea level changes in west Lancashire and their implications for the 'Hillhouse Coastline'. *Geological Journal*, **11**, 137–152.

—— 1978. *Sea level Changes in Northwest England during the Flandrian Stage*. Clarendon Press, Oxford.

—— 1982. Sea level changes in northern England. *Proceedings of the Geologists' Association*, **93**, 43–51.

—— 1985a. Sea-level changes and coastal morphology in northwest England. *In:* JOHNSON, R. H. (ed.) *The Geomorphology of Northwest England*. Manchester University Press, Manchester, 94–121.

—— 1985b. Climate, sea-level and coastal changes. *In:* TOOLEY, M. J. & SHEAIL, G. M. (eds) *The Climatic Scene*. Allen & Unwin, London, 206–234.

—— & KEAR, R. 1977. *The Isle of Man, Lancashire Coast and Lake District. Guidebook for Excursion 4, 10th INQUA Congress, Birmingham*. Geo Abstracts, Norwich.

TRAVIS, C. B. 1926. The peat and forest bed of the southwest Lancashire coast. *Proceedings of the Liverpool Geological Society*, **14**, 263–277.

VINCENT, P. 1986. Differentiation of modern beach and coastal dune sands — a logistic regression approach using the parameters of the hyperbolic function. *Sedimentary Geology*, **49**, 167–176.

WILSON, P. 1991. Buried soils and coastal aeolian sands at Portstewart, Co. Londonderry, Northern Ireland. *Scottish Geographical Magazine*, **107**, 198–202.

——, BATEMAN, R. M. & CATT, J. A. 1981. Petrography, origin and environment of deposition of the Shirdley Hill Sand of Southwest Lancashire, England. *Proceedings of the Geologists' Association*, **92**, 211–229.

WRAY, D. A., COPE, W. F., TONKS, L. H. & JONES, R. C. 1948. Geology of Southport and Formby. *Memoirs of the Geological Survey of Great Britain*. HMSO, London.

WRIGHT, J. E., HULL, J. H., McQUILLIAN, R. & ARNOLD, S. E. 1971. *Irish Sea Investigations 1969–71*. Institute of Geological Sciences Report 71/1.

WRIGHT, P. 1976. *The Morphology, Sedimentary Structures and Processes of the Foreshore at Ainsdale, Merseyside, England*. PhD thesis, University of Reading.

—— 1984. Facies development on a barred (ridge and runnel) coastline: the case of southwest Lancashire (Merseyside), U.K. *In:* CLARK, M. (ed.) *Coastal Research: U.K. Perspectives*. Geo Books, Norwich, 105–118.

Progressive vadose diagenesis in late Quaternary aeolianite deposits?

RITA A. M. GARDNER & SUSAN J. McLAREN

Department of Geography, King's College, Strand, London WC2R 2LS, UK

Abstract: Since the mid-1960s a number of models (as exemplified by Gavish & Friedman (1969) and Land *et al.* (1967)) have been developed which suggest that a progressive sequence of morphological, mineralogical and chemical changes occurs during carbonate diagenesis in the vadose zone. The models raise the possibility that the degree of cementation and diagenetic alteration can be used as a guide to interpreting the relative, and even the absolute, age of deposits. A study of the early vadose diagenesis of a wide range of late Quaternary age aeolianites from Oman, South India, the Bahamas, Tunisia and Mallorca has been undertaken to assess the wider applicability of the models. The results presented in this paper indicate that the rigid application of such models to areas outside those for which they have been developed is inappropriate. In the aeolianites analysed, spatial and temporal heterogeneity is the key characteristic. All tend towards an end point of low-Mg calcite allochems and meniscus, rim and pore-filling cement types but at different rates in different places, both locally and regionally, and via several different diagenetic routes.

Detrital carbonate grains are a common component of aeolian sand bodies. They are usually derived from a coastal sand input, and many deposits are located close to that source. In cases of abundant supply, strong onshore seasonal winds and aridity, the sediments may be transported several hundred kilometres inland, as in the aeolianite deposits of NW India and Oman. The carbonate particles include bioclastic fragments (molluscs, echinoderms, algae, and foraminifera in particular) and, less commonly, ooliths which are mostly found in aeolianites along the north African coasts and in the Caribbean.

The preservation potential of Pleistocene aeolian sand bodies can be enhanced in two main ways: through climatic changes resulting in stabilization and vegetation; by post-depositional diagenetic changes leading to cementation of the sands, usually by low-Mg calcite. The latter form is arguably of greater use to the palaeoenvironmentalist as it provides sufficient lithification for stratigraphic sections to be exposed and it results in the preservation of bedding structures. Aeolianites are widely found within tropical and subtropical latitudes (Gardner 1981).

The lithification itself, under freshwater vadose, mixed and/or phreatic conditions, can also be of use in the interpretation of palaeo-environments, from delimiting former levels of the water table (Land *et al.* 1967; Budd 1984, 1988*a,b*), to inferring climatic changes (Ward 1975) and assessing age of deposits (Hutto & Carew 1984). However, as in all palaeoenvironmental analysis, its value depends largely upon the level of knowledge and understanding of processes and rates of carbonate diagenesis, and on the applicability of the empirical models developed out of this understanding.

Models developed in the field of vadose diagenesis of carbonates, which is that of most relevance to aeolian sand bodies, fall into two broad groups. The first concerns cement morphologies in various combinations of vadose/phreatic and freshwater/marine environments (see McLaren 1993), and the second, the subject of this paper, deals with the progression of diagenetic changes over time. The call for a model relating diagenetic changes to climate (Reeckmann & Gill 1981) has so far gone unheeded.

The idea of progressive and ordered diagenetic change in the freshwater vadose environment was first formulated in the mid-1960s (Friedman 1964) and subsequently developed and applied through the work of Land *et al.* (1967) in Bermuda, Gavish & Friedman (1969) in Israel, Ristvet (1972) in Bermuda, and Reeckmann & Gill (1981) in Australia. Despite the fact that much of the research dates back over 20 years, the models remain generally accepted today, as exemplified by Morse & Mackenzie (1990) who state that there is 'remarkable similarity between the trends in diagenetic texture, mineralogy and geochemistry with progressive geologic age exhibited by Bermudan biosparites and those of the carbonate sediments of the Mediterranean coast of Israel'. Examples of recent applications of the models to 'date' sediments include those of Titus (1984) and Hutto and Carew

From Pye, K. (ed.), 1993, *The Dynamics and Environmental Context of Aeolian Sedimentary Systems*.
Geological Society Special Publication No. 72, pp. 219–234.

Table 1. Models of progressive vadose diagenesis

Author/location	Diagenesis	Stage 1	Stage 2	Stage 3	Stage 4	Stage 5	Stage 6
Friedman (1964) Israel	a. product		meniscus and rim cement	low Mg-calcite allochems	secondary porosity	moulds infilled	pore-filling cement
	b. process		early aragonite dissolution	loss of Mg^{2+}	aragonite dissolution	low-Mg calcite precipitation	low-Mg calcite precipitation
	c. porosity						
	d. time						
Land et al. (1967) Bermuda	a. product	unconsolidated sediments	meniscus cement (external source)	low-Mg calcite allochems	pore-filling cement	total loss of aragonite	
	b. process			loss of Mg^{2+}	dissolution of aragonite and re-precipitation		
	c. porosity						
	d. time						
Gavish & Friedman (1969) Israel	a. product	unconsolidated sediments	meniscus and rim cement (external)	pore-filling cement	fully stable	fully stable	
	b. process		loss of Mg^{2+}	dissolution of aragonite			
	c. porosity	porosity 35%	porosity 20%	porosity 10%	porosity 3%		
	d. time		7–10 ka	80–130 ka	>130 ka		
Reeckmann & Gill (1981) Australia	a. product	unconsolidated sediments	meniscus cement (external)	pore-filling cement	total occlusion of pores by low-Mg calcite		
	b. process		loss of Mg^{2+}	dissolution of aragonite			
	c. porosity						
	d. time		90 ka	400–700 ka	>700 ka		

(1984), both of whom used lack of significant lithification to propose Holocene ages for the sediments under consideration.

The aim of this paper is to undertake an evaluation of the models of progressive vadose diagenesis in order to determine whether they can be sensibly applied to deposits of an aeolian and sandy nearshore character in other areas, and whether, as a result, they have wider applicability in aeolian palaeoenvironmental and geochronological studies.

Models of vadose diagenesis

The models have three components: the diagenetic product; the diagenetic processes proposed to account for the product; and the rates of diagenesis (Table 1). In terms of diagenetic product, all the models propose progressive, sequential change. With the exception of Gvirtman & Friedman (1977) this commences from an original unlithified sandy sediment in which the bioclasts comprise high-Mg calcite and aragonite. In all models the end point, reached after three or more stages, comprises a fully stable, low-Mg calcite, lithified biocalcarenite with all or almost all of the pore space occluded by drusy calcite cement (Table 1). In the intervening stages low-Mg calcite rim and meniscus cements develop first. Gavish & Friedman (1969) quote a decrease in porosity from 35% in the original sediments to 20% after the formation of rim and meniscus cements. This first diagenetic stage may be accompanied by the

transformation of high-Mg bioclasts to low-Mg calcites. However, Friedman (1964) and Land et al. (1967) indicate that the loss of magnesium occurs as a separate stage. This is followed by the dissolution of aragonite and the precipitation of low-Mg calcite, pore-filling, cement crystals over one or more stages until porosity is occluded. All models conform broadly to the same sequential development of diagenetic products although this occurs over a varying number of stages.

There are more differences between the models when diagenetic processes are considered, and in particular those associated with the supply of carbonates. While most researchers suggest that rim and meniscus cements are derived from an external supply of carbonates because they occur prior to evidence for in situ dissolution of aragonite, Friedman (1964) argues for re-precipitation of low-Mg calcite following early aragonite dissolution. All the models see a substantial part of the pore-filling cements (estimated 40% in the case of Gavish & Friedman; Fig. 1) being derived locally from dissolution of aragonite allochems and re-precipitation of low-Mg calcite. Friedman (1964) in particular, and to a lesser extent Gavish & Friedman (1969), stress the development of secondary (mouldic) porosity as a result of aragonite bioclast dissolution. This is subsequently infilled during the precipitation of pore-filling cements.

Thus, most of the models agree on the following points. First, meniscus and rim cements are

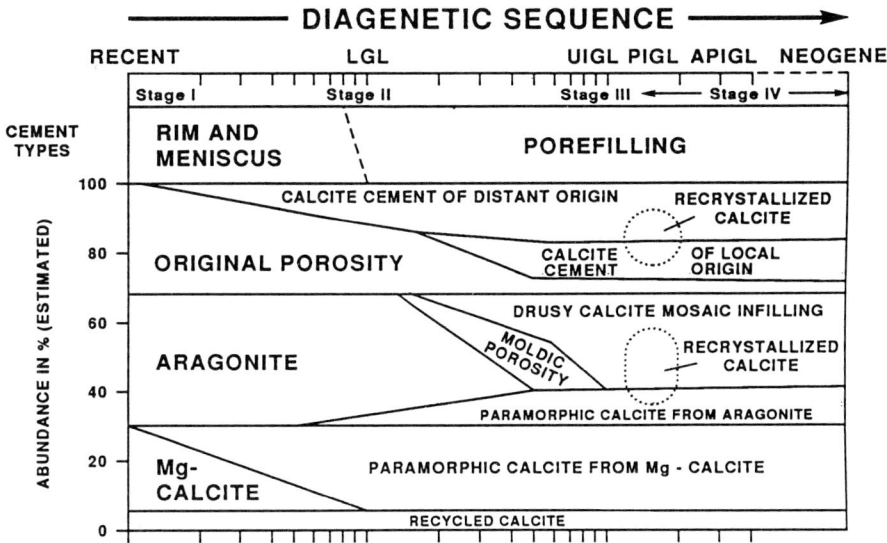

Fig. 1. The model of progressive vadose diagenesis proposed by Gavish & Friedman (1969), amended to include cement types.

the earliest forms and they are derived from an external source. Secondly, the loss of magnesium from high-Mg calcites occurs before the dissolution of aragonite. Thirdly, a substantial component of the pore-filling cement is supplied by the re-precipitation of dissolved aragonite, which involves an increase in volume of approximately 8% (Bathurst 1975). Fourthly, porosity falls over time until it is finally occluded, or remains stable at low levels. Lastly, the identification of separate stages implies that local variability in diagenetic product and process is low.

The rates of progressive diagenesis are less well-documented and rely heavily upon the Israeli sequence (Gavish & Friedman 1969) which has been dated using radiometric methods (radiocarbon) for the Holocene deposits. The ages of last glacial and older deposits have only been determined by stratigraphical means. Although some questions have been raised regarding the accuracy of the dates (Buchbinder & Friedman 1980), it remains the most widely used guide to rates of early subaerial diagenesis. The formation of rim and meniscus cements takes place within 7–10 ka. Dissolution of aragonite occurs over the period 10–50 ka and by stage 3 (80–130 ka), the formation of pore-filling cement has occluded the secondary pores and much of the primary porosity resulting in an overall porosity of only 10%. This subsequently falls to close to 3% in stage 4.

This rate of diagenesis is approximately 10 times faster than that recorded by Reeckmann & Gill (1981) in a skeletal grainstone in Victoria, Australia. The marine calcarenites and aeolianites were dated using radiocarbon techniques for the Holocene deposits and uranium/thorium assays combined with stratigraphical techniques for the younger Pleistocene sediments (Gill 1976). No evaluation of likely contamination is presented. Despite the fact that the sediments contain 90–95% carbonates and the climate is wetter than in Israel (700–750 mm rainfall per year), the rate of diagenesis appears to be unusually slow when compared with most other aeolianites. Mg^{2+} has been lost from high-Mg calcites and only meniscus cements have developed in deposits of 90 ka (stage 1). Meniscus cement continues to be precipitated in stage 2, during which aragonite is slowly dissolved. By stage 3 (400–700 ka) calcite is being precipitated as pore-filling cement and no aragonite remains.

As yet there is no explanation for the apparent wide differences in rates of diagenesis between the Israeli and Australian sequences. The well-studied Bermuda aeolianites (Land *et al.* 1967;

Ristvet 1972; Schroeder 1973) are of little help because dating has proved difficult and the diagenetic rates do not appear to be uniform.

Controls on diagenesis

The idea of progressive diagenetic change in a dated sequence places great importance on time as a controlling factor in diagenesis. This is particularly so if the models are then applied as interpretive tools to other deposits. The relevance of some other factors has been illustrated by earlier research in which individual components such as climate or texture have been found to be important in affecting diagenesis in a particular formation. Only rarely have a number of factors been considered together, and a systematic evaluation of all the recognized possible controls on diagenesis has long been overdue (McLaren 1991).

The controls on diagenesis can be viewed at three spatial scales — macro-, meso- and micro-. The level of explanation of the diagenetic processes increases as the scale focuses on smaller areas (Table 2). At the macro-scale, the regional spatial scale, the tectonic and sea-level history of an area is an enabling factor, and one that determines in part the length of exposure of a sediment body in the vadose zone. The same is true of climate and climatic change in that the products of vadose carbonate diagenesis, as identified above, appear to be almost entirely restricted to warm semi-arid to seasonally humid areas. The original environment of deposition of the sediment, through its influence on sedimentological and lithological properties is a further macro-factor. Lastly, there is time, the factor that the general diagenetic models under discussion have concentrated upon.

Meso-scale controls are clearly linked to the macro-factors but they can vary within, as well as between, sediment bodies. Texture is the most commonly cited influence at this scale, including particle size and shapes and the associated packing, porosity and the size of pore spaces. Its

Table 2. *Potential controls on vadose diagenesis*

Macro-scale	Meso-scale	Micro-scale
Tectonics	Texture	Pore water
Sea-level	Allochems	Chemistry
Climate	Sea spray	Particle
Lithology	Nearness to surface	Mineralogy
Time	Vegetation Water table	Microsolution

importance lies particularly in affecting the rate of water movement through deposits (Evans & Ginsburg 1987; Martin *et al.* 1986; Harris & Matthews 1968; Morrow 1971). In general, finer-grained sediments (Land 1970) and those with poorer sorting (Morrow 1971) are better cemented. The presence or absence of biogenic carbonate fragments has also been cited as an important control (Harris & Matthews 1968; Gavish & Friedman 1969; Steinen 1974; Pingitore 1976). The shells act both as nucleii for initial precipitation and as potential carbonate sources.

Two further meso-scale factors can be identified: the location within a sediment body and the influence of plants and organisms. Location is important in relation to carbonate addition, leaching or localized redistribution. Thus, location in the zone of capillary rise, and nearness to exposed surfaces in receipt of precipitation, surface runoff, sea spray or prone to high evaporation rates is likely to affect process, product and rate of diagenesis. There are contrasting reports of the resulting diagenetic impact (e.g. Longman 1980; McLaren 1991), and in the case of capillary rise of carbonate-charged groundwaters little is known. While the role of plants and organisms in diagenesis has been recognized, the processes are poorly understood. Their primary positive role is in enhancing $CaCO_3$ precipitation at the sediment–root interface leading to the formation of rhizoliths. Whether the process is evapotranspiration affecting pore water concentrations (Marion *et al.* 1985), symbiotic reactions between roots and soil bacteria to produce carbonate (Klappa 1978), or the ability of bacteria associated with blue-green algae to precipitate $CaCO_3$ (Jones & Kahle 1986), or another cause, remains unclear.

Micro-scale controls operate within sediment bodies, usually at the scale of individual particles, pores and groups of pores. They relate closely to the processes of dissolution of aragonite and precipitation of low-Mg calcite and thus concern the chemistry of the substrate and the pore water, its composition, its saturation with respect to carbonate minerals, and its movement. For example, aragonitic bioclasts appear to dissolve more readily than ooliths (Evans & Ginsburg 1987; McLaren 1991); the rate and location of initial crystal development varies according to the mineralogy of the substrate and the presence of adsorbed organic molecules on it (Bathurst 1975). Micro-scale spatial variability in the products of diagenesis have been noted by Buddemeier & Oberndorfer (1986) and James & Choquette (1984). The latter suggested that micro-scale dissolution and replacement may occur either when pore waters are only slightly undersaturated with respect to aragonite or when the rate of throughflow is slow.

The existence of multiple controls on diagenesis operating at different spatial scales, i.e. a nested hierarchy of controls, suggests that diagenetic products, processes and rates should exhibit levels of variability at all spatial scales (both within and between deposits), unless one, or a clearly defined group, of the controls has a dominant influence. Spatial and temporal heterogeneity in diagenesis would severely limit the wider applicability of diagenetic models and the use of diagenesis in palaeoenvironmental analysis.

Methodology

This study is restricted largely to aeolianites of late Quaternary age for three reasons. First, this is the period over which, according to the Gavish & Friedman model and most other dated deposits, much of the vadose diagenetic change occurs; second, aeolianites of this age are moderately well dated, in contrast to the older aeolianites; and third, they have experienced less in the way of climatic and sea-level changes and thus their history of exposure is less complex. Aeolianite deposits were selected from a wide range of modern environments, including arid (Oman), semi-arid (S. Spain and Tunisia), mediterranean (Mallorca) and seasonally humid (S. India and the Bahamas where rainfall exceeds 1500 mm per year).

The aeolianites, and sandy raised beach samples where appropriate, were examined primarily in thin section. Oriented samples were impregnated with araldite under vacuum prior to sectioning. Over 1000 sections were examined in the course of a larger study. Point counting was used to establish proportions of each diagenetic fabric and porosity (550 counts per section). Staining of sections with Fiegl's solution (Fiegl 1937; aragonite), Alizarin red S and potassium ferricyanide (Friedman 1959; Dickson 1966; calcite/dolomite/ferroan calcite) and Titan yellow (Choquette & Trusell 1978; high-Mg calcite) facilitated differentiation between carbonate minerals. This was further supported by SEM analysis (Hitatchi S-450 and Hitatchi 2500) with energy dispersive X-ray micro-analysis (Link Systems AN 10,000) on selected samples.

The sampling framework was designed to accommodate possible macro-, meso- and micro-scale variability in diagenesis by examining deposits from very different regions, sampling individual beds from several sections within each deposit, and, in a few locations, close sampling within individual beds. The choice of sample sections was restricted to available exposures, but care was taken to sample as far back as possible from the exposed surface to eliminate any case-hardening effect. The selection from the range of available exposures was made in order to accommodate the likely diagenetic controls. Thus, for example, locations within one deposit with varying exposure to sea spray were examined in Mallorca. Comparison with documented sea-level histories of the field areas was used to ensure that all the sediments sampled had been subjected to vadose, rather than phreatic, freshwater conditions. All cement types discussed are low Mg-calcite.

Details of the locations and field stratigraphy can be found in Gardner (1981, 1987) and McLaren (1991). The ages of the deposits examined are shown in Table 3.

Progressive, ordered vadose diagenesis?

In all of the samples examined, with the exception of one biocalcarenite section in SE India, meniscus and rim cements were found to pre-date pore-filling cements. This is not surprising as unlithified sediments will be free-draining in the vadose zone and any water held within the pores by capillary forces will be present as meniscii between grains and as films on the grain surfaces. As the pores become smaller and less well connected in the course of diagenesis, then the capacity for more water to remain longer within the primary pores is likely to increase. Furthermore, the rim and meniscus cements were generally found to have formed prior to substantial aragonite dissolution, thus supporting the view that the carbonates in these cements were derived from an external source.

In the Holocene aeolianite from North Point, the Bahamas, there is evidence of some loss of aragonite bioclasts but no loss of high-Mg calcite bioclasts. Thus it is not always the case that high-Mg calcite is dissolved out prior to aragonite.

One of the most interesting findings was that pore-filling cementation may occur without substantial loss of aragonite and that the loss of aragonite need not occur early on in the diagenetic history of a deposit. Oolitic aeolianites from the Bahamas generally contain more than 80% of original aragonitic grains, and in some cases over 95% remain unaltered. For example, ooliths comprise 40% of the deposit at White Bay Cay and 48% of the deposit at Watlings Quarry, the remainder being predominantly carbonate mud pellets. Of these, on average, 89% of the ooliths retain their aragonitic mineralogy while at the same time the deposits have 19% pore-filling cement (remaining porosity is 14%) (Fig. 2). As a result, there are poor correlations between the percentage of allochems (as defined by Bathurst 1975 p. 565, and to include bioclasts and ooliths) in the original sediment and the percentage of cement. Pearson's 'r' values of 0.031 and 0.053 were calculated for a total of 41 samples at White Bay Cay and 20 samples at Watlings Quarry. The lack of aragonite dissolution is further supported by the low values of secondary porosity; mean value at White Bay Cay is 2.2%.

Table 3. *The ages of the deposits studied*

Location	Age	Method	Reference
Roquetas, S. Spain	Stage 5	U-series	McLaren 1991
			Zazo *et al.* 1981
Amoladares, S. Spain	Stage 5	U-series	Zazo *et al.* 1981
Mallorca	Stage 5e	U-series	Hearty 1986
			McLaren 1991
Tunisia	Stage 5	Amino acid	Miller *et al.* 1986
		Stratigraphy	McLaren 1991
Bahamas	Holocene	^{14}C	Carew & Mylroie 1986
		U-series	McLaren 1991
	Stage 5	U-series	Carew & Mylroie 1986
			McLaren 1991
SE India	Holocene	^{14}C	Gardner 1986
	Stage 5	Stratigraphy	Gardner 1986
Wahiba Sands, Oman	Holocene	^{14}C	Gardner 1987
	Stage 5	Stratigraphy	Gardner 1987
	> Stage 5	Stratigraphy	Gardner 1987

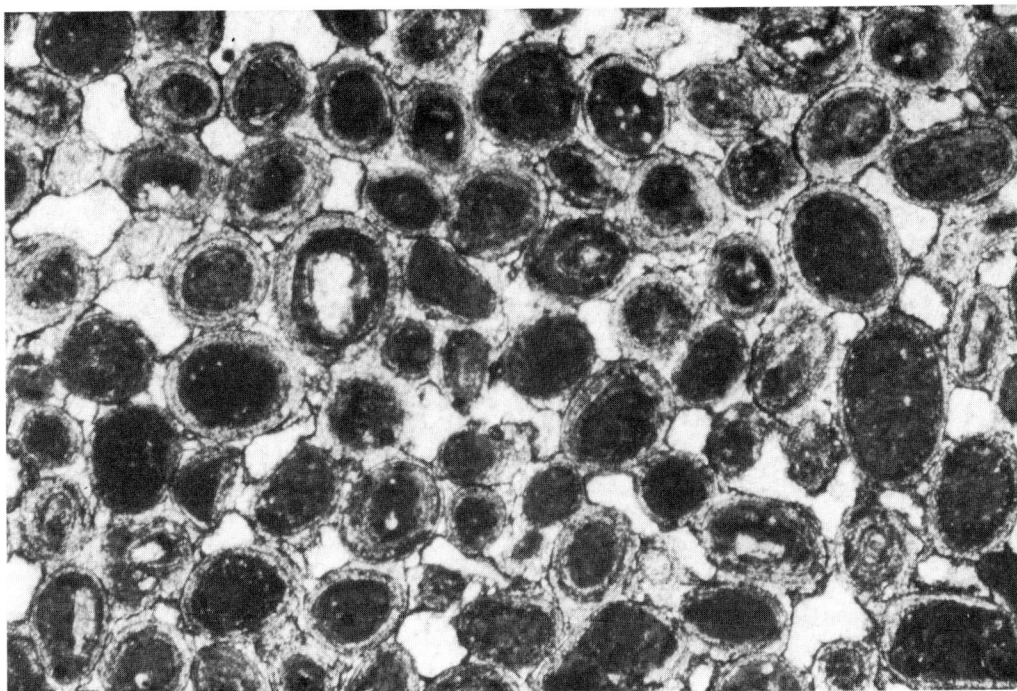

Fig. 2. Patchily developed pore-filling cement in an oolitic aeolianite from White Bay Cay, The Exuma Cays, the Bahamas (PPL ×20).

Tunisian aeolianites were found to be similar in that many aragonite (oolitic and bioclastic) grains remained in well-cemented aeolianites (Fig. 3). The last interglacial deposits at Mahdia, for example, comprise over 90% aragonitic allochems and are sufficiently well cemented (porosity 13%) to be used locally as a building stone.

The frequency with which oolitic grains remained as unaltered aragonite, led to a comparison being made between oolitic and bioclastic deposits of similar age and position within one climatic area in the Bahamas. An examination of 107 samples, revealed that in general the oolitic deposits tend to retain a higher proportion of their aragonite allochems and to have low values of secondary porosity whereas the bioclastic deposits show greater aragonite dissolution (average of 30% allochems dissolved out) and have higher values (5%) of secondary porosity. Thus, it would appear either that the textural differences between the two types of deposit affect the leaching potential, or that the dissolutional processes preferentially affect the bioclasts while the ooliths are more resistant. In support of the latter idea, the cortex of the

ooids is composed of alternating layers of clear aragonite and darker cryptocrystalline aragonite that are dense in comparison with the often more porous shell structures.

These results support the work of Robinson (1967), who found aragonite ooids with 'considerable calcite cement' in the Bahamas and Florida, and Evans & Ginsburg (1987) who found differences in dissolution between bioclasts and ooliths in the Pleistocene Miami limestone. In the coarse shelly grainstone most of the skeletal components were recrystallized to low-Mg calcite whereas in the oolitic units many of the allochems (both ooliths and bioclasts) retained their original mineralogy, including bryozoans and *Halimeda* spp. Aragonite shells and ooliths in last interglacial or older deposits from Florida and the Caribbean have also been identified by Multer & Hoffmeister (1968), Multer (1971), James (1972), Harrison (1977), Kahle (1977) and Chafetz *et al.* (1985).

Thus, in oolitic aeolianites (but also in some of the bioclastic aeolianites) in particular pore-filling cement is not for the most part supplied from dissolution of aragonite allochems. An elusive external cement source still awaits

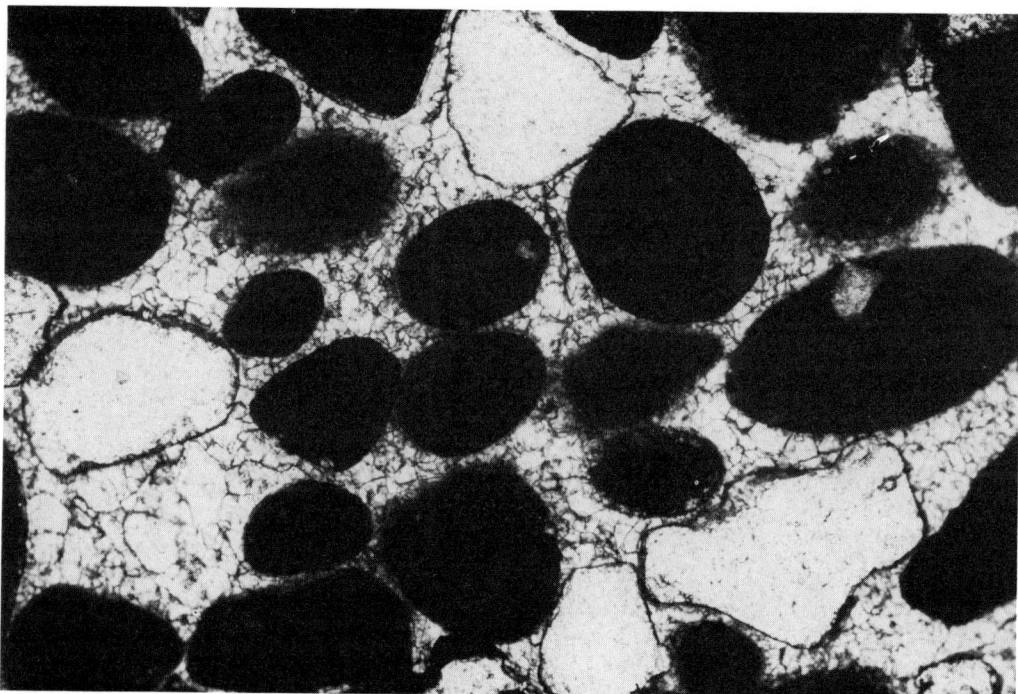

Fig. 3. Occluded aeolianite from Mahdia, Tunisia. The allochems have been stained black with Feigl's solution showing that they are aragonitic (PPL ×40).

identification in the aeolianites of the Caribbean. In bioclastic aeolianites, aragonite may be dissolved much earlier in the diagenetic history and may therefore contribute significantly to the pore-filling cement in some instances, as proposed in the models. Generally, however, in situ dissolution is incapable of accounting for more than 4% of pore-filling cement. For example, take an extreme case of aeolianite with 40% porosity and 60% particulates, of which 80% are aragonite allochems and 20% lithoclasts. Dissolution and complete reprecipitation of the aragonite allochems locally within a closed system will only supply an additional 3.8% volume of calcite. Assuming not all the dissolution moulds are infilled and approximately 6% secondary porosity remains, then the amount of pore-filling cement supplied will be about 10%.

While in most areas diagenesis was found to progress towards occlusion of pores by cement, there were two examples where a negative trend towards higher porosity was observed at a late stage in the diagenesis. In some exposed parts of the Older Aeolianite in the eastern Wahiba Sands, Oman, congruent dissolution of the aragonitic allochems has resulted in high intraparticle secondary porosity (preserved by micritic envelopes) in a well-cemented aeolianite (Fig. 4). The samples examined contained an average of 30.5% calcite cement and had an average porosity of 18.4%. However, approximately half of the porosity was secondary. Similarly, in the bioclastic aeolianite from Campo de Tiro, Mallorca, secondary porosity is high. In this case, the high secondary porosity can be related to the influence of sea spray, and a similar effect is observed in raised marine biocalcarenites in Mallorca (Table 4). In Oman the presence of a saline playa close to the aeolianite with high secondary porosities deserves further investigation.

Local variability

Detailed studies of the range of diagenetic fabrics found within deposits gives information on meso-scale variability. At Arenal, Mallorca, five sections were studied from a last interglacial raised beach. Table 4 shows the mean and range of the percentages of bioclasts, total cement, primary porosity and secondary

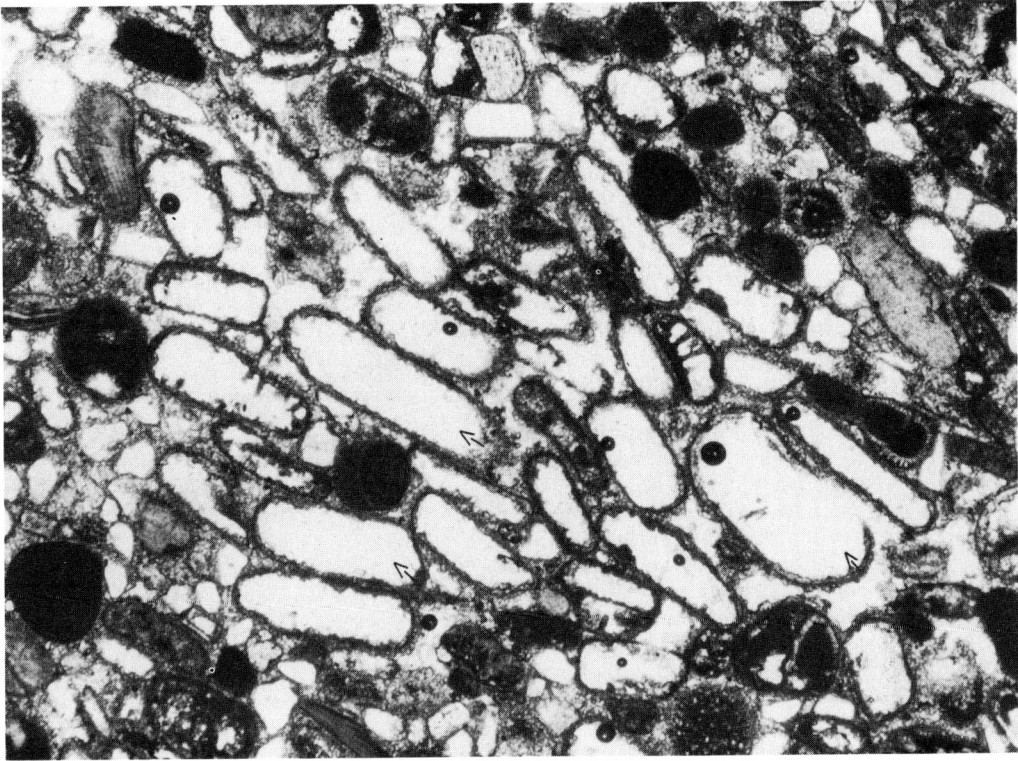

Fig. 4. Intraparticle secondary porosity (arrows) following dissolution of aragonite. The Older Aeolianite, Oman (PPL ×20).

porosity of all five sections. The range in values is high in all cases, indicating high micro-variability associated with textural differences between laminae. However, a definite trend in mean values can also be discerned, indicating meso-scale variability between sections. The distinctive feature of these sections is that they are subject to differing exposure to sea spray. Section 4 is in the most sheltered position followed by sections 1, 2, 3 and finally section 5 which suffers constant sea spray and occasional marine wash during storms. With an increase in the intensity of sea spray and marine wash, a number of diagenetic changes can be detected.

Table 4. *Percentage mean values and ranges of allochems, total cement and porosity in sections exposed to varying amounts of sea spray, Arenal, Mallorca*

Section	Sample Size	Shells		Total cement		PP*		SP**	
		Mean	Range	Mean	Range	Mean	Range	Mean	Range
4	10	41.4	27–46	24.2	18–49	23.3	8–32	1.2	0–5
1	14	41.5	28–52	24.9	21–39	23.3	15–35	4.2	0–10
2	14	34.9	18–49	30.1	10–46	22.5	16–36	5.8	2–16
3	11	32.2	12–45	28.9	13–51	17.8	10–32	8.4	3–17
5	5	30.1	22–44	37.2	33–45	14.0	11–18	8.5	7–18

Note: From section 4 at the top of the table to section 5 at the bottom there is a gradual increase in the amount of sea spray and marine wash affecting the sediments.
PP* primary porosity.
SP** secondary porosity.

First, there is a gradual increase in the percentage of total cement (low-Mg calcite); secondly, and clearly related, there is a decrease in primary porosity. Thirdly, there is a decrease in the percentage of bioclasts and, fourthly, an increase in secondary porosity. Therefore, in the beach at Arenal, sea spray appears to exert an important control on diagenetic fabric in deposits of the same age, height and overall textural characteristics, leading to meso-scale variability within one deposit. Furthermore, coastal aeolianites and raised marine sediments normally experience differential exposure to sea spray.

The second common meso-scale variation in diagenetic change concerned vertical variation downwards from the exposed surface of a deposit, normally to a depth of between 2 and 3 metres. Decalcification associated with palaeosols has been widely reported (Klappa 1983), but in this instance exposed surfaces without palaeosols were examined. A substantial increase in cementation (particularly pore-filling cements) and degree of diagenetic alteration, and decrease in porosity, towards the top surface was noted in sections in Spain, Mallorca, S. India, Oman, and the Bahamas (in some dunes from Oman however, there has been so much wind erosion that the tops of the deposits have not been preserved). This is often, but not always, linked to the formation of a surface calcrete. For example, at Sandy Point, the Bahamas, there is a gradual increase in cement values from an average of 28% at a depth of 1 m to 51% just below a thin calcrete on the exposed surface. The details of one representative section are shown in Fig. 5. Similar meso-scale variations are found in Mallorca, southern Spain and south India.

Texture is a further meso-scale variable. Clearly, this will be greater in marine sediments than in aeolianites, however, the marked effects of textural differences across aeolianite laminae and small beds on patterns of cementation are well reported (Gardner 1988). The results of earlier work were supported by the findings from this study.

The effects of organisms such as fungi, algae, bacteria, and vascular plants on diagenesis was studied primarily in the Bahamas. Localization of cements around former roots resulted in observable differences in diagenetic fabric both within beds and between beds that contain differing concentrations of rhizoliths. The redistribution of the available $CaCO_3$ and its concentration around roots, leaves the parts of the deposits that are away from the roots relatively friable. For example, at Crab Cay, the

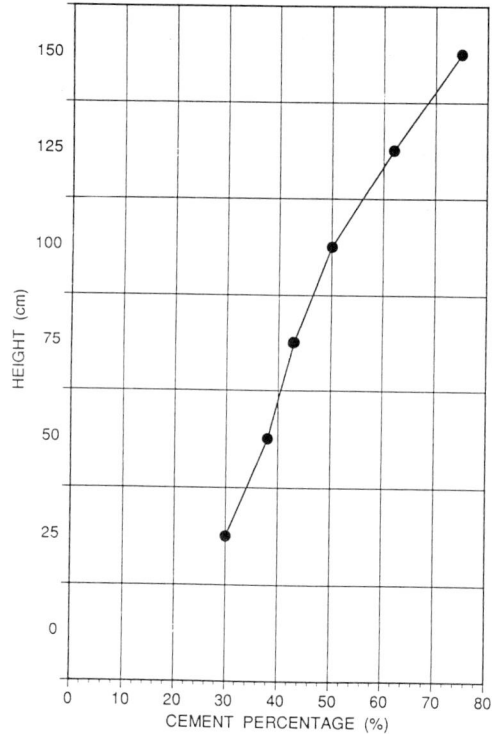

Fig. 5. The variation in total cement values down from the top of a land surface at Sandy Point, San Salvador Island, the Bahamas, determined by point counting.

Bahamas, there is on average, 85% of cement around the root moulds, 32% total cement a few centimetres from the roots, but only 17% cement in the deposits away from the roots (Fig. 6).

Thus, local variability was found to be normal in all of the aeolianites and raised marine sediments examined. The degree of meso-scale variability observed seriously questions the (implicit) assumption of low spatial variability within deposits of the same age that exists in the progressive diagenesis models. At best these findings suggest that a good idea of the environmental history of the deposit is needed and a large number of samples should be carefully selected for examination from each deposit.

Rates of diagenesis

The lack of well-dated deposits has been a great hindrance to the study of rates of diagenesis, and until a way is found of more precisely dating

Fig. 6. Rhizoliths developed within a biolithic aeolianite at Crab Cay, the Bahamas.

aeolianites of up to 150 000 years old, this is likely to continue. However, it is possible to compare deposits of similar (last interglacial) age from different areas, to determine whether diagenetic development (and hence rates) vary in general between different climatic regions (Table 5).

In Oman, the Younger Aeolianites are thought to be of late interglacial/early last glacial age, and are thus slightly younger than the deposits reported from other areas. They are

very friable and exhibit the least overall diagenetic development: pore-filling cement is low and restricted to the finer-grained laminations. Secondary porosity is almost absent and the original high-Mg calcite and aragonite mineralogies are preserved in many allochems (Fig. 7).

In Almeria, deposits also show a low degree of diagenetic alteration. Rim and meniscus cements alone occur most near the base of the outcrops, while there is better development of

Table 5. *Diagenetic patterns in deposits of similar age from different areas*

Location	Sample size	Meniscus cement (%)	Rim cement (%)	Pore-filling cement (%)	Secondary cement (%)	Secondary porosity (%)	Total cement
S. Spain	59	3.4	9.5	8.8	0.3	0.9	21.9
Mallorca	76	3.3	6.9	11.2	4.5	3.7	25.9
Tunisia	111	1.5	6.1	21.9	0.1	0.1	29.6
Bahamas	97	7.5	4.0	16.1	0.6	1.1	28.2
Oman	15	10.6		5.0	0.3	0.6	15.9
SE India	60	9.1		14.8	3.6	2.5	28.2

pore-filling cement near the top. In general, the deposits at the base of the outcrops would fall into stage II but those near to the top of the deposits would fall into stage III of Gavish & Friedman's (1969) model. Amounts of secondary porosity and secondary cement are low in the basal beds but increase upwards towards the top of the stratigraphic sequences. Most of the original aragonite and high-Mg calcite allochems

are preserved in the lower parts of the outcrop, but the amount of dissolution of the allochems increases up profile.

In Mallorca, secondary cements are more abundant than elsewhere in the Mediterranean and pore-filling cements are characteristically patchy in nature. Amounts of meniscus and rim cements, although variable, are generally low. Except for the deposits in the sea spray zone,

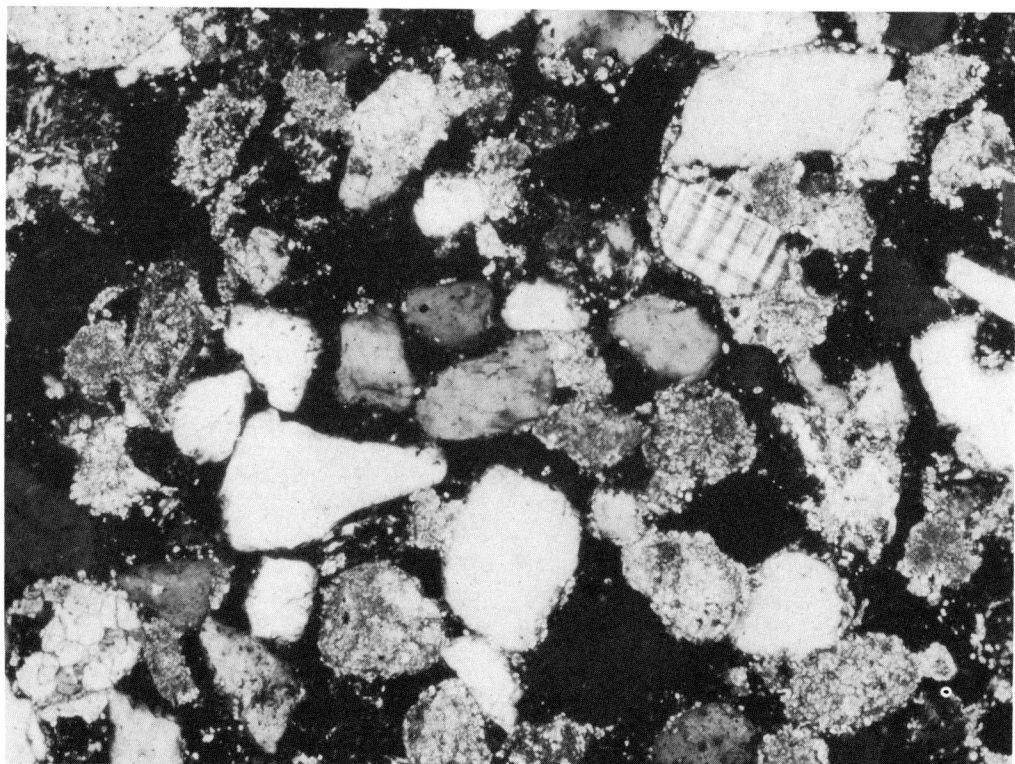

Fig. 7. Coastal aeolianite from Oman showing only poorly developed rim and meniscus cements and high primary porosity (XPL ×20).

all contain aragonitic and some high-Mg calcite allochems, and individual beds/units range between being friable to well cemented. Secondary porosity and secondary cement values are variable, being far higher in the deposits in the sea spray zone.

In Tunisia and the Bahamas the dominant cement forms present may be a function of a textural control as they range from quite well to very well cemented. In the fine-grained aeolianites pore-filling cements are most abundant, although again are patchy in nature. Frequent occlusion of pores with cement has meant that rim and meniscus cements are often less readily discernible. Also in the Bahamas, the amounts of meniscus cements are higher than from any of the other areas, probably owing largely to the smooth roundness of the ooliths. The allochem content in the deposits from the Bahamas and Tunisia is nearly 100% aragonitic. Secondary porosity and secondary cement values in both areas are characteristically low.

In South India the aeolianites exhibit relatively high levels of pore-filling cements (remaining porosity is less than 20%) and almost total loss of both Mg^{2+} and aragonite. Secondary porosity is low. The South Indian aeolianites overall show one of the highest levels of diagenetic alteration.

Diagenetic variability, therefore, appears to exist on a regional scale as well as the mesoscale, in so far as the deposits are all of broadly last interglacial age and all have been exposed to diagenetic changes over the intervening period. This implies that rate of diagenesis does vary in different climatic conditions, a view first proposed by Ward (1975). In some instances the diagenetic fabrics observed do not fit into the Gavish & Friedman (1969) model at all, and where they do, the diagenetic development observed encompasses a range from the beginning of stage II to the end of stage III.

None of the deposits studied that are 130 000 years in age or older (Older Aeolianite in Oman, and Mallorca), have reached the stage of complete stabilization, nor is there total occlusion of pore spaces with pore-filling cement. Also, none of the following researchers have found evidence of total or even near-total occlusion of pore spaces in deposits of last interglacial age: Evans & Ginsburg (1987) in Florida; Matthews (1971) in Barbados; Ward (1973, 1975) in Mexico; or Pingitore (1970) and Schroeder (1973) in Bermuda. Indeed, the low percentages of pore-filling cement found in Oman and southern Spain (less than 8% in many instances) are less than has been recorded in Holocene sediments from the Caribbean.

Well-cemented Holocene deposits

A study of the types and rates of early mineral inversion as well as cementation of a Holocene aeolianite from North Point, San Salvador, the Bahamas, shows that not all Holocene deposits are uncemented or very friable. This deposit comprises mostly aragonitic allochems with a low percentage of high-Mg calcite. The sediments have undergone very little mineralogical stabilization and secondary porosity values are very low (less than 1% on average) and yet they are well cemented with abundant pore-filling cement. In total, the average amount of cement present is 34% (standard deviation 5.4%) and average porosity is 10%. However, the amount of cement in individual pores is highly variable with some pores totally occluded by cement and other pores containing very little, or no, cement (Fig. 8).

In terms of cement, the deposit is close to stage III in Gavish & Friedman's model whereas in terms of mineralogical stability, it lies in stage II of the model. Halley & Harris's (1979) and Budd's (1984, 1988a, 1988b) work, also in the Bahamas, supports these findings. Budd found oolitic sands in the Schooner Cays with between

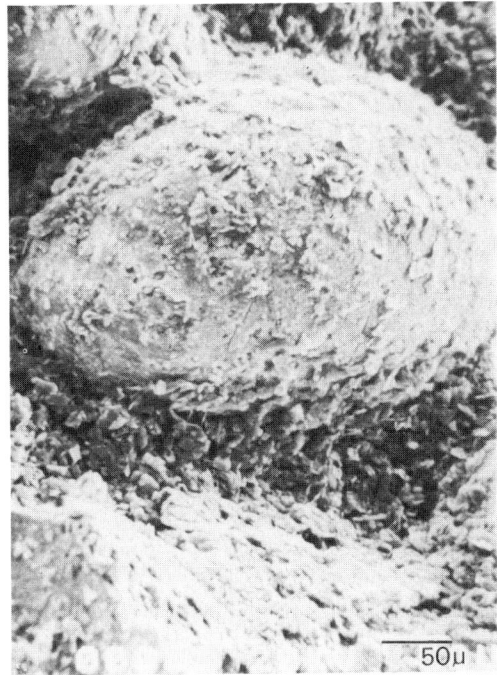

Fig. 8. SEM micrograph showing a well-cemented Holocene aeolianite from North Point, San Salvador, the Bahamas.

28–39% vadose cement, including abundant pore-filling cement. The sands have been radiocarbon dated to 2910 years BP. Thus pore-filling cement is not necessarily only found in post-stage II aged deposits. Owing to the limited amount of work on Holocene aeolianites, it is not yet possible to determine how common substantial diagenetic change is in very young deposits.

Conclusions

A detailed, systematic analysis of aeolianites (and some sandy beach deposits) of last interglacial age from Oman, southern India, the Bahamas, Tunisia, Mallorca and southern Spain has questioned the application of progressive vadose diagenetic models outside the areas from which the models were developed. First, the results presented show that progressive stages in the development of low-Mg calcite products can be identified, although stages in the removal of aragonite could not be identified. Rim and meniscus cements do pre-date pore-filling cements, and in most cases porosity does decrease with carbonate diagenesis.

Second, however, there is far less evidence for sequential and predictable stages in terms of process. Pore-filling cements may develop without substantial loss of aragonite in bioclastic sediments, and commonly do so in oolitic sediments. The progressive diagenetic models all suggest that the loss of aragonite occurs early on in the diagenetic history of a deposit and that its dissolution acts as the source for second stage pore-filling cements. The last interglacial aeolianites studied which were well cemented have retained their aragonitic allochems to varying degrees. Thus, in some cases at least, the allochems within the deposit as a whole are not supplying the CaCO$_3$ for the pore-filling cements.

Third, while some models have associated rates of diagenesis in the vadose zone to the progressive sequences, the rates of diagenesis observed from the current analyses are highly variable. The variability in diagenetic fabric and rate, both within deposits and between deposits of similar ages, is seen to result from the many different combinations of macro-, meso- and micro-scale factors that affect diagenesis. There is no evidence as yet that time, or indeed any other single factor, or particular combination of factors, exerts a general and predictable over-riding influence upon diagenetic change in the vadose zone. The role of climate in this context needs further examination as more independent evidence of Quaternary climate change becomes available.

Thus, the applicability of models of progressive vadose diagenesis to areas outside of those for which they were developed is highly questionable. Spatial and temporal heterogeneity at all scales (macro-, meso- and micro-) exists; the carbonate deposits examined all tend towards an end point of low-Mg calcite allochems and cement but at highly variable rates and via various routes. Hence, the use of the models as tools in palaeoenvironmental and geochronological analyses is not to be recommended. More rigorous and reliable dating of deposits and a better understanding of changing environmental contexts and diagenetic processes may allow greater scope for refined, regional models to be developed in the future.

McLaren's research was supported by a NERC research studentship (GT4/87/GS/64). The authors also wish to acknowledge the assistance of the Geomaterials unit at Queen Mary and Westfield College, University of London, the Royal Geographical Society Oman Wahiba Sands Project (RG), Major Chris Griffiths (RG), and the drawing office in the Department of Geography at King's College.

References

BATHURST, R. G. C. 1975. *Carbonate Sediments and Their Diagenesis.* (Second edition). Developments in Sedimentology, **12**. Elsevier, Amsterdam.

BUCHBINDER, L. G. & FRIEDMAN, G. M. 1980. Vadose, phreatic and marine diagenesis of Pleistocene-Holocene carbonates in a borehole: Mediterranean coast of Israel. *Journal of Sedimentary Petrology*, **50**, 395–407.

BUDD, D. A. 1984. *Freshwater diagenesis of Holocene ooid sands, Schooner Cays, Bahamas.* PhD Dissertation, the University of Texas at Austin, Texas.

—— 1988a. Petrographic products of freshwater diagenesis in Holocene ooid sands, Schooner Cays, the Bahamas. *Carbonates and Evaporites*, **3**, 143–163.

—— 1988b. Aragonite-to-calcite transformation during freshwater diagenesis of carbonates: insights from porewater chemistry. *Geological Society of America Bulletin*, **100**, 1260–1270.

BUDDEMEIER, R. W. & OBERNDORFER, J. A. 1986. Internal hydrology and geochemistry of coral reefs and atoll islands: key to diagenetic variations. *In*: SCHROEDER, J. H. & PURSER, B. H. (eds) *Reef Diagenesis*. Springer, Berlin, 91–111.

CAREW, J. L. & MYLROIE, J. 1986. A refined chronology for San Salvador Island. *In*: CURRAN, H. A. (ed.) *Proceedings of the Third Symposium on the Geology of the Bahamas*. Don Heuer, USA, 35–45.

CHAFETZ, H. S., WILKINSON, B. H. & LOVE, K. M. 1985. Morphology and composition of non marine carbonate cements in near surface settings. In: SCHNEIDERMANN, N. & HARRIS, P. (eds) Carbonate Cements. SEPM Special Publication, 36, SEPM, Tulsa, 337–348.

CHOQUETTE, P. W. & TRUSELL, F. C. 1978. A procedure for making the titan-yellow stain for making Mg-calcite permanent. Journal of Sedimentary Petrology, 48, 639–641.

DICKSON, J. A. D. 1966. Carbonate identification as revealed by staining. Journal of Sedimentary Petrology, 36, 491–505.

EVANS, C. C. & GINSBURG, R. N. 1987. Fabric selective diagenesis in the late Pleistocene Miami Limestone. Journal of Sedimentary Petrology, 57, 311–318.

FIEGL, F. 1937. Qualitative Analysis by Spot Tests. Nordemann, New York.

FRIEDMAN, G. M. 1959. Identification of carbonate mineral by staining methods. Journal of Sedimentary Petrology, 29, 57–97.

—— 1964. Early diagenesis and lithification in carbonate sediments. Journal of Sedimentary Petrology, 34, 777–813.

GARDNER, R. A. M. 1981. Geomorphology and Environmental Change in South India and Sri Lanka. Unpublished DPhil. Thesis, Oxford.

—— 1986. Quaternary coastal sediments and stratigraphy, south east India. Man and Environment Indian Society for Prehistoric and Quaternary Studies, X, 51–72.

—— 1988. Aeolianites and marine deposits of the Wahiba Sands: character and palaeoenvironments. The Journal of Oman Studies Special Report, 3, 75–95.

GAVISH, E. & FRIEDMAN, G. M. 1969. Progressive diagenesis in Quaternary to late Tertiary carbonate sediments: sequence and time scale. Journal of Sedimentary Petrology, 39, 980–1006.

GILL, E. D. 1976. Quaternary Warrnambool-Port Fairy District. In: DOUGLAS, J. G. & FERGUSON, J. A. (eds) Geology of Victoria. Geological Society of Australia Special Publication, 5, 299–304.

GVIRTMAN, G. & FRIEDMAN, G. M. 1977. Sequence of Progressive Diagenesis in Coral Reefs. American Association of Petroleum Geologists Studies in Geology, 4, 357–380.

HALLEY, R. B. & HARRIS, P. M. 1979. Freshwater cementation of a 1,000-year-old oolite. Journal of Sedimentary Petrology, 49, 969–988.

HARRIS, W. H. & MATTHEWS, R. K. 1968. Subaerial diagenesis of carbonate sediments: efficiency of the solution-reprecipitation process. Science, 160, 77–79.

HARRISON, R. S. 1977. Caliche profiles: indicators of near surface subaerial diagenesis, Barbados, West Indies. Canadian Petroleum Geologists Bulletin, 25, 123–173.

HEARTY, P. J. 1986. An inventory of last interglacial deposits from the Mediterranean basin: a study of isoleucine epimerisation and uranium series dating. Zeitschrift fur Geomorphologie Supplement Band, 62, 51–69.

HUTTO, T. & CAREW, J. L. 1984. Petrology of eolian calcarenites, San Salvador Island, Bahamas. In: TEETER, J. W. (ed.) Proceedings of the Second Symposium of the Geology of the Bahamas. Don Heuer, USA, 197–209.

JAMES, N. P. 1972. Holocene and Pleistocene calcareous crust profiles: criteria for subaerial exposure. Journal of Sedimentary Petrology, 42, 817–836.

—— & CHOQUETTE, P. W. 1984. Diagenesis 9 — limestones — the meteoric diagenetic environment. Geoscience Canada, 11, 161–194.

JONES, B. & KAHLE, C. F. 1986. Dendritic calcite crystals formed by calcification of algal filaments in a vadose environment. Journal of Sedimentary Petrology, 56, 217–227.

KAHLE, C. F. 1977. Origin of subaerial Holocene calcareous crusts: role of algae, fungi, and sparmicritisation. Sedimentology, 24, 413–435.

KLAPPA, C. F. 1978. Biolithogenesis of Microcodium: elucidation. Sedimentology, 25, 489–522.

—— 1983. A process response model for the formation of pedogenic calcretes. In: WILSON, R. C. L. (ed.) Residual Deposits: Surface Related Weathering Processes and Materials. Geological Society, London, Special Publication, 11, 211–221.

LAND, L. S. 1970. Phreatic versus meteoric diagenesis of limestones: evidence from a fossil water table in Bermuda. In: BRICKER, O. P. (ed.) Carbonate Cements. Johns Hopkins Press, Baltimore, 133–137.

——, MACKENZIE, F. T. & GOULD, S. J. 1967. Pleistocene history of Bermuda. Bulletin of the Geological Society of America, 78, 993–1006.

LONGMAN, M. W. 1980. Carbonate diagenetic textures from near surface diagenetic environments. Bulletin of the American Association of Petroleum Geologists, 64, 461–487.

MCLAREN, S. J. 1991. Vadose Diagenesis of Late Quaternary Coastal Sediments. Unpublished PhD Thesis, London University.

—— 1993. Use of cement types in the palaeoenvironmental interpretation of coastal aeolian sedimentary sequences. In: PYE, K. (ed.) The Dynamics and Environmental Context of Aeolian Sedimentary Systems. Geological Society, London, Special Publication, 72, 235–244.

MARION, G. M., SCHLESINGER, W. H. & FONTEYN, P. J. 1985. Caldep: a regional model for soil calcium carbonate (caliche) deposition in southwestern deserts (USA). Soil Science, 139, 468–481.

MARTIN, G. D., WILKINSON, B. H. & LOHMAN, K. C. 1986. The role of skeletal porosity in aragonite neomorphism: Strombus and Montastrea from the Pleistocene Key Largo Limestone, Florida. Journal of Sedimentary Petrology, 56, 194–203.

MATTHEWS, R. K. 1971. Diagenetic environments of possible importance to the explanation of cementation fabric in subaerially exposed carbonate sediments. In: BRICKER, O. P. (ed.) Carbonate Cements. Johns Hopkins Press, Baltimore, 127–133.

MILLER, G. H., PASKOFF, R. & STEARNS, Ch. E. 1986. Amino acid geochronology of Pleistocene littoral deposits in Tunisia. *Zeitschrift fur Geomorphologie Supplement Band*, **62**, 197–207.

MORROW, N. R. 1971. Small scale packing heterogeneities in porous sedimentary rocks. *American Association of Petroleum Geologists Bulletin*, **55**, 514–522.

MORSE, J. W. & MACKENZIE, F. T. 1990. *Geochemistry of Sedimentary Carbonates*. Developments in Sedimentology, **48**, Elsevier, Amsterdam.

MULTER, H. G. 1971. Cementation and recrystallisation of Miami limestone ooids. *In*: BRICKER, O. P. (ed.) *Carbonate Cements*. Johns Hopkins Press, Baltimore.

—— & HOFFMEISTER, J. E. 1968. Subaerial laminated crusts of Florida Keys. *Bulletin of the Geological Society of America*, **79**, 183–192.

PINGITORE, Jr. N. E. 1970. Diagenesis and porosity modifications in *Acropora palmata*, Pleistocene of Barbados, West Indies. *Journal of Sedimentary Petrology*, **40**, 712–721.

—— 1976. Vadose and phreatic diagenesis: processes, products and their recognition in corals. *Journal of Sedimentary Petrology*, **46**, 985–1006.

REECKMANN, S. A. & GILL, E. D. 1981. Rates of vadose diagenesis in Quaternary dune and shallow marine calcarenites, Warrnambool, Victoria, Australia. *Sedimentary Geology*, **30**, 157–172.

RISTVET, B. L. 1972. The progressive diagenetic history of Bermuda. *Geological Society of America Annual Meeting: Abstracts and Programs*, **4**, 638–639.

ROBINSON, R. B. 1967. Diagenesis and porosity development in Recent and Pleistocene oolites from southern Florida and the Bahamas. *Journal of Sedimentary Petrology*, **37**, 355–364.

SCHROEDER, J. H. 1973. Submarine and vadose cements in Pleistocene Bermuda reef rock. *Sedimentary Geology*, **10**, 179–204.

STEINEN, R. P. 1974. Phreatic and vadose diagenetic modification of Pleistocene limestone: petrographic observations from subsurface of Barbados, West Indies. *American Association of Petroleum Geologists Bulletin*, **58**, 1008–1024.

TITUS, R. 1984. Physical stratigraphy of San Salvador Island, Bahamas. *In*: TEETER, J. W. (ed.) *Proceedings of the Second Symposium of the Geology of the Bahamas*. Don Heuer, USA, 209–229.

WARD, W. C. 1973. Influence of climate on the early diagenesis of carbonate eolianites. *Geology*, 1, 171–174.

—— 1975. *Petrology and Diagenesis of Carbonate Eolianites of North Eastern Yucaton Peninsula*. American Association of Petroleum Geologists Studies in Geology, **2**, 500–571.

ZAZO, C., GOY, J. L., HOYOS, M., DUMAS, B., PORTA, J., MARTINELL, J., BAENA, J. & AGUIRRE, E. 1981. Ensayo de sintensis sobre el Tirreniense peninsular espanol. *Estudios Geologie*, **37**, 257–262.

Use of cement types in the palaeoenvironmental interpretation of coastal aeolianite sequences

SUSAN J. McLAREN

Department of Geography, Kings College, Strand, London WC2R 2LS, UK

Abstract: The interpretation of palaeodune sequences is an important component in palaeoenvironmental analyses. Studies of the diagenetic alteration of palaeodunes is one of the main ways of determining post-depositional environmental conditions. This paper questions the long-held and generally accepted ideas that cement types/fabrics are a 'major method of recognising the original mineralogy of the cements and by inference their origin under marine, vadose or phreatic conditions' (Arthur *et al.* 1982). A detailed study of last interglacial aeolianites from Mallorca, Tunisia and the Bahamas has shown that although many cement types may be characteristic of the meteoric vadose zone, none are unique. No single cement type is diagnostic of the vadose zone.

Cement distribution patterns are commonly thought to be a function of the diagenetic environment in which the cements have formed. This paper presents the findings of the types of cement that have resulted from diagenesis in the meteoric vadose environment (a few of the deposits are subjected to some sea spray). A wide range of coastal aeolianites of last interglacial age have been examined to see which, if any, cement types are diagnostic and may therefore provide indicators of the vadose zone.

The changes that affect a sediment after deposition may be an important indicator of post-depositional environmental conditions. These changes are collectively known as diagenesis and may involve changes in the mineralogy, geochemistry, texture, and fabric of the sediment. The mineralogical changes associated with the diagenesis of carbonate-rich aeolian deposits include alteration of unstable aragonite and high-Mg calcite and precipitation of relatively low-Mg calcite. Geochemical changes are associated with the mineralogical changes and result in the loss of Mg and Sr ions. Formation of cement and changes in porosity values (primary and secondary) comprise the textural and fabric changes. The last interglacial aeolianites examined in this study from Mallorca, Tunisia and the Bahamas have all been exposed solely to meteoric vadose (but not case hardening) conditions. The types of cement that develop in more texturally variable raised beach deposits are also discussed.

Traditional views on typical vadose cements

Previous work has suggested that a number of factors control the development of different primary carbonate cement mineralogies, crystal sizes and morphologies in the vadose zone; these are discussed below.

Mineralogy

Low-Mg calcite tends to precipitate out in the vadose environment as it is far more stable at surface temperatures and pressures than aragonite and high-Mg calcite. However, dolomite, or (more rarely) high-Mg calcite may precipitate out preferentially in the subaerial environment subject to the influence of sea spray.

Crystal size

Crystal sizes often increase away from the initial substrate and commonly there is an increase towards the centre of the pores. Thus, the larger the pores, the greater the potential for coarser crystals. The rate of crystal growth is a second important control over crystal size. Rapid precipitation is favoured where saturation is high and nuclei are abundant. This results in fine crystal sizes of less than 5 μm (Given & Wilkinson 1985).

Cement morphology

Traditional views state that in the phreatic zone pore spaces between sand grains are completely filled with water; any cements derived from the interstitial waters would be isopachous in nature (Muller 1971). However, in the vadose zone both water and air fill pore spaces giving an 'uneven' cement distribution. The typical cements that are thought to develop in the vadose zone are as listed below.

From Pye, K. (ed.), 1993, *The Dynamics and Environmental Context of Aeolian Sedimentary Systems*. Geological Society Special Publication No. 72, pp. 235–244.

(a) **Pendant/gravity cements** result from calcite precipitation from saturated waters held on the underside of grains, owing to the force of gravity. Such gravity cements have been described as being common in, or diagnostic of, the vadose environment (Longman 1980; Moore 1989; Tucker & Wright 1990; Morse & MacKenzie 1990).

(b) **Drapestone cements** are thought to occur on the upper surfaces of grains due to the dripping of water from overlying grains during drainage. Badiozamani *et al.* (1977) carried out experiments to simulate cementation from evaporation and found that droplet-like cement formed on the upward side of grains.

(c) **Meniscus cements** are concentrated where capillary water is present at grain contacts. Binkley *et al.* (1980) noted that the crystals often have a bladed form (crystals with a length-to-width ratio between 1.5 : 1 and 6 : 1). Uneven cements precipitated within a liquid meniscus results in pore rounding (Scoffin 1987).

(d) **Rim cementation** is rarely complete and the cement crystals gradually decrease in size around the grain and disappear in the zone where the pore space is filled with air. Average sizes of crystals range from about 5–50 μm in diameter. Cement crystals are commonly bladed, equant or syntaxial in shape. Meniscus and rim cements are traditionally considered to be the dominant forms found in the vadose zone where pore-filling cements are common with few or no rim cements and no meniscus cements. This general explanation relies on the idea that the vadose zone is an area of dissolution (Buchbinder & Friedman 1980; Longman 1980) with little precipitation. Land (1966, 1967, 1970, 1971) originally put forward the idea that the vadose zone, where pore spaces are mostly filled with air but water is localized around the points of grain contact, is poorly cemented. The source of the water is generally rain water which is undersaturated with respect to $CaCO_3$, and consequently carbonates within the sediment start to dissolve (Buchbinder & Friedman 1980). Production of CO_2 by root respiration and microbial breakdown of organic material in the soil zone may also enhance solutional processes (Flugel 1982).

In the vadose zone, in situ dissolution of aragonite and reprecipitation of calcite mostly occurs concurrently, although at a rate dependent upon percolation of pore fluids (Harris & Matthews 1968). This means that there is no change in gross porosity values other than as a result of the looser atomic structure and lower density of calcite, which accounts for approximately an 8% decrease in porosity.

(e) **Elongate crystal habits** (length-to-width ratio of more than 6 : 1) are found in vadose settings where CO_2 degassing generates CO_3^{2-} ions by disassociation of HCO_3^-; the system is carbonate-rich, *c*-axis growth is enhanced and elongate crystals result (Given & Wilkinson 1985). These are often termed **needle fibre** or **whisker cements** and are fine and long with respect to their width. Needle fibre cement is not widely reported in the literature, but, increasingly, it is becoming associated with rhizoliths and soils in the vadose zone (e.g. Klappa 1979; Wright 1986).

Questioning the traditional views

In the past, a variety of studies have been undertaken in different climatic environments but often the cement morphology is not the main emphasis of the research. Very few researchers, indeed, give any quantitative assessment of the relative abundances of the different cement types that they find; nor do they give any indication about the distribution of the various cement types within profiles. Also, with many of the deposits studied there is uncertainty as to whether the deposits have always been solely in the vadose zone. The types of cement that have been described in detail by a number of researchers are shown in Table 1 and their global distribution can be seen in Fig. 1. (It should be noted here that there is a small potential problem related to the fact that thin sections are two-dimensional not three; this may lead to complications in identifying rim from meniscus or gravity cements although this should be avoidable by careful study of the cement crystals.)

Very few researchers have actually witnessed gravity cement, and no detailed systematic research has been carried out on these cements. Where observed, they appear to be in dune deposits from arid or semi-arid environments. Ward (1973, 1975) identified this cement type in Holocene dunes from Mexico; Buchbinder & Friedman (1980) described them in Pleistocene dune deposits in Israel; and Gardner (1988) found a little gravity cement in dunes from the Wahiba Sands, Oman.

Contrary to the experimental work of Badiozamani *et al.* (1977), drapestone cement does not appear to be a characteristic cement type of the

Table 1. *The number of researchers who have identified particular cement types in the vadose, mixed and phreatic zones*

	Number of researchers					
Cement type	Vadose		Mixed		Phreatic	
Gravity	4	P–R	2	C	0	–
Drapestone	1	R	0	–	0	–
Meniscus	20	C	3	P	0	–
Rim	26	C	8	C	9	C
Pore-filling	25	C	8	P	10	C
Needle fibre	5	C	0	–	0	–
Secondary	6	P	2	P	0	–

Generalized abundance of different cement types: C, common; P, present; R, rare.

vadose zone. It has only been described in the literature by Buchbinder & Friedman (1980), who also identified gravity cement.

Recent work (e.g. Budd 1988*a,b*) questions the idea that rim and meniscus cements form only in the vadose zone and more importantly that pore-filling cement is indicative of the phreatic zone. Table 1 shows that meniscus and rim cements do appear to be ubiquitous in the meteoric vadose setting. Often the meniscus cements form as a first stage development at points of grain contact where waters have concentrated following seepage or evaporation. However, neither meniscus nor rim cements appear to be restricted to vadose conditions; both have been identified in mixing zone deposits such as beachrocks (Stoddart & Cann 1965; Binkley *et al.* 1980; Strasser & Davaud 1986). Rim cements appear to be common in phreatic deposits as well (Table 1). Rim and meniscus cements, therefore, are not restricted to the vadose environment and cannot be thought of as diagnostic of the meteoric vadose zone.

Table 1 shows that vadose cemented deposits contain not only rim and meniscus cements but also pore-filling cement. In fact, 79% of all the deposits reviewed in this study contain pore-filling cement. This suggests that the vadose zone is often an area of cementation and not just one of dissolution. In fact, the relatively high frequency of pore-filling cement in the vadose zone leads one to question the generally accepted notion that pore-filling cement is diagnostic and characteristic of the phreatic zone.

Published data, therefore, provide no conclusive evidence concerning which cement types are characteristic of particular environments. Thus, a detailed petrographic study of vadose cements has been undertaken to identify what,

if any, are the typical cements that develop in the vadose zone.

Materials and methods

A wide range of last interglacial aeolianites from Mallorca, Tunisia and the Bahamas, as well as raised beaches from Almeria, were examined in thin section (see McLaren 1991 for a detailed description of the sites and sections discussed). A total of 342 samples were studied. All hand specimens were taken from freshly exposed faces or sampled well into the face to avoid any case-hardening influences. All samples collected in the field had their vertical orientation recorded and this information was transferred to all thin sections which were cut transverse to bedding.

The abundances of primary cements (gravity, drapestone, meniscus, rim, needle fibre and pore-filling) were all measured in thin section using a point counter. A total of 550 point counts evenly distributed across each thin section was made to determine the percentages of the various constituents present in the rock.

Results

Gravity and drapestone cements

No evidence of either gravity or drapestone cement was found in any of the samples studied from the meteoric vadose zone. This finding supports work by Schroeder (1973) who found that gravity cements were 'notably absent' in Pleistocene vadose cemented deposits from Bermuda. A wide variety of sediments with different textures were analysed to see if texture acted as a control on the presence/absence of gravity and drapestone cements. Sediments with

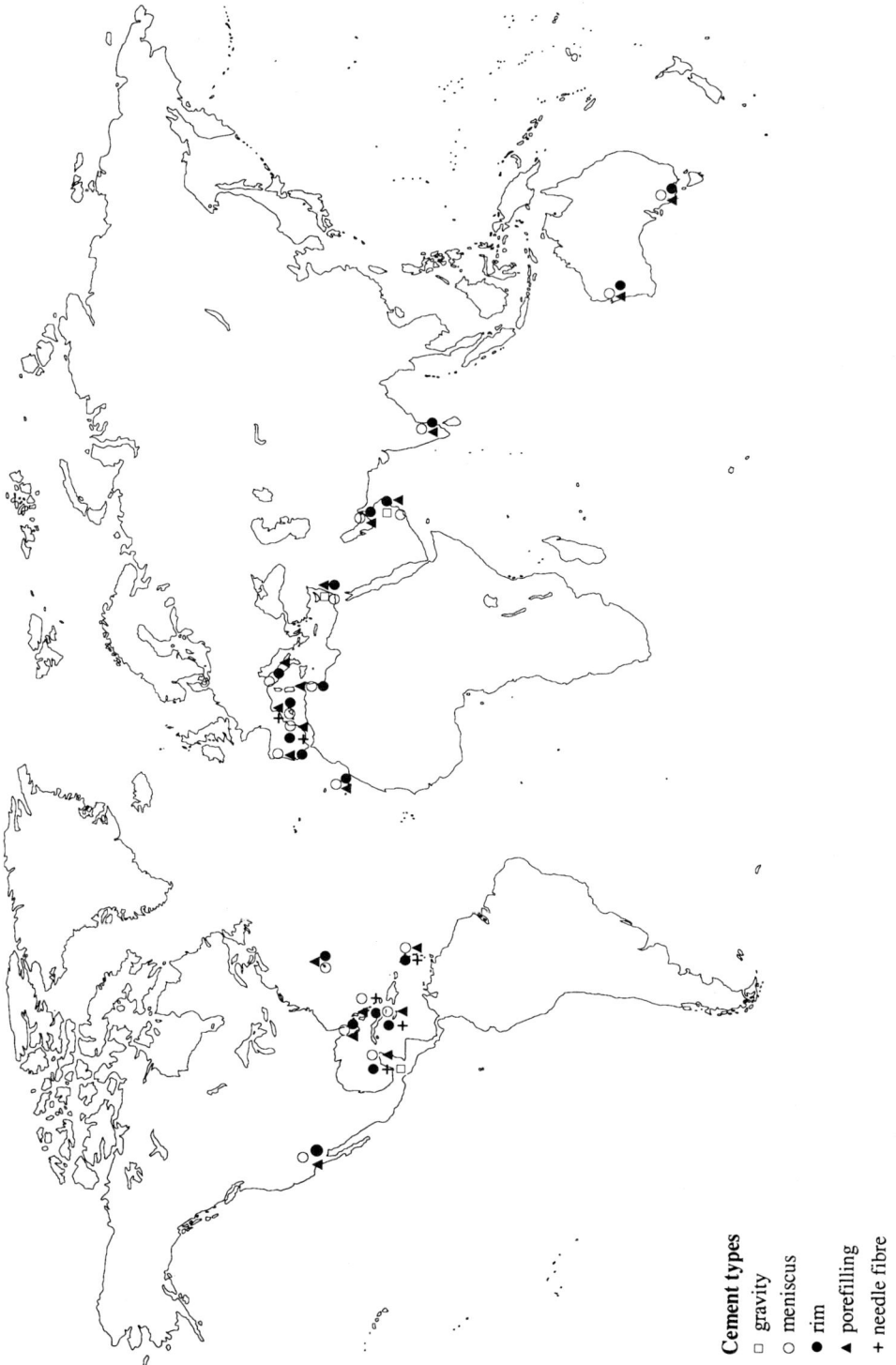

Cement types
□ gravity
○ meniscus
● rim
▲ porefilling
+ needle fibre

Fig. 1. Location of late Quaternary deposits where cement morphologies have been studied in detail.

different packing, sorting, size and shape (platy-to-spherical) were all studied. These included a raised beach at Arenal, Mallorca that contained some beds that were comprised of coarse sand-sized (500 μm to 2 mm) platy shell fragments; at Watlings Quarry, San Salvador, the Bahamas, a medium sand-sized (250–500 μm), very well-sorted oolitic (spherical grain shape) aeolianite; at Amoladares, southern Spain, an angular, poorly-sorted coarse sand to fine pebble (1–8 mm) raised beach; and at Aghir, Tunisia, very fine to fine dune sands (62–250 μm) that are moderately well to well sorted; as well as a number of other types of textures. Even though such a wide range in textures was observed, no textural control was found. This finding, however, disagrees with the work by Schroeder (1973) who suggested (but could not demonstrate) that a textural control was important in the development of gravity cements.

Such a widespread absence of gravity and drapestone cements in deposits with a large range in textures and from a number of different environments, when combined with limited reporting in the recent literature, seriously questions the idea that gravity and drapestone cements are 'typical' vadose cement types.

Meniscus cement

Meniscus cements are present in most of the deposits studied, although evidence of some may have been masked by later addition of pore-filling cement (but often the meniscus form can still be identified). Individual crystals are mostly blunt-ended with the exception of the earliest crystals which are predominantly small and equant. Meniscus cement crystals range from about 5–25 μm in length and the resultant shapes of the meniscii are available, ranging from gently to strongly concave. Table 2 summarizes the amount of meniscus cement (bulk rock volume) in the four main areas studied. Overall, in the Bahamas there is more meniscus cement than in the other areas; the Mallorcan and Spanish deposits contain similar amounts of meniscus cement and Tunisia contains the least. Thus, there appears to be spatial variation in the relative abundance of meniscus cement both within and between sections. This seems to be related to two factors: (1) Meniscus cements appear to be more abundant in the relatively more friable deposits. For example, in some of the dune deposits from Mallorca (Campo de Tiro), meniscus cements are more abundant in the lower portions of the outcrops, where the beds are more friable. In the better-cemented upper dune units towards the surface of the outcrops, pore-filling cements often mask any meniscus development. (2) The development of meniscus cements also appears to be related to texture. This cement type tends to be weakly developed or absent in the more poorly sorted beaches at Amoladares and Roquetas, southern Spain. In Tunisia, at most sites, meniscus cements make up less than 1.5% (bulk rock volume) on average, but at Khniss sections 1 and 2, meniscus cements make up 3.2% and 4.4% (mean bulk rock volume), respectively. The deposits at Khniss are different from elsewhere in Tunisia as they are relatively coarser grained. At most sites grain sizes are very fine to fine sands (62–250 μm) that are well packed, whereas at Khniss grain sizes range between coarse sands to very fine pebbles (500 μm to 4 mm).

In contrast, in the Bahamas, meniscus cements are much more common compared to all the other sites studied (Table 2) and they are particularly well seen in oolitic deposits (Figs 2a and 2b). The oolitic aeolianites of the Bahamas are medium sand-sized (250–500 μm), well rounded, very well-sorted sediments. The development of meniscus cements may be related to the smooth, convex surface of the ooliths and the packing characteristics that leave a suitably sized gap for water meniscii to develop. The Bahamas also represent a wetter environment compared to the other (Mediterranean and semi-arid) areas, and suffer much

Table 2. *Diagenetic patterns in deposits of similar age from different areas*

Area	Sample size	Mean % meniscus	SD	Mean % rim	SD	Mean % pore-filling	SD
Almeria	59	3.4	1.8	9.5	1.7	8.8	5.4
Mallorca	76	3.3	1.2	6.9	1.2	11.2	5.0
Tunisia	110	1.5	0.5	6.1	1.8	21.9	7.4
Bahamas	97	7.5	2.2	4.0	1.2	16.1	7.5

SD: Standard deviation.

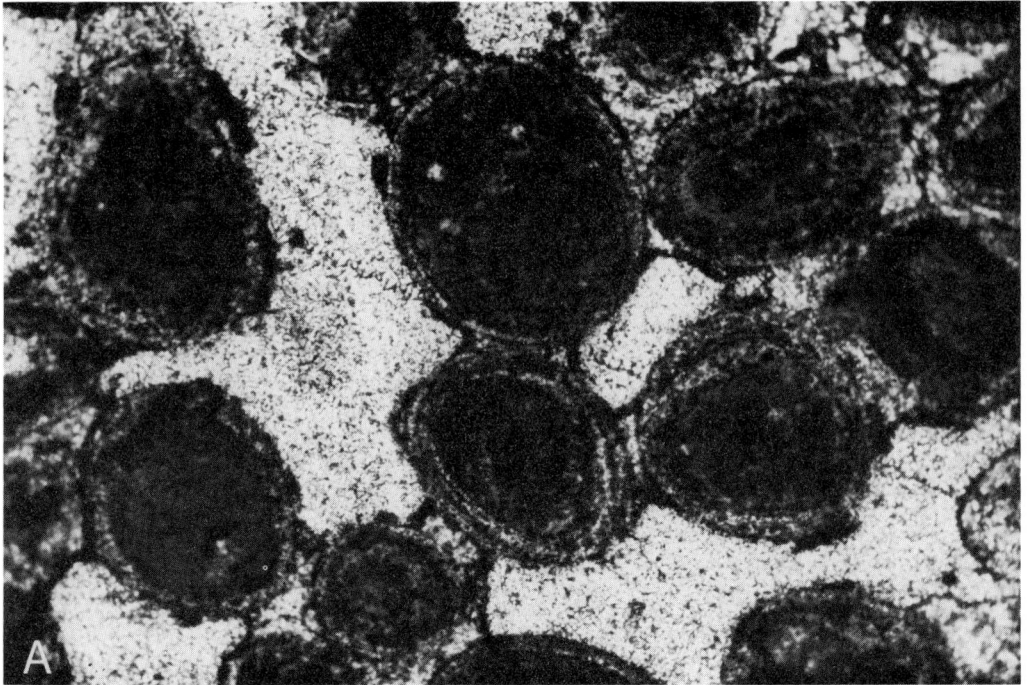

Fig. 2a. Clearly developed meniscus cement in an oolitic aeolianite from Watlings Quarry, San Salvador Island, the Bahamas (× 40).

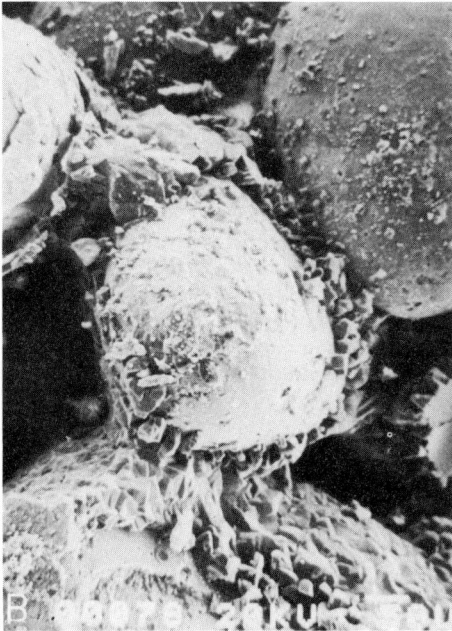

Fig. 2b. SEM micrograph showing meniscus and rim cement developed in an oolitic aeolianite from Watlings Quarry.

more flushing of water through the sediments, both now and in the past (Tarbox 1986). An abundant, but fluctuating supply of water rapidly moving through vadose deposits may result in waters concentrating at points of grain contact allowing meniscus cements to precipitate.

Rim cement

Two varieties of rim cement are evident, the first of which is syntaxial rim cement that is precipitated in optical continuity with the crystals of the host (echinoderm) grain. Syntaxial rim cement appears to be fairly common in most of the deposits studied, but it is not necessarily always present as its development is dependent on the type and abundance of suitable allochems to act as nucleii.

The second type of rim cement is far more common in all areas and mostly occurs either in the form of rhombic blocks or scalenohedral spar ('dog tooth crystals'). Commonly, early generations of rim cements consist of rhombohedral crystals and later generations of scalenohedra (Fig. 2c). In general, the crystals that develop as rims tend to be slightly larger than meniscus cement crystals and range in size from

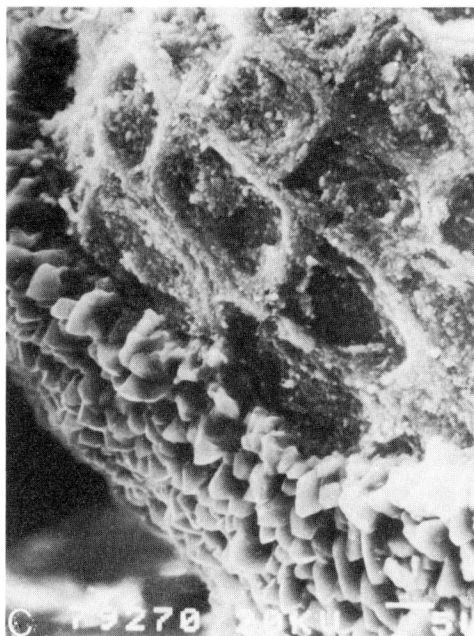

Fig. 2c. Scalenohedral rim cement developed in a raised beach deposit from Arenal, Mallorca.

patchily developed, often with pores devoid of any filling cement adjacent to occluded pores. Frequently, finer-grained areas contain more pore-filling cement than pores in coarser-grained areas (Fig. 2d).

The amount of pore-filling cement varies significantly between the four study areas (Table 2). It is most abundant in the last interglacial deposits from Tunisia. The Bahamas have, on average, the next highest pore-filling cement values. In Mallorca there is generally much less pore-filling cement and in the Almerian beach deposits there is least of all. In Almeria the average amounts of pore-filling cement increase up-profile from 0 to 24% (of the bulk rock volume); it may be totally absent in the lower units. Both equant rhombic crystals and scaleno-hedral spar create a blocky mosaic.

In Mallorca pore-filling cement is highly variable ranging from 5 to 35.3% (of the mean bulk rock volume). The amount of pore-filling cement tends to be lowest in quantity near the base of many sections and overall is highest in some of the aeolianites that are subjected to sea spray and/or occasional marine wash.

In Tunisia and the Bahamas pore-filling

10–50 μm. However, crystal sizes in all the deposits are also related to the size of the original pore which can limit the size of crystal growth.

Rim cements are, overall, common in the last interglacial deposits studied. In Almeria, rim cements are relatively abundant throughout the profile. Compared to the other areas, Almeria is relatively arid and it may mean that as less water is available for dissolution, potentially less total cement develops. What does form tends to develop preferentially as first generation rim cement. In the Bahamas and Tunisia mainly small pore spaces appear to have resulted in relatively rapid infilling of many pores by pore-filling cement often masking any initial rim cements.

Pore-filling cement

Pore-filling cement crystals range in size between 10–200 μm in length but generally do not exceed 100 μm. Crystal sizes often increase away from the initial site of precipitation (i.e. host substrate) towards the centre of pores. The resultant blocky cement mosaic only sometimes completely occludes pore spaces (15–20% of all pores studied are occluded) and tends to be

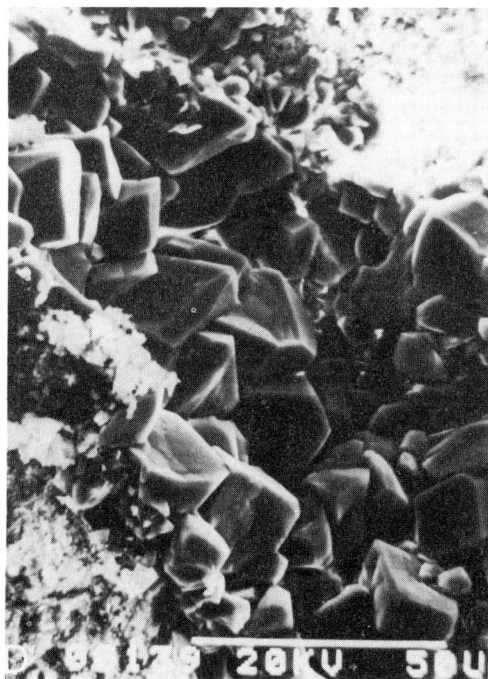

Fig. 2d. Pore-filling cement developed in an aeolianite from Aghir, Tunisia. Note the increase in crystal size towards the centre of the pores.

cements make up 21.9% and 16.1% of the mean bulk rock volume, respectively. The pore-filling cement content ranges on average from 5.3 to 33.7% and is characteristically patchily developed and not present within all pore spaces. Pore-filling cement is most abundant in finer-grained deposits such as those from Mahdia and Salakta in Tunisia. Such finer-grained deposits tend to have smaller pore spaces that require less cement in order to become occluded.

The crystal sizes of pore-filling cement also vary between the four areas. In general, the largest crystal sizes are found in the deposits from Almeria (crystal sizes range between 10–175 μm in length). This may be because of a number of reasons. The nature of the pore spaces in terms of their size and regularity may be of importance. Pore spaces are generally larger in Almeria than elsewhere allowing larger crystals to develop. Also some of the beds in the Almerian deposits are relatively poorly sorted and thus do not drain as freely as deposits that are better sorted, allowing longer residence times for solutions. Alternatively, some pore waters may have longer residence times because there is no constant flushing of water through the deposits in a semi-arid climate, allowing slower, coarser crystallization.

Cement crystals tend to be relatively smaller in the Bahamas and Tunisia because these deposits generally contain smaller pores than in the Spanish deposits, which restrict the length that the crystals can grow. In Mallorca, crystal size is variable tending to be finer in the aeolianites subjected to sea spray. Possibly rapid evaporation of the sea spray limits crystal development. On the whole, vadose pore-filling cements are very common in the deposits studied.

Needle fibre cements

Needle fibre cements are not ubiquitous in the vadose zone. They were found in deposits sampled near to the top of the relevant stratigraphic sequences (generally the upper 50 cm), and were not found at greater depth. Needle fibre cements are usually found in close association with rhizoliths. Both straight and wavy-edged needles exist (Fig. 2e), ranging in length between 20–180 μm and in width between 0.5–4 μm and they are randomly oriented. They are found in aeolianites from Campo de Tiro and Son Mosson in Mallorca and Crab Cay, San Salvador, the Bahamas. At Crab Cay, as well as needle fibre cement, grain coating needle mats are also found in close juxtaposition. These

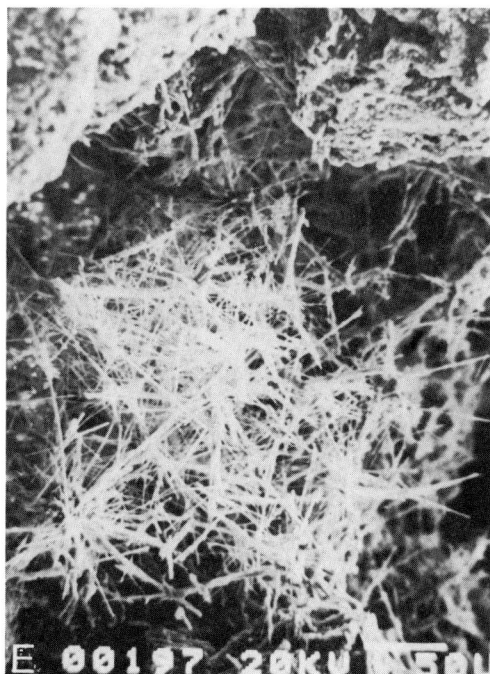

Fig. 2e. Needle fibre cement developed in a raised beach deposit from Roquetas, southern Spain.

mats are a dense array of needle fibres with little void space (Fig. 2f). The needles are randomly oriented and ropy in nature with serrated edges and surface relief.

In the vadose zone, needle fibre cements are associated with microbial activity and probably reflect the site of ectomycorrhizal sheaths of fungi (Wright 1986). Needle fibre cements are thus typical in the upper vadose zone if rhizoliths are present or plant roots were formerly present. When identified, needle fibre cements and grain coating needle mats are good indicators of the nearness to an original exposure surface (although the nearness is dependent on the depth of root growth).

Conclusions

It has been shown that diagenetic change in the four areas (the Bahamas, Tunisia, Mallorca, and Almeria, southern Spain) is similar in scope but not in degree. Gravity cements, contrary to traditional ideas that this cement type is 'typical' of the vadose diagenetic zone, are generally uncommon and often absent. This evidence is supported by its significant absence in other

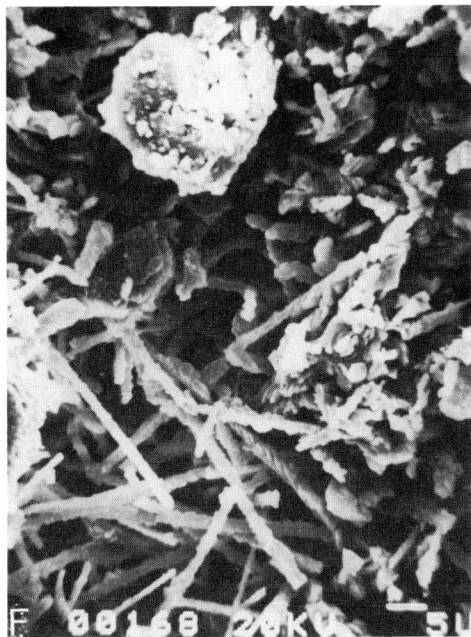

Fig. 2f. Grain coating needle fibre mat, Crab Cay, San Salvador Island. Note the ropy nature of the fibres.

areas such as Bermuda (Schroeder 1973) (Schroeder 1973).

Meniscus cements are common in the vadose diagenetic zone. They are an early phase cement that is clearly evident in the friable deposits where a lack of pore–filling cement allows the easy identification of the meniscus forms. There appears to be a textural (grain size and packing) control with meniscus cements developing preferentially in spherical, medium sand size, well-sorted grains such as many of the oolitic deposits of the Bahamas.

Rim cement crystals are generally abundant in the vadose zone although they are relatively less common in fine–medium sand deposits. In these deposits small pore spaces seems to result in their rapid infilling by pore-filling cements. Syntaxial rim cements are precipitated in optical continuity with the echinoderm grains. The development of this cement type is dependent upon the nature and abundance of contained allochems, thus syntaxial rims may not be present in all vadose cemented deposits. However, where suitable host grains exist, syntaxial rim cements can be used as a characteristic feature of the vadose zone.

Pore-filling cements were found to be very common and characteristic of the meteoric vadose zone deposits studied. The pore-filling cement is typically patchily developed, often with pores devoid of any pore-filling cement adjacent to occluded pores.

Cement types are generally highly variable in amount and distribution, both vertically and horizontally throughout a deposit. For example, in the upper vadose zone near the top of exposed sequences, pore-filling cements are common. Rim and meniscus cements are more common lower down the sequence. This is contrary to Longman's (1980) proposal that the upper vadose zone is a site of solution and lower down is a zone of precipitation.

At any one time, each pore has reached its own distinct stage of diagenetic change. Some pores may contain a thin rim cement while other pores may be devoid of rim cement and yet full of pore-filling cement. This finding disagrees with the conclusions of Land (1967) and Friedman (1964) who both suggest that first generation rim and meniscus cements are followed by second generation pore-filling cements. These two generations need not be separated by time (see Gardner & McLaren 1993). Some sub-areas within the vadose diagnetic zone may also have characteristic cement types. For example, needle fibre cements are characteristics of the upper vadose zone especially when found with rhizoliths and other calcified filaments. Needle fibre cements are strongly related/associated with rhizoliths and, when found, are evidence of proximity to a former erosion suface (dependent on the depth of root growth).

Therefore, there are characteristic cement patterns that develop in each area but there also exists a high degree of variability both within and between deposits. Although many cement types (together) are characteristic of the vadose zone, none are unique.

I would like to thank Dr R. A. M. Gardner for her helpful guidance during the course of this study. Thanks also go to the technicians in the Geography Department, Kings College and in the Geomaterials unit at Queen Mary and Westfield College. This study was funded by a NERC Research Training Award, no. GT4/87/GS/64.

References

ARTHUR, M. A., SCHOLLE, P. A. & HALLEY, R.B. 1982. Stable isotopes in sedimentary geology. *Journal of Sedimentary Petrology,* **52**, 1039–1045.

BADIOZAMANI, K. MacKENZIE, F. T. & THORSTENSON, D. C. 1977. Experimental carbonate cementation: salinity, temperature and vadose-phreatic effects. *Journal of Sedimentary Petrology,* **47**, 529–542.

BINKLEY, K. L., WILKINSON, B. H. & OWEN, R. M. 1980. Vadose beachrock cementation along a south-eastern Michigan marl lake. *Journal of Sedimentary Petrology*, **50**, 953–962.

BUCHBINDER, L. G. & FRIEDMAN, G. M. 1980. Vadose, phreatic and marine diagenesis of Pleistocene–Holocene carbonates in a borehole: Mediterranean coast of Israel. *Journal of Sedimentary Petrology*, **50**, 395–407.

BUDD, D. A. 1988a. Petrographic products of fresh-water diagenesis in Holocene ooid sands, Schooner Cays, Bahamas. *Carbonates and Evaporites*, **3**, 143–163.

—— 1988b. Aragonite-to-calcite transformation during freshwater diagenesis of carbonates: insights from pore-water chemistry. *Geological Society of America Bulletin*, **100**, 1260–1270.

FLUGEL, E. 1982. *Microfacies Analysis of Limestones*. Springer, Berlin.

FRIEDMAN, G. M. 1964. Early diagenesis and lithification in carbonate sediments. *Journal of Sedimentary Petrology*, **34**, 777–813.

GARDNER, R. A. M. 1988. Aeolianites and marine deposits of the Wahiba sands: character and palaeoenvironments. *Journal of Oman Studies. Special Report 3*, 75–95.

—— & McLAREN, S. J. 1993. Progressive vadose diagenesis in late Quaternary aeolianite deposts? *In*: PYE, K. (ed.) *The Dynamics and Environmental Context of Aeolian Sedimentary Systems*. Geological Society, London, Special Publication, **72**, 219–234.

GIVEN, R. K. & WILKINSON, B. H. 1985. Kinetic control of morphology, composition and mineralogy of abiotic sedimentary carbonates. *Journal of Sedimentary Petrology*, **55**, 109–119.

HARRIS, W. H. & MATTHEWS, R. K. 1968. Subaerial diagenesis of carbonate sediments: efficiency of the solution-reprecipitation process. *Science*, **160**, 77–79.

KLAPPA, C. F. 1979. Calcification and significance of soil filamentous micro-organisms in Quaternary calcrete profiles from eastern Spain. *Bulletin of the American Association of Petroleum Geologists*, **63**, 480.

LAND, L. S. 1966. *Diagenesis of Metastable Skeletal Carbonates*. Unpublished PhD Thesis, Lehigh University, Bethlehem, P.A.

—— 1967. Diagenesis of skeletal carbonates. *Journal of Sedimentary Petrology*, **37**, 914–930.

—— 1970. Phreatic versus meteoric diagenesis of limestones: evidence from a fossil water table. *Sedimentology*, **14**, 175–185.

—— 1971. Phreatic versus meteoric diagenesis of limestones: evidence from a fossil water table in

Bermuda. *In*: BRICKER, O. P. (ed.) *Carbonate Cements*. Johns Hopkins Press, Baltimore, 133–137.

LONGMAN, M. W. 1980. Carbonate diagenetic textures from near surface diagenetic environments. *Bulletin of the American Association of Petroleum Geologists*, **64**, 461–487.

McLAREN, S. J. 1991. *Vadose Diagenesis of Late Quaternary Coastal Sediments*. Unpublished PhD Thesis, London University.

MOORE, Jr, C. H. 1989. *Carbonate Diagenesis and Porosity*. Developments in Sedimentology **46**. Elsevier, Amsterdam.

MORSE, J. W. & MACKENZIE, F. T. 1990. *Geochemistry of Sedimentary Carbonates*. Developments in Sedimentology **48**. Elsevier, Amsterdam.

MULLER, G. 1971. Gravitational cement, an indicator for the vadose zone of the subaerial diagenetic environment. *In*: BRICKER, O. P. (ed.) *Carbonate Cements*. Johns Hopkins Press, Baltimore, 301–303.

SCHROEDER, J. H. 1973. Submarine and vadose cements in Pleistocene Bermuda Reef Rock. *Sedimentary Geology*, **10**, 179–204.

SCOFFIN, T. 1987. *An Introduction to Carbonate Sediments and Rocks*. Blackie, Glasgow.

STODDART, D. R. & CANN, J. R. 1965. Nature and origin of beachrock. *Journal of Sedimentary Petrology*, **35**, 243–247.

STRASSER, A. & DAVAUD, E. 1986. Formation of Holocene limestone sequences by progradation, cementation and erosion: two examples from the Bahamas. *Journal of Sedimentary Petrology*, **56**, 422–429.

TARBOX, D. L. 1986. Occurrence and development of water resources in the Bahama Islands. *In*: CURRAN, H. A. (ed.) *Proceedings of Third Symposium on the Geology of the Bahamas*. Don Heuer, USA, 139–144.

TUCKER, M. E. & WRIGHT, V. P. 1990. *Carbonate Sedimentology*. Blackwell, Oxford.

WARD, W. C. 1973. Influence of climate on the early diagenesis of carbonate eolianites. *Geology*, **1**, 171–174.

—— 1975. *Petrology and diagenesis of carbonate eolianite of northeastern Yucactan Peninsula, Mexico*. American Association of Petroleum Geologists Studies in Geology, **2**, 500–571.

WRIGHT, V. P. 1986. The role of fungal biomineralization in the formation of Early Carboniferous soil fabrics. *Sedimentology*, **33**, 831–838.

Temperate and cold climate continental dunes

Genesis and sedimentary structures of late Holocene aeolian drift sands in northwest Europe

EDUARD A. KOSTER, ILONA I. Y. CASTEL & RON L. NAP

Department of Physical Geography, University of Utrecht, PO Box 80.115,
3508 TC Utrecht, The Netherlands

Abstract: In northwest Europe local resedimentation of terrestrial deposits by wind from the Neolithic to the present has resulted in widespread accumulation of (aeolian) drift sands. These deposits consist mainly of fine, well-sorted and well-rounded unimodal sands; their mineralogical composition usually reflects local provenance. Palynological studies combined with radiocarbon analyses of the topmost part of buried peat sections reveal no distinct regional phases in drift sand accumulation: ages mainly vary from AD 500 to 1700. Small remnants of current actively moving drift sands provide evidence of sub-recent deflation and accumulation processes in relation to relief, topsoil and aerodynamic conditions. These are reflected in the sedimentary structures. Principal structures show dune foreset cross-bedding as well as even or wavy sub-horizontal lamination, while occasional strings of granules or small pebbles represent deflation phases. Primary structures resulting from tractional deposition (mainly sub-critical climbing ripple migration) and grainfall deposition, secondary structures (mainly slump, scour-fill and adhesion structures) and non-aeolian structures (mainly caused by rain impact, convolution structures, wetted sand crusts, and foot- or hoof-imprints) have a distinct appearance in lacquer peels and in SEM micrographs. Transport and accumulation of these temperate aeolian sands appear to have occurred under dry, moist and wet conditions.

Widespread aeolian formations of Weichselian to Recent age are found at the surface in the 'sand-belt' of the Northwest and Central European Lowlands, extending from Belgium and The Netherlands through Germany, Denmark, and Poland into the Baltic States (Koster 1978). The total area within the 'sand-belt' covered by surficial aeolian sands, including extensive sheets of Weichselian cover sands and Holocene drift sands not showing a dune relief, is of the order of several tens of thousands of square kilometres (Ruegg 1983; Niessen *et al.* 1984; Koster 1988). A minor proportion of this region is occupied by inland and river dunes. The terminology and lithostratigraphic division used in describing aeolian sand deposits and related relief forms is confused by the fact that geomorphological, stratigraphical and pedological concepts are used indiscriminantly (Koster 1982).

Local resedimentation by wind of terrestrial (mainly aeolian) deposits occurred on a large scale from the beginning of the Neolithic up to the present (Castel *et al.* 1989) mainly in the western part of this region. This resulted in the accumulation of so-called drift sands, which are scattered throughout northern Belgium (areal extent 200–300 km²), The Netherlands (approx. 950 km²), northwestern Germany (1400–

2000 km²), and Denmark (450–550 km²) in many relatively small patches. Similar deposits covering small areas are found in England, particularly in East Anglia (Williams 1975). The majority of these drift sand areas have been stabilized by afforestation since the end of the nineteenth century, but small remnants are still actively forming, mainly in the central Netherlands.

The morphogenesis, sediment characteristics, provenance, and age of these late Holocene drift sands have been studied by Pyritz (1972), Koster (1978), Van Mourik (1988), and Castel (1991*a*). Field experiments and simulation studies on processes of wind erosion and sedimentation in active drift sand areas have been presented by De Ploey (1977, 1980), Brugmans (1983), and Castel (1988). As a follow-up to the overview by Castel *et al.* (1989), the first objective of this paper is to present new data on the age of the first phases of sand drifting, the reliability of radiocarbon ages obtained from peat layers which have been buried by aeolian sand, and the causes of sand drifting with special attention to the influences of climate and human activity.

The sedimentary structures in late Holocene drift sand deposits have so far not been studied. The second objective of this paper is, therefore, to describe specific sedimentary structures

From Pye, K. (ed.), 1993, *The Dynamics and Environmental Context of Aeolian Sedimentary Systems.*
Geological Society Special Publication No. 72, pp. 247–267.

present in the active drift sand areas and to demonstrate the value of these structures for interpreting the genesis of these deposits.

Nature of the drift sands

In this paper the terms 'drift sand deposits' or 'drift sands' are reserved for those aeolian sands which originate through relatively recent (Holocene) resedimentation of terrestrial Pleistocene deposits. They usually display a chaotic dune relief and are sometimes still actively forming. The drift sands, which are mainly of local origin, are characterized by a light yellow–greyish colour (10 YR 5/2–6/2), relatively loose grain-packing, the presence of layers rich in organic (sometimes clastic) material, and an absence of periglacial structures (Koster 1978; Castel *et al.* 1989). Although the sediment attributes of drift sands usually reflect their local provenance quite well, they mainly consist of very fine to moderately fine, non-loamy, well-sorted and well-rounded unimodal sands. Most drift sands in the investigated regions are non-calcareous quartz sands with feldspars and a variable but mainly low content of heavy minerals. Differences in heavy mineral composition of the various source materials are preserved during local resedimentation. Regional differences in the degree of roundness are small. Despite the fact that most drift sands are extremely poor to very poor in humus (< 1.5% organic matter), one of its field characteristics is the occurrence of thin, humic, partly clastic layers. Such layers are usually absent in cover sands. Notwithstanding the often seemingly chaotic drift sand relief, it is generally accepted that these aeolian sands of northwest Europe have been transported predominantly by south-westerly to northwesterly winds (Pyritz 1972; Koster 1978). A major proportion of the drift sand regions displays similar characteristics to the 'low-angle aeolian sand sheet deposits' described by Fryberger *et al.* (1979).

On the basis of palynological analyses and radiocarbon dating of peat layers covered by drift sand from various locations in the Veluwe area (central Netherlands), the Emsland in Niedersachsen (Germany) and southern Jutland (Denmark), Castel *et al.* (1989) concluded that, in general, major sand drift phases did not occur before AD 950. Additionally, the beginning of many sand drifts appears to be closely related, both in time and space, to the strong expansion of arable fields of the so-called 'plaggen' soil type in the Late Middle Ages in northwest Europe. Earlier phases of sand drifting have been documented by Van Gijn & Waterbolk (1986) and Van Mourik (1988), but these probably never covered a large area.

Historical sources point to an important growth of drift sand areas in northwestern Europe, particularly in the eighteenth and nineteenth centuries, during which sand drifting locally became a severe threat to agricultural fields and settlements (Pyritz 1972; Meyer 1984).

Palaeoenvironmental control

In order to investigate the local and regional vegetation development just before and during the initiation of sand drifting, and to detect the causes of sedimentation of these sands, palaeo-ecological analyses were carried out. Research focused on the so-called plateau-dunes (Fig. 1) which originated through relief inversion of the

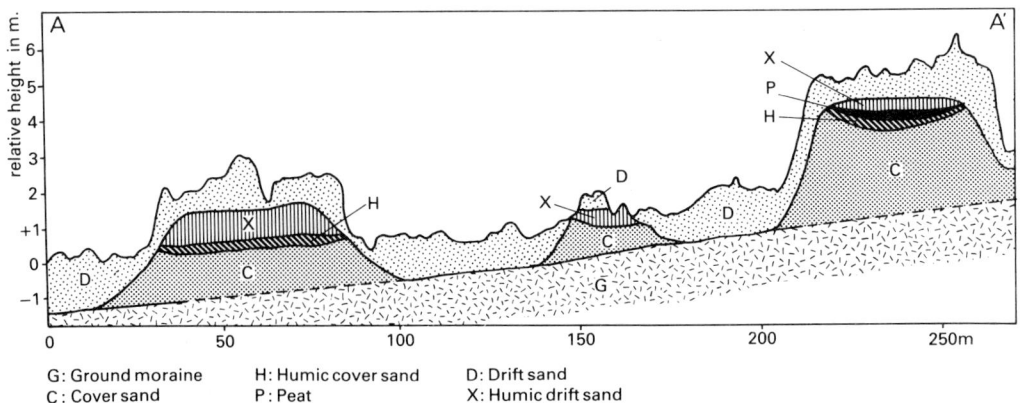

G: Ground moraine   H: Humic cover sand   D: Drift sand
C: Cover sand   P: Peat   X: Humic drift sand

Fig. 1. Cross-section through a drift sand area in the province of Drenthe (The Netherlands), showing three plateau-dunes resulting from relief inversion (from Castel 1991*a*).

Fig. 2. General map showing localities of the investigated sites (lacquer peels). 1: approximate areal extent of plaggen soils in northwest Europe (Pape 1970); 2: loess boundary (west of the Rhine), 200 m contour line (east of the Rhine).

originally gently undulating cover sand topography. They represent the former wet depressions were peat development, mainly consisting of *Eriophorum*, *Sphagnum*, *Scheuchzeria* and sometimes *Carex* vegetation, often occurred. During the subsequent phases of blowing sand, these wet spots preferentially trapped the sand and, after some time, the originally low-lying areas were built up into dunes (Schelling 1957). Palynological evidence from peat layers in plateau-dunes from the Veluwe region (Koster

1978), the province of Drenthe (Castel 1991*a*) and the Emsland and southern Jutland (Castel *et al.* 1989), (for locations see Fig. 2) suggested the following points.

(1) Near the end of the period of peat formation, and before the beginning of the drift sand accumulation, most pollen diagrams show a marked increase in the proportion of Ericaceae. At the same time, or shortly thereafter, an increase in Gramineae and cultural indicators such as *Plantago lanceolata*, *Rumex*, and the crops *Secale*, Cerealia, *Fagopyrum* and *Cannabis* occurs (Fig. 3). The amount of clastic material in the peat sections increased more or less simultaneously. These data indicate a change in land use which initiated the formation of local sand drifts.

(2) *Fagus*, *Carpinus* and *Alnus* decrease at the same time, while *Tilia*, *Ulmus* and *Fraxinus* virtually disappear. The decline of woodland species may have been caused by human exploitation for valuable leaf fodder, causing a retreat, at least in the flowering, of these species. The woods may have been cut in order to open up forests for grazing or to favour the development of agricultural fields.

(3) Surprisingly, at the time of cessation of peat formation and the beginning of (pure) aeolian sand accumulation no significant changes in pollen assemblages were found. The values of the cultural-indicator species increase in most diagrams, suggesting a continuing intensification of agriculture. The depth range between the first occurrence of sand within the peat, representing the

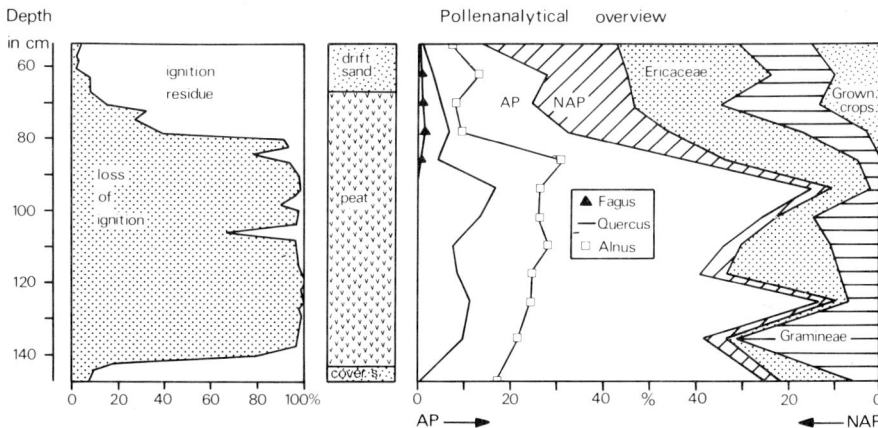

Fig. 3. Simplified pollen diagram illustrating the changes at the transition from peat to drift sand (compiled after Castel 1991*a*).

initiation of sand drifts, and the widespread and intense drift sand sedimentation, which was vigorous enough to terminate local peat growth, is quite variable at different locations.

(4) The percentages of cultural indicators usually do not reach high values, pointing to fairly long distances between the arable fields and pastures and the sampling sites. The areal distribution of drift sands and soils of the 'plaggen' type in northwest Europe (Pape 1970) (Fig. 2) justifies the conclusion that the beginning of extensive sand drifts is related, both in time and space, to the strong expansion of arable fields, together with the use of plaggen fertilizer and exploitation of heathlands by sheep grazing and heath mowing, mainly after AD 950.

(5) In several areas increasingly humid local conditions seem to be reflected in the upper part of the peat, indicated by the presence of algae and fungal types such as Zygnemataceae (a.o. *Debarya*), *Tilletia* and *Botryococcus* and also by the presence of pollen derived from plants favouring moist habitats, such as *Typha, Sparganium, Oenanthe* and Ranunculaceae. Consequently, it is assumed that the removal of trees necessary for the expansion of heathland and arable fields, resulted in a significant decrease in evapotranspiration such that local ground water levels were raised considerably (Castel 1991a).

In the central Netherlands a relationship between sand drifting and a drier climate in the tenth century has been suggested by Heidinga (1984). In spite of the fact that a somewhat warmer and drier period in western Europe between AD 950 and 1250 has been repeatedly documented (Folland *et al.* 1990), no evidence for this climatic oscillation has been found in the analysed peat sections, all of which encompass this period. On the contrary, the above mentioned increased wetness of the peat sections suggests that, in general, the beginning of sand drifting should not be interpreted as a natural response to the 'climatic optimum' of the Middle Ages.

Age of sand drifting

In order to determine a '*post quem*' age of drift sand accumulation, radiocarbon ages of the topmost part of buried peat sections and of buried horizons from humic podzol soils at several locations have been obtained by Koster (1978), Van Mourik (1988, 1991), Van Mourik & Odé (1990), Castel *et al.* (1989) and Castel (1991a). These are summarized in Table 1, which also includes data about the beginning of peat formation at the investigated sites. Several problems concerning interpretation of the radiocarbon ages and additional results need to be highlighted. The dates have been calibrated using the computer programme of the Centre for Isotope Research of the University of Groningen (Van der Plicht & Mook 1990). The peat and soil samples, which cover a large number of years, required a calibration curve which is 'smoothed' (100-year moving average) (Mook 1983). When ages in years BP are given, ^{14}C ages measured from AD 1950 are indicated. The ages are corrected for $\delta^{13}C$-values deviating from $-25‰$. For the calibrated ages, a notation with BC or AD is used.

Organic components can be subdivided into three fractions according to their solubility in acids and in alkalis:

- fulvic acids, soluble in alkalis and in acids;
- humic acids, soluble in alkalis, insoluble in acids;
- residual fraction (humine), insoluble in acids and in alkalis.

In the study by Castel (1991a), conventional radiocarbon ages have been determined from the residual fraction and from an (alkali) extract, which is an equivalent of the humic acid fraction. In most cases, the age of the extract appears to be older than the age of the residual fraction, which is in contrast to the fractionated radiocarbon analyses performed on soil profiles by Van Mourik (1988). Van Mourik (1988) found that the ages determined from the humic acid fraction were usually younger than those from the residual fractions. In order to ascertain which data are the most reliable, several additional radiocarbon ages were obtained by means of AMS dating. To this purpose individual seeds were collected from the same sections (and depths) in the province of Drenthe, The Netherlands by Castel (1991a), and analysed according to the methods described by Van der Borg *et al.* (1987). The AMS ages appeared to correspond most closely with those obtained from the residual fractions. Consequently, it is assumed that the organic residue fractions provide the most reliable results. Therefore, only the data obtained by Castel (1991a) from the residual fractions are given in Table 1.

During the inventory of buried peat deposits in the province of Drenthe, an abrupt change from the peat into the overlying aeolian sand was observed in most cases, for which a hiatus (non-

Table 1. *Radiocarbon ages of the maximum age of drift sand accumulation (nos 1–30) and of the beginning of peat formation (nos 31–42)*

	Location		Material	Sample	^{14}C years BP	Calibrated BC/ AD with 2σ	Reference
1	Kootwijk 1	(Veluwe, The Netherlands)	peat	GrN-6822	900 ± 70	AD 1005–1245	Koster 1978
2	Leuvenum 1	(Veluwe, The Netherlands)	peat	GrN-5986	760 ± 30	AD 1225–1278	Koster 1978
3	Coldenhove	(Veluwe, The Netherlands)	peat	GrN-7395	755 ± 30	AD 1228–1280	Koster 1978
4	Woeste Hoeve	(Veluwe, The Netherlands)	peat	GrN-6827	695 ± 30	AD 1255–1325	Koster 1978
5	Geeste 1	(Niedersachsen, Germany)	peat	GrN-9166	1115 ± 45	AD 820–1000	Castel *et al.* 1989
6	Herzlake 1	(Niedersachsen, Germany)	peat	GrN-9168	1295 ± 55	AD 635–845	Castel *et al.* 1989
7	Fröslev 1	(Jutland, Denmark)	peat	GrN-9164	845 ± 50	AD 1065–1260	Castel *et al.* 1989
8	Defensiedijk 1.1	(Noord-Brabant, The Netherlands)	soil, humic acids	GrN-12804	410 ± 45	AD 1420–1590	Van Mourik 1988
9	Defensiedijk 1.2	(Noord-Brabant, The Netherlands)	soil, humic acids	GrN-12805	1365 ± 25	AD 638–676	Van Mourik 1988
10	Defensiedijk 1.3	(Noord-Brabant, The Netherlands)	soil, humic acids	GrN-12806	3615 ± 35	2095–1905 BC	Van Mourik 1988
11	Defensiedijk 2.1	(Noord-Brabant, The Netherlands)	soil, bulk	GrN-13511	1395 ± 35	AD 600–672	Van Mourik 1988
12	Defensiedijk 2.2	(Noord-Brabant, The Netherlands)	soil, bulk	GrN-13512	3140 ± 35	1495–1360 BC	Van Mourik 1988
13	Defensiedijk 3	(Noord-Brabant, The Netherlands)	soil, bulk	GrN-13514	1225 ± 25	AD 725–860	Van Mourik 1988
14	Boshoverheide 1.1	(Noord-Brabant, The Netherlands)	soil, humic acids	GrN-12869	390 ± 25	AD 1445–1540	Van Mourik 1988
15	Boshoverheide 1.2	(Noord-Brabant, The Netherlands)	soil, humic acids	GrN-12870	615 ± 45	AD 1285–1405	Van Mourik 1988
16	Tungeler Wallen 1	(Noord-Brabant, The Netherlands)	soil, bulk	GrN-14347	945 ± 25	AD 1020–1140	Van Mourik 1988
17	Tungeler Wallen 2	(Noord-Brabant, The Netherlands)	soil, bulk	GrN-14346	2140 ± 30	300–100 BC	Van Mourik 1988
18	Herperduin	(Noord-Brabant, The Netherlands)	peat, bulk	GrN-16535	410 ± 30	AD 1430–1520	Van Mourik & Odé 1990
19	Peelterbaan 1	(Noord-Brabant, The Netherlands)	soil, humic acids	GrN-17644	770 ± 110	AD 1030–1385	Van Mourik 1991
20	Peelterbaan 2	(Noord-Brabant, The Netherlands)	soil, humic acids	GrN-17643	1100 ± 65	AD 775–1040	Van Mourik 1991
21	Peelterbaan 3	(Noord-Brabant, The Netherlands)	soil, humic acids	GrN-17642	1555 ± 65	AD 370–620	Van Mourik 1991
22	Boschoord 1	(Drenthe, The Netherlands)	peat, residue	GrN-16157	760 ± 110	AD 1040–1395	Castel 1991*a*
23	Berkenheuvel 1	(Drenthe, The Netherlands)	peat, residue	GrN-16156	710 ± 70	AD 1185–1390	Castel 1991*a*
24	Kraloo 1	(Drenthe, The Netherlands)	peat, residue	GrN-16162	500 ± 80	AD 1295–1560	Castel 1991*a*
25	Ter Horsterzand 1	(Drenthe, The Netherlands)	peat, residue	GrN-16161	240 ± 60	AD 1495–1865	Castel 1991*a*
26	Ter Horsterzand 3	(Drenthe, The Netherlands)	peat, residue	GrN-16160	510 ± 90	AD 1285–1580	Castel 1991*a*
27	Heuvingerzand 1	(Drenthe, The Netherlands)	peat, residue	GrN-16159	120 ± 70	> AD 1640	Castel 1991*a*
28	Sleenerzand 1	(Drenthe, The Netherlands)	peat, residue	GrN-16163	370 ± 60	AD 1435–1640	Castel 1991*a*
29	Emmerdennen 1	(Drenthe, The Netherlands)	peat, residue	GrN-16158	245 ± 40	AD 1525–1725	Castel 1991*a*
30	Zeegse Duinen 1	(Drenthe, The Netherlands)	peat, residue	GrN-16164	285 ± 45	AD 1490–1675	Castel 1991*a*
31	Kootwijk 3	(Veluwe, The Netherlands)	peat	GrN-6824	1580 ± 80	AD 295–620	Koster 1978
32	Leuvenum 2	(Veluwe, The Netherlands)	peat	GrN-5987	1910 ± 50	25 BC–AD 205	Koster 1978
33	Geeste 2	(Niedersachsen, Germany)	peat	GrN-9167	2260 ± 50	405–210 BC	Castel *et al.* 1989
34	Herzlake 2	(Niedersachsen, Germany)	peat	GrN-9169	2150 ± 55	360–70 BC	Castel *et al.* 1989
35	Fröslev 2	(Jutland, Denmark)	peat	GrN-9165	1375 ± 50	AD 580–740	Castel *et al.* 1989
36	Boschoord 3	(Drenthe, The Netherlands)	peat, residue	GrN-17061	4925 ± 90	3950–3540 BC	Castel 1991*a*
37	Berkenheuvel 4	(Drenthe, The Netherlands)	peat, residue	GrN-17066	3875 ± 50	2490–2215 BC	Castel 1991*a*
38	Ter Horsterzand 7	(Drenthe, The Netherlands)	peat, residue	GrN-17057	5060 ± 60	3990–3745 BC	Castel 1991*a*
39	Heuvingerzand 3	(Drenthe, The Netherlands)	peat, residue	GrN-17062	1780 ± 60	AD 105–380	Castel 1991*a*
40	Sleenerzand 6	(Drenthe, The Netherlands)	peat, residue	GrN-17063	1870 ± 60	5 BC–AD 280	Castel 1991*a*
41	Emmerdennen 6	(Drenthe, The Netherlands)	peat, residue	GrN-17059	2650 ± 50	910–775 BC	Castel 1991*a*
42	Zeegse Duinen 4	(Drenthe, The Netherlands)	peat, residue	GrN-17065	4080 ± 80	2885–2475 BC	Castel 1991*a*

depositional or erosional) between the peat and overlying drift sand cannot be excluded (Castel 1991*a*). However, if such a hiatus exists, this would have consequences for the interpretation of the radiocarbon ages as well as for the estimated ages of the pollen zone boundaries. The presence or absence of a hiatus has been investigated by micromorphological study of the transitional zones. The macroscopically observed transition zones varied in thickness from less than 1 cm to 8–10 cm. The results of the micromorphological analyses (Castel 1991*b*) show that features indicative of pedogenesis never occurred in the upper part of the peat zone. Layers with highly humified peat were always found at a certain depth in the peat zone, while modexi, root channels and organans (cutans) were only found in the drift sand part of the thin sections. The lack of pedogenetic features at the transition from peat into drift sand therefore points to an absence of a chrono-

stratigraphic hiatus between peat formation and drift sand sedimentation. In Fig. 4 the main results of the micromorphological analyses have been summarized. Two examples of sections with an abrupt change from peat to sand in the field are presented. Since both the curves for the decomposed part of the organic material (humicol, cutans and modexi) and for the total amount of organic material are given, this means that the amount of basically undecomposed organic material (humiskel) is inferred. In the Berkenheuvel section the analysis of this transition clearly shows that the amount of mineral grains gradually increases upwards with a pronounced increase at the boundary itself. In the upper centimetres of the peat zone, approximately 10% consists of mineral grains, more or less horizontally layered between the peat. The marked change in character of the material at the transition is accentuated by the change in character of the organic material from humiskel

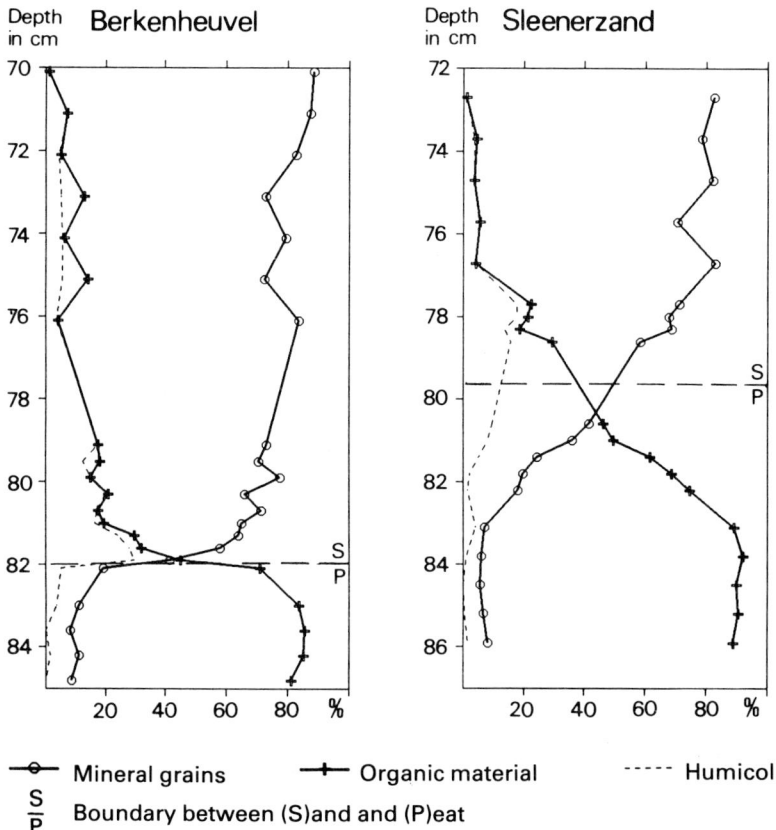

Fig. 4. Percentages of mineral grains and organic material at the transition from peat to drift sand, determined by micromorphological analysis of thin sections. Organic material includes both undecomposed and decomposed material. Locations in the province of Drenthe (The Netherlands) (from Castel 1991*a*).

to humicol, but it is clear that no hiatus exists at this boundary.

Although the Sleenerzand section also showed an abrupt change from peat to drift sand in the field, the micromorphological study points to a more gradual transition. In Fig. 5, the grain size distribution as determined by analyses of the thin sections is given for the same sections. Although the number of grains counted in the upper part of the peat zone is relatively small, it appears that grains with a size of 50–105 μm dominate; the second most abundant size fraction is 15–50 μm. At the transition itself no sudden change in grain size is evident, but rather a gradual coarsening upwards is found in the drift sand zone.

In summary, it can be stated that no hiatus occurs at the transition from peat into aeolian sand. Near the transition, frequently alternating layers of peat and fine sand grains have been observed (Fig. 6), indicating synchronous peat formation and drift sand sedimentation during certain periods. The depth range over which the transition occurs is variable, which can be explained by a varying rate of peat formation and/or by a difference in the supply of drift sand. Consequently, it is concluded that the radiocarbon ages obtained from the topmost one or two centimetres of peat accurately represent the age of the beginning of local sand drifting.

Table 1 also includes calibrated ages of the beginning of rather extensive drift sand sedimentation at several locations in the province of Drenthe. The data clearly illustrate that the deposition of drift sands did not occur at the same time all over this province. It has been seen that, within the upper part of the peat, sand was always deposited over a certain depth range, the

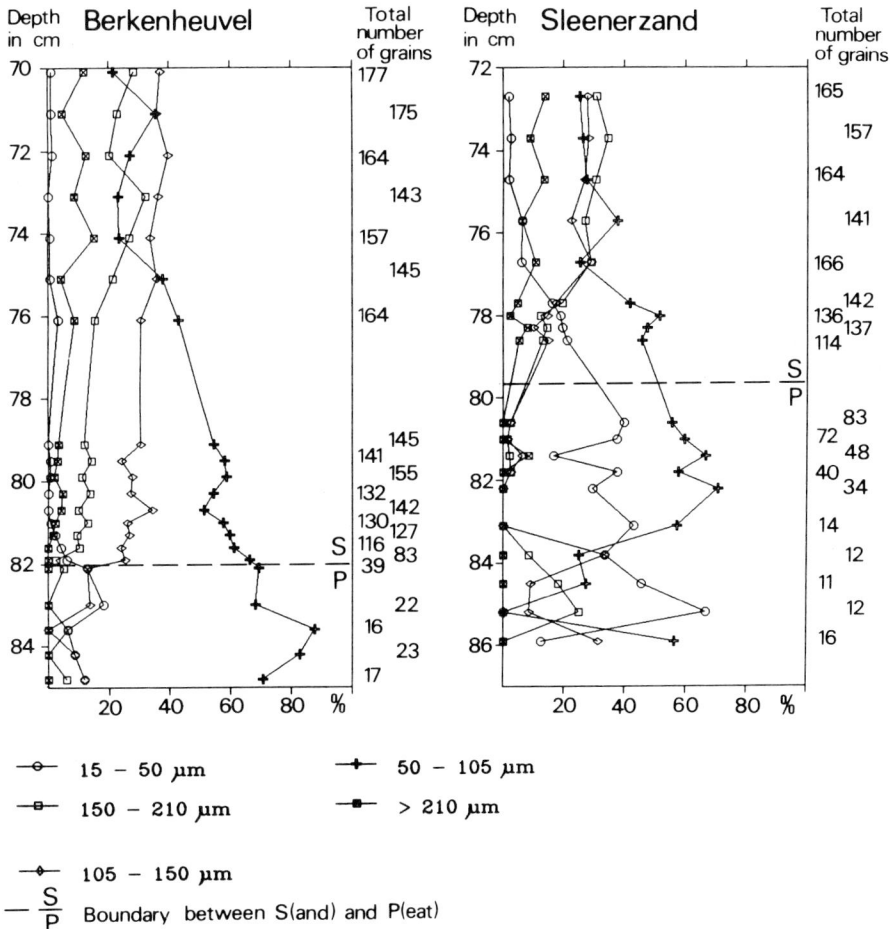

Fig. 5. Grain size distributions for the sections shown in Fig. 4.

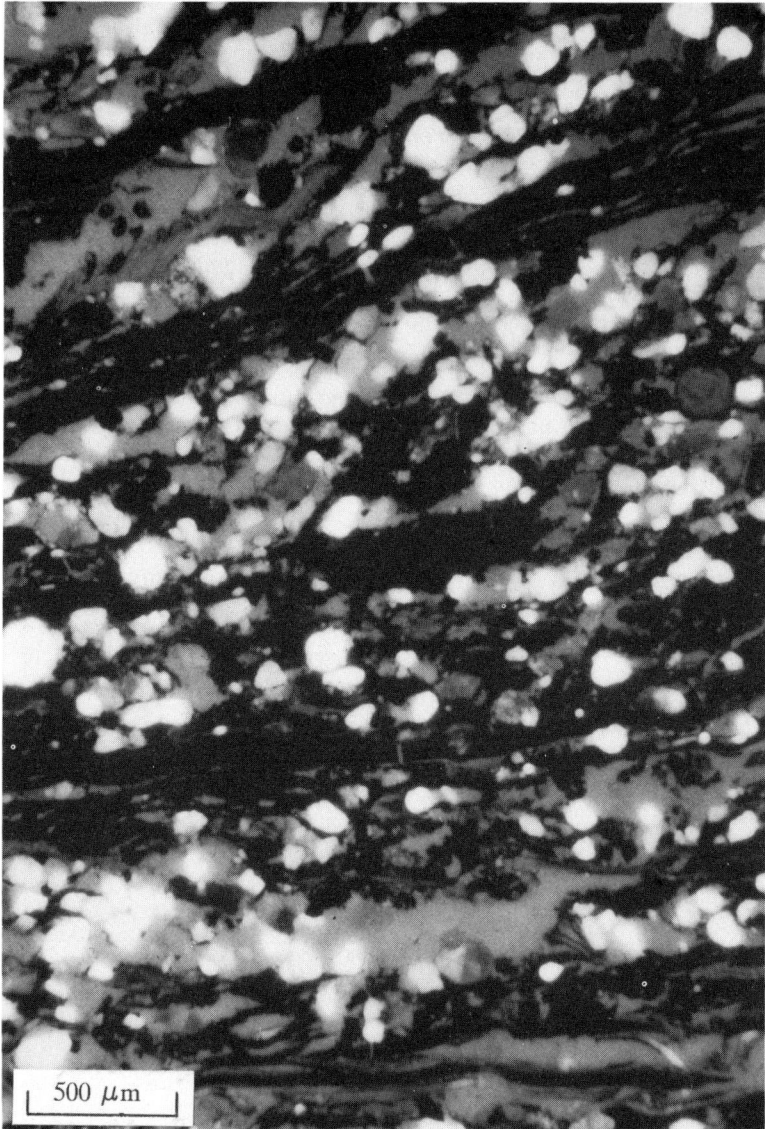

500 μm

Fig. 6. Thin section showing alternating layers of peat and fine sand grains in the transition zone peat–drift sand.

extent of which is dependent on local conditions. The maximum depths of sand occurrence have been deduced from the ignition residues, which have been determined from all peat sections. For these depths, an age was estimated by means of interpolation between the radiocarbon ages of the lower and upper part of the peat sections. In Table 2, the interpolated ages of the first phase of sand drifting at the various locations are given. Again the diachronism of these events is striking.

Comparison of the data in Tables 1 and 2 leads to the inevitable conclusion that a prolonged interval of many centuries must have occurred between the first onset of sand drifting and the more or less extensive drift sand accumulation at a specific location.

If one compares the previously obtained data on the age of drift sands from other regions in The Netherlands, northwestern Germany and southern Jutland (Table 1) with those from Drenthe, the diachronism is even more evident.

Table 2. *Interpolated ages of the first appearance of sand drifting at the locations in the province of Drenthe as listed in Table 1 (from Castel 1991a)*

Location	Interpolated age ^{14}C years BP
Boschoord	1630
Berkenheuvel	1280
Ter Horsterzand	600
Heuvingerzand	255
Sleenerzand	1315
Emmerdennen	285
Zeegse Duinen	2140

Van Mourik (1988) was able to distinguish three phases of drift sand sedimentation in the province of Noord-Brabant (Fig. 2): the first one is ascribed to the Bronze Age and early Iron Age (after *c.* 3500 BP); the second one took place during the Middle Ages (after *c.* 1300 BP); and the third one occurred after the Middle Ages (after *c.* 400 BP). These phases are based upon palynological data combined with fractionated radiocarbon dating of the well-developed podzolic profiles which separate the units. Van Gijn & Waterbolk (1986) present documentation for the occurrence of fully developed double and even triple podzolic profiles in drift sands in the province of Drenthe, spanning the time interval from the early Bronze Age to the early Medieval period. However, they also came to the conclusion that there was no synchronous occurrence of drift sand sedimentation phases in this region. It, therefore, seems possible to distinguish several chronostratigraphic units in drift sands within a specific, limited area; however, this cannot be justified on the basis of the compiled information from the various investigated regions in northwest Europe. Thus far all available evidence seems to confirm the conclusion of Castel *et al.* (1989); the majority of drift sand formations originated during the early part of the Late Middle Ages, but minor drift sand sedimentation occurred throughout the late Holocene. In all cases, a causal relationship seems to exist between human cultural activities and changing local environmental conditions.

Preliminary results of thermoluminescence dating of cover sands and drift sands by Dijkmans & Wintle (1991) and Dijkmans *et al.* (1992) seem promising with respect to distinguishing the sand accumulation phases themselves, instead of conventional dating of only the stabilization phases, represented by interbedded organic horizons, soil profiles, and peat formations.

Sedimentary processes and facies description

In the central Netherlands, sand is still actively moving in only about 5% of the area covered by drift sands (Fig. 7); in other parts of The Netherlands and adjacent countries this is considerably less. Under present-day climatic conditions and land use, only limited sand transport can be expected in these areas, and natural growth of vegetation tends to stabilize the aeolian sands completely (Castel 1988). Consequently, the mobility of the dunes can only be guaranteed by special management measures such as locally removing all vegetation. Presently, sand transport mainly takes place during sandstorms, which according to the field measurements by De Ploey (1980) from a drift sand area in northwestern Belgium, only occur about 10 times per year. Brugmans (1983) (who studied wind ripple initiation in an active drift sand region of the central Netherlands), reported a similar relatively small number of sandstorms per year. Wind directions during episodes of transport vary from west-northwest to south-southwest (particularly during autumn and winter) and from northeast to east (during spring and summer), but resulting transport directions are mainly from southwest to northeast (Koster 1978; Castel 1988). Remarkably, intense sandstorms also occur during heavy winter rainstorms (minimum wind speeds 7–8 m sec^{-1}) when the dunes are completely wet, and when fine gravel (2–5 mm) may also be transported. Widespread formation of granule ripples (Sharp 1963) has been frequently observed. Similarly, De Ploey (1977, 1980) observed simultaneous movement over the dune surface of a dense basal flow, consisting of fine gravel, coarse sand, fine sand and silt, and an optimum saltation fraction of medium fine to medium coarse sand. If the infiltration capacity of dune sand is exceeded during heavy rainfall, or if the dune surface is covered by runoff from higher ground, the surface sand layers may become saturated and move as a liquified flow (Pye and Tsoar 1990). Contrary to the data available on temperate coastal dunes (Rutin 1983; Jungerius & Van der Meulen 1988), geomorphic processes such as rain wash, rain splash and splash drift have not yet been quantified for inland drift sand areas. However, observations of temperate coastal dunes suggest that these features also play a role in remodelling drift sand surfaces. This is especially true on topsoils with relatively low infiltration rates caused by water repellency in possible combination with enriched organic material or colonization of algae and fungi (Koster 1978; De Ploey 1977; Witter *et al.* 1991).

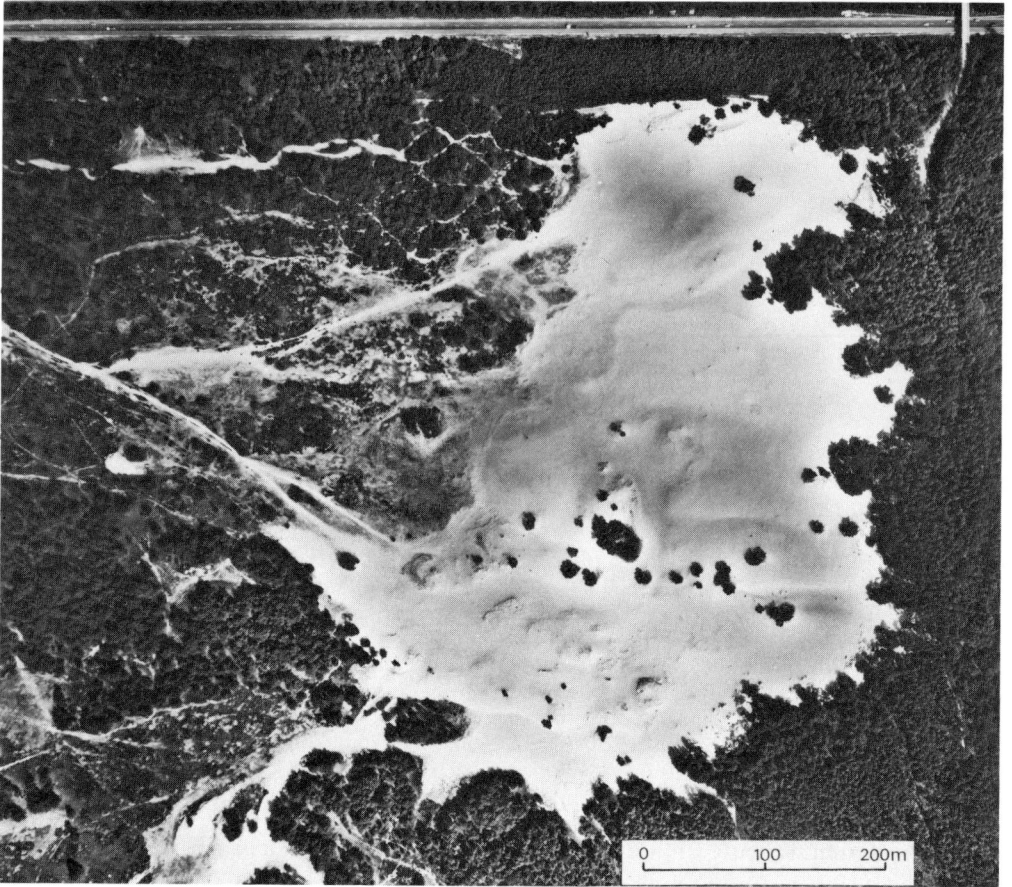

Fig. 7. Active drift sand area, location Hulshorster Zand, Veluwe, The Netherlands. Original scale 1 : 4000 (July 1985).

In order to describe the sedimentary structures in these sands and to relate these structures to processes of sand drifting, more than 20 nitrocellulose lacquer peels have been collected from the windward and leeward slopes and at the crest of various partly active dune types. The collections were made at the following locations (Fig. 2): Kootwijk and Hulshorst (central Netherlands), Geeste and Haken-graben (western Niedersachsen, Germany). The drift sands at Kootwijk and Hulshorst are derived from cover sands, and to a minor extent, from fine, sandy snow meltwater deposits of Weichselian age (Koster 1982). The drift sands at Haken-graben and Geeste originate from Weichselian fluvial deposits along the river Ems as well as aeolian deposits (Schwan 1987). The positions of some of the lacquer peels, which are discussed further in the next section, are indicated in Fig. 8. Drift sands in which sedimentary structures are strongly or completely obliterated by soil formation and bioturbation processes (Ahlbrandt *et al.* 1978) are common but are not discussed here.

The following sedimentary structures are distinguished: (1) primary aeolian structures (terminology as proposed by Hunter (1977*a* and *b*) and Ahlbrandt & Fryberger (1982)); (2) secondary aeolian structures (terminology according to McKee *et al.* (1971) and Kocurek & Fielder (1982)); (3) composite structures (McKee & Bigarella 1972); and (4) non-aeolian structures (McKee & Bigarella 1972; McKee *et al.* 1971).

Kootwijk

Figure 8a illustrates the cross-section of a dome-shaped dune with an almost circular outline, which is partly stabilized by grasses and pine

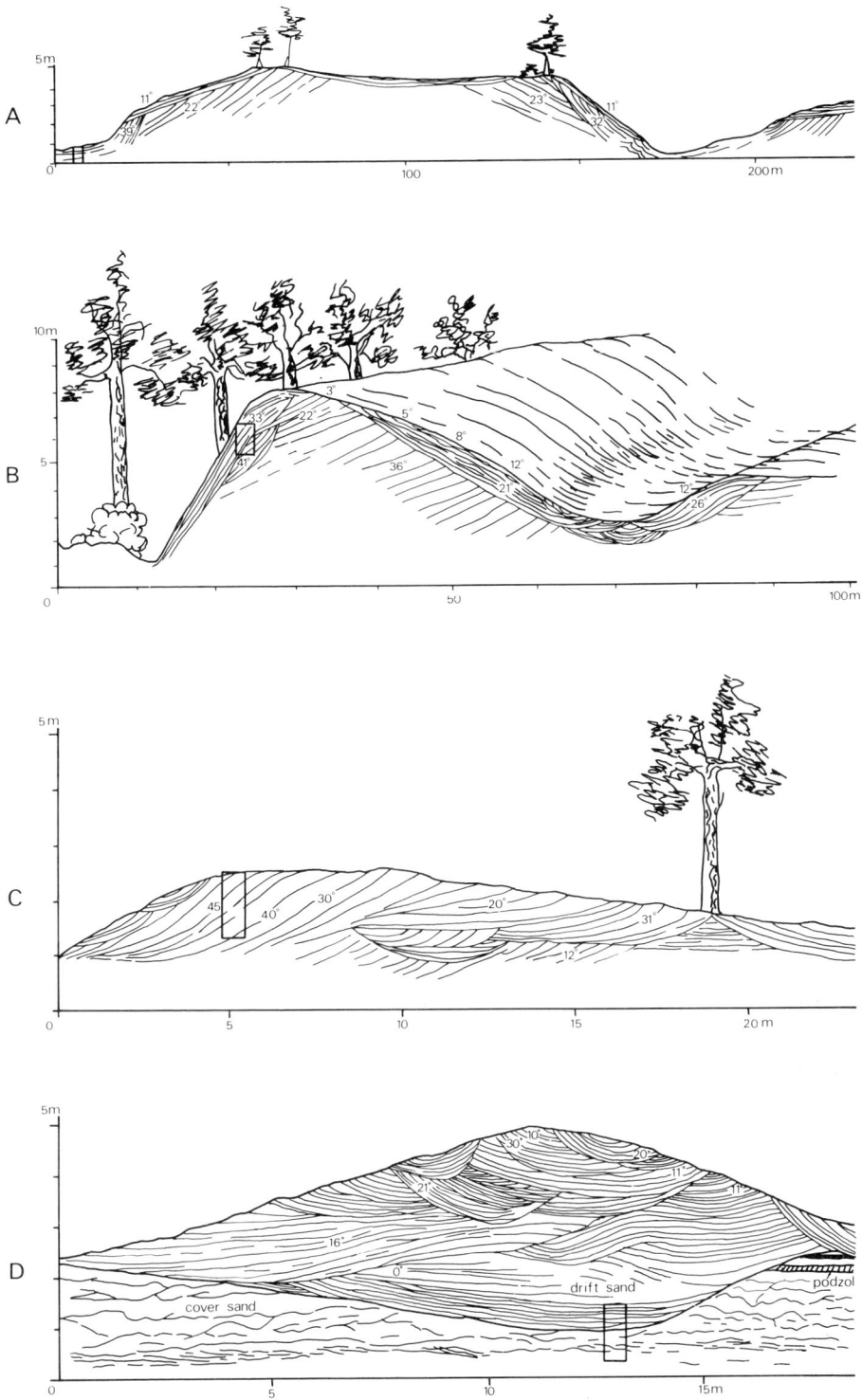

Fig. 8. Cross-sections of dunes showing internal stratification; the positions of the described lacquer peels are indicated. A: location Kootwijk; B: Hulshorst; C: Haken-graben; D: Geeste (for locations see Fig. 2).

trees, and is surrounded by a large, partially active interdune area. This dune type occurs regularly and reflects highly variable sand transport directions (McKee 1966). The top of this dune has recently been deflated and subsequently thick tabular sets with even, parallel lamination have accumulated. Internal structures, with foreset cross-stratification dip directions spreading over 180°, demonstrate that this dune is the remnant of the head of a parabolic dune's initial stage (McKee & Bigarella 1979). The arms have been eroded, the head itself is strongly modified and presently the dune has the appearance of a complex dome dune.

Figure 9 depicts the structures frequently observed in the large, relatively flat interdune areas. The laminae in this lacquer peel are very distinct, caused by a strong grain segregation within the sets, with fine-grained laminae exhibiting positive relief and the somewhat

coarser laminae exhibiting positive relief and the somewhat coarser laminae having negative relief. The following 12 sets have been distinguished:

(1) even to wavy parallel lamination in fine sand;
(2) even, parallel lamination in coarse sand, crinkly (convoluted) laminae at the top;
(3) even, parallel lamination alternating with cross-lamination in fine sand, shallow scours in the middle;
(4) even, parallel lamination in coarse sand and fine gravel, crinkly laminae at the bottom, lower erosive boundary;
(5) even, parallel lamination alternating with crinkly laminae in fine sand, distinct deformation structures at the top;
(6) even, parallel lamination (low lateral continuity) in fine to coarse sand, erosion planes present;
(7) even, parallel lamination (high lateral continuity) in fine sand;
(8) wedge-shaped beds with even, parallel lamination in fine sand, slightly curved erosion planes present, lower boundary erosive, crinkly laminae at the top;
(9) even, parallel lamination in fine sand, distinct deformation structures, fine gravel at the top;
(10) even, parallel lamination alternating with crinkly laminae and contorted lamination near the top in fine sand;
(11) strongly contorted lamination in fine to coarse sand;
(12) even, parallel lamination alternating with crinkly laminae in fine sand, lower boundary is partly erosive.

Even, sometimes wavy, parallel lamination prevails in nearly all sets and is interpreted as the result of sub-critical climbing ripple migration (Hunter 1977a and b). It is thought that wind gusts are responsible for the regular appearance of shallow, composite, scour-fill structures as well as for the sometimes low lateral continuity of the laminae. Erosive boundary planes separate some sets, which reflect different environmental conditions. Thin layers of coarse sand and fine gravel represent deflation lags (top of sets 9 and 11). The very distinct deformation structures (Fig. 10) appear as depressions, primarily filled by coarse sand or even, fine gravel (top of sets 5, 9 and 11). These imprints, which strongly contort some sets, show a bending down of the laminae near the lower cutting planes. These non-aeolian structures are interpreted as foot- or hoof-print structures (Van der Lingen & Andrews 1969; Lewis &

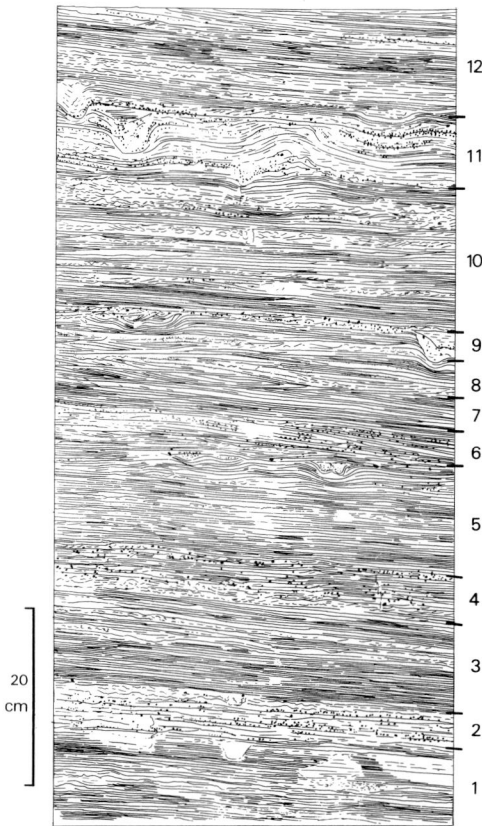

Fig. 9. Sedimentary structures in an interdune area, location Kootwijk. The original drawing was made from a lacquer peel at a scale 1 : 1 (for description see text).

Fig. 10. Lacquer peel showing deformation structures. Detail from the middle part of Fig. 9.

Titheridge 1978). Furthermore, root growth structures which have deformed primary structures are not uncommon in drift sands. The single crinkly (convoluted) laminae represent moist conditions at the accumulation surface such as those that are caused by light rain. The crinkly laminae in combination with coarse sand and fine gravel layers (sets 2, 4 and 11), indicate strong deflation, and are probably related to more severe rain storms. Adhesion ripple structures, observed in other lacquer peels, point to similar conditions (Hunter 1973; Kocurek & Fielder 1982); however, adhesion plane bedforms (Kocurek & Fielder 1982) have not been identified. The low dip angles of the laminae are consistent with the situation in a relatively flat interdune area, where vertical accretion is dominated by deposition of sand, due to migrating sub-critical climbing ripples.

Hulshorst

Figure 8b illustrates a so-called bordering dune ridge (or precipitation ridge), which is a type of transverse dune with a highly irregular front migrating over the edge of a dense forest. The lower windward slope is modified by erosion from an elongated blowout. This blowout is moving almost perpendicular to the dune axis and the top of the windward slope has also been recently eroded. Large-scale cross-stratification on the lee side indicates slow northeasterly migration into the forest. The irregular front of the dune ridge implies dune migration which is

strongly influenced by complex secondary wind flows. The absence of a clear slipface on the lee side results in convex foreset cross-stratification near the dune crest, where the topsets are connected to the steeply dipping foresets. Figure 11 illustrates some details of primary and secondary structures at the upper part of the lee side of this dune. The upper left part of the lacquer peel shows wavy parallel lamination alternating with more or less structureless layers, and is interpreted as a combination of sub-critical ripple migration and grainfall lamination (compare Fryberger & Schenk 1981). In the middle part, a strongly deformed unit occurs with rotated blocks at varying angles, very contorted laminae, and structureless parts. Obviously, this indicates massive slump structures and grainflow structures caused by avalanching on the upper lee side of the dune. McKee *et al.* (1971) and McKee & Bigarella (1972) have also shown that cohesive stratified sand can be deformed to a degree so as to show well-developed fracture planes with little or no bending of the laminae. It is true that the organic matter content of the drift sand is low, but it may just be sufficient to cause some cohesion of the sand. In the lower right part of the peel, parallel lamination prevails alternating with crinkly laminae indicative of moist conditions at the surface during deposition. Other lacquer peels collected in the Hulshorst area frequently display inclined sets up to 15 cm thick with distinct adhesion ripple structures, also indicating wet surface conditions during sand accumulation.

Fig. 11. Lacquer peel from the upper leeward side of a bordering dune ridge, location Hulshorst (for description see text).

Haken-graben

Figure 8c shows the presently inactive remnant of a type of low shadow dune behind an initial sand accumulation around a pine tree. Dune migration has been repeatedly interrupted by erosional phases, which is indicated by large-scale concave erosion planes. Eventually, when dune accumulation became more regular, large-scale, very steeply inclined (up to 45°) cross-strata were produced. Figure 12 illustrates a sequence of four thick sets marked by large-scale cross-stratification. The very distinct laminae all dip in the same direction but show large differences in dip angles (varying from 22° to 45°). Set boundaries are sharp and gradational with the exception of the erosion plane between sets 3 and 4. The lamination density is very high at the gradational boundaries, which makes it difficult to decide whether these boundaries are erosional or not. At the base of sets 2 and 3 wedge-shaped structureless subsets occur, which quickly taper out to the left. Within these subsets a fining-upwards trend is apparent. Many laminae, particularly in the somewhat coarser-grained parts of sets, have a crinkly appearance. An incipient podzolic profile (micropodzol) has developed at the surface during recent decades, destroying the original structures. Based on the characteristics of the laminae, it is suggested that the propagation of this dune is caused mainly by the migration of climbing ripples moving perpendicular to the dip of the lee slope, a situation that was recognized earlier by Sharp (1963). The direction of ripple migration is possibly caused by flow separation of the wind during passage over the dune, converging again on the lee side (Hesp 1981; Clemmensen 1986). Many of these relatively low and small dunes in drift sand areas are dome shaped without well-developed slipfaces, making this kind of propagation plausible. The wedge-shaped structureless features at the base of sets 2 and 3 are interpreted as the result of grainfall deposition. The presence of crinkly laminae as well as the abundance of very steeply dipping laminae probably indicate dune migration under relatively moist environmental conditions. However, with respect to the steeply dipping laminae, which often exceed the angle of repose of these well-sorted and well-rounded sands, the absence of clear avalanche structures is remarkable (Fig. 12). It has been suggested that steep angles of dip up to 46° in aeolian deposits are the result of the addition of very fine particles, high moisture content of the deposit, salt-spray coating, partial cementation, or are the result of sand trapped on vegetated surfaces (McKee &

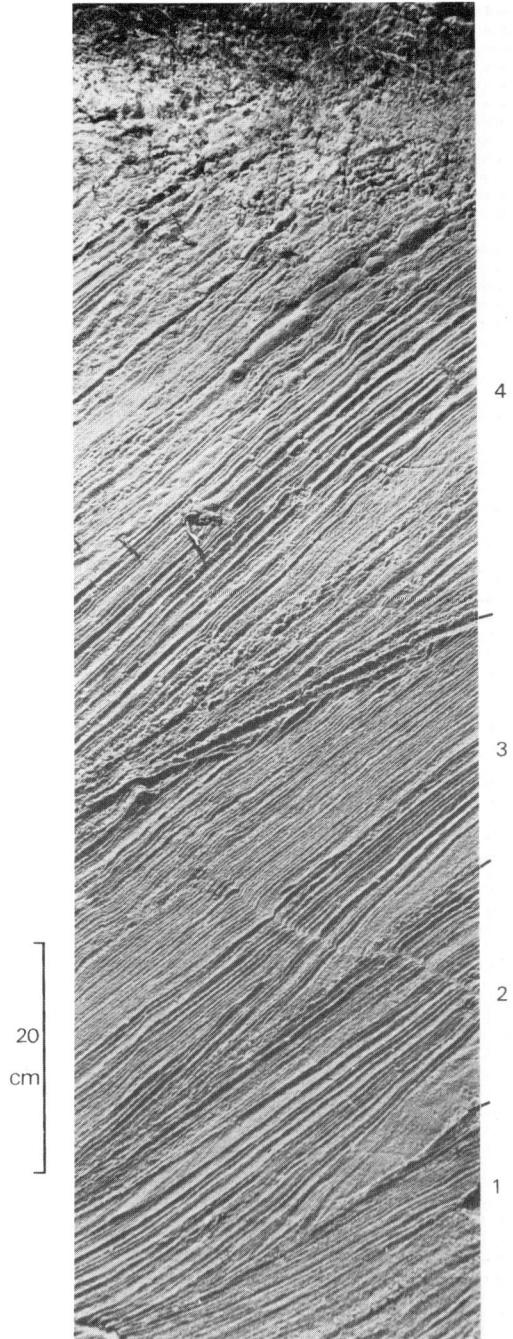

Fig. 12. Lacquer peel illustrating high-angle, large-scale cross-stratification, location Haken-graben (for description see text).

Bigarella 1972; Koster 1978; Hesp 1988; Pye &
Tsoar 1990). Stoutjesdijk (1959) has reported
slope angles of up to 47° on vegetated dune
surfaces. However, in this particular case the
angle of repose has been exceeded under
apparently dry conditions without evidence of
vegetation during deposition. The steeply
inclined laminae show characteristics indicating
the migration of ripples perpendicular to the lee
side of the dune. It appears that this depositional
mechanism can create better grain packing than
other lee side depositional mechanisms such as
grainfall and grainflow. This agrees with the fact
that the packing density of grainfall laminae is
intermediate between loosely packed grainflow
cross-strata and the more tightly packed climbing ripple laminae (Pye & Tsoar 1990).

Geeste

The dune cross-section shown in Fig. 8d is an
example of a former small blowout in cover sand
on top of which a complex dune form has
developed. This form of relief inversion
frequently occurs in drift sand areas. The
position of the original cover sand surface is
shown by the podzolic profile at the right-hand
side of the cross-section. The drift sand layers,
which show a strong admixture of organic
matter, in contrast to the underlying cover sand,
taper out towards both sides where cover sand
occurs at the surface. The distinction between
the brownish, finely-laminated drift sand and
the truncated, yellowish cover sand, in which
structures are strongly disturbed by soil formation, is clearly visible (Fig. 13). Moreover, the
cover sand exhibits a dense grain packing in contrast to the drift sand which is characterized by
a uniform, loose grain packing from the bottom
to the top (Koster 1978). Sedimentary structures
of Weichselian cover sands, which usually
underly the drift sands, have been described and
interpreted by Ruegg (1983) and Schwan (1986,
1987, 1988). Dune structures are dominated by
large-scale trough cross-bedding (composite
structures), becoming more pronounced and
complex towards the top. The vegetation, which
was responsible for catching the drift sand
in the former blowout in cover sand, may have
played an increasingly important role as the
dune grew upwards, by influencing the secondary wind flow (cf. Hesp 1988). The relatively
small, but deep scours in the upper part of the
profile probably represent deflation around and
between clumps of grass or other vegetation. Eventually a period with dominant
erosion changed the overall appearance of
the dune, which is once again stabilized by
vegetation.

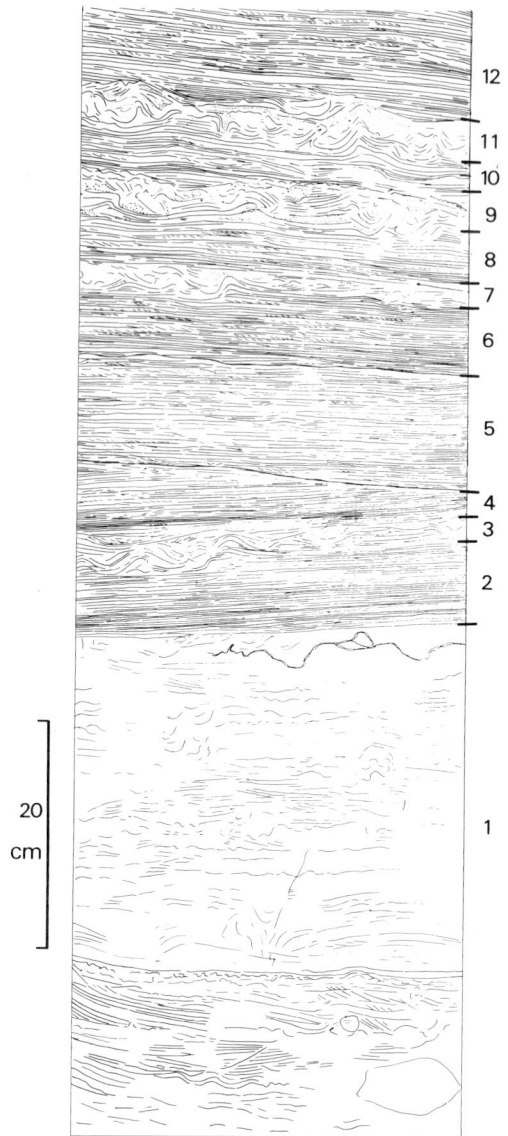

Fig. 13. Sedimentary structures exemplifying the
distinction between cover sand (set 1) and drift sand
(sets 2–12), location Geeste. The original drawing was
made from a lacquer peel at a scale 1 : 1 (for
description see text).

Figure 13 serves to illustrate the internal structures; 12 sets are distinguished:

(1) — structures are strongly deformed by soil
    formation, roots, and burrowing animals;
    remnants of even and wavy parallel lamina-
    tion, cross-lamination, and composite,
    scour-fill structures are still recognizable;
    humus fibres are present near the top, which

indicate that not more than about one metre of cover sand has been removed in this blow-out;

(2) & (4) — even, parallel lamination (high lateral continuity) in fine sand;

(3), (7), (9) & (11)— strongly deformed wavy parallel lamination in moderately coarse sand to fine gravel;

(5), (6), (8), (10) & (12) — even, parallel lamination alternating with very distinct small-scale cross-lamination in fine sand.

Set boundaries are either erosional (e.g. between sets 4, 5 & 6) or gradational, which is particularly evident at the transition from sets 2, 6, 8 & 10 to the overlying strongly deformational structures. The parallel laminated sets represent a regular migration of sub-critical climbing ripples. The common appearance of foreset cross-lamination within the laminae indicates their origin by migrating ripples and implies a uniform wind direction during deposition (from left to right). Convolution structures are interpreted as being caused by liquefaction of the sediment during rain storms. The upward bending of the parallel lamination on top of these deformed sets might indicate that deformation was active during sand deposition. Curiously, no other structures pointing to wet surface conditions, such as adhesion structures, are visible in this case. Under conditions of heavy rainfall on bare drift sand surfaces, especially in interdune areas, it has been observed that the top few centimetres become completely saturated, temporarily blocking further infiltration of water. Consequently, a reversed density stratification resulting from trapped air in the surficial sand layers may occur. This process may explain the origin of convolute lamination in specific cases, possibly similar to that described by De Boer (1979) in intertidal sediments.

Sedimentary microstructures

In order to study particular structures and fabric — defined as the spatial arrangement (orientation and packing) of the clastic components — on a microscale, the use of a scanning electron microscope (SEM) proved very useful. For this analysis samples of c. 2 cm in diameter were cut from leftover strips after framing the lacquer peels. Special attention was given to examples of distinct sub-critical climbing ripple lamination, crinkly (convoluted) laminae and wetted sand crusts. For a first analysis of the texture and structure of the samples a stereo-microscope was used. The main advantages of using the SEM, in contrast to the use of a stereo-microscope or a petrographic microscope, are the large depth of focus possible at high magnifications, the ability to examine relatively large samples at low and high magnification, and the relative simplicity of sample preparation. Samples from several lacquer peels obtained at the locations of Kootwijk, Hulshorst, and Geeste were investigated at magnifications varying from 20× to 1000×.

Sub-critical climbing ripple lamination

Tabular sets with even parallel lamination occur in the lower part of the drift sand at Geeste (Fig. 13). These laminae, which vary in thickness from 0.5–3.0 mm, are supposed to result from deposition by sub-critical climbing ripples. Figure 14a shows three distinct laminae, each marked by the presence of a thin, fine-grained part at the bottom and a relatively thick coarse-grained part at the top. The variation in laminae thickness can be ascribed to the variable thickness of the coarse-grained parts. The fine-grained part of the lamina, as illustrated by Fig. 14b, is very distinct, however. This is partly caused by contamination from the lacquer. The lamina boundaries are fairly sharp; the fine grains are draped on top and partially fill the interstices of the larger grains of the underlying laminae. These fine-grained laminae probably represent the so-called pin stripe lamination described by Fryberger & Schenk (1988). The laminae generally are characterized by sharp boundaries, strong grain segregation and inverse grading, confirming the formation by sub-critical climbing ripples as defined by Hunter (1977a and b) and described in eolian deposits by Fryberger & Schenk (1988).

In Fig. 14c the boundary between the coarse-grained upper part of one lamina and the overlying fine-grained part of the next lamina is somewhat obscured by the infilling of the pores with fine grains between the exceptionally large grains. Nevertheless, these laminae also show inverse grading. Remarkable in this case is the orientation of coarse, as well as fine, grains which are all vaguely dipping to the right, thereby forming micro-cross-stratification. This structure is interpreted as ripple-foreset cross-lamination (Hunter 1977a), indicating ripple migration from left to right.

The upper part of the lacquer peel from Hulshorst (Fig. 11), was typified by alternating wavy parallel lamination with structureless or diffusely laminated layers, interpreted as the result of the combined effect of sub-critical ripple migration and the deposition by grainfall

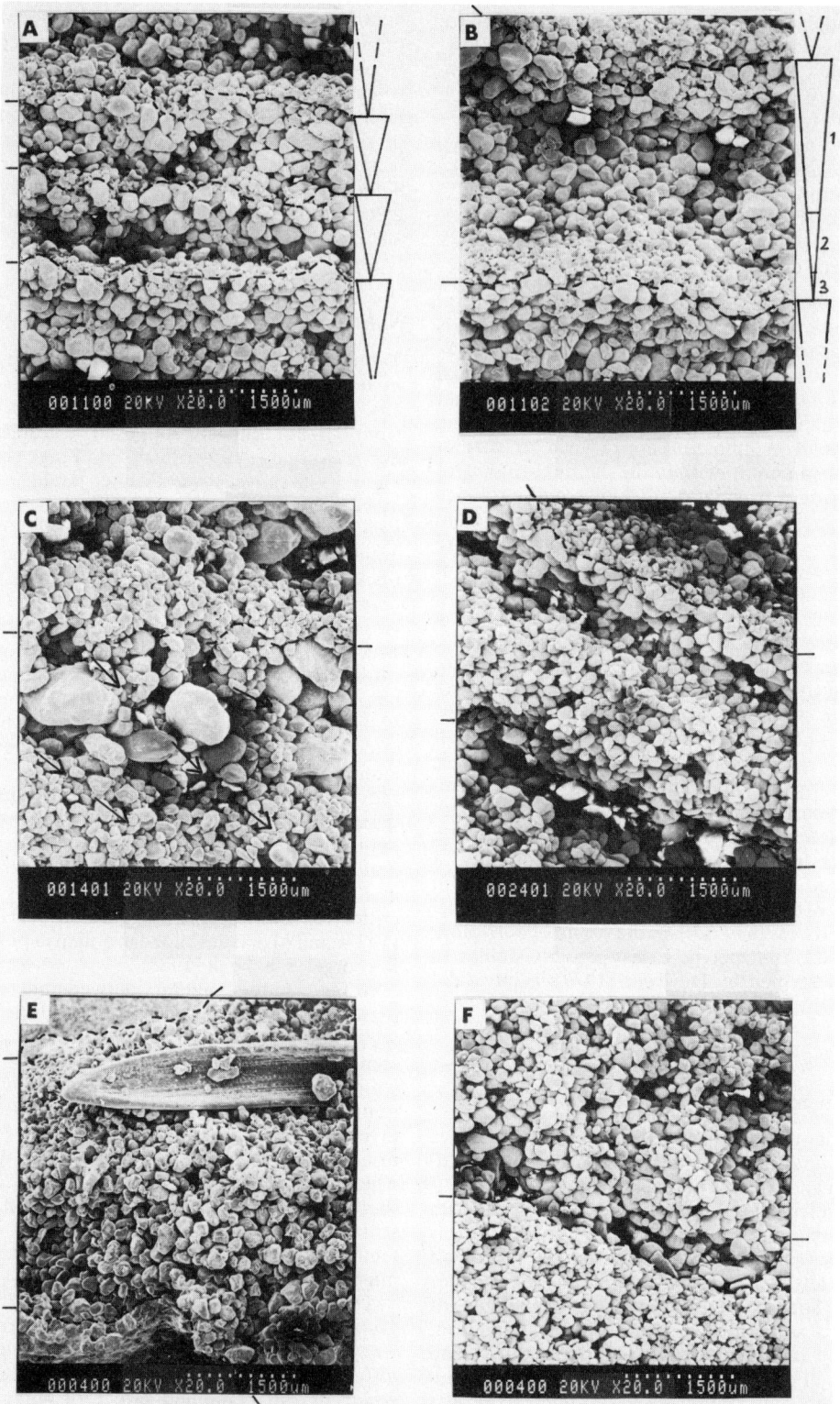

on the lee side of a dune. Figure 14d shows this feature in more detail. The inclined laminae boundaries are wavy and, as in the previously described cases, they separate the coarse-grained part of the laminae from the fine-grained part of the overlying laminae, suggesting inverse grading. However, the main part of the lamina in the middle of the picture consists of moderately fine grains without any grading, indicative of grainfall deposition by the wind blowing over the crest of the dune. Near the laminae boundaries there seems to be a preferential orientation of the coarse grains parallel to the inclined boundaries (tractional deposition).

,rinkly laminae and wetted sand crusts

The samples exemplifying a wetted sand crust (Figs 14e and f) were collected from a lacquer peel taken near the surface in an interdune area at Kootwijk. These pictures represent the many so-called crinkly layers shown in Figs 9 and 10. In the left-hand corner of Fig. 14e, above the pine needle, a rain impact crater at the present surface is just visible. Rain had created a crust about 5 mm thick, which had completely dried out by the time this lacquer peel was taken. The sediment is very homogeneous in grain size, displays a dense grain packing, and shows no preferred orientation of the grains. The impact of rain drops seems to pull the grains in a best-fit position, creating a minimum porosity, which is preserved after drying out. The irregular lower boundary of the crust indicates that the wetting front does not penetrate as a plane surface.

Figure 14f represents a buried sand crust, presently several decimetres below the surface. The rain impact crater at the lower right-hand side is striking and there is sharp contrast between the closely packed, fine-grained crust underneath the irregular boundary and the loosely packed coarse-grained material on top. The sand above the boundary shows indistinct normal grading. The wave boundaries as seen in the lacquer peels are in fact a series of connected depressions formed by raindrop impact. The impact creates a thin veneer of compacted fine

grains. Sand grains dislocated by splash fill up the original depressions. The repetition of this process together with continuing sedimentation during the rainfall event will result in crinkly bands of variable thickness which are left after a rainstorm.

The effect of compaction is particularly important in laminated sand deposited under alternating wet or moist and dry conditions. It is not yet fully understood why wetted sand crusts retain their relatively high degree of compaction after drying (McKee et al. 1971). Wet sand layers or sand layers which were once wet tend to fracture and break as rigid bodies and will move downhill by slumping (cf. Fig. 11) rather than by grainflow (McKee & Bigarella 1972).

Conclusions

Several conclusions can be drawn from the observations and data represented above.

- The beginning of extensive sand drifting in northwestern Europe is related, both in time and space, to the rapid expansion of agriculture, together with the use of plaggen fertilizer and exploitation of heathlands, mainly after AD 950.
- In general, the beginning of sand drifting should not be interpreted as a natural response to the 'climatic optimum' of the Middle Ages; in all cases a causal relationship seems to exist between human cultural activities and changing local environmental conditions.
- In many cases, a prolonged interval of many centuries occurred between the first initiation of sand drifts and the more or less widespread accumulation of drift sands at particular locations.
- The majority of drift sand formations originated during the early part of the Late Middle Ages, but minor drift sand occurrences have taken place throughout the late Holocene. An elaborate dating programme cannot distinguish specific drift sand phases of more than local significance.

Fig. 14. SEM micrographs illustrating specific sedimentary structures in drift sands. (**a**) and (**b**) demonstrate inverse grading (shown by triangles — 1: 210–420 μm; 2: 105–210 μm; 3: 50–105 μm) in sub-critical climbing ripple lamination, location Geeste; (**c**) shows sub-critical climbing ripple lamination with internal foreset structures (see arrows), location Kootwijk; (**d**): combination of sub-critical climbing ripple lamination and grainfall lamination, location Hulshorst; (**e**) and (**f**): normal grading in wetted sand crust (pine needle near the top) and rain impact structure (depression outlined at lower right of (**f**), location Kootwijk (explanation see text). The size of particular features can be inferred from the scale bar at the lower right of the figures (length of the dotted line = 1500 μm).

- Apart from dune foreset cross-bedding in individual dunes, large parts of the relatively low-relief drift sand regions are characterized by evenly or wavy (sub)horizontal lamination.
- Primary structures resulting from tractional deposition (mainly sub-critical climbing ripple migration) and grainfall deposition, secondary structures (mainly slump, scour-fill and adhesion structures) and non-aeolian structures (mainly caused by rain impact, convolution structures, wetted sand crusts, and foot- or hoof-imprints) predominate in the drift sands.
- The study of sedimentary structures in very small remnants of still actively moving (temperate) drift sands has confirmed that sub-recent processes of deflation and accumulation take place under dry, moist, and wet conditions.

Professor E. Derbyshire (University of Leicester) is thanked for placing SEM facilities at our disposal and for his instruction of the third author. We also thank Dr W. Eisner (Byrd Polar Research Center, Ohio), Dr J. H. Van den Berg (University of Utrecht), Dr P. Wilson (University of Ulster) and an unknown referee for their comments and for correcting the English text.

References

AHLBRANDT, T. S., ANDREWS, S. H. & GWYNNE, D. T. 1978. Bioturbation in eolian sediments. *Journal of Sedimentary Petrology*, **48**, 839–848.

—— & FRYBERGER, S. G. (1982). Introduction to eolian deposits. *In*: SCHOLLE, P. A. & SPEARING, D. (eds) *Sandstone Depositional Environments*. American Association of Petroleum Geologists, Tulsa, USA, 11–47.

BRUGMANS, F. 1983. Wind ripples in an active drift sand area in The Netherlands: a preliminary report. *Earth Surface Processes and Landforms*, **8**, 527–534.

CASTEL, I. I. Y. 1988. A simulation model of wind erosion and sedimentation as a basis for management of a drift sand area in The Netherlands. *Earth Surface Processes and Landforms*, **13**, 501–509.

—— 1991a. *Late Holocene Eolian Drift Sands in Drenthe (The Netherlands)*. Thesis, University of Utrecht, Netherlands Geographical Studies 133.

—— 1991b. Micromorphology of the transition peat-Holocene drift sand deposits in the northern Netherlands. *Zeitschrift für Geomorphologie Supplement Band*, **90**, 29–43.

——, KOSTER, E. A. & SLOTBOOM, R. T. 1989. Morphogenetic aspects and age of Late Holocene eolian drift sands in Northwest Europe. *Zeitschrift für Geomorphologie*, **33**, 1–26.

CLEMMENSEN, L. B. 1986. Storm-generated eolian sand shadows and their sedimentary structures, Vejers strand, Denmark. *Journal of Sedimentary Petrology*. **56**, 520–527.

DE BOER, P. L. 1979. Convolute lamination in modern sands of the estuary of the Oosterschelde, the Netherlands, formed as the result of entrapped air. *Sedimentology*, **26**, 283–294.

DE PLOEY, J. 1977. Some experimental data on slopewash and wind action with reference to Quaternary morphogenesis in Belgium. *Earth Surface Processes*, **2**, 101–115.

—— 1980. Some field measurements and experimental data on wind-blown sands. *In*: DE BOODT, M. & GABRIELS, D. (eds) *Assessment of Erosion*, Wiley, Chichester, 541–552.

DIJKMANS, J. W. A., VAN MOURIK, J. M. & WINTLE, A. G. 1992. Thermoluminescence dating of aeolian sands from polycyclic soil profiles in the southern Netherlands. *Quaternary Science Reviews*, **11**, 85–92.

—— & WINTLE, A. G. 1991. Methodological problems in thermoluminescence dating of Weichselian cover sand and late Holocene drift sand from the Lutterzand area, E. Netherlands. *Geologie en Mijnbouw*, **70**, 21–33.

FOLLAND, C. K., KARL, T. R. & VINNIKOV, K. Ya. 1990. Observed climate variations and change. *In*: HOUGHTON, J. T., JENKINS, G. J. & EPHRAUMS, J. J. (eds) *Climate Change the IPCC Scientific Assessment*. Cambridge University Press, Cambridge, 195–238.

FRYBERGER, S. G., AHLBRANDT, T. S. & ANDREWS, S. 1979. Origin, sedimentary features, and significance of low-angle eolian 'sand sheet' deposits, Great Sand Dunes National Monument and vicinity, Colorado. *Journal of Sedimentary Petrology*, **49**, 733–746.

—— & SCHENK, C. J. 1981. Wind sedimentation tunnel experiments on the origins of aeolian strata. *Sedimentology*, **28**, 805–821.

—— & —— 1988. Pin strip lamination: a distinctive feature of modern and ancient eolian sediments. *Sedimentary Geology*, **55**, 1–15.

HEIDINGA, H. A. 1984. Indications of severe drought during the 10th century AD from an inland dune area in the Central Netherlands. *Geologie en Mijnbouw*, **63**, 241–248.

HESP, P. A. 1981. The formation of shadow dunes. *Journal of Sedimentary Petrology*, **51**, 101–112.

—— 1988. Morphology, dynamics and internal stratification of some established foredunes in southeast Australia. *Sedimentary Geology*, **55**, 17–41.

HUNTER, R. E. 1973. Pseudo-crosslamination formed by climbing adhesion ripples. *Journal of Sedimentary Petrology*, **43**, 1125–1127.

—— 1977a. Basic types of stratification in small eolian dunes. *Sedimentology*, **24**, 361–387.

—— 1977b. Terminology of cross-stratified sedimentary layers and climbing-ripple structures. *Journal of Sedimentary Petrology*, **47**, 697–706.

JUNGERIUS, P. D. & VAN DER MEULEN, F. 1988. Erosion processes in a dune landscape along the Dutch coast. *Catena*, **15**, 217–228.

KOCUREK, G. & FIELDER, G. 1982. Adhesion structures. *Journal of Sedimentary Petrology*, **52**, 1229–1241.

KOSTER, E. A. 1978. *De stuifzanden van de Veluwe: een fysischgeografische studie. (The eolian drift sands of the Veluwe (Central Netherlands): a physical geographical study).* Thesis, University of Amsterdam, Publicaties van het Fysisch Geografisch en Bodemkundig Laboratorium van de Universiteit van Amsterdam 27.

—— 1982. Terminology and lithostratigraphic division of (surficial) sandy eolian deposits in The Netherlands: an evaluation. *Geologie en Mijnbouw*, **61**, 121–129.

—— 1988. Ancient and modern cold-climate aeolian sand deposition: a review. *Journal of Quaternary Science*, **3**, 69–83.

LEWIS, D. E. & TITHERIDGE, D. G. 1978. Small scale sedimentary structures resulting from foot impressions in dune sands. *Journal of Sedimentary Petrology*, **48**, 835–838.

MCKEE, E. D. 1966. Structures of dunes at White Sands National Monument, New Mexico (and a comparison with structures of dunes from other selected areas). *Sedimentology*, **7**, 1–69.

—— & BIGARELLA, J. J. 1972. Deformational structures in Brazilian coastal dunes. *Journal of Sedimentary Petrology*, **42**, 670–681.

—— & —— 1979. Sedimentary structures in dunes. With sections on the Lagoa Dune Field, Brazil. *In*: MCKEE, E. D. (ed.) *A Study of Global Sand Seas.* US Geological Survey Professional Paper **1052**, 83–134.

——, DOUGLASS, J. R. & RITTENHOUSE, S. 1971. Deformation of lee-side laminae in eolian dunes. *Geological Society of America Bulletin*, **82**, 359–378.

MEYER, H. H. 1984. Jungdünen und Wehsande aus historischer Zeit im Gebiet nördlich des Dümmers. *Oldenburger Jahrbuch*, **84**, 403–436.

MOOK, W. G. 1983. [14]C calibration curves depending on sample time-width. *PACT*, **8**, 517–525.

NIESSEN, A. C. H. M., KOSTER, E. A. & GALLOWAY, J. P. 1984. *Periglacial sand dunes and eolian sand sheets. An annotated bibliography.* US Geological Survey Open-File Report 84–167.

PAPE, J. C. 1970. Plaggen soils in The Netherlands. *Geoderma* **4**, 229–255.

PYE, K. & TSOAR, H. 1990. Aeolian Sand and Sand Dunes. Unwin Hyman, London.

PYRITZ, E. 1972. *Binnendünen und Flugsandebenen im Niedersächsischen Tiefland.* Göttinger Geographische Abhandlungen 61.

RUEGG, G. H. J. 1983. Periglacial eolian evenly laminated sandy deposits in the Late Pleistocene of NW Europe, a facies unrecorded in modern sedimentological handbooks. *In*: BROOKFIELD, M. E. & AHLBRANDT, T. S. (eds) *Eolian Sediments and Processes.* Developments in Sedimentology **38**, Elsevier, Amsterdam, 455–482.

RUTIN, J. 1983. *Erosional processes on a coastal sand dune. De Blink, Noordwijkerhout, The Netherlands.* Thesis, University of Amsterdam, Publicaties van het Fysisch Geografisch en Bodemkundig Laboratorium van de Universiteit van Amsterdam 35.

SCHELLING, J. 1957. Herkunft, Aufbau und Bewertung der Flugsande im Binnenlande. *Erdkunde*, **11**, 129–135.

SCHWAN, J. 1986. The origin of horizontal alternating bedding in Weichselian aeolian sands in northwestern Europe. *Sedimentary Geology*, **49**, 73–108.

—— 1987. Sedimentologic characteristics of a fluvial to aeolian succession in Weichselian Talsand in the Emsland (F.R.G.). *Sedimentary Geology*, **52**, 273–298.

—— 1988. The structure and genesis of Weichselian to Early Holocene aeolian sand sheets in western Europe. *Sedimentary Geology*, **55**, 197–232.

SHARP, R. P. 1963. Wind ripples. *Journal of Geology*, **71**, 617–636.

STOUTJESDIJK, Ph. 1959. *Heaths and inland dunes of the Veluwe. A study on some of the relations existing between soil, vegetation and microclimate.* Thesis, University of Utrecht, Wentia 5, 1–96.

VAN DER BORG, K., ALDERLIESTEN, C., HOUSTON, C. M., DE JONG, A. F. M. & VAN ZWOL, N. A. 1987. Accelerator mass spectrometry with ^{14}C and [10]Be in Utrecht. *Nuclear Instruments and Methods in Physics Research*, **B29**, 143–145.

VAN DER LINGEN, G. J. & ANDREWS, P. A. 1969. Hoofprint structures in beach sand. *Journal of Sedimentary Geology*, **39**, 350–357.

VAN DER PLICHT, J. & MOOK, W. G. 1990. Calibration of radiocarbon ages by computer. *Radiocarbon*, **31**, 805–816.

VAN GIJN, A. L. & WATERBOLK, H. T. 1986. The colonization of the salt marches of Friesland and Groningen: the possibility of a transhumant prelude. *Palaeohistoria (Acta et Communicationes Instituti Bio-Archaeologici Universitatis Groninganae)*, **26**, Balkema, Rotterdam, 101–122.

VAN MOURIK, J. M. (ed.) 1988. *Landschap in beweging: Ontwikkeling en bewoning van een stuifzandlandschap in de Kempen.* Netherlands Geographical Studies 74.

—— 1991. Zandverstuivingen en plaggenlandbouw. Het bodemarchief van de Peelterbaan. *Historisch Geografisch Tijdschrift*, **3**, 88–95.

—— & ODÉ, B. 1990. Her Herperduin. *Geografisch Tijdschrift*, **24**, 160–167.

WILLIAMS, R. B. G. 1975. The British climate during the last Glaciation; an interpretation based on periglacial phenomena. *In*: WRIGHT, A. E. & MOSELEY, F. (eds) *Ice Ages: Ancient and Modern.* Geological Journal Special Issue **6**, 95–120.

WITTER, J. V., JUNGERIUS, P. D. & TEN HARKEL, M. J. 1991. Modelling water erosion and the impact of water repellency. *Catena*, **18**, 115–124.

Climbing and falling sand dunes in Finnish Lapland

MATTI SEPPÄLÄ

Department of Geography, University of Helsinki,
Hallituskatu 11, SF-00100 Helsinki, Finland

Abstract: This paper presents the first description of climbing and falling sand dunes from Finnish Lapland. They are located on the slopes of a glaciated valley containing an esker. The largest climbing dune is 1800 m long and 2–5 m high. The slope of the fell dips 7 degrees on average. The material comprising the dune is typical aeolian sand with a median grain size about 0.23 mm but not very well sorted. The dunes were formed just after deglaciation under periglacial conditions.

Sand dunes are not exceptional features in regions which have experienced glaciation in the Quaternary (e.g. Seppälä 1971, 1972; Aartolahti 1973, 1976; Sollid *et al.* 1973; David 1977; Bergqvist 1981). Deglaciation produced great amounts of sorted glaciofluvial deposits suitable for aeolian transportation (Seppälä 1980).

Most sand dunes in Finnish Lapland are parabolic in shape. Some transverse and longitudinal inland dunes have also been discovered. Normally they are located on rather horizontal, flat surfaces. Falling and climbing sand dunes have not previously been reported from Finnish Lapland, although there is one reported observation of a small sand dune of this type climbing up from the Oulanka river valley, just south of the Arctic Circle, in eastern Finland, which was filled by a valley train during the deglaciation

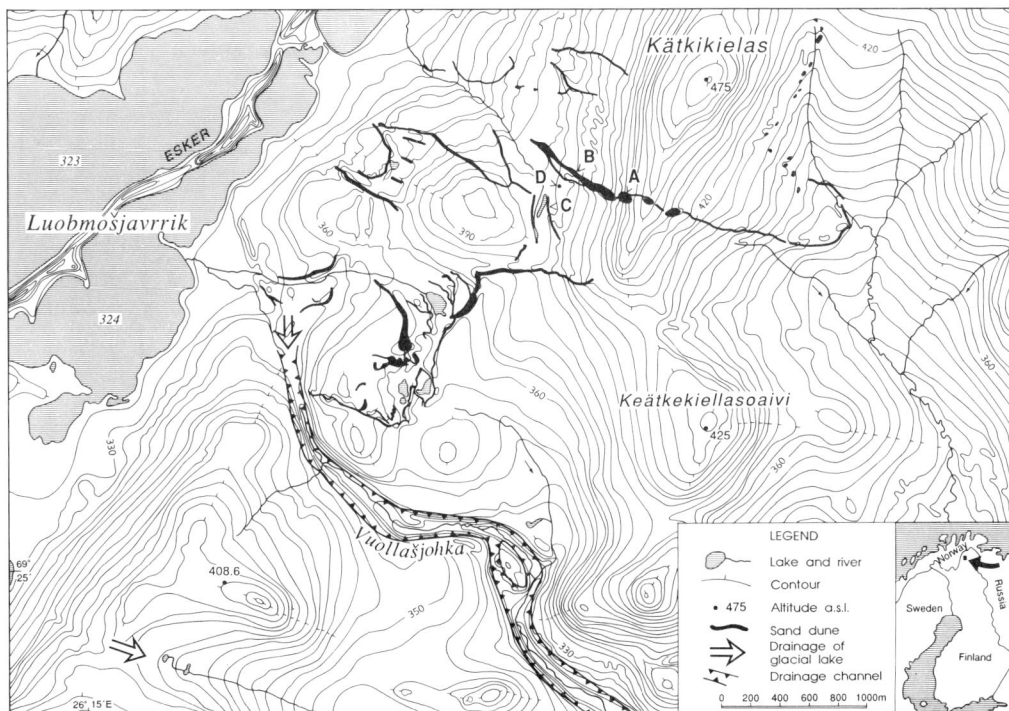

Fig. 1. Topography of the studied region redrawn from the topographic map of Finland nr. 3913 04. Contour interval 5 m. Sand dunes interpreted from the aerial photographs 61170 : 270–274. A, B, C and D are sampling sites (Table 1).

From Pye, K. (ed.), 1993, *The Dynamics and Environmental Context of Aeolian Sedimentary Systems*. Geological Society Special Publication No. 72, pp. 269–274.

(Koutaniemi 1979). Most previously published reports of climbing dunes come from warm desert environments (e.g. Evans 1962; Pye & Tsoar 1990).

This paper is based on an unpublished larger report about deflation in Luobmošjavrrik area (Seppälä 1964).

Study area

The studied area is located at 69° 27'N 26°22'E in Finnish Lapland about 80 km south of the Tromsø-Lyngen ice margin of the Younger Dryas substage (Sollid *et al*. 1973, pp. 301–304) and about 160 km north of the last ice divide located at about latitude 68°N (see e.g. Hirvas 1991). According to Ruuhijärvi (Radiocarbon 7 1965, p. 3), the region became deglaciated about 9800 years ago. He obtained a date of 9800 ± 250 ^{14}C years BP from some mire bottom peat at Petsikko, located about 35 km east of the sand dunes, and Seppälä (1971) dated similar material with an age 9740 ± 220 ^{14}C years BP from Suttisjoki valley, located about 25 km SE of the studied dune region.

Topographically, the study area forms a wider continuation of a 50 km long, narrow and some 200 m deep fault-influenced valley of the Kevojoki river, oriented in the direction ENE–WSW. At its southern end the valley is filled by the Luobmošjavrrik lakes which are crossed by a 12 km long and up to 30 m high mid-valley esker running in a SW to NE direction (Figs 1 & 2). Hills rise on both sides of the valley to a height of 100–200 m above the lake level which is 323 m above sea-level. The highest summit is the Ruohtir fell, 552 m above sea-level, located 3 km NE of the sand dunes.

During the deglaciation, much glaciofluvial sand and gravel was deposited in the valley which was occupied by an ice-dammed lake. Its level, as recorded by the glaciofluvial delta terrace found on the western side of the valley and the deep Vuollašjohka overflow channel (Fig. 1) facing towards the SE through a saddle at an elevation of 343 m, was some 20 m above the present lake level. Meltwaters from the Luobmošjavrrik valley could have drained through this channel for a short time during the deglaciation.

The bare sand deposited in the valley was then drifted by wind and deposited as rather unique sand dunes on the eastern slope of the valley (Fig. 3), overlying slopes covered by tills and summital block fields formed mainly of granulite.

The aim of this paper is to describe these periglacial sand dunes and to discuss the conditions under which they formed.

Fig. 2. General view of the valley with sand dunes and blow-outs. Photographed by the author.

Fig. 3. Part of the northern slope of the longest climbing sand dune on the Kätkikielas slope with large blow-out on the back. Photographed by the author.

Morphology of the sand dunes

North of the previously mentioned overflow channel on the slope occur several sand dune ridges (Fig. 1). The southern dunes are parabolic in shape. The ridges are 400–1200 m long and from 2–4 m high. Intervals between ridges are about 130 m. The ridges dam several small ponds and are bordered by mires and block fields. The southeastern slopes are steeper than the west-facing slopes. In some trench exposures the dip of the sand strata was observed to be 25° to the SE. The altitudinal difference between the lowest and highest parts of the long ridge is 50 m. Some of the dunes are so eroded by subsequent deflation, rain and melt-waters running down the slopes that the original shape is impossible to determine. In some places residuals only about 1 m high are left.

North of this dune area there is a longitudinal dune winding up the slope of Kätkikielas and crossing this fell at an elevation of about 450 m above sea-level. The dune is about 1800 m long and from 2 to 5 m high with a maximum width of up to 100 m on the western slope of the fell. It climbs some 100 m up from the valley floor (Figs 1 & 4) and then falls through about 90 m after crossing the fell. It has also been partly destroyed by erosion and deflation (Figs 2, 3 & 5). On the western slope of the fell the sand dunes are bordered by coarser sand layers, but on the eastern slope the sand dune margin comprises block fields and till. The strata dip towards both sides of the ridge (Fig. 6). East of this long climbing and falling dune we find a parabolic dune which falls down the slope towards the east. Parts of it are still 4–5 m high.

North of the long dune ridge on the western slope of Kätkikielas occur two more climbing sand dunes, each 400–500 m long and 3–4 m high (Fig. 1). On the eastern slope are some small amounts of aeolian sand fill, the upper ends of several former meltwater channels on the slope. They are formed on the lee side of the fell as downwind accumulations. Insufficient sand has been available to produce falling dunes in this case. To the east of these deposits are two discontinuous curving sand ridges up to 6 m high. They have been interpreted as poorly developed parabolic dunes.

Dune sediments

The dune ridges are formed of fine sand. Sample A in Table 1 provides a typical example of the material found in the large deflation basin on the slope (Figs 3 & 4). The median grain size of sample A is 0.24 mm and it is rather well sorted (sorting coefficient (So) = 1.39).

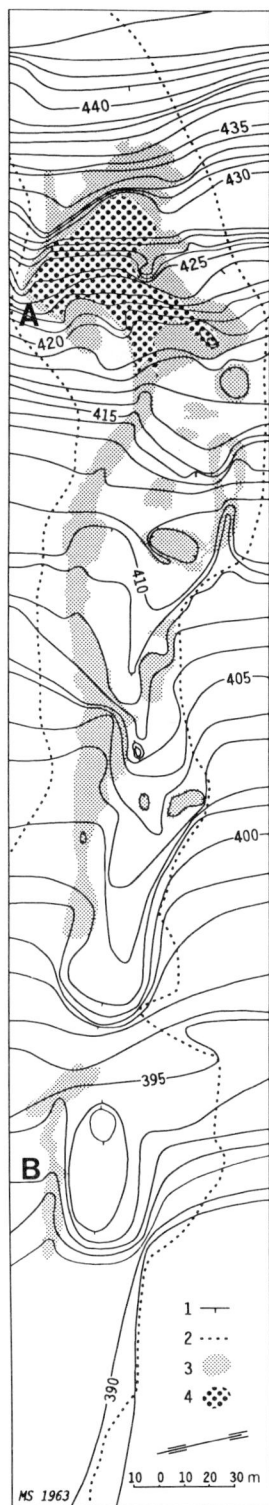

Sample B was collected from the lower part of the climbing dune and it represents a rather fine layer of material with a median grain size of 0.13 mm and poorer sorting (So = 1.58).

Sample C (Fig. 1) represents a coarser sand layer found between fine sands similar to those in sample A. The median grain size is 0.38 mm and sorting coefficient is 1.56.

Sample D (Fig. 1) was taken from the depression between the sand dunes. It is very similar to sample A and could be the source material for the dunes.

The grain size of the samples analysed is typical of Lappish sand dunes (cf. Seppälä 1971) but is not as well sorted in these climbing dunes as usual. This might mean that deposition took place quickly and that the material was transported only short distances by the wind. The esker in the middle of the valley is partly formed of similar fine sand and the sand sources could not be further away. The maximum distance for sand drift has, therefore, been less than 3 km (Fig. 1).

Discussion and conclusions

The lowest level with sand dunes in Luobmuš-javrrik valley is above the highest level of the former ice-dammed lake at about 345–350 m above sea-level. It is postulated that from here strong winds transported sorted glaciofluvial sand up the slope and built ridges.

The orientation of the climbing and falling dunes indicates that the effective direction of sand transporting winds was NW–WNW, which corresponds very well with the conclusions of former studies from Fennoscandia (Seppälä 1971, 1972, 1973; Sollid et al. 1973; Aartolahti 1976), which interpreted wind directions from the orientation of parabolic dune axes.

Wind velocities were very strong because they formed longitudinal ridges climbing up the slope with an average dip of 7° and dunes also exist far beyond the summit on the lee side of the fell.

In one place, post-depositional dislocations were observed in the stratified aeolian sand which can be explained by sand accumulation on ice, with subsequent burying of some ice, as in

Fig. 4. Part of the longest climbing sand dune on the slope of Kätkikielas fell (See A and B in Fig. 1.). Key: 1, contours with 1 m interval; 2, edge of the sand dune; 3, blow-outs in the sand; 4, deflation basin with stones at its bottom (cf. Fig. 3). The map is based on field measurements made with a Kern tachymeter (Kipregel, 1963).

Fig. 5. Deflated southern slope of the longest climbing sand dune at its lower course with typical birch forest. Photographed by the author.

many kames. No organic layers have been found under the dunes which means that they were formed just after the deglaciation or during it. Thus, the environment was unvegetated during the deposition of the climbing dunes. Later, at the end of the period of dune formation, the dunes were partly covered by vegetation and a parabolic shape was established.

Fig. 6. Characteristic stratified material in the climbing sand dune. The shovel is 50 cm long. Photographed by the author.

Table 1. *Grain size distribution of selected samples in weight per cent, grain size in mm (Q_{75}, Md and Q_{25}), sorting (So), skewness (Sk), and kurtosis (K): (in moment measures)*

Sample	<0.06	<0.2	<0.6	0.6–2	Q_{75}	Md	Q_{25}	So	Sk	K
A	2.4	51.2	46.4	–	0.33	0.24	0.17	1.39	0.97	0.28
B	18.3	73.5	8.2	–	0.20	0.13	0.08	1.58	0.95	0.26
C	1.6	26.0	71.5	0.9	0.56	0.38	0.23	1.56	0.89	0.28
D	3.2	64.4	32.4	–	0.34	0.23	0.16	1.46	1.03	0.22

At present, deflation and erosion is destroying the dunes. The present forest limit is about 360–370 m above sea-level in this region (Fig. 1), and although deflation is most effective above this limit it also occurs at lower levels because the vegetation is very scattered and the birch trees are usually less than 6 m high. On the slope of Kätkikielas fell the largest blow-out has dimensions of 90 m by 120 m and its edges are up to 3 m high. Its floor is covered by big stones and blocks showing that the deposition of aeolian sand took place outside of the source area.

In the field I was assisted by Mr Mikko Harri. Final drawings of the figures were made by Mrs Kirsti Lehto and the photographs printed by Miss Leena Heiskanen. Dr Kenneth Pye kindly revised the English of the manuscript.

References

AARTOLAHTI, T. 1973. Morphology, vegetation and development of Rokuanvaara, an esker and dune complex in Finland. *Fennia*, **127**, 1–53.

—— 1976. Lentohiekkaa Suomessa (Aeolian sand in Finland.) (In Finnish only.) *Suomalainen Tiedeakatemia, Esitelmät ja pöytäkirjat*, 83–95.

BERGQVIST, E. 1981. Svenska inlandsdyner. Översikt och förslag till dynreservat. (Ancient inland dunes in Sweden. Survey and proposals for dune reserves.) *Statens naturvårdsverk Rapport* (The national Swedish Environmental Protection Board), **PM 1412**, 1–109.

DAVID, P. P. 1977. *Sand dune occurences of Canada.* Indian and Northern Affairs, National Parks Branch, Contract 74–230, 1–183.

EVANS, J. R. 1962. Falling and climbing sand dunes in the Cronese ("Cat") Mountain area, San Bernardino County, California. *Journal of Geology*, **70**, 107–113.

HIRVAS, H. 1991. Pleistocene stratigraphy of Finnish Lapland. *Geological Survey of Finland, Bulletin*, **354**, 1–123.

KOUTANIEMI, L. 1979. Late-glacial and post-glacial development of the valleys of the Oulanka river basin, North-eastern Finland. *Fennia*, **157**, 13–73.

PYE, K. & TSOAR, H. 1990. *Aeolian Sand and Sand Dunes*. Unwin Hyman, London.

RADIOCARBON 7. 1965. *The American Journal of Science*.

SEPPÄLÄ, M. 1964. *On the wind erosion in Luomushjärvi region, Utsjoki*. (In Finnish only.) Unpublished BSc thesis. Department of Geography, University of Turku.

—— 1971. Evolution of eolian relief of the Kaamasjoki-Kiellajoki river basin in Finnish Lapland. *Fennia*, **104**, 1–88.

—— 1972. Location, morphology and orientation of inland dunes in northern Sweden. *Geografiska Annaler*, **54A**, 85–104.

—— 1973. On the formation of periglacial sand dunes in northern Fennoscandia. *Ninth Congress International Union for Quaternary Research, Abstracts, Christchurch*, 318–319.

—— 1980. Deglaciation and glacial lake development in the Kaamasjoki river basin, Finnish Lapland. *Boreas*, **9**, 311–319.

SOLLID, J. L., ANDERSEN, S., HARME, N., KJELDSEN, O., SALVIGSEN, O., STUROD, S., TUVEITÄ, T. & WILHELMSEN, A. 1973. Deglaciation of Finnmark, North Norway. *Norsk Geografisk Tidsskrift*, **27**, 233–325.

Dust and loess

Contrasting origin and character of Pleistocene and Holocene dust falls on the Canary Islands and southern Morocco: genetic and climatic significance

GENEVIÈVE COUDÉ-GAUSSEN[1] & PIERRE ROGNON[2]

[1] *Centre de Recherches en Géographie Physique de l'Environnement, Université de Caen, France*

[2] *Département de Géodynamique des Milieux Continentaux, Université P. & M. Curie, Paris, France*

Abstract: Various surficial calcitizations and laminar calcretes overlie non-carbonate basement rocks in southern Morocoo (Ifni and Anti-Atlas areas) and basalt outcrops or biodetrital sands in the eastern Canary Islands (mainly Fuerteventura). Geomorphological, sedimentological and micromorphological data provide evidence for their origin by deposition of Pleistocene calcitic dust and syn- to post-depositional colluvial reworking. These dusts were derived from the comminution of marine and local biodetrital sands mobilized by the westerlies on the emergent offshore shelves during times of Pleistocene low sea-level.

An opposite aeolian system related to Holocene and present-day Saharan dusts explains the formation of silts resting on the slopes of Fuerteventura and the fine fraction of the colluvial soils of southern Morocco. These deposits often lie unconformably on the calcitic bodies and their sedimentological and micromorphological characters are similar to those of the modern Saharan dusts. This suggests opposite major climatic conditions for the two aeolian dust systems: the calcitic dust system was controlled by a southerly shift of the mid-latitude circulation in the Pleistocene; the later Saharan dust contribution seems to be related to the re-establishment of the trade wind circulation and Saharan air outbreaks, particularly in the early Holocene when the continuance of blocking situations in the mid-latitude Atlantic may have been notable.

Currently, general aspects of the production, mobilization, transport and sedimentation of aeolian dusts are well-documented and a recent review of the literature addressing these aspects can be found in Pye (1987). One of the most important problems concerning the dust contribution to soils and surficial deposits is the characterization of the aeolian input by pedological, geomorphological, stratigraphical and sedimentological criteria (Coudé-Gaussen 1989a; Coudé-Gaussen 1991). Another main question concerns the environmental conditions of the dust deposition and their general climatic significance.

An attempt to evaluate such a dust contribution has been made during several field trips to the eastern Canary Islands (Fuerteventura and Lanzarote) and southern Morocoo (Ifni and Anti-Atlas areas), where calcitic coatings on rocks and siliceous silty colluvial soils have been observed (Fig. 1). Saharan air outbreaks and siliceous dust falls are frequent today in these areas (Coudé-Gaussen *et al.* 1987; Bergametti *et al.* 1989), and so it has been relatively easy to compare these desert dusts with the silty soils and rock coatings.

The calcitic coatings lie unconformably on non-carbonate rocks in both areas. Their occurrence relates to the general problem of the genesis of calcitizations and calcretes and their possible formation by deposition and alteration of calcitic dusts, as previously hypothesized by Blümel (1981). However, in the eastern Canary Islands and southern Morocco, the geographical origin of this calcite is questionable because the modern Saharan dusts have a mostly siliceous composition. This suggests that the calcitic dusts are derived from an extra-Saharan source and that, during their emplacement as surficial deposits, the climatic conditions were different in these areas from those of today.

Evidence for calcitic dust deposition

Field data

1. Types of calcitic coatings. In the two areas, various types of calcitizations have been dis-

From Pye, K. (ed.), 1993, *The Dynamics and Environmental Context of Aeolian Sedimentary Systems.*
Geological Society Special Publication No. 72, pp. 277–291.

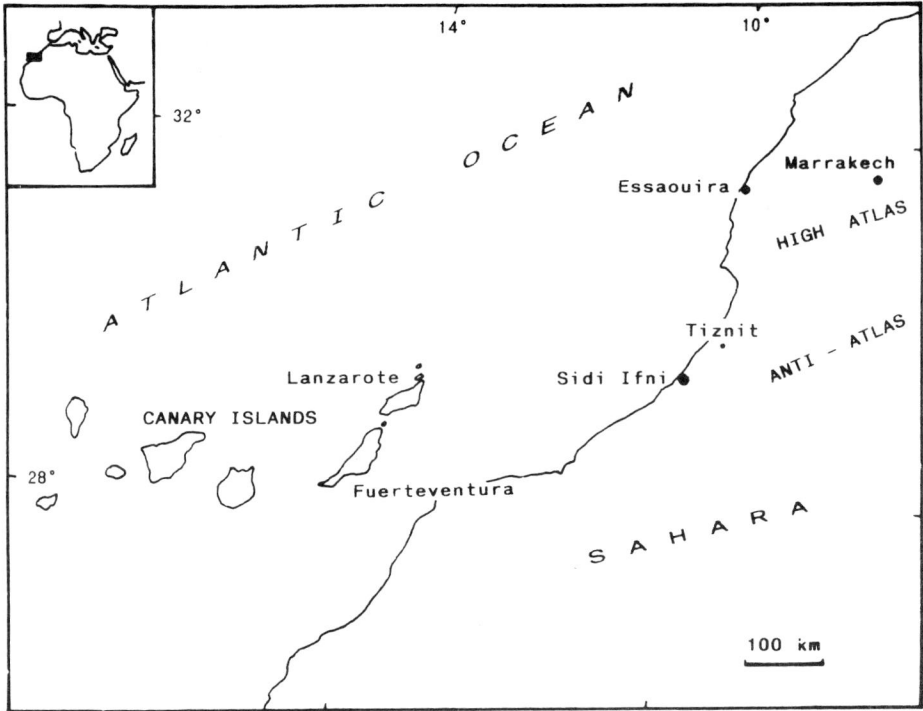

Fig. 1. General location map.

tinguished according to their topographical locations, sedimentological characteristics and micromorphological situations with regard to the bed-rock:

- *pellicular calcitization* constitutes a thin (1–5 mm) surficial coating of harder carbonate chips capping the rocks. This sheet may also cover the walls of large open cracks within the rock;
- *crack-filling calcitization* fills the rock cracks entirely and often has a powdery texture;
- *penetrating calcitization* fills up all the pores and microcracks in the rock. Sometimes this pervasive calcitization has entirely modified the mineralogical components by epigenesis;
- *powdery calcitization* fills and covers the irregularities in the rock surface with a soft, light-coloured, sometimes tens of centimetres thick, layer of powdery material.

These calcitizations may be overlaid by *calcretes* which often have a laminar microstructure. At other sites, the calcretes can rest directly upon the rock.

2. General location and mineralogical differ-

ences. Most of the carbonate coatings are found on topographical highs excluding their initial emplacement by processes other than aeolian dust deposition. This is not incompatible with possible syn- or post-depositional reworking by run-off or slope processes. Moreover, the calcitizations and calcretes often overlie non-carbonate-bearing intrusive, volcanic and metamorphic rocks. This mineralogical difference is further indisputable evidence for their aeolian origin.

3. Significant sites and field sections. In southern Morocco, the calcitizations and calcretes cap the granitic hills of the Ifni and Kerdouss massifs. They thin and become more discontinuous eastwards from the Atlantic shores.

A good section is exemplified by a granitic slope near Ifni which exhibits various calcitic accumulations (Fig. 2):

- *pervasive calcitization* penetrating into the heart of deeply weathered rock between less altered woolsack-like boulders;
- a light, fine-textured *powdery calcitization.*
- surficially sealed by a *laminar calcrete* in

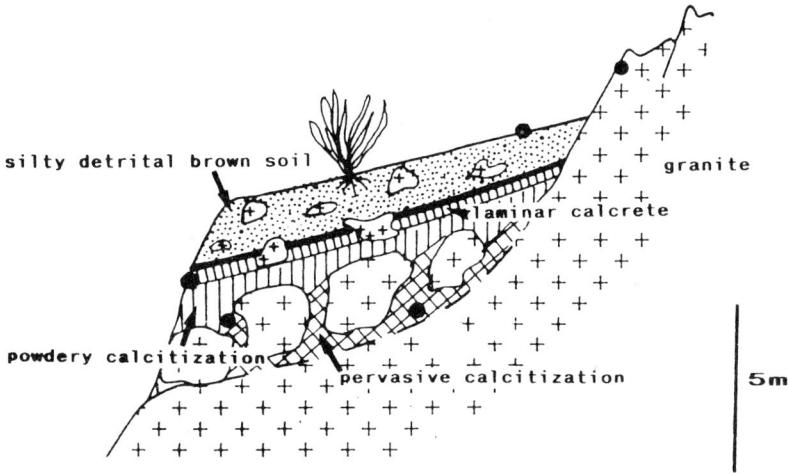

Fig. 2. A field section on a western granitic slope (Ad Bousgaou, 10°03W/29°25N/alt. 630 m), near Ifni (southern Morocco).

which the inclination and distribution of granitic gravels and stones exhibit the influence of reworking by slope processes;
- a *silty soil* forming the top of the section.

Comparable data can be observed on the nearby Precambrian and Palaeozoic metasedimentary outcrops. On schistose substrata, the layered microstructure and rock cracks are filled by white calcite. On the Kerdouss massif, slopes of

Precambrian siliceous sandstone are frequently covered by various types of calcitizations (Fig. 3).

Calcitic coatings also occur on volcanic rocks in the both areas. In southern Morocco, Precambrian rhyolite and andesite are overlain and penetrated by calcite (Fig. 4). In some complex slope sections, it is obvious that the soft calcitic bodies (powdery calcitizations) have been affected by colluvial processes, as testified by

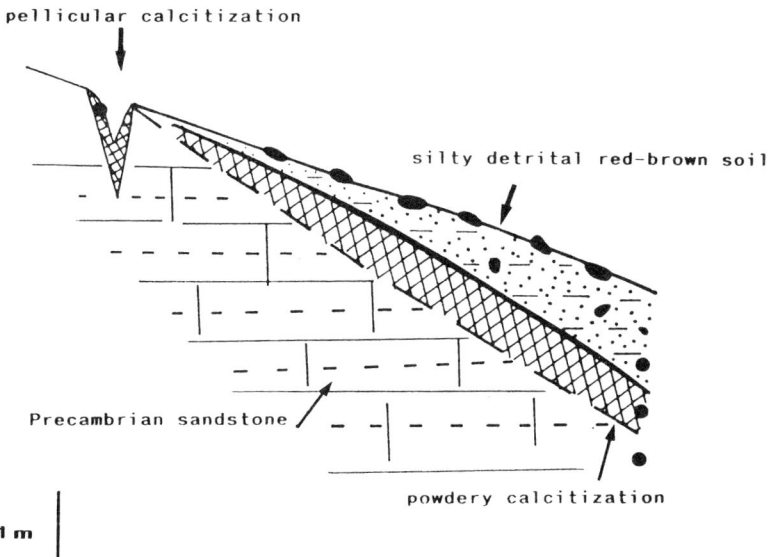

Fig. 3. A field section on a sandstone slope (north of Tafraoute, 9°W/29°48N/alt. 1300 m), near Kerdouss pass (southern Morocco).

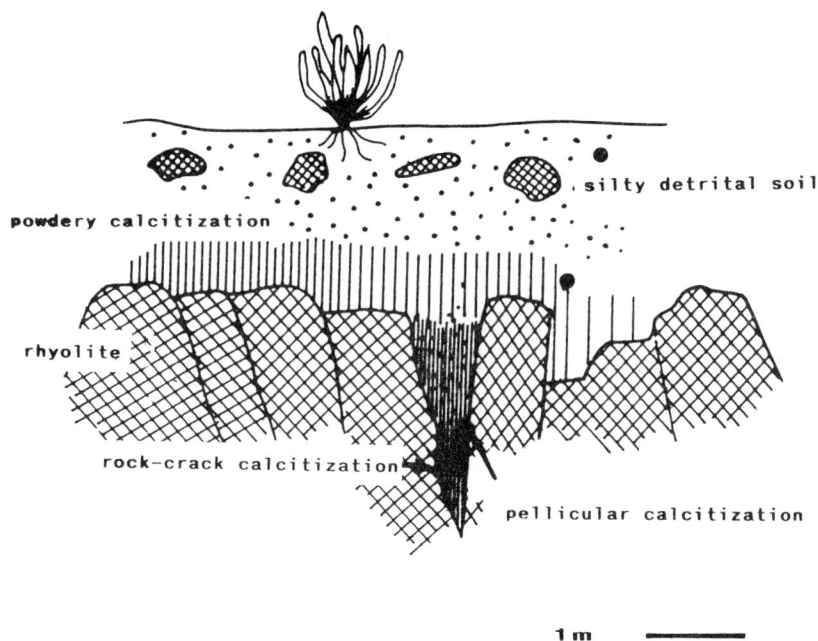

silty detrital soil

powdery calcitization

rhyolite

rock-crack calcitization

pellicular calcitization

1 m

Fig. 4. A field section on a western rhyolitic slope (Djebel Asijar, 10°11W/29°20N/alt. 170 m), southwest of Ifni (southern Morocoo).

their downslope thickening and numerous inter-stratified rhyolitic gravel and stone lines. Horizons of laminar calcrete are covered by a silty detrital soil on the top of such sections.

In the eastern Canary Islands, particularly in Fuerteventura, the geomorphological setting of the surficial calcitic bodies (often thick, hard crusts) is also very significant. They are located on the summits and upper slopes of late Cenozoic to Pleistocene basaltic traps and volcanoes (Fig. 5). On the steep slopes, the calcretes have been incised by run-off to form chevron patterns. In the lower areas, calcitic crusts occur also on the Tertiary gabbroic basement and its Pleistocene biodetrital sand cover (Coudé-Gaussen & Rognon 1988a). Such

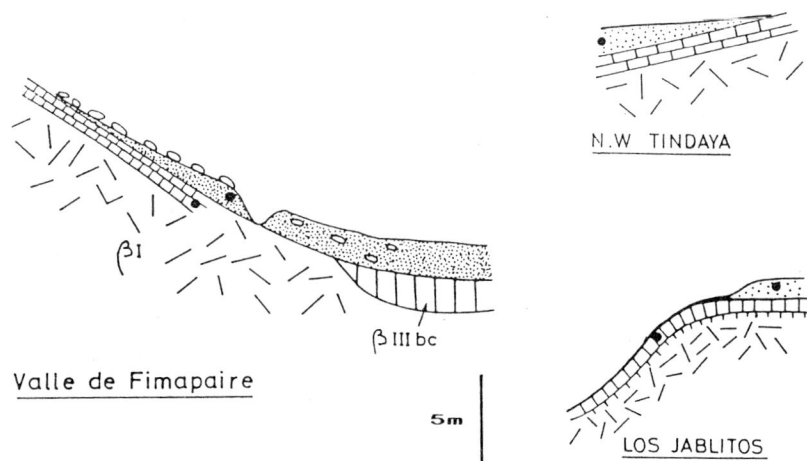

N.W TINDAYA

β1

β III bc

Valle de Fimapaire

5 m

LOS JABLITOS

Fig. 5. Typical field sections with calcretes and silty deposits on basalt outcrops in Fuerteventura, Canary Islands (national grid: Valle de Fimapaire FS0963/alt. 110 m; NW Tindaya ES9864/alt. 150 m; Los Jablitos, FS1070/alt. 140 m).

calcitizations are less developed on the most recent (Holocene) volcanic cones, suggesting a pre-Holocene age for their emplacement on the other older sites.

Analogous calcitizations and calcretes also occur on carbonate terrains in southern Morocco, such as in the serrated dolomitic Akhass Highlands (east of Ifni) and the Pliocene calcareous lacustrine deposits of the Tiznit plain. However, mineralogical evidence of their allochthonous origin is less directly evident on such substrata.

Sedimentological and mineralogical data

Grain-size analyses showed that the samples (≤2 mm fraction) are generally very poorly sorted, indicative of limited reworking by slope processes immediately after deposition.

Moreover, the mineralogical bulk sample and ≤2 μm X-ray diffraction (XRD) analyses and chemical data confirm the difference between the calcitic coatings and the underlying substrata.

1. Granitic substrata. The main difference is the calcite abundance in the coatings and its absence in the granite (Table 1). On the other hand, quartz, feldspar, plagioclase and mica are less important in the calcitic bodies. In the clay fraction, the abundance of palygorskite in the calcitizations can be related to epigenesis of primary granitic minerals by calcite as described by Millot *et al.* (1977). The chemical analysis data also show the opposite trends in concentration of silica and calcium between the rock and calcitic coatings. It is noticeable that the loss on ignition is lower than 2% in solid granite, but ranges from 16–35% in the calcitic bodies.

In the granite, potassic feldspars are dominant while the plagioclases are mainly sodic, in agreement with the chemical data. It confirms that the calcitic surficial bodies cannot be the result of in situ calcium concentration by the weathering of the underlying rock.

Internal differences exist between the calcitic coatings. The contamination by autochthonous (or subautochthonous if slope processes are taken into account) minerals issued from the

Table 1. *Mineralogical and chemical composition of calcitic bodies and underlying granitic substratum*

	Paly	Clay	Q	Pla	Kf	Mic	Cal
		XRD Bulk sample mineralogy (%)					
Granite	—	2–28	35–37	19–32	16–25	0–11	—
Pellicular calcite	—	—	20	7	—	2	71
Crack-filling calcite	—	—	29	11	5	2	53
Penetrating calcite	0–2	0–23	7–39	2–8	0–5	8–9	27–70
Powdery calcite	0–3	—	7–13	8–13	5–8	0–7	57–68
Calcrete	—	—	17	21	—	3	59

	Kao	Sme	Int. G	Ill	Chl	Int	Paly	Q
		XRD Clay mineralogy (%)						
Granite	0–25	0–40	—	11–70	9–10	0–8	—	—
Crack-filling calcite	10	—	20	20	—	—	50	—
Penetrating calcite	5	0–95	0–5	0–15	0–5	—	0–70	tr
Powdery calcite	5	0–15	0–10	0–25	0–15	—	45–85	tr
Calcrete	15	—	25	45	15	—	—	tr

	SiO_2	Al_2O_3	Fe_2O_3	MgO	CaO	K_2O	Na_2O	TiO_2	MnO	H_2O
		Chemical composition (%)								
Granite	60–72	14–15	2–6	0.3–2	0.5–2	3–5	3–4	≤1	≤0.1	≤1
Pellicular calcite	15	4	2	1	39	≤1	≤1	0.1	≤0.1	0.9
Crack-filling calcite	32	8	4	1.5	25	2	1.5	0.5	≤0.1	0.8
Penetrating calcite	12–46	3–12	1–2	1–1.5	15–42	≤1.5	≤0.5	≤0.1	≤0.1	1–4
Powdery calcite	20–23	5–6	2.5	1–2	32–34	1–2	1	0.2	≤0.1	1–2
Calcrete	25	6	2	1	32	2	1	0.2	≤0.1	0.6

Paly: palygorskite; Q: quartz; Kf: K-feldspar; Mic: mica; Cal: calcite; Kao: kaolinite; Sme: smectite; Int. G: interlayered illite–smectite; Ill: illite; Chl: chlorite; Int: interlayered chlorite.

granitic substratum is less important for the most surficial calcitizations (pellicular and powdery) than for the deepest ones (crack filling and penetrating), considering the various quartz, feldspar and mica contents. With regard to the penetrating calcitizations, the very wide range of values corresponds with the varying degree of calcitic epigenesis. Surprisingly, the laminar calcrete is not the most calcitic level but is richer in basement minerals: this agrees with its mode of emplacement as is described later.

2. *Volcanic substrata.* Similar data and conclusions result from the mineralogical and chemical analyses of the calcitic coatings on rhyolitic and basaltic outcrops (Tables 2 & 3).

In the rhyolitic areas of southern Morocco, the differences in mineralogical and chemical composition between the substratum and calcitic coatings are obvious, with high percentages of quartz in the former but highest percentages of calcite in the coatings. The presence of significant amounts of smectite suggests its neogenesis

Table 2. *Mineralogical and chemical composition of calcitic bodies and underlying rhyolitic substratum*

| | **XRD Bulk sample mineralogy (%)** | | | | | | |
	Paly	Clay	Q	Pla	Kf	Mic	Cal
Rhyolite	—	—	70–74	0–7	2–30	0–17	—
Pellicular calcite	—	—	5	—	—	—	95
Crack-filling calcite	—	2	3	3	—	—	92
Penetrating calcite	—	3	13	—	—	4	80
Powdery calcite	0–9	0–3	7–15	0–6	1–6	—	76–82
Calcrete	0–6	—	10–16	—	—	—	78–90

| | **XRD Clay mineralogy (%)** | | | | | | | |
	Kao	Sme	Int. G	Ill	Chl	Int	Paly	Q
Rhyolite	—	—	—	100	—	—	—	—
Pellicular calcite	—	70	—	20	10	—	—	—
Crack-filling calcite	10	85	—	5	—	—	—	—
Penetrating calcite	tr	—	—	30	—	70	—	—
Powdery calcite	5–15	—	0–40	10–30	0–15	0–5	80–85	tr
Calcrete	5–10	—	—	20–25	0–5	—	65–70	—

| | **Chemical composition (%)** | | | | | | | | | |
	SiO$_2$	Al$_2$O$_3$	Fe$_2$O$_3$	MgO	CaO	K$_2$O	Na$_2$O	TiO$_2$	MnO	H$_2$O
Pellicular calcite	13	4	2	1.3	40	1	1	0.2	⩽0.1	0.9
Crack-filling calcite	6	2	1	1.3	46	0.3	0.8	0.1	⩽0.1	1.7
Powdery calcite	11–21	3–7	1–3	1–2	31–44	1	⩽1	⩽0.3	⩽0.1	1–3

Table 3. *Mineralogical composition of calcitic bodies on basalt substratum*

| | **XRD Bulk sample mineralogy (%)** | | | | | | |
	Paly	Clay	Q	Pla	Kf	Mic	Cal
Pellicular calcite	—	—	2	6	—	—	83
Crack-filling calcite	—	5	2–9	3–9	—	—	77–78
Powdery calcite	—	—	tr	tr	tr	—	80
Calcrete	—	—	tr	tr	tr	—	80

| | **XRD Clay mineralogy (%)** | | | | | | | |
	Kao	Sme	Int. G	Ill	Chl	Int	Paly	Q
Pellicular calcite	5	20	—	—	15	10	55	—
Crack-filling calcite	5	—	—	—	—	80	15	—
Powdery calcite	—	—	—	tr	25	10	65	—
Calcrete	—	0–5	—	—	0–70	0–30	0–95	—

in the richest calcitic context (pellicular and crack-filling calcitizations) and high contents of palygorskite in the powdery calcitizations and calcretes are probably indicative of epigenetic formation.

The question of calcitizations and calcrete formation on basalt and gabbroic outcrops in Fuerteventura has been considered in earlier publications (Rognon & Coudé-Gaussen 1987; Chamley *et al.* 1987; Coudé-Gaussen & Rognon 1988*a*). These studies showed that calcitizations and thick calcretes cover the Plio-Pleistocene 'old basalts' and Tertiary gabbros which contain < 10–12% carbonate.

These coatings are characterized by the highest proportions of calcite. The occurrence of quartz (even in trace amounts) is surprising in regard to its absence in the volcanic substratum of the island and suggests an allochthonous contribution. The calcitic bodies contain abundant fibrous minerals (palygorskite and sepiolite), with mainly significant chlorite values on gabbroic substrate. These data may suggest deep and ancient weathering of the basalts and gabbros but they do not explain the primary origin of the calcite in the overlying coatings.

3. Metamorphic substrata. The calcitizations sampled on the old basement sandstones and schists of southern Morocco also have a clear mineralogical and chemical dissimilarity (Tables 4 & 5).

The calcite values are higher in the calcitizations than in the siliceous metasediments. The most internal calcitizations have, as usual, a lower calcitic content because of either a moderate calcite input in the less altered underlying substratum or a contamination by the local detrital minerals during the process of crackinfilling. The observations and analyses in the schist area are illustrative. The abundance of smectite and palygorskite in the calcitizations reflects the explanations given previously. Correspondingly, the loss on ignition value reaches 30–40% in the calcitizations but only 1–2% in the metamorphic rocks.

Micromorphology

Thin sections were examined by XPL or PPL optical microscopy in order to characterize the mineral components and the relationships between coatings and underlying rocks, according to standard methods of soil thin section description (Bullock *et al.* 1985). Extra-large sections (15 × 10 cm) have been used to examine the internal microstructures and fabrics which are particularly significant in providing evidence of the emplacement and reworking in these calcitic bodies.

The calcitic coatings mainly consist of white to very light-brown micritic matrix. In the samples characterized by a porous microstructure, the calcite has often been recrystallized as microsparite spiculae in the internal voids, with some acicular front to micro-geode-like patterns around the main opened vesicles.

Table 4. *Mineralogical and chemical composition of calcitizations and underlying sandstone substratum*

| | **XRD Bulk sample mineralogy (%)** | | | | | | |
	Paly	Clay	Q	Pla	Kf	Mic	Cal
Sandstone	—	5	71	11	—	9	2
Pellicular calcite	—	—	15	2	—	2	81
Powdery calcite	—	—	8	—	—	2	90

| | **XRD Clay mineralogy (%)** | | | | | | | |
	Kao	Sme	Int. G	Ill	Chl	Int	Paly	Q
Sandstone	—	—	45	35	20	—	—	tr
Pellicular calcite	10	55	—	30	5	—	—	tr
Powdery calcite	5	50	—	15	5	—	25	—

| | **Chemical composition (%)** | | | | | | | | | |
	SiO_2	Al_2O_3	Fe_2O_3	MgO	CaO	K_2O	Na_2O	TiO_2	MnO	H_2O
Sandstone	76	10	4	1.5	≤1	2	1	0.4	≤0.1	0.1
Pellicular calcite	13	4	2	0.7	41	≤1	≤1	≤1	≤0.1	1.1
Powdery calcite	7	2	1	0.6	49	≤1	≤1	≤0.1	≤1	0.5

Table 5. *Mineralogical and chemical composition of calcitizations and underlying schist*

	XRD Bulk sample mineralogy (%)						
	Paly	Clay	Q	Pla	Kf	Mic	Cal
Schist	—	2–3	42–57	14–37	3–26	2–15	—
Pellicular calcite	—	—	10	12	—	6	72
Crack-filling calcite	—	2–5	13–34	4–29	—	4–22	13–70
Penetrating calcite	0–3	0–5	28–37	4–6	0–2	4–16	34–59

	XRD Clay mineralogy (%)							
	Kao	Sme	Int. G	Ill	Chl	Int	Paly	Q
Pellicular calcite	20	25	—	55	5	—	—	—
Crack-filling calcite	15–25	0–80	0–35	5–25	0–30	—	—	tr
Penetrating calcite	5	0–55	0–5	20	5	—	15–70	tr

	Chemical composition (%)									
	SiO_2	Al_2O_3	Fe_2O_3	MgO	CaO	K_2O	Na_2O	TiO_2	MnO	H_2O
Schist	73–76	10–14	3–4	0.5	≤1	3–5	1–3	≤1	≤0.01	0.1
Pellicular calcite	22	9	2	0.7	33	3	≤1	≤1	≤0.01	0.5
Crack-filling calcite	21–53	6–15	2–6	1	7–35	1–2	1–3	≤1	≤0.01	1–2
Penetrating calcite	32	9	3	1.4	25	2	≤1	≤1	≤0.01	1–2

1. Deepest internal calcitic features. Inside the rock, the relationships between the calcite and substratum minerals vary from moderate and thinnest crack-filling to pervasive alteration.

A good example of the first situation is given by an apparently unaltered granite from the Ifni area (southern Morocco). The microcracks are weekly penetrated by narrow linear micritic edges, but without any modification of the primary minerals (Fig. 6a).

Another example from a schist outcrop in southern Morocco shows a more elaborate internal calcitization with wedge-shaped penetration by micrite in the phyllitic structure. Here, a selective micro-epigenesis has occurred where some phyllitic laminae are unaltered and others more or less replaced by micrite. In the large voids, the calcite filling is totally microcrystallized (Fig. 6b).

2. Surficial calcitic features. The calcitic fillings of the large open cracks and the surficial soft calcitic rock covers are dominated by micritic texture, often with recrystallized microsparitic features in the planar and vesicular voids. Secondary red-brown laminated clayey infillings may occur in the pores.

The beige micritic matrix may have a subangular, blocky microstructure where the primary fine material is divided into rough aggregates by planar voids. It suggests an incipient mechanical disintegration of the previous homogeneous micritic filling.

The reworked calcite in the large crack fillings or rock covers may be more complex (e.g. Fig. 6c, from a powdery calcitization sampled on a granitic area). The primary micritic aggregates are separated by large open passages filled by a secondary heterogeneous material with high porosity. It is composed of small detrital particles (quartz, feldspars) and variously sized globular to oblate, mainly micritic, nodules. Some nodules are composed of the same light-coloured homogeneous micrite as the large blocky aggregates. Others have a mixed constitution with detrital components in a micritic matrix. Composite meganodules also have been examined, consisting of elementary nodules and detrital contaminant grains in a micritic matrix.

This nodular microstructure occurs frequently in the most surficial calcitizations. Observations of thin sections from basaltic cracks in Fuerteventura have shown two successive infillings: the first forming a calcitic coating on the crack walls with a largely recrystallized sparitic structure containing local mono- or polymineral particles and chips: and the second composed of isolated or joined homogeneous and heterogeneous nodules in a general light-brown micritic matrix. Microparticles of quartz and orange-coloured concentric-shaped clayey grains (palygorskite/sepiolite) are commonly seen in such basaltic crack fillings.

3. Laminar calcretes. The micritic microstructure of the laminar calcretes consists of

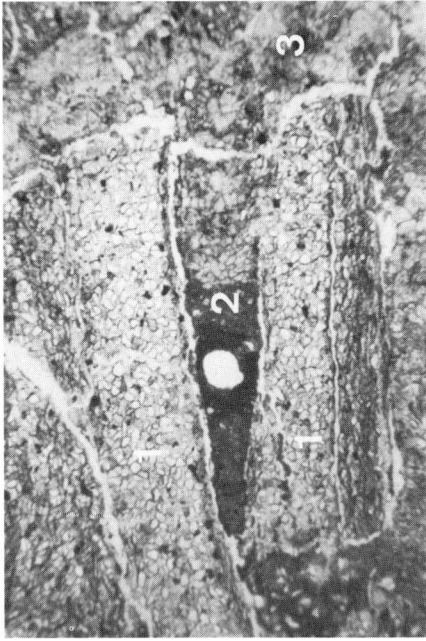

(a)

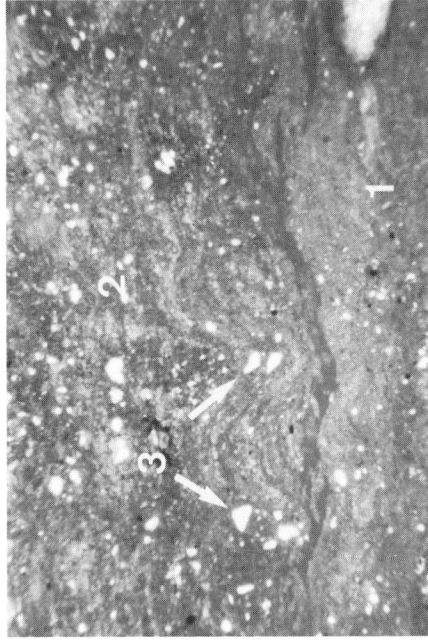

(b)

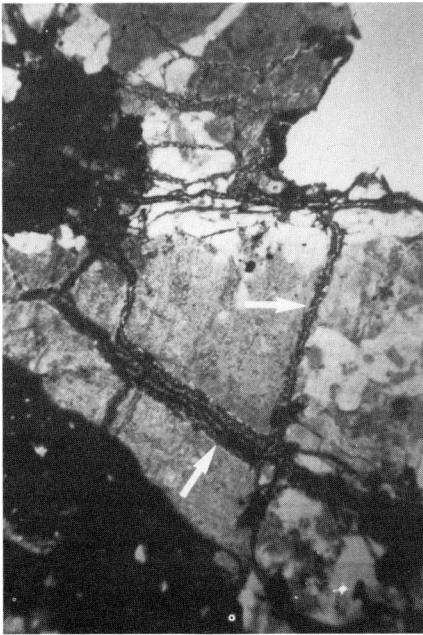

(c)

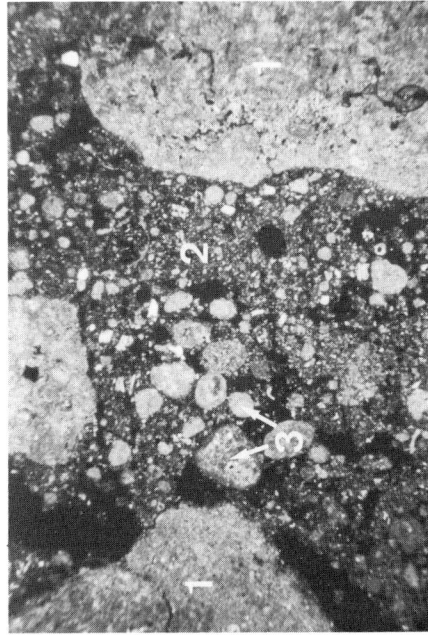

(d)

Fig. 6. Thin section descriptions: (**a**) incipient linear calcitizations along rock micro-cracks from the granite of the Ad Bousgaou section (e.g. Fig. 2) (XPL; frame length 480 μm); (**b**) selective calcitic epigenesis in a schistose microstructure (from a rock sampled near Aït Cherif, 10° 22W/29° 52N) — (1) unaltered phyllitic laminae, (2) partially micritic epigenetized laminae, (3) recrystallized micritic infilling of a large void (XPL; frame length 1.5 mm); (**c**) micromorphology of a powdery calcitization from the rhyolitic Djebel Asijar section (e.g. Fig. 4) — (1) homogeneous micritic blocky aggregates, (2) secondary heterogeneous infilling, (3) micritic nodules (PPL; frame length 750 μm); (**d**) micromorphology of a laminar calcrete from the granitic Ad Bousgaou section (e.g. Fig. 2) — (1) intermediate homogeneous micritic layer with some quartz particles, (2) laminar micritic stromatolitic growth features, (3) detrital particles trapped in the depressions (XPL; frame length 1.2 mm).

parallel bands of nodular structure with inter-layed stromatolites. Some intermediate bands have a homogeneous micritic matrix where vesicles and planar voids are frequently filled by recrystallized microsparite.

The nodules in the nodular bands are com-parable with those in the calcitizations: some small ones consist of homogeneous brown micrite but meganodules have a composite con-stitution with small nodules and detrital chips wrapped in a light-coloured micritic matrix. In the stromatolitic layers, the micritic laminae have a crescentic shape where small detrital particles (quartz, feldspars) have been trapped in the depressions between each convexity (Fig. 6d).

These micromorphological features indicate calcitic penetration downwards into the rock but suggest a rather moderate epigenesis of the rock minerals by the calcite except where it has pene-trated deepest into a previously weathered sub-stratum. The micromorphological data also indicate reworking after the initial deposition of the carbonate as aeolian dust on the rock. The disintegration of the first-formed homo-geneous micritic fillings and coatings into rough blocky aggregates, the occurrence of successive crack infillings, the rounded shape of the secon-dary micritic nodules and the contamination by local or allochthonous detrital particles are indicative of post-depositional slope processes in a wetter environment.

Finally, the layered micritic microstructure of the laminar calcretes is derived from reworking responsible for the nodular and detrital bands and from stabilization phases when micro-organisms were able to form stromatolitic laminated bands. Such stromatolitic growth phases have only occurred subaerially according to Vogt (1983) and Adolphe (1987), during times when the rate of calcite dust deposition slowed down.

Discussion

On Fuerteventura, almost all the calcitizations and calcretes have formed on Pleistocene or older volcanic rocks. An important calcrete, dated to $13\,850 \pm 200$ [14]C years BP, has stabilized the biodetrital dune sand cover of the lowlands (Rognon et al. 1989). It suggests that the calcitic dust falls responsible for the calcitiz-ations and calcretes in Fuerteventura and, more generally, eastern Canary Islands, are of Pleisto-cene and Late-Glacial age.

During Pleistocene low sea-level stands, the lowlands and topographic saddles of Fuerte-ventura were invaded by dune sands mobilized from a continental shelf, 4–9 km wide, to the west of Fuerteventura (Rognon & Coudé-Gaussen 1987; Coudé-Gaussen & Rognon 1988a; Rognon et al. 1989). These marine and shore sands, derived from shells and marine organisms, were carbonate rich (70–90%). During their migration, they were fragmented and comminuted, in a similar manner to that described near the Cap Sim in western Morocco (Coudé-Gaussen et al. 1982b). Calcitic dust was produced by reworking of the biodetrital sand and blown by winds in the direction of sand transport from west to east, blanketing the interior fixed dunes and volcanic relief. Several lines of evidence support this contention, includ-ing the gradual loss of Mg-calcite and aragonite (marine markers) from seashore sands to interior fixed dunes, and the spatial transition of calcretes upon sands to calcretes upon volcanic rocks (Chamley et al. 1987). There is a corresponding easterly increase of low-Mg calcite and smectite (whose formation is prob-ably associated with the magnesium lost by the Mg-calcite during its transformation to secon-dary low-Mg-calcite). Moreover, δ^{13}C and δ^{18}O values of the bulk carbonate samples are positive in the seashore and western mobile sands, reflecting direct marine influence, while they are intermediate in the weakly lithified dunes and negative in the calcretes. These isotopic data are comparable with those obtained from Holocene and Pleistocene sands in Bermuda (Friedman 1964). Further, micromorphological examin-ation provides evidence for a marine origin of the calcite constituting the calcretes on the highest slopes, far away from the biodetrital sand source. Some thin sections show more or less preserved (sometimes only ghost shapes, but sometimes easy to identify) fragments of red algae, lamellibranchiata and foraminifera.

In the case of southern Morocco, similar con-clusions can be drawn based on observational evidence. The thickest calcitic coatings occur near the Atlantic shore and are often located on the west-facing slopes. Thus, the source areas would also be the Pleistocene emergent con-tinental shelf west of the present seashore and, in all probability, the limestone and marl areas of the Tiznit and Massa plains (Fig. 7).

Evidence for a siliceous dust contribution

In both areas, siliceous soils and deposits often uncomformably overlie the previously described calcitic bodies, or rest directly upon various substrata.

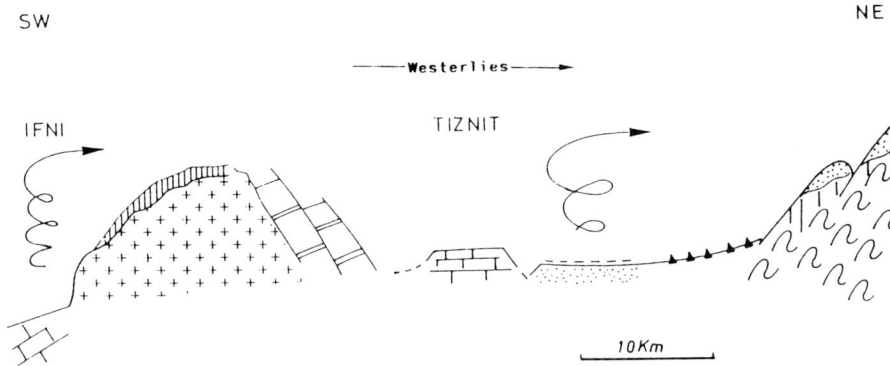

Fig. 7. The Pleistocene calcitic dust system in southern Morocco.

General character and location

In southern Morocco, red-brown detrital soils commonly occur on the hill slopes. They are several decimetres thick but they can attain several metres in low areas. They contain coarser detrital gravels and stones derived from local bed-rock. These soils are seen on the granitic and rhyolitic hills near Ifni (Figs 2 & 4) and on the schist and sandstone outcrops of the Kerdouss massif (Fig. 3). Comparable soils occur on the dolomitic hills of the Akhass ridge and on the lacustrine limestones of the Tiznit plain.

Table 6. *Mineralogical and chemical composition of Moroccan red-brown soils overlying various substrata*

| | **XRD Bulk sample mineralogy (%)** | | | | | | |
	Paly	Clay	Q	Pla	Kf	Mic	Cal
Soils on:							
Granite	—	2	53	32	8	4	—
Rhyolite	—	7	61	17	11	4	—
Sandstone	—	8	63	8	4	8	6
Schist	—	6	59	9	—	—	26
Dolomite	—	5	63	3	2	2	25
Limestone	—	2	64	15	4	—	15

| | **XRD Clay mineralogy (%)** | | | | | | | |
	Kao	Sme	Int. G	Ill	Chl	Int	Paly	Q
Soils on:								
Granite	20	—	10	55	15	—	—	+
Rhyolite	17	—	7	61	10	3	—	+
Sandstone	20	—	35	40	5	—	—	—
Schist	10	—	40	35	15	—	—	—
Dolomite	25	—	5	50	18	2	—	+
Limestone	10	—	10	25	10	—	45	+

| | **Chemical composition (%)** | | | | | | | | | |
	SiO_2	Al_2O_3	Fe_2O_3	MgO	CaO	K_2O	Na_2O	TiO_2	MnO	H_2O
Soils on:										
Granite	62	15	2	1	3	2	$\leqslant 1$	$\leqslant 0.1$	2	6
Rhyolite	61	14	6	2	1	3 2	$\leqslant 1$	2	6	—
Sandstone	58	15	6	1	3	2	$\leqslant 1$	$\leqslant 1$	$\leqslant 0.1$	4
Schist	55	13	4	1	9	3	$\leqslant 1$	$\leqslant 1$	2	11
Dolomite	49	13	5	1	9	3	$\leqslant 1$	1	0.1	1

On Fuerteventura, light-brown silty deposits, typically a few decimetres thick, frequently occur upon the volcanic rocks, the calcretes and the Pleistocene sand cover. As they are seen even on the summits and slopes of the most recent Holocene volcanoes, it suggests their very recent emplacement.

Sedimentological, micromorphological and isotopic data

The sedimentological character of these siliceous soils and deposits is very similar, suggesting an allochthonous source.

The mineralogical and chemical data relating to the Moroccan soils and their various substrata are shown in Table 6. The ≤ 2 mm fraction of these soils contains similar contents of quartz, feldspars and micas, but calcite occurs only in some samples. Many of these were obtained from deposits on carbonate bed-rocks and can be explained by local contamination due to mixing by slope processes. On sandstones and schists, the occurrence of carbonate in the silty deposits seems related to the contamination from underlying calcitizations.

The mean mineralogical data relating the silty deposits and soils upon basalt outcrops on Fuerteventura are presented in Table 7. The dominance of quartz (reaching 55–60% in some samples) in the ≤ 2 mm fraction is noteworthy considering the absence of this mineral in the volcanic rocks of Fuerteventura. It is also likely that some of the feldspars (especially K-feldspars) are allochthonous. Calcite also has a notable importance (40–50%). It can be

explained by local contamination from upper calcitized slope surfaces, or by a present-day deposition of calcitic marine and seashore dusts as documented by Coudé-Gaussen et al. (1987) and Coudé-Gaussen (1991). Part of the calcite may also be far-travelled from the limestone areas of the Saharan Atlas. Such an origin has been previously documented by studies of present north Saharan air outbreaks. In the clay fraction, the smectite content may correspond to a local contribution derived from the weathering of basalts. Some of the palygorskite may result from the same process, but some may be related also to an aeolian input.

Grains from these Canarian soils have been examined by SEM and EDAX. Each sample was found to contain smoothed and even-rounded quartz grains (20–350 μm) with high-energy erosive surficial microfeatures. The grains are often covered by an amorphous silica film or adhering particles. The frequent etched microrelief suggests post-depositional pedogenesis, while the surficial frosting suggests reworking and transport by slope processes. The smoothed feldspar grains are very altered and mainly of potassic type. Some mica flakes with typical wind-abraded edges were also noted. The detrital carbonate grains are well rounded and often polycrystalline. The palygorskite and sepiolite clays often occur as puffy and porous aggregates of mixed fibres which suggest a local origin with little reworking. They are also frequent as surficial coatings upon the biggest allochthonous grains or as wind-shaped pellets which are very common in the proximal Saharan dusts (Coudé-Gaussen & Blanc 1985).

Lastly, the very different Sr–Nd isotopic

Table 7. *Mean mineralogical and chemical composition of silty deposits overlying basalt in Fuerteventura, eastern Canary Islands*

	XRD Bulk sample mineralogy (%)						
	Paly	Clay	Q	Pla	Kf	Mic	Cal
Soils on:							
Basalt	—	15	35	10	10	—	20

	XRD Clay mineralogy (%)							
	Kao	Sme	Int. G	Ill	Chl	Int	Paly	Q
Soils on:								
Basalt	20	15	5	30	5	10	15	+

	Chemical composition (%)									
	SiO_2	Al_2O_3	Fe_2O_3	MgO	CaO	K_2O	Na_2O	TiO_2	MnO	H_2O
Soils on:										
Basalt	39	11	6	4	14	2	≤1	1	≤0.1	4

signatures of particles derived from Canarian volcanic rocks and from African continental crust allowed us to identify the source of Holocene Canarian silty deposits: 75% of the particles are of African origin and only the last quarter seems derived from local weathered volcanic rocks (Grousset *et al.* 1992).

Discussion

These observations suggest that the red-brown soils of southern Morocco and the silty beige deposits of eastern Canary Islands cannot be interpreted as upper horizons of carbonated soils of which the overlying calcretes and calcitizations would represent the calcareous lower horizons. The detrital soils and silty deposits frequently rest directly upon bed-rock without any interlayered calcitic bodies. The mineralogical and chemical relationships of these soils and deposits provide the best evidence for their partly extraneous origin. Although the local lithology has exercized some influence in producing notable compositional differences, mainly in their coarse gravelly and stony fraction, the high topographical locations

of the deposits argue for their primary aeolian origin.

The source of this recent allochthonous contribution is apparently the Saharan desert. The mineralogical and chemical assemblages observed are very comparable with those of proximal dusts sampled on the Saharan margins, and of recent dust fall deposits collected on the Canary Islands (Coudé-Gaussen *et al.* 1987; Bergametti *et al.* 1989; Coudé-Gaussen 1989*b*).

Conclusion

The contrasting characters and origins of these two successive dust depositional episodes can be explained by an important climatic change around latitude 28–32°N, at the end of the Pleistocene (Fig. 8).

During the last glacial period, the growth of ice sheets induced a southerly shift of the climatic zones and an increase in the north–south thermal gradient over subtropical latitudes. The westerly cyclonic circulation system, today located over the mid- to northern part of western Europe, moved south over the Mediterranean (Rognon 1976). This can be explained

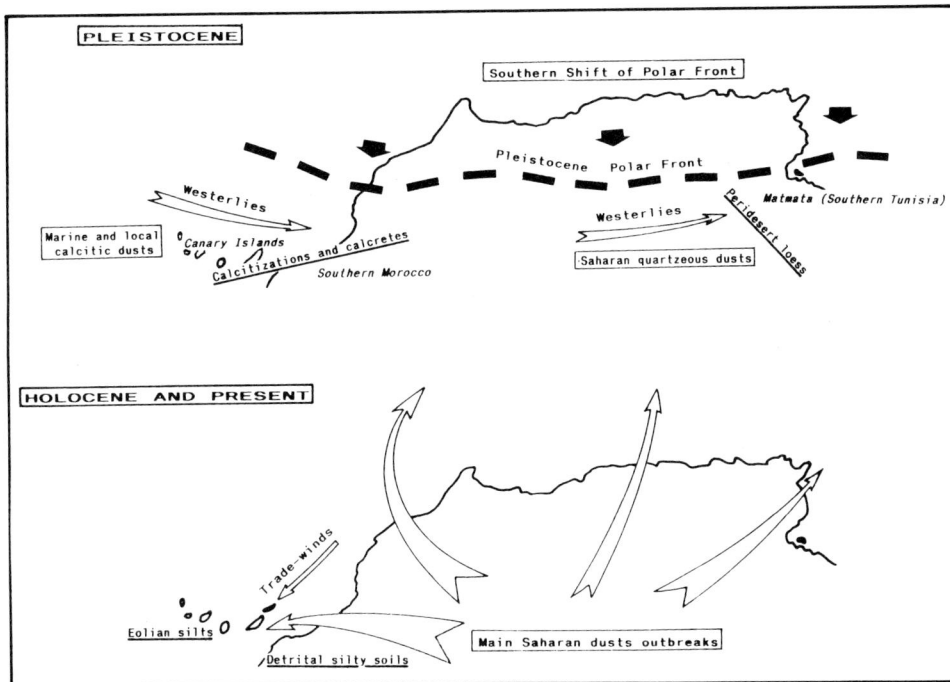

Fig. 8. Opposite climatic situations associated with the Pleistocene dust system and Holocene to Present dust system in the Maghreb.

by greater heat losses over the glaciated areas in Europe and by a more frequent southerly shifting, both in winter and summer, of polar mobile highs.

The best evidence of the prevalence of the westerlies over southern Europe and northern Africa in this period is given by the general sand migration from the emergent continental shelves to the eastern hinterlands. This has been demonstrated here for Fuerteventura in the eastern Canary Islands, but is also documented in the Landes (France), Guadalquivir (Spain), and western Maghreb (Gharb and Souss plains). Further evidence that winds blew from the west is provided by the distribution of volcanic ash deposits from Tenerife (Booth 1973), the pattern of peridesert loess deposition in southern Tunisia (Coudé-Gaussen et al. 1982a; Coudé-Gaussen & Rognon 1988b; Coudé-Gaussen 1990) and, as far as central Sahara, by the firn line location in the Atakor massif (Rognon 1967).

Thus, the deposition of calcitic dust in the eastern Canary Islands and southern Morocco is related to this general late Pleistocene palaeoclimatic situation.

During the Late Glacial, the rapid warming-up of ocean surface waters and the concomitant ice melting in northern Europe induced a northerly shift of the cyclonic trajectories, whereas the continental shelf was gradually submerged, reducing the availability of carbonate sand and dust.

In the early Holocene, the influence of the dynamic subtropical anticyclone (prevalent along the Senegal and Mauritania shore between 20 000 and 12 000 BP) decreased in the south and moved northward over the Atlantic Ocean. Blocking situations related to this continuance of high pressures over the oceanic middle latitudes gave a drier climate in western Europe during the Boreal period (10 000–8000 BP) and induced winds blowing from an eastern sector at latitude 28–32°N.

This easterly influence is corroborated in western Morocco by a dry phase, north of the Atlantic Atlas (Safi, Marrakech) between 10 000 to 7500 BP (Weisrock et al. 1985; Weisrock & Miskovsky 1988) and by the closing of the main Moroccan estuaries, such as Ksob wadi (Weisrock & Rognon 1977; Chahboun 1988).

Prevailing ridge-and-trough structures related to this northerly location of the subtropical Atlantic anticyclone would control frequent air outbreaks issuing from southern Atlas and northern Saharan areas and would be responsible for important dust falls in southern Morocco and eastern Canary Islands. These

siliceous dusts may explain the fine contribution to the detrital soils around Ifni and silty deposits of Fuerteventura which formed during the early Holocene. Moreover, the dynamic Azores anticyclone kept up some influence along the Moroccan coast as indicated by the shore dune ridges, which are oriented NNE–SSW in the Essaouira-Cap Sim area.

During the Atlantic period (7500–5000 BP), an intensive cyclonic circulation system developed in western Europe, associated with the disappearance of the blocking situations. The southerly shift of the Atlantic subtropical anticyclone to its present latitudes induced a strengthening of the near-ground trade-winds along the Canary Islands and the southern Morocco coast. Even if some winds blow today from the southwest or the east during the summer months along the Moroccan coast, or from the west during the winter on the Canary Islands, this NNE–SSW prevalent direction of the trade-winds explains the present decrease in frequency of the Saharan dust falls in both the eastern Canary Islands and southern Morocco. A more accurate estimate of the Saharan dust deposition on the study areas during the Late Holocene and the present period would need a significantly better time resolution.

The field trips have been supported by the CNRS, Paris (France) and CSIC, Madrid (Spain). The sedimentological analyses were undertaken at the Centre de Géomorphologie du CNRS, Caen (France). The authors are particularly grateful to Dr K. Pye, PRIS, University of Reading, for his help and advice in improving the English of this contribution.

References

ADOLPHE, J. P. 1987. Formations carbonatées continentales. In: MISKOVSKY, J. C. (ed.) Géologie de la Préhistoire. Géopré, Paris, 197–224.

BERGAMETTI, G., GOMES, L., COUDÉ-GAUSSEN, G., ROGNON, P. & LE COUSTUMER, M. N. 1989. African dust over Canary Islands: source-regions identification and transport pattern for some summer situations. Journal of Geophysical Research, 94 (D212), 14855–14864.

BLÜMEL, W. D. 1981. Pedologische und geomorphologische Aspekte der Kalkkurstenbildung in Südwestafrika und Südostspanien. Karlsruher Geographische, Hefte 10.

BOOTH, B. 1973. The Granadilla Pumice Deposit of Southern Tenerife, Canary Islands. Proceedings of the Geologists' Association, 84, 353–370.

BULLOCK, P., FÉDOROFF, N., JONGERIUS, A. STOOPS, G., TURSINA, T. & BABEL, U. 1985. Handbook for Soil Thin Section Description. Waine Research Publications, Wolverhampton.

CHAHBOUN, A. 1988. *Les formations sableuses fluviatiles, littorales et éoliennes aux embouchures des oueds Tensift, Ksob et Souss (Atlas atlantique, Maroc)*. Thesis, d'Université P. & M. Curie, Paris.

CHAMLEY, H., COUDÉ-GAUSSEN, G., DEBRABANT, P. & ROGNON, P. 1987. Contribution des aérosols à la sédimentation quaternaire de l'Ile de Fuerteventura (Canaries) *Bulletin de la Societe Géologique de France*, **8**, (3, 5), 939–952.

COUDÉ-GAUSSEN, G. 1989a. *Les Poussières Sahariennes et leur Contribution aux Sédimentations Désertiques et Péridésertiques*. Thesis, Etat Sciences, d'Université, P. & M. Curie, Paris.

—— 1989b. Local, proximal and distal Saharan dusts: characterization and contribution to the sedimentation. *In*: LEINEN, M. & SARNTHEIN, M. (eds), *Palaeoclimatology and Paleometeorology: Modern and Past Patterns of Global Atmospheric Transport*. NATO ASI Series, C. 282. Kluwer Academic Publishers, Dordrecht, 339–358.

—— 1990. The loess and loess-like deposits on either sides of Western Mediterranean Sea: genetic and paleoclimatic significance. *Quaternary International*, **5**, 1–8.

—— 1991. *Les Poussières Sahariennes*. John Libbey Eurotext, Paris.

—— & BLANC, P. 1985. Présence de grains éolisés de palygorsekite dans les poussières actuelles et les sédiments récents d'origine désertique. *Bulletin de la Societe Géologique de France*, **1**, (4), 571–579.

——, MOSSER, C. ROGNON, P. & TOURENQ, J. 1982a. Une accumulation de loess du Pléistocène Supérieur dans le Sud-Tunisien: la coupe de Téchine. *Bulletin de la Societe Géologique de France*, **24**, (2) 283–292.

—— ROGNON, P. 1988a. Origine Eolienne de Certains Encroûtements calcaires sur l'Ile de Fuerteventura (Canaries orientales). *Geoderma*, **42**, 271–293.

—— & ——. 1988b Caractérisations sédimentologique et conditions paléoclimatiques de la mise en place de loess au Nord du Sahara à partir de l'exemple du Sud-Tunisien. *Bulletin de la Societe Géologique de France*, **8**, (4, 6), 1081–1090.

——, ——, BERGAMETTI, G., GOMES, L., STRAUSS, B., GROS, J. M. & LE COUSTUMER, M. N. 1987. Saharan dust on the Fuerteventura Island (Canary Islands): chemical and mineralogical characteristics, air-mass trajectories and probable sources. *Journal of Geophysical Research*, **92**, 9753–9771.

——, —— & WEISROCK, A. 1982b. Evolution du matériel sableux au cours de son déplacement dans un système dunaire — les barkhanes du Cap Sim au Sud d'Essaouira (Maroc). *Comptes Rendus de l'Académie des Sciences Paris*, **295** (2), 621–624.

FRIEDMAN, G. M. 1964. Early diagenesis and lithification in carbonate sediments. *Journal of Sedimentary Petrology*, **34** (4), 777–813.

GROUSSET, F. E., ROGNON, P., COUDÉ-GAUSSEN, G. & PEDEMAY, P. 1992. Origins of peri-Saharan dust deposits traced by their Nd and Sr isotopic composition. *Palaeogeography Palaeoclimatology Palaeoecology*, **93**, 203–212.

MILLOT, G., NAHON, D., PAQUET, H., RUELLAN, A. & TARDY, Y. 1977. L'épigénie calcaire des roches silicatées dans les encroûtements carbonatés en pays subaride: Anti Atlas, Maroc. *Sciences Géologiques*, Strasbourg, **30** (3), 129–152.

PYE, K. 1987. *Aeolian Dust and Dust Deposits*. Academic Press, London.

ROGNON, P. 1967. *Le Massif de l'Atakor et ses bordures (Sahara central). Etude géomorphologique*. CRZA Série Géologie, **9**, CNRS, Paris.

—— 1976. Essai d'interprétation des variations climatiques au Sahara depuis 40 000 ans. *Revue de Géographie physique et de Géologie dynamique* 2–3, **18**, 251–282.

—— & COUDÉ-GAUSSEN, G. 1987. Reconstitution paléoclimatique à partir des sédiments du Pléistocène Supérieur et de l'Holocène du Nord de Fuerteventura (Canaries). *Zeitschrift für Geomorphologie* NF, **31**, 1–19.

——, ——, LE COUSTUMER, M. N., BALOUET, J. C. & OCCHIETTI, S. 1989. Le massif dunaire de Jandia (Fuerteventura, Canaries): évolution des paléoenvironnements de 20 000 BP à l'Actuel. *Bulletin de l'Association Française d'Etudes Quaternaire*, **35** (1), 31–37.

VOGT, T. 1983. *Types et genèse de croûtes calcaires (France méditerranéenne, Afrique du Nord)*. Thesis Etat Sciences, Université P. & M. Curie, Paris.

WEISROCK, A., DELIBRIAS, G., ROGNON, P. & COUDÉ-GAUSSEN, G. 1985. Variations climatiques et morphogenèse au Maroc atlantique (30–33°N) à la limite Pléistocène–Holocène. *Bulletin de la Societe Géologique de France*, **8** (1), 565–569.

—— & MISKOVSKY, J. C. 1988. Nouvelles précisions sur le stratotype holocène de Makhfamane (Haut Atlas occidental, Maroc). *Bulletin de l'Association Française d'Etudes Quaternaire*, **36** (4), 205–215.

—— & ROGNON, P. 1977. Evolution morphologique des basses vallées de l'Atlas atlantique marocain. *Géologie méditerranéenne*, **4**, 313–334.

Occurrence and palaeoenvironmental implications of the Late Pleistocene loess along the eastern coasts of the Bohai Sea, China

PEI-YING LI[1] & LI-PING ZHOU[2]

[1]First Institute of Oceanography, State Oceanic Administration,
Qingdao 266003, China
[2]The Godwin Laboratory, University of Cambridge, Free School Lane,
Cambridge CB2 3RS, UK (Present address: School of Environmental Sciences,
University of East Anglia, Norwich NR4 7TJ, UK

Abstract: Loess is found to blanket various landforms of different altitudes in the eastern coastal areas of the Bohai Sea, from Liaodong Peninsula to the Miaodao Islands and Jiaodong Peninsula in eastern China. Two major lithological units have been recognized: the Penglai Loess and the Dalian Loess. Recent studies reveal a weathered unit within the Dalian Loess and confirm the occurrence of Holocene loess. Random distribution of marine microfossils and concomitant occurrence of contrasting species in the loess rule out the possibility of the coastal loess being a marine sediment. The coastal loess is coarser and contains a higher content of unstable minerals compared to the inland equivalents. Regional variations in texture and heavy mineral assemblage within the coastal loess zone are also observed, suggesting that the primary source of the coastal loess is not northwestern China but is more likely to be from the coastal areas and the adjacent Bohai Sea basin. When the coastline was displaced a few hundred kilometres eastwards during the last glacial period, vast areas of the continental shelf of the Bohai Sea were exposed. Coastal dunefields may have been expanded and perhaps developed locally into a desert shelf environment. With intensified aeolian process, fine particles in the ancient fluvial and marine sediments were transported by small-scale low-level northwesterly and northerly winds and deposited along the eastern coasts of the Bohai Sea. The basal age of the Late Pleistocene Dalian Loess was previously estimated to be 11–25 ka, but preliminary thermoluminescence dating results suggest that it started to accumulate at least 30 ka earlier.

Aeolian deposits are widely distributed in northern China (Fig. 1). Studies have shown that the alternating occurrence of loess and palaeosols in the Loess Plateau, north central China, reflects climate variation in the last 2.6 Ma (e.g. Liu & Yuan 1982; Kukla 1987; Rutter *et al.* 1991). The spatial extent of the loess deposits in China has been expanded through time (Liu 1966). Late Quaternary loess deposits are found in areas outside the Loess Plateau (Liu 1985; Zhang *et al.* 1989) and regional diversity and the palaeoenvironmental implications of these loess records are apparent. In this paper, we focus on the loess found along the eastern coasts of the Bohai Sea and our discussions are particularly concerned with the Late Pleistocene loess.

The present Bohai Sea coastline extends from the southern tip of the Liaodong Peninsula to the northern tip of the Jiaodong Peninsula, with its eastern border at the Laotieshan Strait (Fig. 2). The two peninsulas and the northwest corner of the Bohai Sea are regions of uplift and the Bohai Sea is a river-dominated subsidence basin. Most of the Bohai Sea coast consists of depositional plains but some bed-rock-embayed sections of coast are present on the Liaodong Peninsula and northern coast of the Jiaodong Peninsula (Wang & Aubrey 1987).

Nearly 30 years ago, regional Quaternary investigations (Jin & Zheng 1964) reported widespread occurrence of loess on the Miaodao Islands (Fig. 2). Subsequent studies have mainly attempted to establish the origin of the loess on the islands, and models invoking marine (Jin & Zheng 1964), colluvial–diluvial–alluvial (Zhao 1983) and aeolian (Li and Zhao 1983; Guo *et al.* 1983) processes have been put forward. Recently, Jiao *et al.* (1987) studied loess deposits on the Liaodong Peninsula (Fig. 2) and considered Inner Mongolian deserts (cf. Fig. 1) as the source area, while Lue & Li (1990) suggested the ancient Bohai Sea sediments as an additional source. Han (1987), Li (1987) and Cao *et al.*

From Pye, K. (ed.), 1993, *The Dynamics and Environmental Context of Aeolian Sedimentary Systems.*
Geological Society Special Publication No. 72, pp. 293–309.

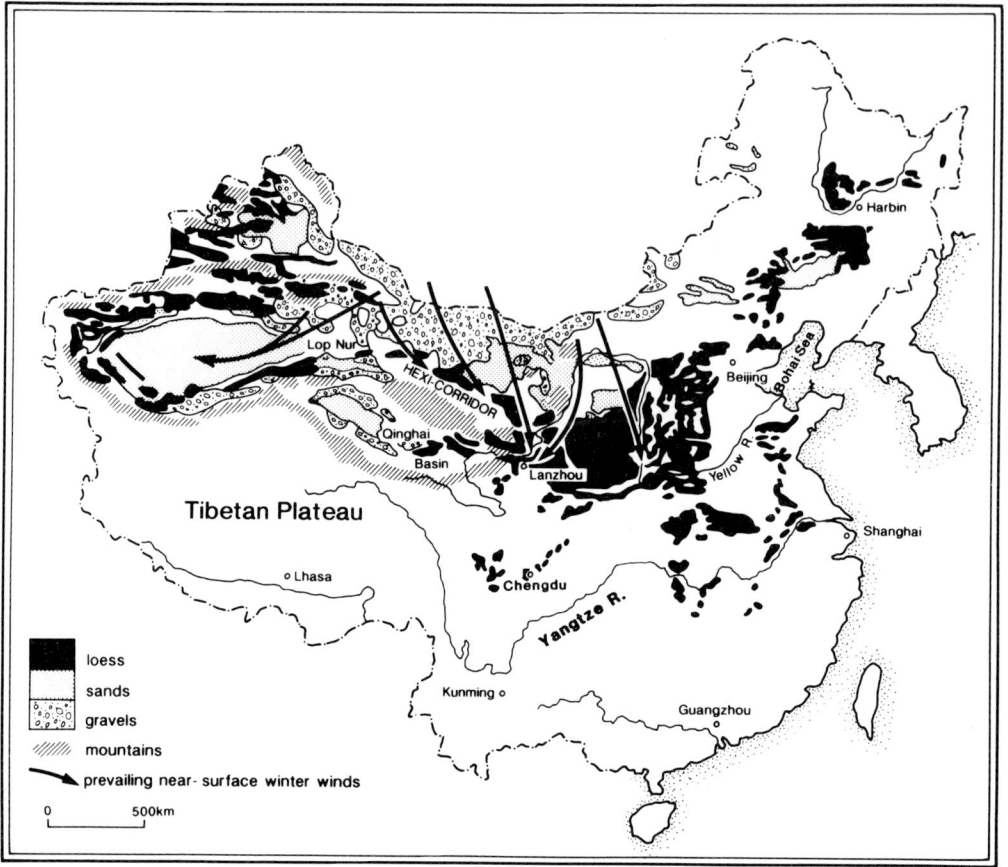

Fig. 1. Map showing distribution of loess and the main desert source regions of dust in northern China. (After Pye & Zhou 1989, based on various sources).

(1988) made systematic investigations on the loess along the northern coast of the Shandong Peninsula (Fig. 2) and on the Miaodao Islands. They concluded that the coastal loess is an aeolian deposit derived primarily from ancient marine sediments of the exposed Bohai Sea during the glacial times. More recent investigations have led to the proposition that loess on the Liaodong Peninsula is the product of desertification in the Liaodong Bay during the last glacial maximum (Li *et al.* 1991, 1992; Xia *et al.* 1991). Here we analyse recent lithological, mineralogical and chronological results and examine the palaeoenvironmental implications of the coastal loess in the eastern Bohai Sea area.

Loess distribution

Figure 2 shows the loess distribution in the

eastern coastal areas of the Bohai Sea. A discontinuous north–south coastal loess belt extends from 36–40.5°N and from 120–124°E, i.e. from the Gaixian, Liaoning in the north to Qingdao, Shandong in the south. This distinct loess zone can be divided into three sub-regions from the north to the south, i.e. the Liaodong Peninsula, Miaodao Islands and the Jiaodong Peninsula (Li, 1992).

Although the loess is generally found to blanket various relief features of different altitudes, its occurrence varies with respect to some large-scale slope orientations. As can be seen in Fig. 2, in the Liaodong Peninsula loess is well expressed along the northwest coast with a 200 km long loess belt running from Gaixian to Lueshun, whereas loess on the southeast coast and on the adjacent islands is rare. The thickness of the loess can attain 20–30 m in the Jinzhou and Lueshun areas with a maximum of up to

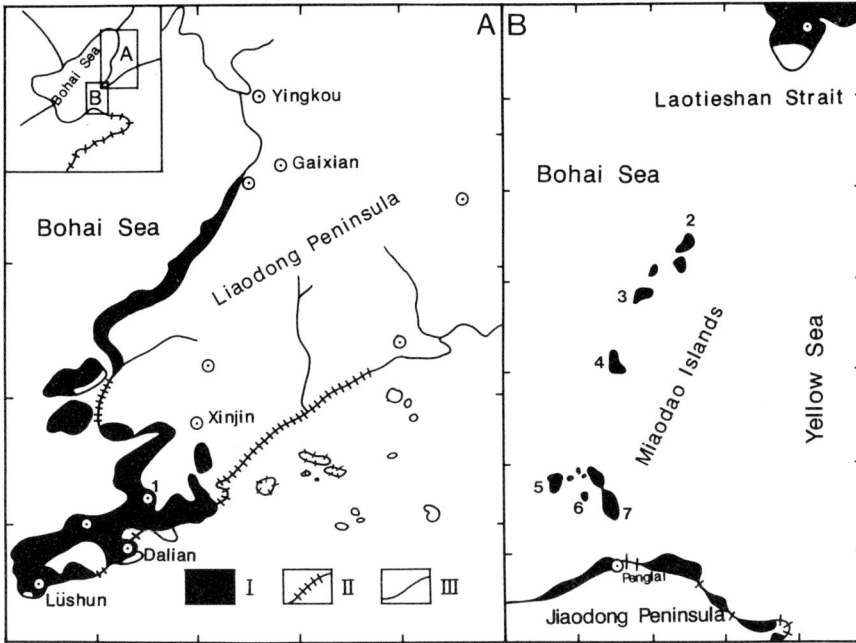

Fig. 2. Loess distribution along the eastern coasts of the Bohai Sea. I: major occurrence; II: sporadic occurence; and III: no loess. 1: Jinzhou; 2: North Hunagcheng Island; 3: Daqin Island; 4: Tuoji Island; 5: Daheishan Island; 6: Miaodao Island; 7: South Changshan Islands.

20 m for the Late Pleistocene loess. Geomorphologically, loess forms terraces or covers the piedmont slopes and relict planation surfaces.

On the Miaodao Islands, loess extends from the shoreline to 100 m above sea-level (asl), covering undulating slopes and denudation surfaces as well as filling the valleys. The older loess unit (see below) is mostly confined to the altitude of 0–30 m asl while the younger loess occurs over a wider altitudinal range.

In the Jiaodong Peninsula, loess is distributed mainly along the northern coast, particularly in the Penglai area. To the west of Penglai, distinctive loess platforms with steep marine-cut edges can be seen. In this area loess also covers a basalt platform and weathered regolith, as well as filling the valleys. The thickness of the loess decreases landward with increasing altitude.

Stratigraphy

In most places, two major loess units have been recognized. The lower unit is brownish with no stratification but vertical joints. It contains one or more dark-brown palaeosols with the top one being the most developed. Carbonate nodules of various sizes (1–10 cm diameter) are scattered

in the loess or below the soils. This loess unit is best expressed in the Penglai area, Jiaodong Peninsula, (Fig. 3) and in some southern islands of the Miaodao Islands. Beneath this loess unit is colluvium or Tertiary Red Loam.

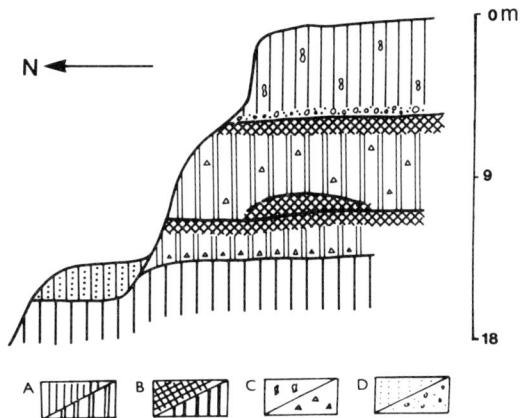

Fig. 3. Loess profile at Lingezhuang, Penglai, showing well-expressed Penglai Loess with interbedded palaeosols: (**a**) young loess/old loess; (**b**) palaeosol/red loam; (**c**) carbonate nodule/rock debris; and (**d**) sand/gravel.

The upper unit is lighter in colour and more porous, forming a distinct vertical cliff. At places where the palaeosol is missing, this unit is separated from the lower unit by a clear erosional surface. This loess is best exposed in the Dalian area, Liaodong Peninsula as well as on the Miaodao Islands. At some places in Dalian, it is underlain by a medium-fine sand (Lue & Li 1990).

As will be shown later, the upper loess unit may be correlated with the Malan Loess in north-central China and the lower unit with part of the Upper Lishi Loess. For convenience of discussion of these loess deposits with distinct regional characteristics, we use the Penglai Loess for the lower unit and the Dalian Loess for the upper one.

There has been some uncertainty as to the subdivision of the Dalian Loess. At Miaodao brickyard on the Miaodao Islands, we observed a sub-unit, over 3 m thick, within the Dalian Loess which is slightly more brownish than the loess above and below and is characterized by increased magnetic signals (Fig. 4a). Based on mineral magnetic measurements (Thompson & Oldfield 1986) of samples from this section and from other sites in the coastal area (Zhou, unpublished data), we interpret the increase in magnetic susceptibility to be the result of relatively intensified weathering or initial pedogenesis during the formation of this sub-unit. The degree of the magnetic enhancement is less pronounced compared with samples from the central and western Loess Plateau (Zhou et al. 1990), thus it seems reasonable to infer continuous loess formation for the time period concerned. While the base of the Dalian Loess and the underlying palaeosol are to be determined, the exposed sequence seems to suggest a possible subdivision of the Dalian Loess, similar to that of the Malan Loess in north China (An & Lu 1984). However, the local variation in stratigraphy is considerable and a complete sequence may be found only at a few favourable sites. At another section 2 km to the north on the same island, the loess is much thinner and contains many carbonate nodules (Fig. 4b). A dark-brownish palaeosol is well exposed and underlain by a diluvial–colluvial deposit. Such a variation is probably related to the local geomorphic conditions which determine both accumulation and preservation of the loess deposits.

At some sites, e.g. at the Miadao brickyard (Fig. 4a) and on the western coast of South Changshan Island, a dark-brownish palaeosol is seen at the top of the Dalian Loess. Above this soil is a light-yellowish loess, up to 1.5 m thick. In most places, however, this loess is often

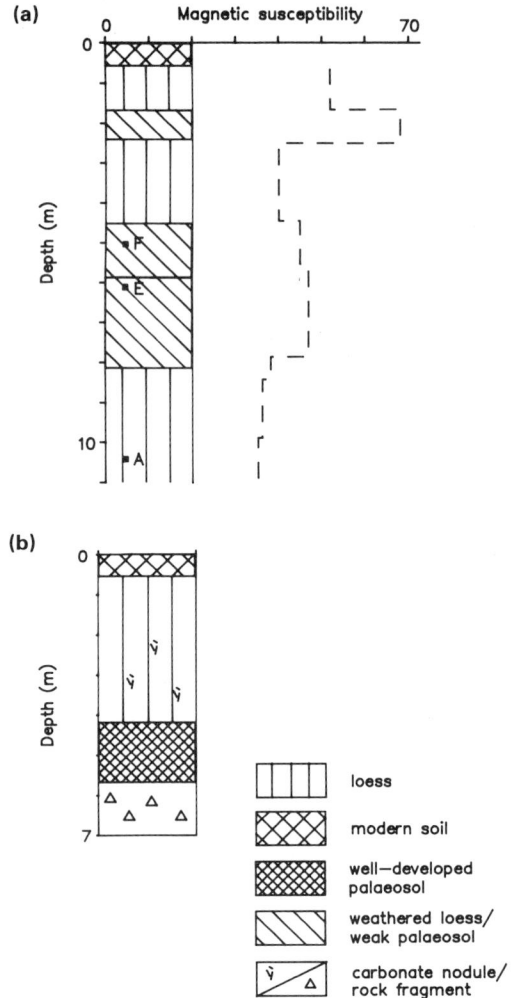

Fig. 4. Loess sections on the Miaodao Islands, showing stratigraphical variation in space: (**a**) at Miaodao brickyard, a weathered sub-unit is found within the loess. Magnetic susceptibility: 10^{-8} m^3 kg^{-1}. TL sampling positions are indicated with A, E and F; (**b**) at a site 2 km to the north of (a).

difficult to recognize due either to recent erosion or present-day cultivation. At a loess section near Penglai, Guo et al. (1983) described a dark-greyish palaeosol at a depth of 5 m. Fragments of Neolithic pottery were found in the soil, suggesting that the loess above is of Holocene age. On the southwest coast of the South Changshan Islands (Fig. 2), we observed a 0.5 m thick shell midden within the loess 3 m below present ground surface.

Rock debris and pebbles of various sizes can be found in some loess sections. However, their occurrence, which probably reflects temporary intervention of alluvial and colluvial processes during loess formation, is sporadic and varies considerably in space.

Chronology

Palaeomagnetism

Palaeomagnetic measurements on the coastal loess samples from various sites indicate that the Penglai Loess and the Dalian Loess were formed in the Brunhes Epoch (Han 1987; Li 1987). Lue & Li (1990) observed an apparent brief polarity departure at a depth of 5 m in a loess section at Jinzhou, which was interpreted as the Blake event. However, the 15 mT AF demagnetization they employed may not be particularly suitable for the sediments studied and further magnetic measurements are needed to confirm their finding.

Radiocarbon dating

Over a dozen ^{14}C dates have been obtained on secondary carbonates mostly from the lower part of the Dalian Loess at various sites (Li, 1987; Lue & Lie 1990). The distribution of these ^{14}C dates which mostly vary between 10 and 25 ka is unlikely to represent stages of loess deposition as the carbonates must have been formed after loess deposition. The assumption that the small carbonate nodules selected for ^{14}C dating may be formed shortly after loess deposition cannot be justified. The time interval between loess accumulation and nodule formation is often difficult to determine as both organic and inorganic carbon in the nodules may be affected by dissolved old detrital or biogenic carbonates and/or addition of recent carbon through precipitation and groundwater. The latter process explains the much younger dates of 3972 ± 160 ^{14}C years BP and 6868 ± 200 ^{14}C years BP for the carbonate nodules from the Dalian Loess at Lueshun reported by Jiao et al. (1987). It appears to be reasonable to suggest that the published ^{14}C dates of secondary carbonates from the Dalian Loess have little significance for establishing absolute chronology for the loess deposition. Therefore, the previous view based on these dates that the accumulation of the Dalian Loess coincides with the last glacial maximum needs reconsideration.

Thermoluminescence dating

In the absence of organic matter, thermoluminescence dating provides the only means of determining the time of loess deposition (Wintle 1981, 1987, 1990; Wintle & Huntley 1982; Wintle et al. 1984). TL dating studies of the coastal loess are few. Sun (1985) reported that quartz grains from the coastal loess showed different TL behaviour compared to that of loess in central China. This was attributed to some uneven zeroing due to the lack of 'heating' prior to deposition. Jiao et al. (1987) reported six TL dates ranging from 22 to 60 ka for the Dalian Loess at three sections in the Liaodong Peninsula (Table 1). The lack of experimental details makes the evaluation of these dates impossible.

In contrast to Sun's (1985) observation, we found that both polymineral fine grains and quartz in coastal loess display TL properties similar to those for loess from the Loess Plateau (Zhou, unpublished data). While detailed discussion on TL dating of Chinese loess will be given elsewhere, we present here preliminary results of a TL dating study on three samples from the Miaodao brickyard (Fig. 4a, Table 1). Samples were prepared following standard procedures (Aitken 1985) and some experimental information can be found in Table 1.

A TL date is determined by a number of physical variables (cf. Wintle & Huntley 1982) and may be strongly influenced by the laboratory procedures employed, including choice of grain size, laboratory bleaching condition, irradiation strategy, preheat treatment, optical filters, and methods of data analysis (Debenham 1985; Wintle 1987, 1990; Berger 1988; Rendell & Townsend 1988; Forman 1989; Zoeller et al. 1988). It is worth noting that TL dates close to true ages may be obtained when errors associated with factors determining a TL age unintentionally cancel one another out.

The most interesting point in Table 1 is the age of sample QTL110A from the base of the exposure. The age of 56 ± 5 ka is significantly older than the ages obtained by radiocarbon dating of secondary carbonates. Taking into account current problems in dating Chinese loess (e.g. Lu et al. 1988; Zhou & Wintle 1989; Wintle 1990; Forman 1991), this TL age estimate may be on the younger side rather than on the older side of the true age. While future study may provide more confidence in this date, we consider that it represents a more realistic age estimation for early Dalian Loess deposition than did the previous radiocarbon dates.

Fossils

Fauna

The coastal loess contains both terrestrial and marine fossils. Ostrich eggs and deer horns were

Table 1. *Thermoluminescence dating results for the coastal loess*

Sample No./Site	Depth (m)	Th[a] (ppm)	U[a] (ppm)	K$_2$O[b] (%)	H$_2$O (%)	a-value	D (mGy a^{-1})	ED[c] (Gy)	Age (ka)
110F	5.0	6.77	2.89	1.79	13	0.10	3.67	98 ± 7	27 ± 3 [d,e,f]
110E	6.1	9.12	2.56	1.49	14	0.09	3.56	102 ± 23	29 ± 5 [e,g,h]
110A	10.4	7.12	2.66	1.87	15	0.10	3.63	125 ± 4	34 ± 3 [d,e,h]
						0.11	3.49	196 ± 6	56 ± 5 [d,h,i]
								191 ± 4	55 ± 5 [d,f,i]
Miaoxi	0.8								34 ± 3 [j]
Miaoxi	4.7								60 ± 5 [j]
Miaoxi II	3.0								51 ± 5 [j]
Muchengyi Reservoir	6.5								22 ± 2 [j]
Muchengyi Reservoir	13.3								44 ± 4 [j]
Zhoujiagou	3.0								36 ± 4 [j]

[a] Obtained by thick source alpha counting. No radon loss was indicated in the comparison of sealed count rates and unsealed count rates.
[b] Obtained by atomic absorption spectroscopy.
[c] Averaged values of equivalent dose (ED) for TL signals above 280°C.
[d] Polymineral fine grains, preheat 16 h at 140°C.
[e] Corning 5-58 + HA3 filters.
[f] ED by the regeneration method, 5 h SOL2 solar simulator for bleaching.
[g] Fine-grained quartz obtained by 72 h treatment with hydrofluosilicic acid. No preheat.
[h] ED by the total bleach method, 5 h SOL2 solar simulator for bleaching.
[i] Corning 7-59 + HA3 filters.
[j] From Jiao *et al.* (1987). Miaoxi, Muchengyi Reservoir and Zhoujiagou all lie in the southern part of the Liaodong Peninsula. TL dates were obtained by the TL laboratory of the Institute of Archaeology, Chinese Academy of Social Sciences, Beijing.

found in the Dalian Loess on the islands and bones of mammoth and rhinoceros recovered from the adjacent seabed (cf. Cao *et al.* 1988).

Recently, over thirty species of foraminifera have been identified in the Penglai Loess and Dalian Loess (Fig. 5) and among the common species are *Cribrononion incertum*, *Elphidium magellanicum*, *Buccella frigida*, *Ammonia beccarii* vars, *Cribrononion subincertum*, *Cribrononion gnythosuturatum*, *Brizalina striatula*, *Elphidium nakanokawaense*, *Protelphidium* sp., *Lagana* sp., *Fissurina* sp., *Fissurina aradasii* and *Nonion glabrum*.

It is interesting to note that the coastal loess also contains radiolaria, mostly spumellaria (Fig. 5). Species such as *Hexacontium enthacanthum* Jorgensen, *Dictyocoryne profuna* Ehrenberg, *Stylodictya* sp., *Heliodiscus* sp., *Xiphosphaera* sp., *Hexastylus* sp. and *Cyclastrum trifastigiatum* sp. nov. have been recorded from the Dalian Loess on the islands. Foraminifera are rare in the Penglai Loess but radiolaria including *Heliodiscus* sp., *Acrosphaera spinosa* Haeckel, *Cenosphera* sp., *Hexastylus* sp. and *Spongocore polyacantha* Popofsky are present.

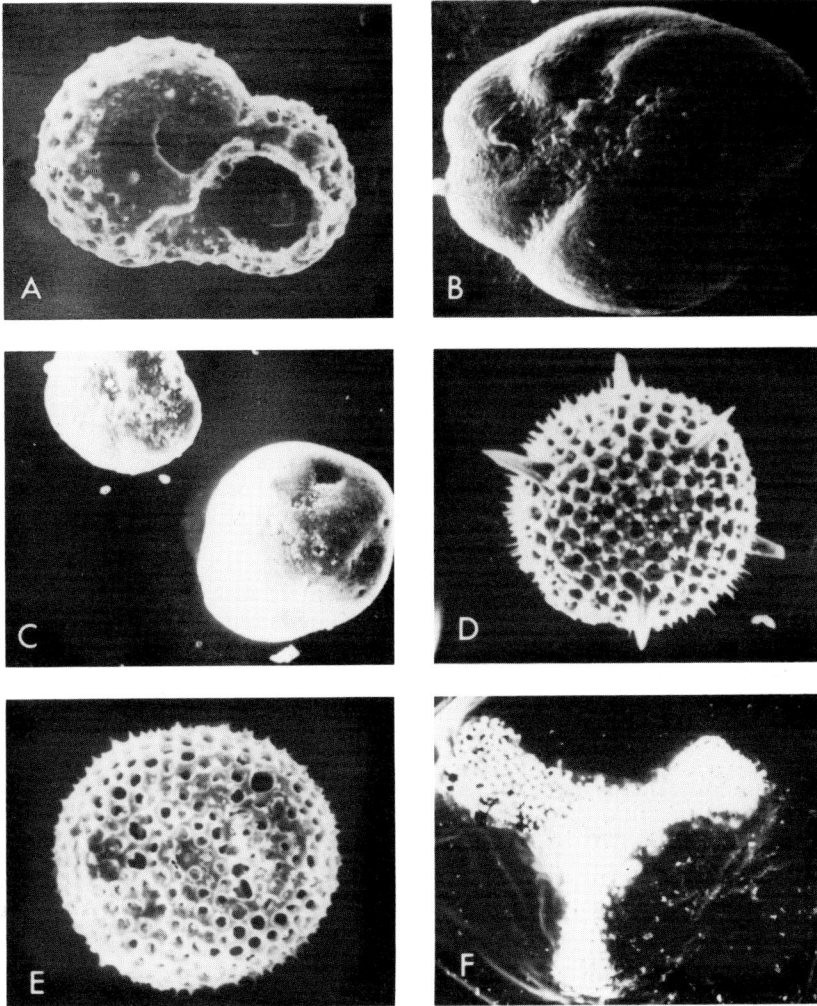

Fig. 5. SEM micrographs of selected marine microfossils found in the coastal loess. (**a**) *Globigerinoides* sp. ×241; (**b**) *Elphidium magellanicum* ×265; (**c**) *Buccella frigida* ×210; (**d**) *Hexacontium enthacanthum* Jorgensen ×225; (**e**) *Cenosphera* sp. ×305; (**f**) *Euchitonia triangulum* (Ehrenberg) ×158.5.

Foraminiferal tests are seen to have been more-or-less evenly abraded and only small numbers of them show some signs of mechanical damage (Figs 5a–c). By comparison, radiolaria samples have nearly intact surface features (Figs 5d–f). Assuming similar modes of transport for them, such a difference is probably due to the lower resistance to abrasion for the calcareous foraminifera or, alternatively, caused by prior transport or post-depositional dissolution of the foraminifera.

The microfauna described above suggest diverse marine environments, ranging from deep water indicated by common Pacific radiolaria to near-shore conditions represented by foraminifera such as *Ammonia beccarii* vars, and from temperate–warm water planktonic *Globigerinoides* sp. to temperate–cold water species such as *Buccella frigida* (Cushman). These fossils are found to have been well mixed and no definite association of certain species with particular loess units has yet been conclusively established. It should be pointed out that these fossils are related to the sea water conditions of different time periods before rather than during the loess formation, and their palaeoecology does not indicate the nature of sedimentary environments during loess accumulation. Despite the great variety of the species, the content of the microfossils in the loess is very low, ranging from 10 to 250 tests per 100 g of dry loess, which may point to a poor preservation condition in the loess. The random distribution of these microfossils and the concomitant occurrence of contrasting species in the loess rules out the possibility of the coastal loess being a marine sediment. Most of the microfossils found in the coastal loess are smaller than 100 μm (Han 1987; Li 1987). Due to their small specific weight, they must have been very susceptible to aeolian transport, consistent with Goudie & Sperling's (1977) observation that foraminifera have been carried

as far as 800 km inland by onshore winds in northwest India.

Flora

Pollen in the coastal loess is generally scarce and is dominated by herb taxa. Lue & Li (1990) and Li *et al.* (1992) reported widespread occurrence of *Artemisia* in the loess of the Liaodong Peninsula. The palaeoflora also comprised Chenopodiaceae, Compositae, Pottiaceae, Humulus and *Pinus*. This vegetation cover is different from the present temperate deciduous broad-leaved forest.

Lithology and mineralogical composition

Granulometry

Table 2 compares the textural characteristics of the Dalian Loess and the Malan Loess in Luochuan, central Loess Plateau and Zibo which lies to the west of the Shandong Peninsula. These loess deposits show some common features such as a dominant silt fraction, moderately good sorting and positive skewness. However, a significant difference in the grain-size distribution is also noted. The Dalian Loess is coarser than the inland Malan Loess. A similar trend is revealed when the Upper Lishi Loess and the Penglai Loess are compared (Han 1987; Lue & Li 1990). The fine sand content attains an average of 32.5% for the Dalian Loess which is significantly higher than 4.8% and 7.5% for the Malan Loess at Luochuan and Zibo. The coastal loess can, therefore, be described as sandy loess, and is inconsistent with the general trend of textural variation in Chinese loess which shows a southeastwards decrease in mean grain size (Liu 1966).

Table 2. *Comparison of granulometry and textural parameters between the coastal Dalian Loess and the inland Malan Loess*

| Location | Textural composition (%) | | | | Statistical parameter (φ) | | | |
	Median sand <2φ	Fine sand 2–4φ	Silt 4–9φ	Clay >9φ	M_z	M_d	σ	S_k
Coastal area	2.4	32.5	49.2	16.0	5.37	4.35	2.16	0.63
Luochuan[a]		4.8	76.8	18.4	6.64	6.00	2.15	0.40
Zibo[b]	0.5	7.5	73.5	18.5	6.35	5.50	2.14	0.46

[a] From Liu (1966).
[b] From Yuan (1959).

On the other hand, it is interesting to note that within the coastal loess zone the mean size decreases systematically from north to south. As may be seen in Fig. 6, loess on the Jiaodong Peninsula is finer than that on the Liaodong Peninsula. The fine sand fraction content decreases southwards from 57.9% on the Liaodong Peninsula to 34.7% on the Miaodao Islands and 11.7% on the northern Jiaodong Peninsula.

It may be concluded from such a marked variation in loess texture within the limited coastal loess zone, together with the evidence of loess, that (1) the coastal loess has a significant component derived from a proximal source(s); (2) a local-scale wind system was responsible for the aeolian transport; and (3) the palaeo-wind was predominantly northerly or northwesterly. On the other hand, as sedimentation in the Bohai Sea is controlled by fluvial influx from four major and many smaller rivers, each discharging characteristic sediments to various parts of the sea shore, it is possible that the grain-size distribution and mineral composition (see below) of the coastal loess have also been influenced by those of the proximal source materials.

Mineralogy

The light mineral fraction, up to 99% of the loess, is dominated by quartz and feldspars. Some of the quartz grains have clean surfaces while others are coated with ferric oxide or alumino-silicate materials. Some of the feldspar grains display brownish surface spots and altered edges which apparently result from weathering.

Although most quartz grains are sub-rounded, sub-angular grains are present in all samples. Surface features of the quartz grains studied by scanning electron microscopy (Sun 1985; Cao *et al.* 1988; Mu *et al.* 1989; Lue & Li 1990) include frosted plates, conchoidal fractures, smooth surfaces, V-shaped pits and dish-shaped depression as well as amorphous silicate coatings, oriented etch pits and etch lines or holes. Most of these surface textural features are common in aeolian sediments while some reflect the combined effects of marine deposition and aeolian reworking as well as weathering.

Heavy minerals, with concentrations ranging from 1 to 5%, are composed mainly of amphiboles, epidote, and opaques (Han 1987; Li 1987; Lue & Li 1990). The amphiboles are dominated by hornblende, and several varieties of pyroxene, mica, actinolite, zoisite, iddingsite, garnet, zircon and magnetite are also present. This mineralogical association bears a general resemblance to those in the inland loess (Wei 1986) except the iddingsite which occurs only at sites in close association with local basalt, e.g. in Penglai area and the Daheishan Island. If the coastal loess shared a source with the inland loess and the loess was primarily transported through a large-scale high-altitude atmospheric pathway, it should have a relatively stable heavy mineral assemblage and there should be little difference between loess in the three sub-regions. However, as we shall show below, this is not the case (Table 3).

Heavy minerals can be classified according to their resistance to chemical alteration as unstable (e.g. hornblende), relatively stable, stable and very stable components. The R value in

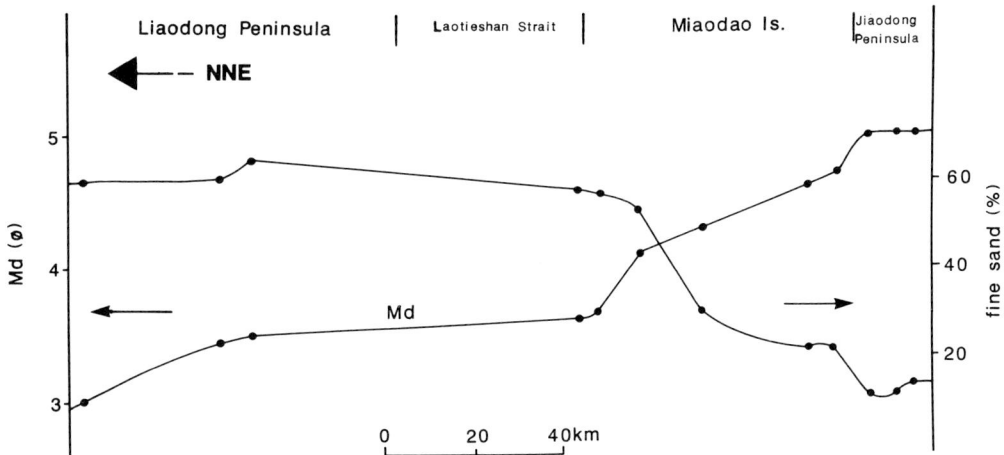

Fig. 6. Textural variations within the coastal loess zone.

Table 3. *Comparison of heavy mineral characteristics between the coastal Dalian Loess and the inland Malan Loess*

Location		Heavy mineral assemblage (%)							
	Unstable	Fairly stable	Stable	Very stable	Amphiboles	Epidotes	Opaques	R
Liaodong Peninsula	66.8	17.7	13.9	1.2	64.6	10.1	12.8	4.45
Miaodao Islands	44.2	24.4	26.2	4.2	38.3	15.9	22.7	1.46
Jiaodong Peninsula	41.9	38.6	15.7	3.0	33.5	26.1	14.5	2.24
Luochuan[a]	31.4	35.5	30.0	3.2	23.7	25.1	27.9	0.94
Zibo[b]	14.2	53.7	26.0	4.4	26.4	30.7	24.5	0.47

[a] From Liu (1966).
[b] From Yuan (1959).

Table 3 is defined as the ratio of the unstable component to stable and very stable components. We consider that the higher R values, i.e. higher contents of unstable heavy mineral species, in the coastal Dalian Loess relative to those in the inland Malan Loess are controlled by the provenance, distance from the source, and the nature of depositional and/or post-depositional weathering, which may be different from those for the inland loess.

Moreover, the relative abundance of unstable heavy minerals seems to show a decreasing trend from the north to the south, i.e. a southwards increase in the stability of the heavy mineral assemblages (Table 3). This is consistent with the spatial variation in loess texture and would also indicate near source deposition under the influence of a palaeo-wind system which transported sand and silt southwards.

A comparison of heavy mineral assemblages provides a further clue about the source area. It may be seen in Fig. 7 that the average contents of unstable and stable minerals for the coastal Dalian Loess are closer to those in Bohai Sea floor sediments (Chen *et al.* 1980) than to those for the Malan Loess of Luochuan. Combined with lithological and microfauna evidence presented above, this suggests that the most likely source lies in the region occupied by the present Bohai Sea.

Illite is the main clay mineral but kaolinite, smectite and chlorite are also present. This clay mineralogy is similar to that in the inland loess (cf. Liu 1985). As these clay minerals are common in temperate or cooler environments, the similarity noted here does not necessarily suggest a shared provenance. Nevertheless, some clays may have their origin in north China and have been brought to the coast by rivers.

The relative abundance of these clays varies with lithology, an increase in kaolinite and decrease in smectite and chlorite content being observed in the palaeosols. Differences also exist between the the two major loess units. The chlorite content is significantly higher in the Dalian Loess than in the Penglai Loess. By comparison, spatial variation of clay mineralogy is relatively small.

The percentage of calcium carbonate varies widely and, in general, is lower than that in the Loess Plateau samples. While it may be as high as 8% in the Dalian Loess, many Penglai Loess and palaeosol samples are almost entirely decalcified (< 1%). Higher values are found, however, in secondary carbonate concentrations associated with palaeosols. For the Penglai Loess, carbonate content is slightly higher in the north than in the south. This may be related to the difference in leaching intensity which might have been controlled by local precipitation. However, no such spatial variation has been found for the Dalian Loess.

Palaeoenvironmental implications

Aeolian transport

The high content of fine sand in the coastal loess provides some indication of the mode of transport. As recently discussed by Pye (1987) and Tsoar & Pye (1987), sand grains larger than 70 μm are mainly moved by saltation or modified saltation. Most silt grains of 20–70 μm are transported in short-term suspension while fine silt and clays, i.e. < 15 μm, may be moved in long-term suspension. Therefore, for the coastal loess, the modified saltation and short-term suspension seem to have been two main modes

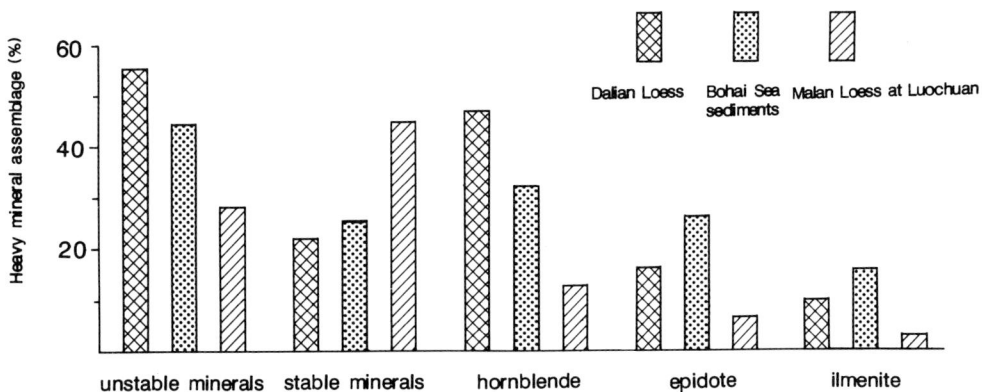

Fig. 7. Comparison of heavy mineral assemblages between coastal loess, inland loess and the Bohai Sea sediments.

of transport. The finest fraction may be related to long-term suspension of materials carried by turbulent wind flows at both low and high levels. As during the dust storms in historical times (Zhang 1982), fine grains from northwest China could have travelled a few thousand kilometers by turbulent winds during the dry and cold periods of the Quaternary. Therefore, it is reasonable to expect some materials from sources of the northwest interior in the coastal loess. This contribution facilitated by high-level winds should be mainly seen in the fine silt and clay fractions. Some of the distal materials may also come from Northeast China and eastern Inner Mongolia which are potential sources deserving further investigation.

The efficacy of the distance-dependent atmospheric transport of dust from central Asia and northwest China is clearly seen in the granulometric variations summarized by Park (1987). The silt content in the red-yellow soil of the southwestern coastal area of Korea is higher than that of Kosa (airborne dust) deposited in Japan, but lower than that in the loess of north China, while a reversed trend is observed in the clay content which ranges between 30 and 40% for the red-yellow soil in Korea. Fine dust ($< 10 \mu$m) is also carried further downwind and has been widely recorded in the deep-sea sediments of the north Pacific (Rea et al. 1985; Hovan et al. 1989). It should be noted that the long-distance transport of dust depends strongly on the frequency and intensity of the vertical dust lifting events (Pye & Zhou 1989).

It is clear that the textural characteristics of the coastal loess which occurs between the mainland (China) and the islands to the east (Korea and Japan) deviate from the distance-controlled granulometric variations for dust originating from northwest China. This may be best explained by the contribution from a proximal source(s) to the coastal loess. The deduced northerly and northwesterly directions of winds responsible for the coastal loess deposition are consistent with present-day prevailing near-surface wind directions for the winter and spring seasons in the region. The height of deflated particles may have been largely confined to the lower 1000–1500 m of the atmosphere. Such a mode of aeolian transport for the coastal loess is consistent with the loess deposition model discussed by Pye & Zhou (1989).

Effects of sea-level and climate change

Several studies (e.g. Cang et al. 1986; Zhao & Qin 1986) have shown that four transgressions occurred in the Bohai Sea area at approximately 85–70, 65–53.5, 39–23 and 8.5 ka BP. The extent of these transgressions varied considerably and current data indicate that the transgression at 39–23 ka was the most extensive one. Between these transgressions were periods with low stands of sea-level.

The sea-level changes in the Bohai Sea may have had a series of effects on the regional sedimentary environments. The lowering of sea-level would have resulted in an extension of major rivers onto the continental shelf, leading to deposition of large amounts of fine-grained clastic sediments. Falling sea-level may have also caused landward sediment transport due to the lowering of wave base. With an increase in shelf area, the progradation of sand dunefields towards the sea may have occurred as fluvial and marine sediments were exposed to deflation. On the other hand, during periods of climate amelioration, increased river influx to the sea would effectively replenish the store of fine materials for aeolian deflation during subsequent regressions. A similar effect of fluvial sedimentation on aeolian dust transport in arid areas has been noted by Pye (1989).

Figure 8 is an example of submarine shallow profiling records obtained to the west of the Liaodong Peninsula. Three different sedimentary structural units, denoted A, B and C, may be distinguished. Note that the chronological framework shown in Fig. 8 is tentative. Discontinuities exist and the sedimentary structures of older sediments might have been disturbed and modified by subsequent marine reworking. The preservation of the original structure depends on several factors, such as the rate of rising or falling of the sea-level, the rate of vegetation colonization and the nature of the lithologies (Glennie 1970). It is interesting to note that unit C displays medium- to high-angle cross-bedding. Such sedimentary structures have been widely recorded in the lag deposits in the northern and central Bohai Sea (Xia et al. 1991).

These lag deposits are mainly composed of medium to fine sands and are found to contain contrasting species of terrestrial and marine fauna and reworked pollen. Thus they have been interpreted as remains of an ancient desert developed on the exposed continental shelf (Li et al. 1991; Xia et al. 1991). The extensive occurrence of such sand deposits with sedimentary structures characteristic of aeolian sands seems to be consistent with this interpretation. While future studies will concentrate on the characterization of these submerged sandbodies including their distribution, morphostratigraphy

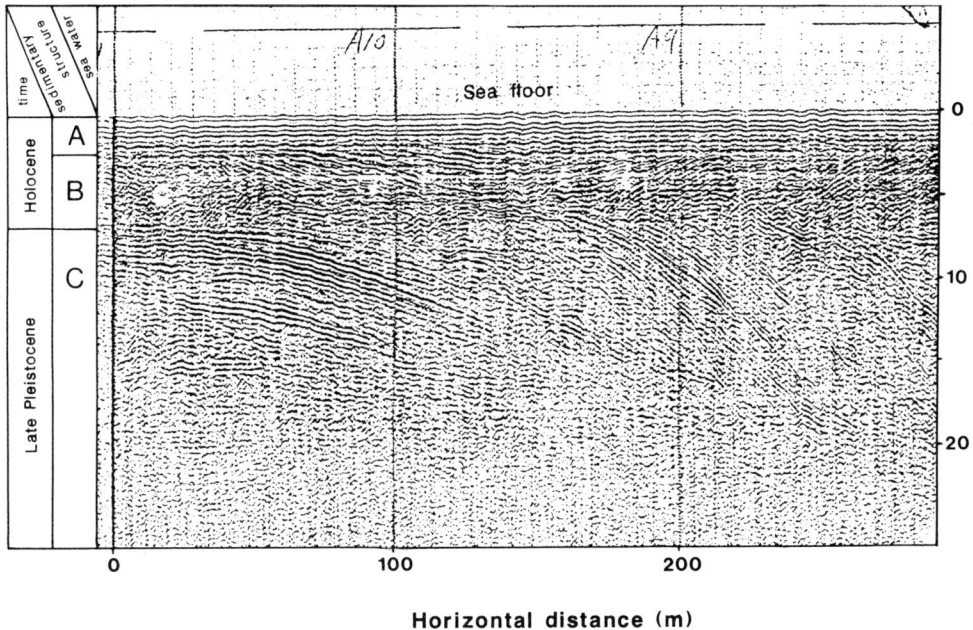

Fig. 8. An example of submarine profiling records commonly observed from the northern Bohai Sea, showing distinguishable sedimentary structures (A, B and C) in the sequence.

and internal geometry, here we examine the palaeoclimatic background which may have determined the sedimentary environments of the ancient Bohai Sea basin.

It is well established that many of the present-day sand seas were enlarged during earlier periods of the Quaternary (e.g. Glennie 1970; Pye & Tsoar 1990). Sarnthein (1978) showed that maximum aeolian activity occurred on a global scale around the time of the last glacial maximum due to combined effect of changes in wind regime, temperature and rainfall (Pye & Tsoar 1990). Increased aeolian activity was accompanied by a lowering of sea-level in most places. Figure 9 shows that the coastline in eastern China was displaced eastwards during the late Pleistocene. Increased loess accumulation in north central China generally coincided with the times of low sea-level (Liu 1985), suggesting that the overall environmental conditions in north China were colder, drier, more windy and more dusty during these periods. A similar trend of climate change may be expected in the Bohai Sea area. There is lack of sand supply and a thick sand accumulation is favoured in the subsiding river-dominated Bohai Sea area. With increased aeolian activity, sediments of various origins exposed on the shelf during glacial times were sorted and deflated by wind. Initially, dunefields similar to those in coastal areas may have been developed (Pye 1983; Goldsmith 1978; Carter et al. 1990). Although the prevailing wind is inferred to have been northerly and northwesterly, seasonal onshore southerly and south-easterly winds might have been primarily responsible for the dune formation. The regional palaeohydrological regime is unclear but rivers and lakes might still have been in existence at this time as seen in the central Bohai Sea area (Cang et al. 1986). Fine materials were transported downwind towards the eastern coasts and deposited to form loess. It should be noted that coastal dunes may be formed under almost all types of climate (Pye 1983; Carter et al. 1990) and the occurrence of such deposits should not always be taken as evidence of increased aridity.

Nevertheless, when the coastline migrated to its extreme position 600–1000 km east of the present coast around the time of the last glacial maximum, some parts of the extended coastal dunefield may have evolved into a shelf desert. Certain sedimentary features characteristic of continental deserts could be observed in these sand deposits. This would not be surprising as the distance between the dunes and the sea at

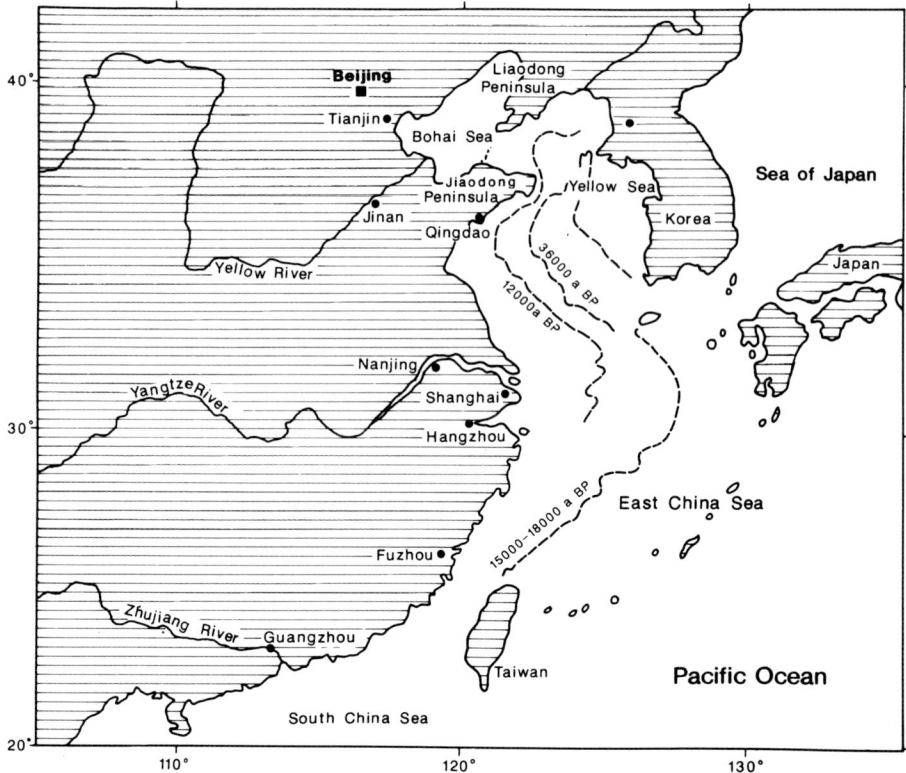

Fig. 9. Late Pleistocene coastline changes in the Bohai Sea and the Yellow Sea (based on Wang & Wang 1980; Liu *et al.* 1987; First Institute of Oceanography, State Oceanic Administration 1989).

that time was of the same order as the interior desert to the present coast in the east. At this time an increase in aridity may have occurred in the ancient Bohai Sea basin, as observed in the coastal areas of Australia (Bowler 1976). Recently, Zhao & Li (1991) also identified aeolian sandy deposits in the Yellow Sea Trough and deduced an intense desertification in the early stage of the regression of the last glacial maximum. This may be related to the increase in global ice volume and the suppression of regional monsoon winds. Although the northwesterly winter monsoonal wind was weaker at this time, loess deposition may have been accelerated as a result of a reduction in vegetation cover in the source areas, particularly on the exposed continental shelf. However, as the development of a desert will depend on the moisture balance as well as temperature and wind intensity, the extremely severe environmental conditions required for the formation of the continental desert may have been confined to certain parts of the basin and only for certain period of time. There is no compelling evidence

to relate loess deposition specifically to any extreme sedimentary environment in the region. The widespread occurrence of Holocene loess on the islands appears to indicate the diversity in loess depositional environments, though all were generally windy, dry and probably cold. The sporadic occurrence of the debris fragments in some of the loess sections also indicates the influence of intermittent and local hillwash during loess formation.

Accumulation rate of the late Pleistocene coastal loess

The thickness of the Dalian Loess varies considerably in space. However, sections over 10 m thick are not uncommon and a maximum thickness of 20 m has been recorded near Lueshun. Such a thick late Pleistocene loess in the coastal area, which is far from the northwestern deserts, is surprising as the inland equivalent Malan Loess rarely exceeds 10 m in most parts of the Loess Plateau and thicker loess has only been found in the western Loess Plateau (Derbyshire 1983). This again lends support to the proximal source model for the coastal loess discussed above.

The stratigraphic sequence at the Miaodao brickyard suggests that the TL age for sample QTL110A should represent the early stage of the Dalian Loess deposition as the base is not fully exposed. In other words, late Pleistocene loess accumulation began earlier than 56 ka ago. While more detailed TL dating studies in the future will provide a more precise constraint on the mode of loess accumulation, the three TL dates suggest an apparent variation in average accumulation rate for the loess at this site which varies between 140 and 220 mm ka^{-1}. The lowest accumulation rate is found in the weathered loess unit during the time interval between 36 and 28 ka constrained by the samples QTL110E and F (Fig. 4a). This is consistent with the marine evidence in which a transgression was recorded between 39 and 23 ka in the western and southern parts of the Bohai Sea (Cang et al. 1986; Zhao & Qin 1986). Intensified aeolian activity may be indicated for the periods represented by unweathered loess above and below this unit. Although this may be associated with an increased dust influx from the northwest interior as a result of enhanced aridity, some local factors related to sea-level changes may have played an important role.

According to the preliminary TL chronology, loess deposition seems to have continued throughout the later part of the last glacial period during which appreciable environmental changes must have taken place. The incomplete nature of the loess sequences in most places may be attributed to unfavourable local topography. This lends further support to the conclusion that coastal loess deposition may occur under a fairly wide range of environmental conditions.

Conclusions

Loess in the eastern coastal areas of the Bohai Sea has a substantial contribution from the deflated marine and fluvial sediments exposed on the coast and shelf of the Bohai Sea during low sea-level stands in the late Pleistocene. Coastal loess deposition seems to have been largely controlled by small- to medium-scale low-altitude atmospheric process. The loess distribution pattern suggests that the predominant palaeo-wind direction was northerly and northwesterly, in agreement with modern winter winds in the region. During the regressions and low stands in sea-level, coastal dunefields probably experienced substantial

expansion and an environment similar to a continental desert may have developed locally when the shoreline stood 600–1000 km east of its present position. Late Pleistocene loess deposition was not confined to the period of last glacial maximum (around 20 ka BP) as previous ^{14}C dates suggested, but started at least 30 ka earlier. Variation in aeolian activity during the last glacial period may be recognized at some sites with well-preserved loess sections. The presence of localized Holocene loess suggests that the coastal loess deposition may have occurred under various environmental conditions.

We thank Professor Cao Jiaxin for introducing us to the coastal loess and for her encouragement throughout the work. Field work has been assisted by Liaoning Normal University, First Regiment of the Changdao Garrison, and local government. Discussions with Han Jingtai, Liu Guohai, Liu Zhenxia, Lue Houyuan, Shi Ning, Wang Yongji, Xia Dongxing and Zhao Songling have been most helpful. Laboratory assistance from Jia Xiufen, Cheng Zhenbo, Lu Kang, Lue Chenggong, Zhuang Yun and Zhang Deyu is gratefully acknowledged. LPZ thanks St John's College, Cambridge for support, and Drs Ken Pye and Ann Wintle for helpful advice. This work was partially supported by Grant No. 48900038 from the State National Sciences Foundation of China.

References

AITKEN, M. J. 1985. *Thermoluminescence Dating.* Academic Press, London.

AN, X. & LU, Y. 1984. A climatostratigraphic subdivision of Late Pleistocene strata named by Malan formation in North China. *Kexue Tongbao*, 29, 1239–1242.

BERGER, G. W. 1988. TL dating studies of tephra, loess and lacustrine sediments. *Quaternary Science Reviews*, 7, 295–303.

BOWLER, J. M. 1976. Aridity in Australia: age, origins and expression in aeolian landforms and sediments. *Earth Science Reviews*, 12, 279–319.

CANG, S. X., HUANG, Q. F., ZHANG, H. C. & ZHAO, S. L. 1986. Transgression and sea-level changes in the late pleistocene in Bohai sea. *In*: CHINA WORKING GROUP, *China Sea Level Changes*. IGCP Project No. 200. China Ocean Press, Beijing, 42–50.

CAO,, J. X., LI, P. Y. & SHI, N. 1988. Study on the loess of Miaodao Islands in Shandong province. *Scientia Sinica* (B), 31, 120–127.

CARTER, R. W. G., NORDSTROM, K. F. & PSUTY, N. P. 1990. The study of coastal dunes. *In*: NORDSTROM, K. F., PSUTY, N. & CARTER, B. (eds) *Coastal Dunes Form and Process*. Wiley, Chichester, 1–14.

CHEN, L. R., LUAN, Z. F., ZHENG, T. M., XU, W. Q. & DONG, T. L. 1980. Mineral assemblages and their distribution patterns in the sediments of the Gulf of Bohai Sea. *Oceanologia et Limnologia Sinica,* **11**, 46–64.

DEBENHAM, N. C. 1985. Use of UV emissions in the TL dating of sediments. *Nuclear Tracks and Radiation Measurements,* **10**, 717–724.

DERBYSHIRE, E. 1983. Environmental change along the Old Silk Road. *Geography,* **69**, 108–118.

FIRST INSTITUTE OF OCEANOGRAPHY, STATE OCEANIC ADMINISTRATION. 1989. *Landform Atlas of China* (Lueshun–Dalian Area). Science Press, Beijing.

FORMAN, S. L. 1989. Applications and limitations of thermoluminescence to date Quaternary sediments. *Quaternary International,* **1**, 47–59.

—— 1991. Late Pleistocene chronology of loess deposition near Luochuuan, China. *Quaternary Research,* **36**, 19–28.

GLENNIE, K. W. 1970. *Desert Sedimentary Environments.* Elsevier, Amsterdam.

GOLDSMITH, V. 1978. Coastal dunes. *In:* DAVIS, R. A. JR (ed.). *Coastal Sedimentary Environments.* Springer-Verlag, New York, 171–235.

GOUDIE, A. & SPERLING, C. H. B. 1977. Long distance transport of foraminiferal tests by wind in the Thar desert, northwest India. *Journal of Sedimentary Petrology,* **47**, 630–633.

GUO, Y. S., HAN, Y. S., YANG, G. F. & ZHANG, M. H. 1983. On the Pleistocene high stands of sea level in the Shandong Peninsula. *Acta Oceanologica Sinica,* **5**, 480–489.

HAN, J. T. 1987. Study on the loess in Penglai district, Shandong Province. *In:* LIU, T. S. (ed.) *Aspects of Loess Research.* China Ocean Press, Beijing, 76–84.

HOVAN, S. A., REA, D. K., PISIAS, N. G. & SHACKLETON, N. J. 1989. A direct link between the China loess and marine delta ^{18}O records: aeolian flux to the north Pacific. *Nature,* **340**, 296–298.

JIAO, Y. N., WEI, C. K. & FU, W. X. 1987. A study on the characteristics of loess in the Liaodong Peninsula. *Scientia Geographica Sinica,* **7**, 231–237.

JIN, L. X. & ZHENG, K. Y. 1964. Preliminary geological survey of the Miaodao Islands. *Oceanologia et Limnologia Sinica,* **6**, 364–369.

KUKLA, G. J. 1987. Loess stratigraphy in Central China. *Quaternary Science Review,* **6**, 191–219.

LI, P. Y. 1987. Late Cenozoic Erathem and environmental vicissitudes of Miaodao archipelago, Shandong Province. *Marine Geology & Quaternary Geology,* **7**, 111–122.

—— 1992. Distribution pattern of coastal loess in the Bohai Strait and Jiaodong and Liaodong Peninsulas. *Journal of Oceanography of Huanghai and Bohai Seas,* **10**, 25–33.

——, CHENG, Z. P., LUE, H. Y. & LIU, G. H. 1992. Loess along the Liaodong coast. *Acta Geologica Sinica,* **66**, 82–94.

——, XIA, D. X. & LIU, G. H. 1991. Approach to the origin of loess in coast of East China and the desertization of Bohai Sea during glacial periods. *In:* LIANG, M. S. & ZHANG, J. L. (eds) *Correlation of Continental and Marine Quaternary in China.* Science Press, Beijing, 50–60.

LI, W. Q. & ZHAO, Y. J. 1983. Preliminary study of unsolidified Quaternary sediments on the Miaodao Islands. *Marine Science,* **3**, 20–23.

LIU, M. H., WU, S. Y. & WANG, Y. J. 1987. *Late Quaternary Sedimentation of the Yellow Sea.* China Ocean Press, Beijing.

LIU, T. S. 1966. *Composition and Texture of Loess.* Science Press, Beijing.

—— 1985. *Loess and Environment.* China Ocean Press, Beijing.

—— & YUAN, B. 1982. Quaternary climatic fluctuation — A correlation of records in loess with that of the deep sea core V28-238. *In: Research on Geology* (I). Institute of Geology, Academia Sinica, Culture Relics Publishing House, Beijing, 113–121.

LU, Y., ZHANG, J. Z. & XIE, J. 1988. Thermoluminescence dating of loess and palaeosols from the Lantian section, Shaanxi Province, China. *Quaternary Science Reviews,* **7**, 245–250.

LUE, J. F. & LI, Z. M. 1990. Loess and its sedimentary environment in Liaodong Peninsula. *Scientia Geographica Sinica,* **10**, 97–106.

MU, Y. Z., LI, X. M. & LIU, Z. B. 1989. Surface features and environmental significance of quartz grains from the loess deposits of the Dalian coast. *Liaoning Normal University Journal (Science series),* **3**, 57–60.

PARK, D. W. 1987. The loess-like red yellow soil of the south western coastal area of Korea in comparison with the loess of China and Japan. *Geojournal,* **15**, 197–200.

PYE, K. 1983. Coastal dunes. *Progress in Physical Geography,* **7**, 531–557.

—— 1987. *Aeolian Dust and Dust Deposits.* Academic Press, London.

—— 1989. Processes of fine particle formation, dust source regions and climatic changes. *In:* LEINEN, M. & SARNTHEIN, M. (eds) *Paleoclimatology and Paleometeorology: Modern and Past Patterns of Global Atmospheric Transport.* NATO ASI series no. 282), Kluewer, Dordrecht, 3–30.

—— & TSOAR, H. 1990. *Aeolian Sand and Sand Dunes.* Unwin Hyman, London.

—— & ZHOU, L. P. 1989. Late Pleistocene and Holocene aeolian dust deposition in North China and the Northwest Pacific Ocean. *Palaeogeography, Palaeoclimatology and Palaeoecology* **73**, 11–23.

REA, D. K., LEINEN, M. & JANECEK, T. R. 1985. Geologic approach to the long-term history of atmospheric circulation. *Science,* **227**, 721–725.

RENDELL, H. M. & TOWNSEND, P. D. 1988. Thermoluminescence dating of a 10 m loess profile in Pakistan. *Quaternary Science Reviews,* **7**, 251–255.

RUTTER, N., DING, Z. L. & LIU, T. S. 1991. Comparison of isotope stages 1–61 with the Baoji-type pedostratigraphic section of north-central China. *Canadian Journal of Earth Sciences,* **28**, 985–990.

SARNTHEIN, M. 1978. Sand deserts during glacial maximum and climatic optimum. *Nature,* **272**, 43–46.

SUN, J. Z. 1985. Some characteristics of the loess on Miao Islands of Shandong Province. *Kexue Tongbao,* **30**, 384–387.

THOMPSON, R. & OLDFIELD, F. 1986. *Environmental Magnetism.* George Allen & Unwin, London.

TSOAR, H. & PYE, K. 1987. Dust transport and the question of desert loess formation. *Sedimentology*, **34**, 139–153.

WANG, J. T. & WANG, P. X. 1980. Relationship between sea-level changes and climatic fluctuations in east China since late Pleistocene. *Acta Geographica Sinica*, **35**, 299–312.

WANG, Y. & AUBREY, D. G. 1987. The characteristics of the China coastline. *Continental Shelf Research*, **7**, 329–349.

WEI, L. 1986. Heavy minerals in the Malan Loess. *Quaternaria Sinica*, **7**, 49–56.

WINTLE, A. G. 1981. Thermoluminescence dating of the late Devensian loesses in southern England. *Nature*, **289**, 479–480.

—— 1987. Thermoluminescence dating of loess. *Catena Supplement*, **9**, 103–115.

—— 1990. A review of current research on TL dating of loess. *Quaternary Science Reviews*, **9**, 385–397.

—— & HUNTLEY, D. J. 1982. Thermoluminescence dating of sediments. *Quaternary Science Reviews*, **1**, 31–53.

——, SHACKLETON, N. J. & LAUTRIDOU, J. P. 1984. Thermoluminescence dating of periods of loess deposition and soil formation in Normandy. *Nature*, **310**, 491–493.

XIA, D. X., LIU, Z. X., LI, P. Y. & ZHAO, S. L. 1991. Inference of ancient desert in the Bohai Sea. *Acta Oceanologica Sinica*, **13**, 540–546.

YUAN, Y. S. 1959. *Petrographical and Mineralogical Study of Loess in the Zibo, Zhangxia and Penglai regions, Shandong.* Unpublished Advanced Certificate thesis, Geology and Geography Department, Peking University.

ZHANG, D. 1982. Analysis of dust rain in the historic times of China. *Kexue Tongbao*, **27**, 294–297.

ZHANG, Z. H., ZHANG, Z. Y. & WANG, Y. S. 1989. *Loess in China.* Geological Publishing House, Beijing.

ZHAO, K. H. 1983. The origin of the loess in the chain islands of Miao-a mineralogic view. *Marine Science*, **1**, 29–32.

ZHAO, S. L. & LI, G. G. 1991. Origin and shallow-layer structure of the sediments in the Yellow Sea trough. *Acta Oceanologica Sinica*, **10**, 106–115.

—— & QIN, Y. S. 1986. Transgressions and sea-level changes in the eastern coastal region of China in the last 300 000 years. *In*: CHINA WORKING GROUP, *China Sea Level Changes.* IGCP Project No. 200 China Ocean Press, Beijing, 122–130.

ZHOU, L. P., OLDFIELD, F., WINTLE, A. G., ROBINSON, S. G. & WANG, J. T. 1990. Partly pedogenic origin of magnetic variations in Chinese loess. *Nature*, **346**, 737–739.

—— & WINTLE, A. G. 1989. Underestimation of regeneration ED encountered in Chinese loess. *In: Long and Short Range Limits in Luminescence Dating.* Research Laboratory for Archaeology and the History of Art, Oxford University, Occasional Paper No. 9, 139–144.

ZOELLER, L., STREMME, H. E. & WAGNER, G. A. 1988. Thermolumineszenz-Datierung an Loss-Palaeoboden-Sequenzen von Nider-, Mittel- und Oberrhein. *Chemical Geology (Isotope Geoscience Section)*, **73**, 39–62.

The magnetic mineralogy of a loess section near Lanzhou, China

TIMOTHY C. ROLPH[1], JOHN SHAW[1], EDWARD DERBYSHIRE[2]
& WANG JINGTAI[3]

[1]*Geomagnetism Laboratory, Oliver Lodge Building, University of Liverpool, PO Box 147, Liverpool L69 3BX, UK*
[2]*Department of Geography, University of Leicester, Leicester LE1 7RH, UK*
[3]*Geological Hazards Research Centre, Gansu Academy of Sciences, Lanzhou, China*

Abstract: A collection of almost 500 hand samples from a 250 m loess section near Lanzhou, Gansu province, have been investigated to determine both the nature of the magnetic mineralogical differences between the loess and inter-bedded soils and the time variance of the magnetic mineralogy. Results indicate that the parent magnetic material, which is aeolian in origin, is cation-deficient magnetite/maghaemite and the strong magnetic susceptibility contrast, particularly in the younger material, between loess and soil, is mostly due to pedogenic enhancement of the magnetic content of soils by in situ formation of ultrafine magnetite, although we should not ignore possible changes in the nature of the available aeolian material during the warmer soil-forming periods. Compared to loess sections further to the east, the degree of enhancement in the soils is small, on average having a ferrimagnetic content which is only 17% greater than that of the loess. Trends in the magnetic mineralogy of the section indicate a possible link with the long-term climate change of Northern China, the average ferrimagnetic grain size increasing with decreasing age in response to increased wind velocity and a colder, drier climate. At a point approximately 110 m above the base there is a dramatic change in the nature and concentration of the magnetic material. This change indicates either a rapid deterioration in the average climate or a sudden change in the provenance of the source material, although at this stage we cannot discount the possibility of some time gap at this position. The loess/palaeosol susceptibility contrast is noticeably reduced below this level, perhaps indicating less extremes of climate.

The loess deposits of central China represent the most complete terrestrial record of climate for the Pleistocene. The deposits, which cover more than 500 000 km^2, consist of alternating layers of loess and palaeosol (fossil soil) which represent a succession of glacial/interglacial cycles. During the glacial intervals, the climate of the area was cold and dry, with strong winds bringing in dust and silt which had been deflated in dry desert areas to the north and west. This material accumulated to form the thick loess deposits. In the interglacial periods, the climate changed to become warm and wet, wind-transport of material decreased and pedogenic processes acted on the loess to form the soils present in the loess deposits (Heller & Liu 1984; Liu *et al.* 1985). Milankovitch, in his classic work *Canon of Insolation and the Ice Age Problem* (1941), showed the link that exists between the glacial/interglacial fluctuations and variations in the Earth's insolation parameters. The parameters are quasiperiodic functions which reflect the obliquity, precession and eccentricity of the Earth's orbit. The quasiperiodicities associated with these orbital parameters are centred at 100 000 years (eccentricity, which also shows a weak quasiperiodicity centred at 400 000 years), 41 000 years (obliquity) and 23 000 and 19 000 years (precession). Spectral analysis techniques have shown that these quasiperiodicities modulate the magnetic susceptibility record of the loess deposits (cf. Wang *et al.* 1990), with high susceptibility in the palaeosols and lower suceptibility in the loess. This indicates the presence of a direct link between the magnetic mineralogy of these deposits and palaeoclimate. The mechanism by which this link occurs is an area of great interest and Kukla *et al.* (1988) proposed a model whereby a constant influx of atmospheric magnetic material undergoes varying degrees of dilution according to whether there is a period of loess accumulation or soil formation. This suggests that the climatic signal in effect modulates the input of the bulk material

From Pye, K. (ed.), 1993, *The Dynamics and Environmental Context of Aeolian Sedimentary Systems.*
Geological Society Special Publication No. 72, pp. 311–323.

and thereby indirectly modulates the magnetic mineral content. This proposal has subsequently been investigated by a number of workers who suggest that the pedogenic enhancement of magnetic material either influences (Zhou *et al.* 1990) or dominates (Maher & Thompson 1991) the susceptibility fluctuations.

The present climate in the loess plateau shows a significant regional variation, with humidity and temperature increasing from northwest to southeast. Evidence for a similar variability in past climate is clearly provided by the loess deposits, with both loess and palaeosols in the southeast showing a higher degree of weathering, the palaeosols standing out as distinctive reddish horizons within yellowish-pink loess. A regional correlation is also apparent for the magnetic susceptibility, which reaches far larger values in the southeast (the loess susceptibility near Xian (SE) is similar to the palaeosol susceptibility near Lanzhou (NW)). Such strong regional variations are likely to be an important factor in preferred explanations for loess/palaeosol magnetic susceptibility contrasts. This may be indicated by the differing conclusions of Maher & Thompson (1991), who prefer a model of dominant pedogenic enhancement (samples from a loess section at Luochuan, eastern loess plateau), and Zhou *et al.* (1990), who suggest only partial pedogenic enhancement (samples from two sites in the western Gansu province). It should be noted that both these studies were limited to loess and palaeosols younger than 700 ka.

The present study is of a 250 m loess section near Lanzhou, Gansu province, which contains more than twenty soils. This number of soils (by comparison with other sections) suggests an age range considerably in excess of 700 ka. Magnetic susceptibility was measured in the field at 10 cm intervals and a collection of almost 500 hand samples was made for detailed laboratory analysis. This detailed coverage was chosen to enable us to look both at differences between loess' and palaeosols and at any longer-term behaviour indicated by trends in the magnetic mineralogy.

Geological setting and sampling procedure

The central loess plateau of China comprises the majority of the country's loess deposits. The city of Lanzhou lies on the dry western margin of the plateau and in this vicinity is found the world's greatest known thickness of loess. Lanzhou, which straddles the Huang He (Yellow River), is at an altitude of more than 1500 m and has a

semi-arid climate. Winters are severe with only limited snow, the majority of the precipitation in the area occurring as heavy showers in July and August. The studied section occurs near the village of Dawan, about 30 km southwest of Lanzhou, overlooking a valley (Xuan Ja Gou) which contains a minor tributary of the Huang He (see Fig. 1). The section is 250 m thick and at the top has an altitude in excess of 2100 m. It is underlain by river gravels which lie in turn upon Cretaceous arenites.

The section is on a steep slope and a series of deep overlapping pits were excavated down the section. Using a Bartington susceptibility meter with field probe, readings were taken at 10 cm intervals for the whole section. Subsequently, a series of hand samples were collected at 1 m intervals for the top 20 m and at 0.5 m intervals for the remaining 230 m. These samples were shipped back to Liverpool University geomagnetism laboratory for further investigation.

Experimental techniques

The use of a series of mineral magnetic techniques provides considerable information about the magnetic mineralogy of a sample, including the type(s) of magnetic mineral(s) and the mass fraction and magnetic grain-size distribution of the ferrimagnetic minerals. For all hand samples the magnetic parameters measured were magnetic susceptibility (X), saturation magnetization (M_s) and saturation remanence (M_{rs}). In addition, twenty samples were further investigated to determine their Curie temperature (T_c), low and high frequency susceptibility (X_L and X_H), low temperature susceptibility behaviour and IRM (isothermal remanent magnetization) acquisition curves. These twenty samples were taken at various levels spanning the whole section and included both loess and palaeosol samples. A number were selected to investigate specific behaviour exhibited during the initial X, M_s and M_{rs} measurements.

Although ferrimagnetic (magnetite/maghaemite) minerals are likely to dominate the magnetic characteristics, other minerals present may be important. The average composition of the Lanzhou loess is quartz (*c.* 60%), carbonate (8–20%) and clay (7–28%; illite > 70%), with haematite and magnetite/maghaemite as minor components (Derbyshire 1983). Quartz and carbonate are weakly diamagnetic (-0.62×10^{-8} and -0.48×10^{-8} m^3 Kg^{-1}) and, therefore, make a very small negative contribution to the susceptibility of loess. However, significant

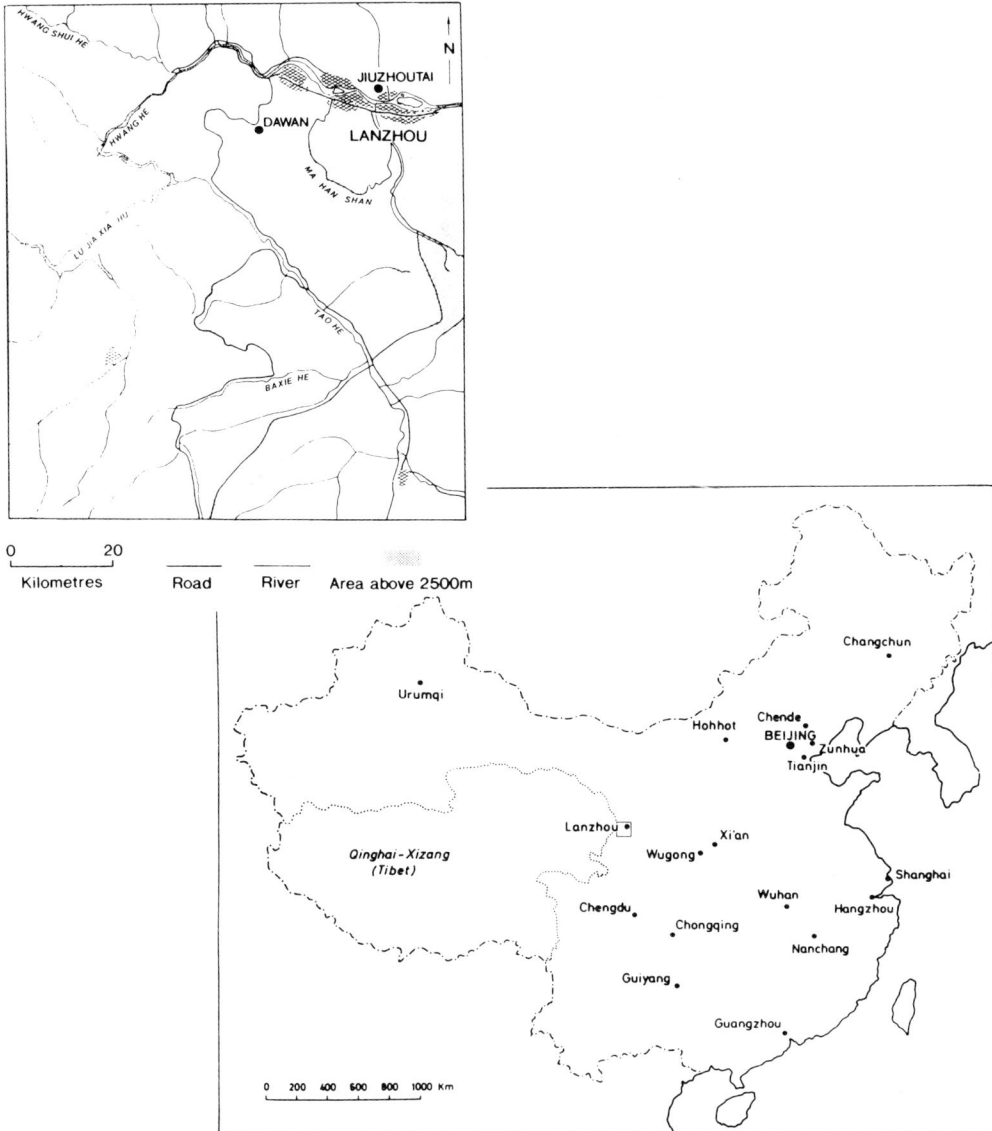

Fig. 1. Location map showing the position of the Dawan section with respect to the city of Lanzhou and the Jiuzhoutai section and its location within China. Adapted from Clarke (1992).

variations in the amount of these two components may be important if they vary at the expense of the more magnetic components. Although clay only has a weak susceptibility ($c. 15 \times 10^{-8}\ m^3\ kg^{-1}$), it occurs in far larger quantities than magnetite/maghaemite and so may make an important contribution to the loess susceptibility. The susceptibility of haematite is weak ($6 \times 10^{-6}\ m^3\ kg^{-1}$) compared to magnetite/maghaemite ($c.\ 5 \times 10^{-4}\ m^3\ kg^{-1}$)

and will be significant only if it occurs in large quantities. If superparamagnetic (SP) magnetite is present it can make a major contribution to a sample's magnetic susceptibility. At room temperature, the magnetic moments of SP magnetite grains are able to align themselves with an applied field. For this reason, SP magnetite has a susceptibility which is considerably stronger than stable magnetite, for which the magnetic moments have an essentially predetermined alignment. The magnetic susceptibility of all hand samples was measured using a Bartington meter with coil system.

To determine M_s and M_{rs}, samples were processed using a Molspin vibrating sample magnetometer (VSM). Small subsamples (< 1 g) are placed in a sample holder which is then placed between the poles of an electromagnet. The magnetic field (H) produced by the electromagnet induces a magnetization (M) in the sample, and M is measured by vibrating the sample, the movement of the sample inducing a voltage in pick-up coils situated either side of the sample. The value of H is cycled in a stepwise manner between ± 1 Tesla, and the value of M measured at each step. For some value of H, which depends on size and type of magnetic mineral, the ferromagnetic minerals reach their saturation magnetization (M_s), beyond which their magnetization no longer increases with increasing H. On reducing H to zero, the magnetization retained by the sample is termed the saturation remanent magnetization (M_{rs}), and depends on the domain state (stable single domain — SSD; multidomain — MD; superparamagnetic — SP) and type of magnetic mineral. As M_s and M_{rs} both show the same dependency on magnetic mineral type and concentration, normalizing M_{rs} by M_s removes this dependency and the ratio M_{rs}/M_s becomes a useful parameter for estimating magnetic grain size. Values of M_s for magnetite and haematite are 92 Am^2 kg^{-1} and 0.5 Am^2 kg^{-1}, respectively (Thompson & Oldfield 1986) and the ratio M_{rs}/M_s is 0.5 for SSD grains, *c.* 0.02 for MD grains and zero for SP grains (SP grains cannot hold a remanence, so M_{rs} is zero). The value of M_s for a sample will reflect the mass fraction of magnetic material, e.g. loess which contains 1% by weight of magnetite will have an M_s value of 0.92 Am^2 kg^{-1} and an M_{rs}/M_s value which reflects the relative contribution of SP, SSD and MD magnetite grains. For part of the measurement cycle the sample is subjected to a field which increases in the opposite sense to the initial magnetizing field. This will gradually reduce the magnetization of the sample until at some field value the magnetization will pass through zero; this field value, H_c, is known as the coercivity of the sample, and has a similar magnetic grain-size/type dependence as M_{rs}, but is independent of the concentration of magnetic minerals. It reflects the magnetic 'hardness' of the sample, i.e. its ability to resist magnetization/demagnetization. The picture is complicated by the presence of paramagnetic minerals such as illite, an important component of loess and palaeosols, for which M increases

linearly with H. This component is subtracted to obtain the value of M_s for the ferrimagnetic component. For each sample, the variation of M with H traces out a hysteresis loop, the shape of which depends on M_s, M_{rs} and H_c, and thus on magnetic mineralogy.

A number of features and trends were indicated by the susceptibility and hysteresis data, and for further study a series of 20 samples were taken from the section and investigated using additional magnetic techniques. The first characteristic investigated was the thermomagnetic behaviour of the samples. This provides us with Curie temperatures and a measure of thermally induced alteration, both of which are useful diagnostic characteristics. The technique involves placing samples in an oven which sits between the pole pieces of an electromagnet. In the presence of a strong magnetic field (*c.* 0.5 T), the sample magnetization is monitored while it is heated to, and cooled from, 700°C. The shape of the heating and cooling curves, the point at which magnetization disappears (Curie temperature, T_c) and the difference between the magnetization before and after heating, all provide important information regarding the type of magnetic minerals present. For example, T_c is 680°C for haematite and 575°C for magnetite. Maghaemite, a low temperature oxidation product of magnetite which may be formed during pedogenesis, has a T_c close to 650°C (maghaemite Curie temperatures are rarely seen due to its alteration to haematite during heating). Transitional states between magnetite and maghaemite (cation deficient or CD magnetite) produce Curie temperatures between the two end-members. The heating curves of samples which contain maghaemite will show a change of slope in the temperature range 250–350°C as the mineral undergoes oxidation, eventually converting to haematite. The saturation magnetization of haematite is much lower than maghaemite or CD magnetite (and furthermore, the applied field of 0.5 T is insufficient to saturate haematite), so this oxidation process results in a cooling curve which returns to a room temperature magnetization value that is much lower than the pre-heating value. Paramagnetic susceptibility is strongly temperature dependent, falling off according to an inverse temperature law, and so the presence of paramagnetic clays superimposes a concave decay/growth on the heating/cooling curve.

Low temperature magnetic susceptibility behaviour was also investigated. If the magnetic susceptibility of a sample is monitored as it warms up from liquid nitrogen temperature

($-196°C$) to room temperature, the susceptibility variation can be used to infer the type and/or domain state of the magnetic mineralogy. For example, as mentioned above, paramagnetic susceptibility shows an inverse relationship with temperature, so as the temperature falls, the paramagnetic susceptibility increases to reach a value at $-196°C$ of three times its room temperature value. The behaviour of SP magnetite is also diagnostic. As the temperature falls, the SP grains pass through the SP/SSD transition, and their susceptibility falls to that of stable SP magnetite. Samples with an appreciable SP content will show a significant fall in susceptibility, this decrease tailing off as all the SP grains eventually pass through the transition. MD magnetite shows a strongly characteristic behaviour which is related to the change in sign of a magnetocrystalline anisotropy constant, which passes through zero at $c. -150°C$. The susceptibility of MD grains is affected because this anisotropy constant plays an important role in fixing the positions of the domain walls. As the constant tends towards zero, wall movements become possible and this allows the susceptibility to increase to a maximum, which occurs as the anisotropy constant passes through zero. The susceptibility then falls as the constant increases with an opposite sign. Thus the presence of MD magnetite is characterized by a peak in susceptibility centred at $c. -150°C$. The behaviour of SSD magnetite is by comparison stable, showing only a small decrease in susceptibility with falling temperature. When interpreting the low temperature susceptibility curves of loess and palaeosol samples is is important to realize that the observed behaviour is likely to be a combination of varying contributions from the clay and SP, SSD and MD magnetite components.

The frequency dependence of magnetic susceptibility is an important diagnostic tool for detecting SP magnetite. The technique involves measuring the magnetic susceptibility at a low and a high frequency in an alternating field (AF) susceptibility system. In the Bartington instrument used, these frequencies are 0.43 kHz (low $- X_L$) and 4.3 kHz (high $- X_H$). The principle of the technique relies on the fact that at low frequencies the magnetic moments of the SP grains can flip between the two AF directions and thus make a large contribution to the susceptibility. As the frequency increases, the SP moments can no longer keep up with the field and their contribution falls to the level of SSD grains. The frequency dependence (X_{fd}) is given by

$$X_{fd} = (X_L - X_H)/X_L$$

and is usually expressed as a percentage. The magnitude of this effect can be used as a measure of the relative contribution of SP grains to the susceptibility record.

The final technique used on the sub-collection of 20 specimens was IRM acquisition. The procedure is a simple one in which the sample is subjected to a set of increasing magnetic fields in the range 0–4 T. The procedure differs from the VSM treatment described earlier in that the sample magnetization is measured after, rather than during, each field application. Also, the maximum field available (4 T) is much higher so that any haematite component will be close to saturation. The magnetizations acquired (IRMs) will increase until the magnetic minerals are saturated; at this point the sample has a saturation (S)IRM. For magnetite this will occur at fields of less than 0.3 T, and is the same as the M_{rs} value obtained from the VSM. The presence of haematite will mean that saturation does not occur until much higher fields. Using a technique developed by D. J. Robertson, of Liverpool University Geography Department, the data are plotted as magnetization against Log_{10} of applied field, and the data are compared to a calculated curve. To calculate the curve we presume that a magnetic mineral phase has a log normal grain-size (and hence coercive force) distribution and the curve is drawn according to the distribution parameters input by the operator. If more than one discrete mineral phase is present then we must input distribution parameters for each phase and calculate a curve which will be a combination of the distributions. These parameters are % of SIRM contributed by the magnetic mineral (which will be less than 100% if more than one phase, i.e. magnetite and haematite, are present) and the mean and standard deviation (both as Log_{10}) of the coercive force distribution. These parameters can be adjusted until the curve fits the data.

In the following section, examples are presented of each of the techniques used and the results briefly discussed.

Results

Figure 2 shows typical hysteresis loops, (a) for loess sample 234.5L and (b) for palaeosol 84.0P. The loess sample has a wider loop which reflects a higher value of H_c and a higher relative value of M_{rs} (magnetization M is normalized to the maximum value in each case). Since M_{rs} and H_c both increase for increasing SSD fractions, and decrease for increasing SP fractions, these loops suggest that the loess has a higher SSD fraction

(a)

(b)

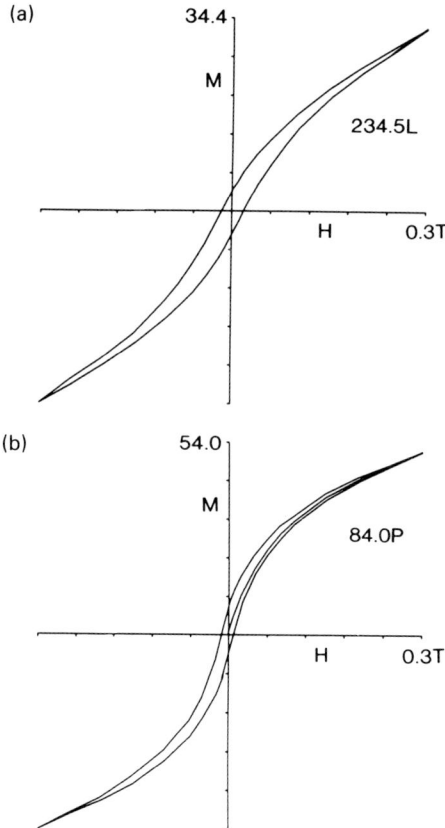

Fig. 2. Hysteresis loops obtained from : (**a**) a loess (234.5L); and (**b**) a palaeosol (84.0P) sample. The maximum field (H) shown is 0.3 T and the magnetization (M) is shown in units of $Am^2 Kg^{-1}$. The points at which the loop cuts the magnetic field and magnetization axes provide estimates of coercivity (H_c) and saturation remanence (M_{rs}), respectively.

or the palaeosol has a higher SP fraction. The peak field ($\pm$ 1 T) of the VSM is not sufficient to saturate any haematite present in the samples. As a consequence the loops are closed at fields above 0.3 T (the saturating field for magnetite), which indicates that paramagnetic clay ($M_{rs} = 0$) dominates the high field behaviour.

The magnetic susceptibility and hysteresis parameters for all samples are presented in Fig. 3. The position of the 20 samples chosen for further analysis is indicated by a sample code, the number of which refers to the depth in the section and the letter to whether the sample is loess (L) or palaeosol (P). The position of these samples is shown in Fig. 3.

Figure 4 shows examples of two Curie curves: (**a**) 67.0L and (**b**) 84.0P. Both have Curie points just above 575°C (CD magnetite) and both show a change of slope centred at *c.* 280°C (inversion of maghaemite). Sample 67.0L shows the larger fall in magnetization on cooling which implies that the loess contained a greater proportion of maghaemite (which changed to weakly magnetized haematite) than the palaeosol sample. All Curie curves have a noticeably concave appearance due to the large fraction of paramagnetic clay minerals and the small amounts of ferrimagnetic minerals present. The low temperature susceptibility behaviour of a number of samples is shown in Fig. 5. We can see that as the temperature increases, some curves show combinations of behaviour such as MD susceptibility 'humps' and paramagnetic decay (e.g. 220.5P) and paramagnetic decay and SP growth (e.g. 44.5P). However, some curves are dominated by a single characteristic behaviour such as MD (142.0L) or paramagnetic (141.5L) behaviour. Figure 6 shows the IRM acquisition behaviour of a palaeosol and loess sample. Sample 32.5P has nearly saturated at 0.3 T, indicating the dominance of a magnetite/maghaemite phase with a small contribution from a haematite phase. In contrast, sample 89.0L had still not saturated at the peak field used (4 T). This is indicated by the behaviour above 0.3 T, with a gradual but clear increase continuing between 0.3 and 4 T. The data can be fitted by a curve which is a combination of magnetite/maghaemite and haematite, with magnetite/maghaemite contributing 90% of the magnetization. For samples which do not saturate, the SIRM is estimated from the trend of the magnetization between 0.3 and 4 T. This is done iteratively until the calculated curve fits the data. In Fig. 6, these calculated curves are shown as solid lines.

Table 1 presents the results of the mineral magnetic experiments for the 20 samples, together with the mean results of the loess and palaeosol sub-groups. The implications of these results are discussed below.

Discussion

Analysed as a group there are clearly differences in the magnetic mineralogy both between loess and palaeosol and with time. Looking first at thermomagnetic behaviour, there is a strong similarity between the curves of the loess and palaeosol samples. All samples have Curie temperatures in excess of 575°C, all heating curves show a change of slope centred around 280°C and all cooling curves return much lower

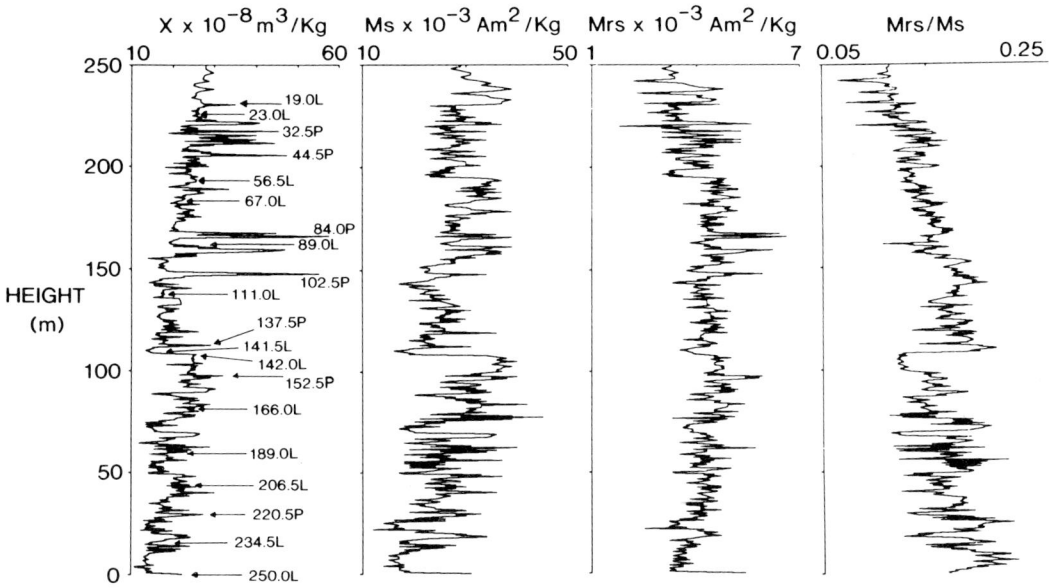

Fig. 3. The variation, with stratigraphic height, of the magnetic parameters susceptibility (X), saturation magnetization (M_s), saturation remanence (M_{rs}) and the ratio M_{rs}/M_s. The 20 samples chosen for further analysis are shown on the susceptibility diagram.

than the heating curves. For the loess and palaeosol samples, the average Curie temperatures are $595° \pm 5°C$ and $593 \pm 4°C$, respectively, temperatures which indicate magnetite with some degree of cation deficiency due to low temperature oxidation. This oxidation can occur during weathering by a mechanism which is thought to involve the migration of Fe^{2+} cations to the surface of the grain where they oxidize to Fe^{3+} (O'Reilly 1983). This will leave vacancies in the lattice and, if the process is completed, produces maghaemite which has the chemistry of haematite but the cubic structure of magnetite and a saturation magnetization close to that of magnetite. Various degrees of oxidation may produce a number of intermediate phases ranging from magnetite grains showing cation deficiency, through magnetite grains with a maghaemite rim to grains which have undergone complete maghaemitization. This will also depend on grain size, with the smaller grains being the first to become fully maghaemitized. On heating, maghaemite inverts to the much more weakly magnetized haematite in the region of $350°C$, and this is reflected by the change of slope at $280°C$ and the post-heating reduction in magnetization. The reduction in magnetization for loess and palaeosol samples averages $42\% \pm 5\%$ and $39\% \pm 4\%$, respectively, and the Curie curves for the loess samples at Dawan show a much more distinctive maghaemite

'hump' during heating than palaeosols (see Fig. 4). Although these differences are not significant in absolute terms, their implication, that loess contains a larger proportion of maghaemite/CD magnetite than palaeosols, is interesting in the context of whether maghaemite is an important pedogenic product. Maher et al. (1991) discuss the presence of maghaemite as a possible pedogenic indicator, formed in soils by oxidation of magnetite. The results discussed here suggest that maghaemite/CD magnetite is more important in the loess at Dawan, suggesting that maghaemite/CD magnetite is a parent (aeolian) material.

Low temperature susceptibility (X_{LT}) behaviour shows distinct groupings within the samples. At $-196°C$, the four youngest palaeosols (32.5P, 44.5P, 84.0P and 102.5P) have susceptibility values within a few percent of their room temperature value. As they warm up, their susceptibility initially falls by a small amount, in all cases falling below the room temperature value. Subsequently the susceptibility shows a gradual increase up to room temperature, the curve showing a slightly concave shape. This indicates that the X_{LT} behaviour of these samples is controlled equally by contributions from SP magnetite (susceptibility increases with temperature) and paramagnetic clay (susceptibility decreases as temperature increases). As sample 152.5P

(a)

67.0L

MAGNETISATION

0　　　TEMPERATURE ˚C　　700

(b)

84.0P

MAGNETISATION

0　　　TEMPERATURE ˚C　　700

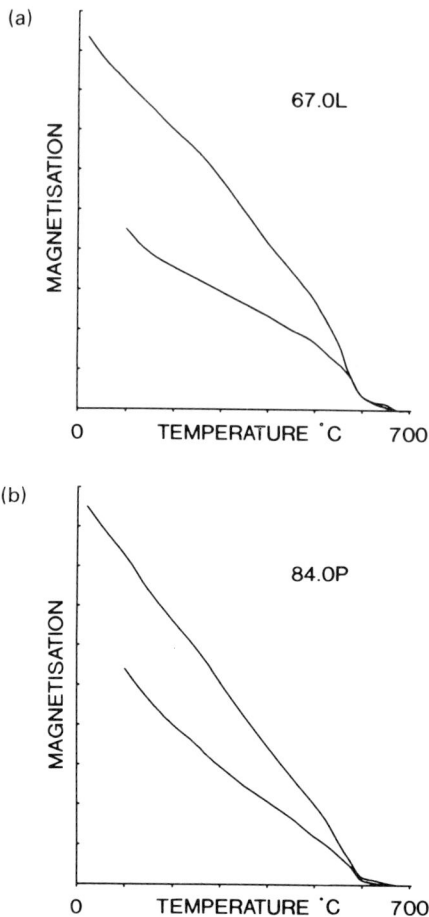

Fig. 4. Thermomagnetic curves for (**a**) a loess (67.0L) and (**b**) a palaeosol (84.0P) sample. Curie temperatures for both are close to 600°C and both show a reduction in magnetization after cooling (lower curve). The heating (upper) curve of sample 67.0L shows a 'hump' in the region of 280°C, indicating the presence of maghaemite.

warms up it simply shows a gradual small decrease to its room temperature value, indicating that clay has slightly more influence than the SP magnetite. Samples 137.5P and 220.5P show considerably different behaviour, appearing as a combination of a gradual small decrease but with a hump, centred at −150°C, superimposed on the curve. This indicates the presence of MD magnetite with contributions from paramagnetic clay and SP magnetite. Looking at the loess samples, they show a range of behaviour. Samples 23.0L, 67.0L, 89.0L, 111.5L, 141.5L, 166.0L and 234.5L all show behaviour dominated by paramagnetic clay, the

curves showing a continuous decrease to room temperature from values at −196°C which range between 120 and 180% of the room temperature value. The remaining loess samples show a combination of paramagnetic clay and MD magnetite, with some showing behaviour dominated by MD magnetite (142.0L and 189.0L), some showing fairly equal contributions (19.0L and 250.0L) and the remainder (56.5L and 206.5L) showing clay to be more important. Small amounts of SP magnetite may play a minor role in the loess samples. The presence of SSD magnetite cannot be determined because, as discussed earlier, its susceptibility shows very little temperature dependence.

IRM acquisition behaviour of all samples indicates a remanence dominated by a magnetite/ maghaemite phase, with minor contributions from haematite. There does not appear to be a consistent difference between loess and palaeosol in terms of the haematite contribution, but in the majority of palaeosol samples, the maghaemite/magnetite phase has a lower mean coercivity than the loess samples, the exceptions being 137.5P and 220.5P (see Table 1). The coercivity of a magnetic mineral is a measure of its magnetic 'hardness', its resistance to magnetization or demagnetization. For magnetite, uniaxial SSD grains have the highest coercivities, with a maximum theoretical value of 0.3 T. MD magnetites are always lower than 0.1 T and SP grains, at room temperature, do not have an effective coercivity. However, magnetite grains on the SP/SSD boundary, while not being in normal terms remanence carriers, may be able to retain a remanence for short periods of time, which may be of the order of the time taken between giving the samples an IRM and measuring it. Thus they may contribute to the IRM, saturating at very low fields, and giving the sample an artificially low mean coercivity. This is the most likely explanation for the lower coercivities associated with the palaeosol samples. Another explanation, that the palaeosols contain greater amounts of MD magnetite, is not supported by the low temperature susceptibility results. The two palaeosol samples which do show low temperature susceptibility curves indicating a significant MD magnetite fraction are 137.5P and 220.5P, and these are the two palaeosol exceptions to low mean coercivity mentioned above.

The frequency dependence (X_{fd}) of magnetic susceptibility is, as discussed in the previous section, highly sensitive to the presence of SP grains. Table 1 shows the frequency effect, expressed as a percentage, for all samples. If we ignore sample 220.5P, the average X_{fd} for the

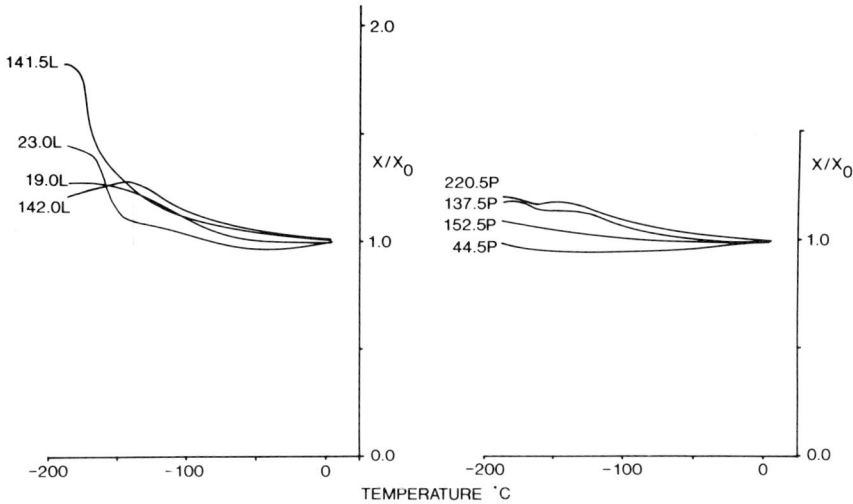

Fig. 5. Low temperature susceptibility curves for a selection of loess and palaeosol samples. Loess samples (suffix L) show behaviour indicative of varying contributions from multidomain (MD) and superparamagnetic (SP) magnetite and paramagnetic clay. The palaeosol samples (suffix P) show behaviour dominated by clay and SP magnetite.

palaeosols is $6 \pm 2\%$ and for the loess samples is $2 \pm 1\%$. Amongst the palaeosols, sample 220.5P is anomalous in that its frequency effect of 0.9% places it within the range of the loess values and indicates little SP magnetite. For the

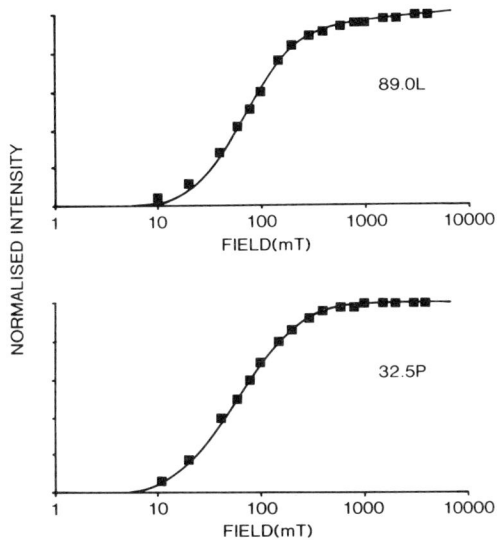

Fig. 6. IRM acquisition curves for a loess (89.0L) and palaeosol (32.5P) sample. The palaeosol sample has saturated by 0.3 T indicating a magnetite/maghaemite mineralogy. The loess magnetization continues to increase after 0.3 T, indicating that the sample also contains haematite.

loess samples, 166.0L gives a frequency effect of 3.4%, significantly higher than the average of the remainder. These results suggest that although frequency dependent susceptibility is in general a good indicator of palaeosols, in some cases the information it provides is not immediately diagnostic.

Considering all the above results, and the information provided by the hysteresis loops, it is possible to make some comments regarding the loess and palaeosols at Dawan. On average the palaeosol samples have higher M_s values, higher susceptibility values, higher frequency effects and a strong SP contribution to low temperature susceptibility. The loess samples tend to show a greater reduction in magnetization after heating, low temperature susceptibility dominated by paramagnetic clay and/or MD magnetite and a larger mean coercivity for the magnetite/maghaemite fraction. These parameters suggest that palaeosols have a larger amount of magnetic material and that this material has a larger fraction of ultrafine SP grains. The loess, in contrast, appears to have a larger fraction of coarser MD grains. These are, however, generalizations, and the results also show that some samples cannot be categorized so easily.

Regarding regional differences, the loess and palaeosols at Dawan have susceptibilities and frequency effects much smaller than equivalent materials further to the east (i.e. Luochuan; Maher & Thompson 1991), and the site contrast

Table 1. *The results from the mineral magnetic experiments are given in terms of saturation magnetization (M_s), the ratio of saturation remanence to M_s (M_{rs}/M_s), magnetic susceptibility (SUS), frequency effect of susceptibility expressed as a percentage (X_{fd}), the mean coercive force (CF), the Curie temperature (T_c) and the percentage decrease in magnetization after heating (%DM). These parameters refer only to the magnetite/maghaemite fraction*

Sample	M_s ($\times 10^{-3}$ Am^2 Kg^{-1})	M_{rs}/M_s	SUS ($\times 10^{-8}$ m^3 Kg^{-1})	X_{fd}	Mean CF (10^{-3} T)	T_c (°C)	%DM
19.0L	39.0	0.044	31.5	1.6	79.4	595	56
23.0L	27.0	0.087	30.2	1.3	79.4	590	46
32.5P	34.0	0.097	52.9	6.0	58.9	595	38
44.5P	35.0	0.093	57.2	5.9	57.5	595	44
56.5L	36.0	0.104	29.3	1.0	75.9	595	39
67.0L	30.0	0.110	25.4	2.8	79.4	595	44
84.0L	39.0	0.104	74.6	9.2	47.9	590	34
89.0L	33.0	0.107	32.0	2.2	72.4	595	37
102.5P	27.0	0.116	53.1	6.6	53.7	600	37
111.5L	27.0	0.127	21.0	2.4	85.1	590	45
137.5P	31.0	0.111	34.9	3.4	72.4	590	38
141.5L	19.0	0.150	18.3	2.2	85.1	590	41
142.0L	33.0	0.105	29.7	1.3	74.1	590	41
152.5P	39.0	0.103	39.9	4.8	66.1	590	38
166.0L	42.1	0.129	26.8	3.4	74.1	590	39
189.0L	35.5	0.122	29.0	1.7	72.4	595	41
206.5L	25.5	0.169	21.1	2.4	75.9	600	39
220.5P	32.9	0.130	32.4	0.9	74.1	590	44
234.5L	18.9	0.208	18.6	2.2	85.1	605	46
250.0L	31.3	0.170	26.4	1.1	83.2	600	36
Mean values							
Loess	31 ± 7	0.13 ± 0.04	26 ± 5	2 ± 1	79 ± 5	595 ± 5	42 ± 5
Palaeosol	34 ± 4	0.11 ± 0.01	48 ± 15	5 ± 3	62 ± 10	593 ± 4	39 ± 4

between loess and palaeosol susceptibility is reduced. Maher & Thompson (1991) have interpreted their SIRM and ARM data as indicating a mean magnetite content at Luochuan of between 0.08 and 0.15% by weight for the loess, and 0.2–0.32% for palaeosols, where all samples come from the part of the section less than $c.$ 500 ka. If we look at the top part of the Dawan section (it was not possible to visibly distinguish palaeosols in the lower part of the section), the equivalent average values are $0.029 \pm 0.005\%$ for the loess samples and $0.034 \pm 0.005\%$ for the palaeosol samples. This would suggest that although X_{fd} in both loess and palaeosol samples tends to be larger at Luochuan, the susceptibility contrast between the two sites mostly reflects the large difference in the total ferrimagnetic content of the two sites. It seems unlikely that such a large contrast in magnetite content can be explained simply by differences in weathering at the two sites, especially as the contrast between loess and palaeosol at Luochuan, which represents climate extremes, is less than a factor of two. It seems more likely that the contrast is related to differences in loess source area and perhaps to differences in mean loess grain size. The loess at Lanzhou is closely linked to the uplift of the Qinghai-Tibet plateau (Derbyshire 1983) and the east–west-trending Gansu corridor. Material at Lanzhou is likely to have a more westerly origin than the material at Luochuan, which is linked to the desert regions to the northwest. The link between loess grain size and susceptibility may be explained by the dilution model of Kukla et al. (1988). The magnetic minerals are associated with the finer fraction and this will be relatively reduced in coarser loess sequences, where the accumulation rate is more rapid and the dilution effect enhanced. The loess in the Lanzhou area falls in the range of medium to coarse silt and so might be expected to show a reduced magnetic content by comparison to the finer-grained loess of Luochuan.

Looking finally at the large M_s, M_{rs} and susceptibility data sets (Fig. 3), there are a number of features and trends present in the data. Firstly, in the upper part of the section the palaeosols are indicated by strong peaks in the magnetic susceptibility, but further down the section the contrast between loess and palaeosol susceptibility is much reduced. Secondly, there appears to be a significant discontinuity in both the susceptibility and M_s (and hence M_{rs}/M_s) data sets at a height of approximately 110 m. Above this break the susceptibility record shows a clear increasing trend and the M_{rs}/M_s data show a decreasing trend as we go up the section.

Below this level any trends are less obvious. The M_s data also show a gradual increase as we go up the section, but strong periodicities present in this data set tend to disguise this trend. M_{rs} shows a different behaviour, with a gradual increase as we go down the top 60 m of the section, at which point the trend stops, with the average M_{rs} value remaining almost constant until a point some 40 m from the base, at which point the trend becomes a decreasing one. This indicates that in the central part of the section there is little change in average remanence. The contrast between the M_s and M_{rs} trends suggests that as the magnetic content increases in the central part of the section there is a change in the nature of the magnetic material. We can imagine two explanations for the observed trends. Firstly, if the increase in magnetic content was restricted to SP grains, this would enhance both M_s and susceptibility but not M_{rs}, as SP grains do not retain a remanence. This scenario seems unlikely, a view which is supported by the frequency effect of susceptibility, which shows no evidence of an increase through the section. The second, and more likely, explanation is that as the magnetic content increases there is a gradual shift in the grain-size distribution to coarser grains, this leading to an increase in the relative (as well as total) contribution from MD magnetite. Although MD and SSD magnetite both have the same value of M_s (per unit mass), the M_{rs} value for MD magnetite is more than an order of magnitude less than SSD magnetite. This means that an increase in MD material would show in the M_s data but not be significant in the M_{rs} data. The change to a decreasing trend at the top of the M_{rs} data suggests that above this point the MD fraction is still increasing while the SSD fraction is now decreasing. Since both M_s and M_{rs} show the same dependence with concentration, the ratio M_{rs}/M_s is a useful tool for estimating the relative proportions of MD and SSD grains, although the presence of SP grains, with a M_{rs}/M_s ratio similar to MD grains, can complicate the picture. However, we cannot find a trend in the SP fraction (from the X_{fd} data), so trends in the M_{rs}/M_s data are likely to represent trends in the SSD/MD ratio, and hence in the magnetic grain size. The M_{rs}/M_s data in Fig. 3 clearly show a change in the magnetic grain-size distribution, the ratio falling almost continually up the section, this trend being particularly obvious above the 110 m discontinuity. If we interpret this as an increase in the larger, MD grains, then this increasing magnetic grain size correlates with an increase in average loess grain size, the Lanzhou loess changing from medium-silt grade to coarse silt as the age decreases (Derbyshire 1983). The discontinuity at 110 m above the base may therefore provide some important climatic information.

To interpret the significance of this abrupt behavioural change, we must first consider what physical parameters are responsible. Samples taken immediately below (142.0L) and above (141.5L) show that there is an almost 50% fall in both M_s and susceptibility for sample 141.5L, but a significant increase in both the M_{rs}/M_s ratio and mean coercivity. This suggests a large decrease in total magnetic content coupled with an increase in the magnetically 'hard' fraction. Both samples show very little frequency dependence of susceptibility indicating a low SP content in both samples, and Curie curves for both samples are identical indicating the magnetic chemistry is consistent. However, low temperature susceptibility behaviour indicates that sample 142.0L has a significant fraction of MD magnetite, whereas sample 141.5L shows behaviour dominated by paramagnetic clay. This discontinuity, therefore, appears to simply reflect a large decrease in magnetic content and, perhaps, a reduction in magnetic grain size, although because of the very low ferrimagnetic content in sample 141.5L it is possible that the relatively large paramagnetic content simply masks all other low temperature susceptibility behaviour. Such a large and abrupt change can be interpreted in a number of ways. Firstly, it could represent a significant increase in loess accumulation, with an accompanying increased dilution of magnetic material as discussed above, and such a change could represent a change to colder, drier conditions (Liu *et al.* 1985). Alternatively, the change could arise from a change in loess source area, perhaps as a result of changing wind patterns and/or topography. Finally, the change may simply represent a depositional hiatus/erosional break in the section, with an accompanying time gap. Although we could not find evidence of such a break in the field, this latter hypothesis cannot be discounted at this stage.

The samples used in this study were not orientated and so it has not been possible to use palaeomagnetism to date the section. However, by considering the climatic and tectonic history of China, we can arrive at one possible explanation for such an abrupt change. At approximately 1.4 Ma BP there was a change from subsidence to uplift with the onset of the Qiantang movements (1.4–1.2 Ma); these movements led to the uplift of the Qinghai-Tibet plateau (Sun & Wu 1985). Subsequent to this the climate became progressively colder and drier (Liu *et al.* 1985), and evidence of this can be seen in our susceptibility record. Prior to the discontinuity, the average susceptibility is slightly higher and the contrast between loess and palaeosols reduced, whereas after the discontinuity the palaeosol horizons, particularly as we move up the profile, have a very distinct susceptibility signature. This could indicate that the early loess deposition was in a climate with a higher average temperature and with less extreme temperature fluctuations, with an abrupt change to a colder climate at 1.4 Ma. The M_{rs}/M_s ratio also indicates that the magnetic grains, in common with the average loess grain size, become coarser as we move upwards, and this can be related to the change in climate which, as well as becoming colder and drier, also showed an increased wind velocity (Liu *et al.* 1985).

Conclusions

The magnetic mineral record of the Dawan section has provided some interesting climatic information. We can conclude that although pedogenesis plays a lesser role in this part of the loess plateau, the magnetic mineralogy of the palaeosols, particularly the younger soils, still shows clear pedogenic enhancement, in agreement with the findings of Zhou *et al.* (1990). This is sufficient to question the theoretical basis of the technique by which Kukla *et al.* (1988) use the susceptibility of a loess or palaeosol unit as an indicator of its duration. However, Xu (pers. comm.) has found that when this technique is applied to the Xifeng section susceptibility record, subsequent spectral analysis provides evidence of the Milankovitch periodicities to a much greater accuracy than otherwise. This suggests that although the technique is based on an over-simplified view of the susceptibility of palaeosols, there is clearly a link between the degree of pedogenic enhancement and the accumulation rate during the soil forming period. The low level of pedogenesis at Lanzhou perhaps partly reflects the fact that in addition to the colder, drier climate, the accumulation rate is much greater in this western part of the loess plateau. The loess source area is closer and the region is therefore likely to receive a greater input of material in the soil forming interglacials than the more eastern loess areas. In addition, the Huang He (Yellow River) flows through Lanzhou and may have provided, through deflation, an important local source of material during interglacials. The loess at Dawan appears to contain a slightly greater proportion of maghaemite (and cation-deficient magnetite) than the palaeosols. This suggests that pedogenic enhancement at Lanzhou is dominantly of the magnetite fraction, as the amount of maghaemite present in the palaeosols is little different from that present in the loess parent material. The presence of maghaemite in the loess indicates that the source material had already undergone weathering, perhaps in the northwestern desert regions during previous interglacial periods. It may also be further evidence that the Huang He provides an important local source of weathered material.

The magnetic record also provides us with information regarding the long-term climate change of northern China, which is closely linked to the recent tectonic history of the area. There is a clear discontinuity in the magnetic mineral record, above which the data suggest a climate which is becoming steadily colder and drier, with increasing wind velocity. This climate trend has been seen elsewhere (Liu *et al.* 1985) and is linked to the uplift of the Qinghai-Tibet plateau (Sun & Wu 1985). At this stage it is not known whether this discontinuity represents an abrupt change in climate or whether there is a time gap in the section. Palaeomagnetic studies of a second section near Lanzhou indicate either that accumulation of loess in the area did not start until approximately 1.3 Ma BP (Burbank & Li 1985) or was much reduced prior to this time (Rolph *et al.* 1989), with this date corresponding to the onset of Qiantang movements which led to the uplift of the Qinghai-Tibetan plateau. Clearly there is a strong tectonic link with loess accumulation in this area and the section at Dawan may provide important information on this subject. The next stage of this project must, therefore, be to provide an accurate time scale for the section.

We would like to express our gratitude to the various members of the Geological Hazards Research Centre, Lanzhou, who collected the samples used in this study. Many of the susceptibility measurements were carried out by Ms C. E. Farren in the course of an undergraduate project. We would also like to thank Michéle Clarke for many useful discussions and two anonymous reviewers whose comments helped to improve the manuscript. The authors acknowledge financial support for the project from Natural Environment Research Council grant GST/02/540.

References

BURBANK, D. W. & LI, J. J. 1985. Age and palaeoclimatic significance of the loess of Lanzhou, North China. *Nature*, **316**, 429–431.

CLARKE, M. 1992. *Formation, depositional history and magnetic properties of loessic silts from the Tibetan front, China*. PhD thesis, University of Leicester.

DERBYSHIRE, E. 1983. Origin and characteristics of some Chinese loess at two localities in China. *In*: BROOKFIELD, M. E. & AHLBRANDT, T. S. (eds) *Eolian Sediments and Processes*. Elsevier, Amsterdam, 95–103.

HELLER, F. & LIU, T. S. 1984. Magnetism of Chinese loess deposits. *Geophysical Journal of the Royal Astronomical Society*, **77**, 125–142.

KUKLA, G., HELLER, F., LIU, X. M., XU, T. S. & AN, Z. S. 1988. Pleistocene climates in China dated by magnetic susceptibility. *Geology*, **16**, 811–814.

LIU, T. S., ZHANG, S. X. & HAN, J. M. 1985. Stratigraphy and environmental changes in the loess of central China. *Quaternary Science Reviews*, **5**, 489–495.

MAHER, B. A., SINGER, M. J. & VEROSUB, K. L. 1991. Maghaemite as a factor in interpreting the palaeoclimatological record from soil/loess sequences. *Abstracts of the XIII INQUA Congress*, 226.

—— & THOMPSON, R. 1991. Mineral magnetic record of the Chinese loess and palaeosols. *Geology*, **19**, 3–6.

MILANKOVITCH, M. M. 1941. *Canon of Insolation and the Ice-Age Problem*. Beograd, Koninglich Serbische Akademie.

O'REILLY, W. 1983. The identification of titanomaghaemites: model mechanisms for the maghaemitization and inversion processes and their magnetic consequences. *Physics of the Earth and Planetary Interiors*, **31**, 65–76.

ROLPH, T. C., SHAW, J., DERBYSHIRE, E. & WANG, J. T. 1989. A detailed geomagnetic record from Chinese loess. *Physics of the Earth and Planetary Interiors*, **56**, 151–164.

SUN, D. Q. & WU, X. H. 1985. Preliminary study of Quaternary tectono-climatic cycles in China. *Quaternary Science Reviews*, **5**, 497–501.

THOMPSON, R. & OLDFIELD, F. 1986. *Environmental Magnetism*. George Allen and Unwin, London.

WANG, Y., EVANS, M. E., RUTTER, N. & DING, Z. 1990. Magnetic susceptibility of Chinese loess and its bearing on palaeoclimate. *Geophysical Research Letters*, **17**, 2449–2451.

ZHOU, L. P., OLDFIELD, F., WINTLE, A. G., ROBINSON, S. G. & WANG, J. T. 1990. Partly pedogenic origin of magnetic variations in Chinese loess. *Nature*, **346**, 737–739.

Index